MULTIVARIATE METHODS IN ECOLOGICAL WORK

STATISTICAL ECOLOGY
Volume 7

a publication from the
satellite program in statistical ecology
international statistical ecology program

Statistical Ecology Series

General Editor: G. P. Patil

*For these first three volumes, contact:
The Pennsylvania State University Press
University Park, PA 16802 USA

For all of the remaining volumes, contact:
International Co-operative Publishing House
P.O. Box 245
Burtonsville, MD 20730 USA

MULTIVARIATE METHODS IN ECOLOGICAL WORK

edited by

LASZLO ORLOCI
Department of Plant Sciences
University of Western Ontario
London, Ontario, Canada

C. RADHAKRISHNA RAO
Indian Statistical Institute
New Delhi, India

WILLIAM M. STITELER
State University of New York
College of Environmental Science and Forestry
Syracuse, New York

International Co-operative Publishing House
Fairland, Maryland USA

Mathematical ecology is moving out of its classical phase carrying with it untold promise for the future, but, as H.A.L. Fisher, the historian, remarks, progress is not a law of nature. Without enlightenment and eternal vigilance on the part of both ecologists and mathematicians there always lurks the danger that mathematical ecology might enter a dark age of barren formalism, fostered by an excessive faith in the magic of mathematics, blind acceptance of methodological dogma and worship of the new electronic gods. It is up to all of us to ensure that this does not happen.

J. G. SKELLAM (1972)
in Mathematical Models in Ecology
edited by J. N. R. Jeffers
Blackwell Scientific Publications
Oxford, England

For top management and general public policy development, monitoring data must be shaped into easy-to-understand indices that aggregate data into understandable forms. I am convinced that much greater effort must be placed on the development of better monitoring systems and indices than we have in the past. Failure to do so will result in sub-optimum achievement of goals at much greater expense.

R. E. TRAIN (1973)
National Conference on Managing the Environment
The United States Environmental Protection Agency
Washington, D.C., USA

International Statistical Ecology Program

ADVISORY BOARD
(Chairman: G. P. Patil)

Berthet, P.	Iwao, S.	Robson, D. S.
Cairns, J., Jr.	Matern, B.	Rossi, O.
Chapman, D. G.	Matis, J. H.	Simberloff, D. S.
Cormack, R. M.	Ord, J. K.	Taillie, C.
Cox, D. R.	Pielou, E. C.	Warren, W. G.
Holling, C. S.	Rao, C. R.	Waters, W. E.

SATELLITE PROGRAM IN STATISTICAL ECOLOGY
(1977-1978)

Director: G. P. Patil   Managing Editor: C. Taillie

COORDINATORS AND EDITORS

Artuz, M. I.	O'Neill, R. V.	Simberloff, D. S.
Cairns, J., Jr.	Ord, J. K.	Smith, W. K.
Chapman, D. G.	Orloci, L.	Solomon, D. L.
Cormack, R. M.	Patil, G. P.	Stiteler, W. M.
Gallucci, V.	Patten, B. C.	Taillie, C.
Grassle, J. F.	Rao, C. R.	Usher, M. B.
Holling, C. S.	Robson, D. S.	Waters, W. E.
Innis, G. S.	Rosenzweig, M. L.	White, G. C.
Matis, J. H.	Shoemaker, C.	Williams, F. M.

HOSTS

Matis, J. H.	Waters, W. E.	Rossi, O.
Gates, C. E.	Noy-Meir, I.	Zanni, R.

ADVISORS

Berthet, P.	Hennemuth, R.	Matern, B.
Engen, S.	Iwao, S.	Seber, G.
Glass, N.	Knox, G. A.	Warren, W. G.

SPONSORS

NATO Advanced Study Institutes Program
NATO Ecosciences Program

The Pennsylvania State University
The Texas A&M University
The University of California at Berkeley

National Marine Fisheries Service, USA
Environmental Protection Agency, USA
Fish and Wildlife Service, USA
Army Research Office, USA

Comunita Economica Europea
Universita degli Studi, Parma, Italy

Consiglio Nazionale delle Ricerche, Italy
Ministero dei Lavori Pubblici,
Affari Esteri, e Pubblica Istruzione, Italy
Societa Italiana di Statistica, Italy
Societa Italiana di Ecologia, Italy

The Participants and
Their Home Institutions and Organizations

PARTICIPANTS

SATELLITE A: COLLEGE STATION AND BERKELEY   July 18-August 13, 1977

Andrews, P. L., Montana
Anthony, R. G., Oregon
Artuz, M. I., Turkey
Bagiatis, K., Greece
Bajusz, B. A., Pennsylvania
Baumgaertner, J., California
Bell, E., Washington
Bellefleur, P., Canada
Berthet, P., Belgium
Beyer, J., Denmark
Bingham, R., Texas
Boswell, M. T., Pennsylvania
Braswell, J. H., Georgia
Braumann, C. A., Portugal
Brennan, J. A., Massachusetts
Bruhn, J. N., California
Cairns, J. Jr., Virginia
Callahan, C. A., Oregon
Cancela da Fonseca, J. P., France
Caraco, T. B., Arizona
Chapman, D. G., Washington
Cho, A., Pennsylvania
Colwell, R. California
Cormack, R., Scotland
Coulson, R. N., Texas
DeMars, C. J., California
Dennis, B., Pennsylvania
Derr, J., Iowa
deVries, P. G., Netherlands
Doucet, P. G., Netherlands
Elterman, A. L., California
Engen, S., Norway
Ernsting, G., Netherlands
Fiadeiro, P. M., Portugal
Flores, R. G., Brazil
Flynn, T. S., California
Folse, L. J., Texas
Ford, R. G., California
Gallucci, V. F., Washington
Gates, C. E., Texas
Gautier, C., France
Gerald, K. B., Texas

Giles, R. H., Virginia
Gokhale, D. V., California
Grant, W. G., Texas
Guardans, R. C., Spain
Hart, D., Virginia
Hazard, J. W., Oregon
Hendrickson, J. A., Pennsylvania
Hennemuth, R., Massachusetts
Hogg, D. B., Mississippi
Innis, G. S., Utah
Janardan, K. G., Illinois
Johnson, D., Texas
Johnson, D. H., North Dakota
Jolly, G. M., Scotland
Kester, T., Belgium
Kie, J. G., California
Kobayashi, S., Japan
Kubicek, F., Czechoslovakia
Labovitz, M. L., Pennsylvania
Lamberti, G. A., California
Lasebikan, B. A., Nigeria
Lindahl, K. Q., California
Laurence, G. C., Rhode Island
Livingston, G. P., Texas
Ludwig, J. A., New Mexico
Ma, J. C. W., Texas
Macken, C. A., Minnesota
Marsden, M. A., Montana
Mason, R., Oregon
Matis, J. H., Texas
Matthews, G. A., Texas
Minello, T. J., Texas
Mizell, R. F., Mississippi
Monserud, R. A., Idaho
Myers, C. C., Illinois
Myers, R. A., Canada
Naveh, Z., Israel
Nebeker, T. E., Mississippi
Neyman, J., California
Norick, N. X., California
O'Neill, R. V., Tennessee
Ord, J. K., England

Overton, S., Oregon
Patil, G. P., Pennsylvania
Pennington, M. R., Massachusetts
Poole, R. W., Rhode Island
Pulley, P. E., Texas
Quinn, T. J., Washington
Rawson, C. B., Washington
Reynolds, J. F., North Carolina
Riggs, L. A., California
Robson, D. S., New York
Roman, J. R., New York
Rosenzweig, M. L., Arizona
Roughgarden, J., California
Roux, J. J. J., South Africa
Sanders, F., Tennessee
Sen, A. R., Canada
Serchuk, F. M., Massachusetts
Shoemaker, C., New York
Singh, K. P., India
Smith, W. K., Massachusetts
Solomon, D. L., New York
Southward, G. M., New Mexico
Stafford, S. G., New York
Steinhorst, R. K., Texas
Stenseth, N. C., Norway
Stiteler, W. M., New York
Stout, M. L., California
Stromberg, L. P., California
Taillie, C., Pennsylvania
Tracy, D. S., Canada
Usher, M. B., England
vanBiezen, J. B., Netherlands
Walter, G. G., Wisconsin
Waters, W. E., California
Wensel, L. C., California
West, I. F., New Zealand
Wiegert, R. G., Georgia
Williams, F. M., Pennsylvania
Wright, J. R., Alabama
Wu, Y. C., California
Yandell, B. S., California
Zweifel, J. R., California

SATELLITE B: PARMA   July 31-September 5, 1978

Arditi, R., France
Azzarita F., Italy
Balchen, J. G., Norway
Bargmann, R. E., Georgia
Barlow, N. D., England
Baxter, M. B., Australia
Behrens, J., Denmark
Beran, H. G., Austria
Berman, M., Maryland
Berryman, A. A., Washington
Boswell, M. T., Pennsylvania
Brambilla, C., Italy
Braumann, C. A., New York
Breitenecker, M., Austria
Buerk, R., West Germany
Callahan, C. A., Oregon
Cancela da Fonseca, J. P., France
Chieppa, M., Italy

Clark, W. G., Italy
Cobelli, C., Italy
Cooper, C., California
Curry, G. L., Texas
DeMichele, D. W., Texas
Derr, J., Iowa
Diggle, P. J., England
Drakides, C., France
Ebenhöh, W., West Germany
Engen, S., Norway
Feoli, E., Italy
Fiadeiro, P. M., Portugal
Fischlin, A., Switzerland
Framstad, E. B., Norway
Frohberg, K., Austria
Gallucci, V. F., Washington
Garcia-Moya, E., Mexico
Gatto, M., Italy

Geri, C., France
Giavelli, G., Italy
Ginzburg, L. R., New York
Gokhale, D. V., California
Goldstein, R. A., California
Granero Porati, M. I., Italy
Greve, W., West Germany
Grosslein, M. D., Massachusetts
Grümm, H. R., Austria
Gulland, J. A., Italy
Gutierrez, A. P., California
Gydesen, H., Denmark
Hanski, I., England
Hanson, B. J., Utah
Hau, B., West Germany
Helgason, T., Iceland
Hendrickson, J. A., Pennsylvania
Hengeveld, R., Netherlands

Hennemuth, R. C., Massachusetts
Hoff, J. M., Norway
Holling, C. S., Canada
Hotz, M. C. B., Belgium
Jancey, R. C., Canada
Kooijman, S., Netherlands
Lamont, B. B., Australia
Levi, D., Italy
Liu, C. J., Kentucky
Marshall, W., Canada
Martin, F. W., Maryland
Matis, J. H., Texas
Menozzi, P., Italy
Meyer, J. A., France
Mohn, R. K., Canada
Mosimann, J., Maryland
Naveh, Z., Israel
Noy-Meir, I., Israel
Olivieri-Barra, S. T., Belgium
Ord, J. K., England
Orloci, L., Canada
Pacchetti, G., Italy
Pagani, L., Italy
Patil, G. P., Pennsylvania
Patten, B. C., Georgia

Pennington, M. R., Massachusetts
Policello, G. E., Ohio
Pospahala, R. S., Maryland
Purdue, P., Kentucky
Radler, K., West Germany
Ramsey, F. L., Oregon
Reyment, R. A., Sweden
Reyna Robles, R., Mexico
Rinaldi, S., Italy
Robson, D. S., New York
Rossi, O., Italy
Russek, E., Washington
Russo, A. R., Hawaii
Sadasivan, G., India
Schaefer, R., France
Shoemaker, C. A., New York
Show, I. T., California
Shuter, B. J., Canada
Simberloff, D., Florida
Slagstad, D., Norway
Smetacek, V. S., West Germany
Smith, W. K., Massachusetts
Sokal, R. R., New York
Soliani, L., Italy
Solomon, D. L., New York

Spremann, K., West Germany
Steinhorst, R. K., Idaho
Stenseth, N. C., Norway
Stiteler, W. M., New York
Subrahmanyam, C. B., Florida
Szöcs, Z., Hungary
Taillie, C., Pennsylvania
terBraak, C. J. F., Netherlands
Torrez, W. C., New Mexico
Tracy, D. S., Canada
Tursi, A., Italy
Vale, C., Portugal
van Biezen, J. B., Netherlands
Vazzana, C., Italy
Wahl, E., West Germany
Walker, B. H., England
Walter, G. G., Wisconsin
Walters, C. J., Canada
Warren, W. G., Canada
Waters, W. E., California
White, G. C., New Mexico
Wise, M. E., Netherlands
Zanni, R., Italy

SATELLITE C: JERUSALEM   September 7-September 15, 1978

Austin, M. P., Australia
Baxter, M. B., Australia
Berthet, P., Belgium
Carleton, T. J., Canada
Eisen, P. A., New York
Galluci, V., Washington
Goldstein, R. A., California
Goodman, D., California
Hanski, I., Finland
Hendrickson, J. A., Pennsylvania
Hengeveld, R., Netherlands

Hennemuth, R. C., Massachusetts
Hirsch, A., Washington, DC
Kempton, R. A., England
Naveh, Z., Israel
Noy-Meir, I., Israel
Odum, H. T., Florida
O'Neill, R. V., Tennessee
Patil, G. P., Pennsylvania
Quinn, T. J., Washington
Resh, V. H., California
Rossi, O., Italy

Rosenzweig, M. L., Arizona
Safriel, U., Israel
Sheshinski, R., Israel
Simberloff, D. S., Florida
Sokal, R. R., New York
Solem, J. O., Norway
Stenseth, N. C., Norway
Subrahmanyam, C. B., Florida
Torrez, W., New Mexico
Waters, W. E., California
Whittaker, R. H., New York

AUTHORS NOT LISTED ABOVE

Adams, J. E., Texas
Barber, M. C., Georgia
Barreto, M., Brazil
Batcheler, C. L., New Zealand
Beuter, K. J., West Germany
Bitz, D. W., Rhode Island
Brown, B. E., Massachusetts
Brown, G. C., Kentucky
Brown-Leger, L. S., Massachusetts
Chaim, S., Israel
Chardy, P. France
Clark, G. M., Ohio
Condra, C., California
Connor, E. F., Florida
Costa, H., Brazil
Dale, M., Australia
De LaSalle, P., France
Duek, J. L., Arizona
Elwood, J. W., Tennessee
Ferris, J. M., Indiana
Ferris, V. R., Indiana
Finn, J. T., Massachusetts
Foltz, J. L., Texas
Gibson, V. R., Massachusetts
Giddings, J. M., Tennessee

Gittins, R., Australia
Godron, M., France
Grassle, J. F., Massachusetts
Green, R., California
Halbach, U., West Germany
Halfon, E., Canada
Helthshe, J. F., Rhode Island
Hildebrand, S. G., Tennessee
Hogeweg, P., Netherlands
Iwao, S., Japan
Jacur, G. R., Italy
Kaesler, R. L., Kansas
Kravitz, D., Massachusetts
Kruczynski, W. L., Florida
Lackey, R., Virginia
Laurec, A., France
Lepschy, A., Italy
Malley, J. D., Maryland
Marcus, A. H., Washington
Matern, B., Sweden
McCune, E. D., Texas
Moncreiff, R., California
Moroni, A., Italy
Nichols, J. D., Maryland
O'Connor, J. S., New York

Paloheimo, J. E., Canada
Perrin, S., California
Pickford, S. G., Washington
Plowright, R. C., Canada
Podani, J., Hungary
Ratnaparkhi, M. V., Pennsylvania
Rescigno, A., Canada
Rickaert, M., France
Rohde, C., Maryland
Rosen, R., Canada
Scott, E., California
Scott, J. M., Hawaii
Seber, G. A. F., New Zealand
Siri, E., Italy
Skalski, J. R., Washington
Stehman, S., Pennsylvania
Steinberger, E. H., Israel
Swartzman, G., Washington
Taylor, C. E., California
Taylor, L. R., England
Tiwari, J. L., California
Watson, R. M., Scotland
Wehrly, T. E., Texas
Wigley, R. L., Massachusetts
Wissel, C., West Germany
Yang, M., Florida

Foreword

The Second International Congress of Ecology was held in Jerusalem during September 1978. In this connection, a Satellite Program in Statistical Ecology was organized by the International Statistical Ecology Program (ISEP) during 1977 and 1978. The emphasis was on research, review, and exposition concerned with the interface between quantitative ecology and relevant quantitative methods. Both theory and application of ecology and ecometrics received attention. The program consisted of instructional coursework, seminar series, thematic research conferences, and collaborative research workshops.

The 1977 and 1978 Satellite Program consisted of NATO Advanced Study Institutes at College Station in Texas, Berkeley in California, and Parma in Italy; NATO Advanced Research Institute at Parma; ISEP Research Conferences, Seminars, and Workshops at College Station, Berkeley, Parma, and Jerusalem; and a Research Conference at Jerusalem.

The Satellite Program has been supported by NATO Advanced Study Institutes Program; NATO Ecosciences Program; National Marine Fisheries Service, USA; Environmental Protection Agency, USA; Fish and Wildlife Service, USA; Army Research Office, USA; The Pennsylvania State University; The Texas A&M University; The University of California at Berkeley; Universita degli Studi, Parma; Consiglio Nazionale delle Ricerche, Italy; Ministero dei Lavori Pubblici, Affari Esteri, e Pubblica Istruzione, Italy; Societa Italiana di Statistica, Italy; Societa Italiana di Ecologia, Italy; Communita Economica Europea; and the participants and their home institutions and organizations.

Research papers and research-review-expositions were specially prepared for the program by concerned experts and expositors. These materials have been refereed and revised, and are now available in a series of ten edited volumes.

HISTORICAL BACKGROUND

The First International Symposium on Statistical Ecology was held in 1969 at Yale University with support from the Ford Foundation and the US Forest Service. The three symposium co-chairmen (G. P. Patil, E. C. Pielou, and W. E. Waters) represented the fields of statistics, theoretical ecology, and applied ecology. The program was well attended, and it provided a broad picture of where statistics and ecology stood relative to each other. While effort was apparent, communication between the two disciplines was inadequate.

It was clear that a focal forum was necessary to discuss and develop a constructive interface between quantifiable problems in ecology and relevant quantitative methods. As a partial solution to fill this need at professional organizations' level, the director of the symposium (G. P. Patil) made certain recommendations to the Presidents of the International Association for Ecology (A. D. Hasler), the International Statistical Institute (W. G. Cochran), and the International Biometric Society (P. Armitage). The International Association for Ecology (INTECOL) took a timely step in creating a section in the organization, namely, the statistical ecology section. The three societies together took a timely step in setting up a liaison committee on statistical ecology. The INTECOL Section and the Liaison Committee together developed the International Statistical Ecology Program. Since its inception in 1970, ISEP (as it has come to be known) has put emphasis on identifying the interdisciplinary needs of statistics and ecology at advanced instructional levels, and also at research conference and workshop levels.

The First Advanced Institute on Statistical Ecology in the United States was organized at Penn State for six weeks in 1972 with support from the US National Science Foundation, the US Forest Service, and the Mathematical Social Sciences Board. The participants of the Institute are all enjoying the benefits of their fruitful participation. With support from the UNESCO program of Man and Biosphere, a six month program was held in Venezuela for participants from Latin America in 1974 under the direction of Jorge Rabinovich, himself a participant in the 1972 Institute. With some initiatives from ISEP, special statistical ecology sessions have been held at the international conferences of the International Statistical Institute and the Biometric Society.

While plans were being made for the Second International Congress of Ecology, the then Secretary General and current President of INTECOL (G. A. Knox), and the ISEP chairman (G. P. Patil) discussed the need and the timeliness of a program in statistical ecology. The Satellite Program in Statistical Ecology took its final shape from this beginning under the care and concern of its director, advisors, coordinators, and sponsors.

SCIENTIFIC BACKGROUND AND PURPOSE

The perceptions of Skellam and Train quoted on page v in this volume recapitulate the cautions, inspirations, and objectives responsible for the Satellite Program. The rigorous formulation of a quantitative scientific concept requires and in a sense creates empirically measurable quantities. Conversely, the scientific validity of the concept is totally dependent upon the measured values of those quantities. This is a capsule version of the feedback process known as the "scientific method." In crude modern terms, we might label the first procedure "modeling" and the second "curve-fitting." For reasons of complexity, historical accident, or whatever, these mutually dependent, complementary components have never become firmly integrated within ecology and its application to environmental studies.

Both procedures involve forms of mathematics: In the "modeling" process, the mathematics is used *relationally* — as a system of logic to ensure rigor and clarity of reasoning. This is in the historical tradition of Volterra and Lotka. Validation is most often based on *qualitative* agreement: the right trend, or the correct shape of the curve. This is as it should be, especially for broad general theories: it is the ideas and understanding that count, not so much the quantitative detail.

In the "curve-fitting" tradition, the mathematics is used *numerically* — as a system for precise measurement and prediction. Validation is most often based on *quantitative* agreement: n-place accuracy, or minimum uncertainty. This is also as it should be, especially for application and management: it is the forecast and ability to act confidently that count, not so much the underlying concept.

It need not be taken as a sign of "physics envy" to assert that as a science matures, these two processes must converge — more quantification must be used in concept validation and more concepts must be incorporated into the methodology of quantification.

The purpose of the Satellite Program is to encourage that convergence within the science of ecology and promote its application in the study of the environment and environmental stress. Monitoring and assessment activities to be meaningful and defensible need: (i) a conceptual and philosophical basis, (ii) a theoretical framework, (iii) methodological support, (iv) a technological toolbox, and (v) administrative management. The ultimate purpose of the program is to help identify and integrate the specifics of these important factors responsible for protecting the environment.

We take as our theme the better melding of fundamental ecological concepts with rigorous empirical quantification. The overall result should be progress toward a stronger body of general ecological theory and practice.

PLANNING AND ORGANIZATION

The realization of any program of this nature and dimension often fails to fully meet the initial expectations and objectives of the organizers. Factors that are both logistic and psychological in nature tend to contribute to this general experience. Logistic difficulties include optimality problems for time and location. Other difficulties which must be attended to involve conflicting attitudes towards the importance of individual contributions to the proceedings.

We tried to cope with these problems by seeking active advice from a number of participants and special advisors. The advice we received was immensely helpful in guiding our selection of the best experts in the field to achieve as representative and balanced a coverage as possible. Simultaneously, the editors together with the referees took a rather critical and constructive attitude from initial to final stages of preparation of papers by offering specific suggestions concerning the suitability, and also the structure, content and size. These efforts of coordination and revision were intensified through editorial sessions at the program itself as a necessary step for the benefit of both the general readership and the participants. It is our pleasure to record with appreciation the spontaneous cooperation of the participants. Everyone went by scientific interests often at the expense of personal preferences. The program atmosphere became truly creative and friendly, and this remarkable development contributed to the maximal cohesion of the program and its proceedings within the limited time period available.

In retrospect, our goals were perhaps ambitious! We had close to 350 lectures and discussions during 50 days in the middle of the summer seasons of 1977 and 1978. For several reasons, we decided that an overworked program was to be preferred to a leisurely one. First of all, gatherings of such dimension are possible only every 5-10 years. Secondly, the previous meetings of this nature occurred some 5-10 years back, and the subject area of statistical ecology had witnessed substantial growth in this time. Thirdly, but most importantly, was the overwhelming response from potential participants, many of whom were to come across the continents!

Satellite A at College Station, Texas, and at Berkeley, California covered a four week period during July 18-August 13, 1977 and had 125 participants. Satellite B at

Parma, Italy had 130 participants spread over a six week period during July 31-September 5, 1978. Satellite C at Jerusalem, Israel had 35 participants. Approximately, one-third of the participants were graduate students, one-half were university faculty, and one-third were agency scientists. Approximately, one-third of the participants had an affiliation with one mathematical science or another, one-half an affiliation with one environmental science or the other, and one-quarter had an affiliation with one environmental management program or another. Thus the group was a good mix of great variety contributing to the effectiveness of the program. Not only what one heard was enlightening, but what one over-heard was equally enlightening!

Professors G. P. Patil, Paul Berthet, J. K. Ord, and Charles Taillie served as scientific directors of the program with Professor Patil assuming the responsibility of its direction from its conception to its conclusion. The inaugural speakers were: Professor H. O. Hartley at College Station, Professor J. Neyman at Berkeley, Professor C. S. Holling at Parma, and Professor G. P. Patil at Jerusalem.

SCIENTIFIC CONTENT AND PUBLICATION

The following summary information on the subjects and corresponding coordinators of the satellite program may be of some interest. The details of each subject and its publication volume are reported elsewhere.

Statistical Distributions in Ecological Work: J. K. Ord, G. P. Patil, and C. Taillie.

Spatial and Temporal Analysis in Ecology: R. M. Cormack and J. K. Ord.

Quantitative Population Dynamics: D. G. Chapman and V. Gallucci.

Sampling Biological Populations: R. M. Cormack, G. P. Patil, and D. S. Robson.

Ecological Diversity in Theory and Practice: J. F. Grassle, G. P. Patil, W. K. Smith, and C. Taillie.

Multivariate Methods in Ecological Work: L. Orloci, C. R. Rao, and W. M. Stiteler.

Systems Analysis of Ecosystems: G. S. Innis and R. V. O'Neill.

Compartmental Analysis of Ecosystem Models: J. H. Matis, B. C. Patten, and G. C. White.

Environmental Biomonitoring, Assessment, Prediction, and Management-Certain Case Studies and Related Quantitative Issues: J. Cairns, Jr., G. P. Patil and W. E. Waters.

Contemporary Quantitative Ecology and Related Ecometrics: G. P. Patil and M. L. Rosenzweig.

Three more subjects were organized in the program.

Scientific Modeling and Quantitative Thinking with Examples in Ecology: G. P. Patil, D. S. Simberloff, and D. L. Solomon.

Conceptual Foundations of Ecological Theory and Applications: M. B. Usher and F. M. Williams.

Optimizations in Ecological Theory and Management: C. S. Holling and C. Shoemaker.

It would be fruitful to reorganize these subjects and add a few more for the next satellite program when it occurs.

It should be mentioned here that the close coordination and cooperation between the coordinators and the authors/speakers of potential contributions to the Proceedings (which are particularly intensive during the satellite program) paid themselves handsomely when the editors were confronted with the technical work related to the publications after the close of the program at Jerusalem. It is therefore very satisfying to report that the edited research papers and research-review-expositions prepared for the program are ready for distribution within 12 months of the conclusion of the program. For purposes of convenience, the contributions are organized in ten volumes in the Statistical Ecology Series published by the International Co-operative Publishing House. Altogether, they consist of an estimated 4,000 pages of research, review, and exposition, in addition to this common foreword in each followed by individual volume introductions. Subject and author indexes are also prepared at the end. Every effort has been made to keep the coverage of the volumes close to their individual titles. May this ten volume set in its own modest way provide an example of synergism.

FUTURE DIRECTIONS

We wish there was no need for a program of this nature and dimension. It would be ideal if the needs of an interdisciplinary program were satisfactorily met in the existing institutions. Unfortunately, universities and governmental agencies have not been able to find effective ways to foster healthy interdisciplinary programs. The individuals attempting to do something in this direction tend to feel disheartened or disillusioned.

The satellite-like-programs help create and sustain enthusiasm, inward strength, and working efficiency of those who desire to meet a contemporary social need in the form of some interdisciplinary work. It should be only proper and rewarding for everyone involved that such programs are planned from time to time.

Plans are being made for a satellite program in conjunction with the next Biennial Conference of the International Statistical Institute and with the next International Congress of Ecology. Care should be exercised that the next program not become a mere replica of the present one, however successful it has been. Instead, the next program should be organized so that it helps further the evolution of statistical ecology as a productive field.

The next program is being discussed in terms of subject area groups. Each subject group is to have a coordinator assisted by small committees, such as a program committee, a research committee, an annual review committee, a journal committee, and an education committee. This approach is expected to respond to the need for a journal on statistical ecology, and also to the need of bringing out well planned annual review volumes. The education committee would formulate plans for timely manuals, modules, and monographs. Interested readers may feel free to communicate their ideas and interests to those involved in planning the next program. With mutual goodwill and support, we shall have met a timely need for today's science, technology, and society.

July 1979   G. P. Patil

Program Acknowledgments

For any program to be successful, mutual understanding and support among all participants are essential in directions ranging from critical to constructive and from cautious to enthusiastic. The present program is grateful to the members of the ISEP Advisory Board, and to the referees, editors, coordinators, advisors, sponsors and the participants for their timely advice and support.

The success of the program was due, in no small measure, to the endeavors of the Local Arrangements Chairmen: J. H. Matis and C. E. Gates at College Station, W. E. Waters at Berkeley, O. Rossi and R. Zanni at Parma, and I. Noy-Meir at Jerusalem. We thank them for their hospitality and support.

And finally those who have assisted with the arduous task of preparing the materials for publication. Barbara Alles has been an ever cheerful and industrious secretary in the face of every adversity. Charles Taillie managed both scientific and non-scientific aspects. Bharat Kapur copyedited and proofread. Bonnie Burris, Bonnie Henninger, and Sandy Rothrock prepared the final versions of the manuscripts. Marllyn Boswell helped with the subject and author indexes. So did Bharat Kapur, Satish Patil, and Rani Venkataramani.

All of these nice people have done a fine job indeed. To all of them, our sincere thanks.

July 1979

G. P. Patil

Reviewers of Manuscripts

With appreciation and gratitude, the program acknowledges the valuable services of the following referees who have served as reviewers of manuscripts submitted to the program for possible publication. The editors thank the reviewers for their critical and constructive reviews.

G. Knott
National Institutes of Health

S. Kotz
Temple University

A. M. Kshirsagar
University of Michigan

S. Kullback
The George Washington University

R. C. Lewontin
Harvard University

B. F. J. Manly
University of Otago

A. H. Marcus
Washington State University

F. Martin
Patuxent Wildlife Research Center

J. H. Matis
Texas A&M University

J. R. McBride
University of California

D. Mollison
Heriot-Wett University

R. V. O'Neill
Oak Ridge National Laboratory

J. Newton
University of St. Andrews

J. K. Ord
University of Warwick

L. Orloci
University of Western Ontario

G. P. Patil
Pennsylvania State University

B. C. Patten
University of Georgia

M. Pennington
National Marine Fisheries Service

S. Pimm
Texas Tech University

K. H. Pollock
University of Reading

R. W. Poole
Brown University

R. S. Pospahala
Patuxent Wildlife Research Center

E. Preston
Environmental Protection Agency

F. Preston
Preston Laboratories

P. Purdue
University of Kentucky

F. L. Ramsey
Oregon State University

C. R. Rao
Indian Statistical Institute

P. A. Rauch
University of California

E. Renshaw
University of Edinburgh

R. Reyment
Uppsala University

D. S. Robson
Cornell University

M. L. Rosenzweig
University of Arizona

O. Rossi
University of Parma

W. E. Schaaf
National Marine Fisheries Service

T. Schopf
University of Chicago

H. T. Schreuder
Forest Service

G. A. F. Seber
University of Auckland

J. Sepkoski
University of Rochester

I. T. Show, Jr.
Science Applications, Inc.

D. Simberloff
Florida State University

D. B. Siniff
University of Minnesota

W. K. Smith
Woods Hole Oceanographic Institution

R. R. Sokal
State University of New York

G. M. Southward
New Mexico State University

S. Stehman
Pennsylvania State University

R. K. Steinhorst
University of Idaho

W. M. Stiteler
Syracuse University

P. Switzer
Stanford University

C. Taillie
International Statistical Ecology Program

C. E. Taylor
University of California

C. Tsokos
University of South Florida

E. Ursin
Danish Institute of Fisheries and Marine Research

D. Vaughan
Oak Ridge National Laboratory

G. G. Walter
University of Wisconsin

W. G. Warren
Western Forest Products Laboratory

W. E. Waters
University of California

S. D. Webb
University of Florida

G. C. White
Los Alamos Scientific Laboratory

R. H. Whittaker
Cornell University

M. E. Wise
Leiden University

S. Zahl
University of Connecticut

J. Zweifel
National Marine Fisheries Service

Contents of Edited Volumes

Determination of Plant Species Diversity in Mediterranean Shrub and Woodland Along Environmental Gradients. V. RESH, Biomonitoring, Species Diversity Indices, and Taxonomy. J. SOLEM, A Comparison of Species Diversity Indices in Trichoptera Communities. W. K. SMITH, V. R. GIBSON, L. S. BROWN-LEGER, and J. F. GRASSLE, Diversity as an Indicator of Pollution: Cautionary Results from Microcosm Experiments. C. B. SUBRAHMANYAM and W. L. KRUCZYNSKI, Colonization of Polychaetous Annelids in the Intertidal Zone of a Dredged Material Island in North Florida. C. E. TAYLOR and C. CONDRA, Competitor Diversity and Chromosomal Variation in Drosophia Pseudoobscura. B. DENNIS and O. ROSSI, Community Composition and Diversity Analysis in a Marine Zooplankton Survey.

Bibliography: B. DENNIS, G. P. PATIL, O. ROSSI, S. STEHMAN, and C. TAILLIE, A Bibliography of Literature on Ecological Diversity and Related Methodology.

MULTIVARIATE METHODS IN ECOLOGICAL WORK
L. Orloci, C. R. Rao, and W. M. Stiteler (editors)   **400 pp. approx.**
R. BARGMANN, Structural Analysis of Singular Matrices Using Union Intersection Statistics with Applications in Ecology. M. DALE, On Linguistic Approaches to Ecosystems and Their Classification. D. V. GOKHALE, Analysis of Ecological Frequency Data: Certain Case Studies. J. HENDRICKSON, Examples of Discrete Multivariate Methods in Ecological Work. R. HENGEVELD and P. HOGEWEG, Cluster Analysis of the Distribution Patterns of Dutch Carabid Species. R. JANCEY, Species Weighting. A. LAUREC, P. CHARDY, P. DE LASALLE, and M. RICKAERT, Use of Dual Structures in Inertia Analysis: Ecological Implications. J. MOSIMANN and J. D. MALLEY, Size and Shape Analysis. L. ORLOCI, Non-Linear Data Structure and Their Description. J. PODANI, A Generalized Strategy of Homogeneity-Optimizing Hierarchical Classificatory Methods. R. REYMENT, Multivariate Analysis in Statistical Paleoecology. E. SCOTT, Spurious Correlation. W. K. SMITH, D. KRAVITZ, and J. F. GRASSLE, Confidence Intervals for Similarity Measures Using the Two Sample Jackknife. R. K. STEINHORST, Analysis of Niche Overlap. W. M. STITELER, Multivariate Statistics with Applications in Statistical Ecology. Z. SZOCS, New Computer Oriented Methods for Structural Investigation of Natural and Simulated Vegetation Patterns. B. LAMONT and K. J. GRANT, A Comparison of Twenty Measures of Site Dissimilarity. R. GITTINS, Ecological Applications of Canonical Analysis.

SPATIAL AND TEMPORAL ANALYSIS IN ECOLOGY
R. M. Cormack and J. K. Ord (editors)   **400 pp. approx.**
J. K. ORD, Time-Series and Spatial Patterns in Ecology. P. J. DIGGLE, Statistical Methods for Spatial Point Patterns in Ecology. R. M. CORMACK, Spatial Aspects of Competition Between Individuals. R. W. POOLE, The Statistical Prediction of the Fluctuations in Abundance in Nicholson's Sheep Blowfly Experiments. W. G. WARREN and C. L. BATCHELER, The Density of Spatial Patterns: Robust Estimation Through Distance Methods. B. MATERN, The Analysis of Ecological Maps as Mosaics. J. A. LUDWIG, A Test of Different Quadrat Variance Methods for the Analysis of Spatial Pattern. S. A. L. M. KOOIJMAN, The Description of Point Patterns. R. HENGEVELD, The Analysis of Spatial Patterns of Some Ground Beetles (Col. Carabidae).

SYSTEMS ANALYSIS OF ECOSYSTEMS
G. S. Innis and R. V. O'Neill (editors)   **425 pp. approx.**
R. K. STEINHORST, Stochastic Difference Equation Models of Biological Systems. R. V. O'NEILL, Natural Variability as a Source of Error in Model Predictions. R. K. STEINHORST, Parameter Identifiability, Validation and Sensitivity Analysis of Large System Models. R. V. O'NEILL, Transmutation Across Hierarchical Levels. R. V. O'NEILL, J. W. ELWOOD, and S. G. HILDEBRAND, Theoretical Implications of Spatial Heterogeneity in Stream Ecosystems. R. V. O'NEILL and J. M. GIDDINGS, Population Interactions and Ecosystem Function: Plankton Competition and Community Production. J. P. CANCELA DA FONSECA, Species Colonization Models of Temporary Ecosystems Habitats. E. HALFON, Computer-Based Development of Large Scale Ecological Models: Problems and Prospects. G. S. INNIS, A Spiral Approach to Ecosystem Simulation.

COMPARTMENTAL ANALYSIS OF ECOSYSTEM MODELS
J. H. Matis, B. C. Patten, and G. C. White (editors)   **400 pp. approx.**
Applications of Compartmental Analysis to Ecosystem Modeling: R. V. O'NEILL, A Review of Linear Compartmental Analysis in Ecosystem Science. G. G. WALTER, A Compartmental Model of a Marine Ecosystem. M. C. BARBER, B. C. PATTEN, and J. T. FINN, Review and Evaluation of Input-Output Flow Analysis for Ecological Applications. I. T. SHOW, JR., An Application of Compartmental Models to Meso-scale Marine Ecosystems.

Identifiability and Statistical Estimation of Parameters in Compartmental Models: C. COBELLI, A. LEPSCHY, G. R. JACUR, Identification Experiments and Identifiability Criteria for Compartmental Systems. M. BERMAN, Simulation, Data Analysis, and Modeling with the SAAM Computer Program. G. C. WHITE and G. M. CLARK. Estimation of Parameters for Stochastic Compartment Models. R. E. BARGMANN, Statistical Estimation and Computational Algorithms in Compartmental Analysis for Incomplete Sets of Observations.

Stochastic Approaches to the Compartmental Modeling of Ecosystems: J. L. TIWARI, A Modeling Approach Based on Stochastic Differential Equations, the Principle of Maximum Entropy, and Bayesian Inference for Parameters. J. H. MATIS and T. E. WEHRLY, An Approach to a Compartmental Model with Multiple Sources of Stochasticity for Modeling Ecological Systems. P. PURDUE, Stochastic Compartmental Models: A Review of the Mathematical Theory with Ecological Applications. A. H. MARCUS, Semi-Markov Compartmental Models in Ecology and Environmental Health. M. E. WISE, The Need for Rethinking on Both Compartments and Modeling.

Mathematical Analysis of Compartmental Structures: G. G. WALTER, Compartmental Models, Digraphs, and Markov Chains. K. B. GERALD and J. H. MATIS, On the Cumulants of Some Stochastic Compartmental Models Applied to Ecological Systems. A. RESCIGNO, The Two-variable Operational Calculus in the Construction of Compartmental Ecological Models.

ENVIRONMENTAL BIOMONITORING, ASSESSMENT, PREDICTION, AND MANAGEMENT — CERTAIN CASE STUDIES AND RELATED QUANTITATIVE ISSUES
J. Cairns, Jr., G. P. Patil, and W. E. Waters (editors)   **450 pp. approx.**
Biomonitoring: J. CAIRNS, JR., Biological Monitoring — Concept and Scope. Z. NAVEH, E. H. STEINBERGER, and S. CHAIM, Use of Bio-Indicators for Monitoring of Air Pollution by Fluor, Ozone and Sulfur Dioxide.

Environmental assessment and prediction: G. P. PATIL, C. TAILLIE, and R. L. WIGLEY, Transect Sampling Methods and Their Application to the Deep-Sea Red Crab. W. E. WATERS, Biomonitoring, Assessment, and Prediction in Forest Pest Management Systems. C. A. CALLAHAN, V. R. FERRIS, and J. M. FERRIS, The Ordination of Aquatic Nematode Communities as Affected by Stream Water Quality. R. A. GOLDSTEIN, Development and Implementation of a Research Program on Ecological Assessment of the Impact of Thermal Power Plant Cooling Systems on Aquatic Environments.

Environmental Management: A. HIRSCH, Ecological Information and Technology Transfer, R. H. GILES, JR., Modeling Decisions or Ecological Systems. R. LACKEY, Appliction of Renewable Natural Resource Modeling in Public Decision-Making Process. F. MARTIN, R. S. POSPAHALA, and J. D. NICHOLS, Assessment and Population Management of North American Migratory Birds. J. E. PALOHEIMO and R. C. PLOWRIGHT, Bioenergetics, Population Growth and Fisheries Management. Z. NAVEH, A Model of Multiple-Use Management Strategies of Marginal and Untillable Mediterranean Upland Ecosystems.

Case Studies and Quantitative Issues: M. D. GROSSLEIN, R. C. HENNEMUTH, and B. E. BROWN, Research, Assessment, and Management of a Marine Ecosystem in the Northwest Atlantic — A Case Study. J. NEYMAN, Two Interesting Ecological Problems Demanding Statistical Treatment. B. DENNIS, G. P. PATIL, and O. ROSSI, The Sensitivity of Ecological Diversity Indices to the Presence of Pollutants in Aquatic Communities. D. SIMBERLOFF, Constraints on Community Structure During Colonization.

CONTEMPORARY QUANTITATIVE ECOLOGY AND RELATED ECOMETRICS
G. P. Patil and M. L. Rosenzweig (editors)   **725 pp. approx.**
Community Structure and Diversity: R. A. KEMPTON and L. R. TAYLOR, Some Observations on the Yearly Variability of Species Abundance at a Site and the Consistency of Measures of Diversity. G. P. PATIL and C. TAILLIE, A Study of Diversity Profiles and Orderings for a Bird Community in the Vicinity of Colstrip, Montana. M. L. ROSENZWEIG, Three Probable Evolutionary Causes for Habitat Selection. O. ROSSI, G. GIAVELLI, A. MORONI, and E. SIRI, Statistical Analysis of the Zooplankton Species Diversity of Lakes Placed Along a Gradient. S. KOBAYASHI, Another Model of the Species Rank-Abundance Relation for a Delimited Community. M. L. ROSENZWEIG and J. L. DUEK, Species Diversity and Turnover in an Ordovician Marine Invertebrate Assemblage.

Patterns and Interpretations: D. S. SIMBERLOFF and E. F. CONNOR, Q-Mode and R-Mode Analyses of Biogeographic Distributions: Null Hypotheses Based on Random Colonization. D. GOODMAN,

Applications of Eigenvector Analysis in the Resolution of Spectral Pattern in Spatial and Temporal Ecological Sequences. R. H. WHITTAKER and Z. NAVEH, Analysis of Two-Phase Patterns. R. R. SOKAL, Ecological Parameters Infered From Spatial Correlograms. M. P. AUSTIN, Current Approaches to the Non-Linearity Problem in Vegetation Analysis. M. GODRON, A Probabilistic Computation for the Research of "Optimal Cuts" in Vegetation Studies. S. IWAO, The m*-m Method for Analyzing Distribution Patterns of Single- and Mixed-Species Populations.

Modeling and Ecosystems Modeling: D. L. SOLOMON, On a Paradigm for Mathematical Modeling. R. W. POOLE, Ecological Models and Stochastic-Deterministic Question. R. ROSEN, On the Role of Time and Interaction in Ecosystem Modelling. E. HALFON, On the Parameter Structure of a Large Scale Ecological Model. G. SWARTZMAN, Evaluation of Ecological Simulation Models. R. WIEGERT, Modeling Coastal, Estuarine and Marsh Ecosystems: State-of-the-Art.

Statistical Methodology and Sampling: J. DERR and J. K. ORD, Field Estimates of Insect Colonization, II. J. A. HENDRICKSON, JR., Analyses of Species Occurrences in Community, Continuum and Biomonitoring Studies. R. SHESHINSKI, Interpolation in the Plane. The Robustness to Misspecified Correlation Models and Different Trend Functions. W. C. TORREZ, The Effect of Random Selective Intensities on Fixation Probabilities. R. GREEN, A Graph Theoretical Test to Detect Interference in Selecting Nest Sites. I. NOY-MEIR, Graphical Models and Methods in Ecology. T. J. QUINN, The Effects of School Structure on Line Transect Estimators of Abundance. J. W. HAZARD and S. G. PICKFORD, Line Intersect Sampling of Forest Residue.

Applied Statistical Ecology: R. C. HENNEMUTH, Man as Predator. V. F. GALLUCCI, On Assessing Population Characteristics of Migratory Marine Animals. P. A. EISEN and J. S. O'CONNOR, MESA Contributions to Sampling in Marine Environments. W. E. WATERS and V. H. RESH, Ecological and Statistical Features of Sampling Insect Populations in Forest and Aquatic Environments. R. L. KAESLER, Statistical Paleoecology: Problems and Perspectives. S. A. L. M. KOOIJMAN and R. HENGEVELD, The Description of Non-Linear Relationship Between Some Carabid Beetles and Environmental Factors. P. E. PULLEY, R. N. COULSON, and J. L. FOLTZ, Sampling Bark Beetle Populations for Abundance.

A Bibliography: B. DENNIS, G. P. PATIL, M. V. RATNAPARKHI, and S. STEHMAN, A Bibliography of Selected Books on Quantitative Ecology and Related Ecometrics.

QUANTITATIVE POPULATION DYNAMICS
D. G. Chapman and V. F. Gallucci, editors   **300 pp. approx.**
D. G. CHAPMAN and V. F. GALLUCCI, Population Dynamics Models and Applications. J. G. BALCHEN, Mathematical and Numerical Modeling of Physical and Biological Processes in the Barents Sea. A. BERRYMAN and G. C. BROWN, The Habitat Equation: A Fundamental Concept in Population Dynamics. C. A. BRAUMANN, Population Adaptation to a "Noisy" Environment: Stochastic Analogs of Some Deterministic Models. L. GINZBERG, Genetic Adaptation and Models of Population Dynamics. M. I. GRANERO-PORATI, Stability of Model Systems Describing Prey-predator Communities. K. J. BEUTER, C. WISSEL, and U. HALBACH, Correlation and Spectral Analysis of Population Dynamics in the Rotifer *Brachionus Calyciflorus Pallas*. G. G. WALTER, Surplus Yield Models of Fisheries Management. SOME MORE PAPERS IN PREPARATION.

Contributors to This Volume

Bargmann, Rolf E.
Department of Statistics and Computer
  Science
University of Georgia

Chardy, P.
Centre Océanologique de Bretagne
Brest, France

Dale, M. B.
Cunningham Laboratory
CSIRO, Australia

De La Salle, P.
Centre Océanologique de Bretagne
Brest, France

Gittins, R.
Department of Biogeography and
  Geomorphology
Australian National University

Gokhale, D. V.
Department of Statistics
University of California, Riverside

Grant, Ken J.
Department of Mathematics and Computing
  Studies
Western Australian Institute of Technology

Grassle, J. Frederick
Woods Hole Oceanographic Institution

Hendrickson, John A., Jr.
Division of Limnology and Ecology
Academy of Natural Sciences

Hengeveld, R.
Department of Geobotany
Catholic University, Nijmegen

Hogeweg, P.
Bioinformatica
University of Utrecht

Jancey, R. C.
Department of Plant Sciences
University of Western Ontario

Kravitz, David
Woods Hole Oceanographic Institution

Lamont, Byron B.
Department of Biology
Western Australia Institute of Technology

Laurec, A.
Centre Océanologique de Bretagne
Brest, France

Malley, James D.
Division of Computer Research and
  Technology
National Institutes of Health

Mosimann, James E.
Division of Computer Research and
  Technology
National Institutes of Health

Orlóci, László
Department of Plant Sciences
University of Western Ontario

Podani, János
Research Institute for Botany
Hungarian Academy of Sciences

Reyment, R. A.
Uppsala University

Rickaert, M.
Centre Océanologique de Bretagne
Brest, France

Scott, E. L.
Department of Statistics
University of California, Berkeley

Smith, Woollcott
Woods Hole Oceanographic Institution

Steinhorst, R. ·Kirk
Institute of Statistics
Texas A&M University

Stiteler, William M.
College of Environmental Science and
  Forestry
State University of New York

Szócs, Zoltán
Research Institute of Botany
Hungarian Academy of Sciences

PREFACE TO THE VOLUME

This volume results from a very productive session on multivariate methods, held at the University of Parma in Parma, Italy in the summer of 1978, as part of the Satellite Program in Statistical Ecology. This session provided a unique opportunity to bring together people with a wide range of backgrounds and interests. The time was right for the Satellite Program.

The past decade has seen a great increase in the level of interest in bringing multivariate techniques to bear on problems in ecology. In some cases the approach has been to simply apply one of the standard multivariate techniques. In other cases the uniqueness of the ecological problem has required that new and innovative techniques be developed. Both of these approaches were represented by the papers at the session.

The papers of the formal session provided the stimulus for many discussions. From these it is apparent that the effects of the Program will continue to be seen in the future as the thinking and research which it stimulated continues to mature.

The contents of the volume are characterized by a diversity which ended all attempts in failure to stratify the subject matter. The papers indeed provide a sample of what is being done in the area, and just as characteristically, of what can and should be done in the future as this important area continues its rapid development.

All of the participants are indebted to ISEP for the opportunity to come together in the stimulating environment at Parma. For their part in making the stay in Parma a truly pleasant experience, we thank O. Rossi, R. Zanni, and the staff of the Istituto di Ecologia. We are particularly grateful to G. P. Patil, who has been the primary force behind the Program. He has accomplished the difficult task with contagious enthusiasm, a task which few would have dared to undertake.

August 1979

L. Orloci
C. R. Rao
W. M. Stiteler

ACKNOWLEDGMENTS

For permission to reproduce materials in this volume, thanks are due to Dr. W. Junk B. V., Publishers, for figures 1 and 2 on page 98.

TABLE OF CONTENTS

MULTIVARIATE METHODS
IN
ECOLOGICAL WORK

L. Orloci, C. R. Rao, and W. M. Stiteler, (eds.),
Multivariate Methods in Ecological Work, pp. 1-9. All rights reserved.
Copyright © 1979 by International Co-operative Publishing House, Fairland, Maryland

STRUCTURAL ANALYSIS OF SINGULAR MATRICES USING UNION-INTERSECTION STATISTICS

ROLF E. BARGMANN

Department of Statistics and Computer Science
University of Georgia
Athens, Georgia 30602 USA

SUMMARY. In many biological and agricultural experiments the number of attributes measured on each experimental unit often far exceeds the number of units. In ecological applications, studies involving description of the enviornment or pattern recognition result in singular matrices of covariances or correlations on the basis of which structural analyses (e.g. factor analysis) must be made. For maximum-likelihood estimation, determinants or inverses of such matrices are required. The union-intersection statistic of internal dependence, however, a function of the characteristic roots of the matrix, can be defined in such a way that it is applicable to singular matrices. Computational algorithms and an example are presented in this paper.

KEY WORDS. structure, analysis, union-intersection, singularity.

1. INTRODUCTION

Factor analysis and statistical analyses of other structural models (Harman, 1967) are gaining increasing importance in ecological research. Examples are the search for characteristics of an environment that attracts rare species of animals, or the description of interactions of chemical compounds in an organism or an ecosystem. Indirectly, these techniques are also useful in pattern recognition and in the pre-processing of categorized data, as occur when three or more qualitative and non-ordinal descriptors (e.g., color of foliage, type of soil) are to be included in a

multivariate analysis. In all these applications it is necessary
to express internal dependence in a sample correlation matrix as
a single index. Many recommendations have been made to define
such a single index of internal dependence (see, e.g., Horst,
1961). Statisticians use the likelihood-ratio test statistic
(Steel, 1949) as such a single measure of dependence. In factor
analysis, the factor loadings can be estimated so as to maximize
the likelihood ratio statistic of partial independence, i.e.,
the determinant of $R_{\sim p}$, the matrix of partial correlations of
the observable variables given the artificial common factors.
The solution thus obtained is identical with the maximum-likelihood estimation of parameters of the covariance structure (or
rather, correlation structure) corresponding to the Factor Analysis
model (Lawley, 1943). In the scaling of categorized data one
can, likewise, estimate the scaled responses in such a way that
the determinant of the correlation matrix of the scaled variables
is a minimum (Chang and Bargmann, 1974). Since all these techniques depend, directly or indirectly, on the determinant or inverse of a correlation matrix, non-singularity of the latter is
required.

Furthermore, it is not usually customary in statistical
optimization procedures to use a likelihood-ratio statistic as an
index, but rather a union-intersection statistic (Roy, 1957). As
an illustration, the dependence between two sets of variables is
expressed as the canonical correlation, which is the square root
of the largest eigenvalue of a matrix $Q_{\sim} = R_{\sim 11}^{-1} R_{\sim 12} R_{\sim 22}^{-1} R_{\sim 12}'$ and not
by the likelihood-ratio test statistic, which is $|I_{\sim} - Q_{\sim}|$. The
reason for this preference is that union-intersection test statistics are global upper bounds of all indices of dependence in
subsets (canonical correlation $\geq$ any multiple correlation $\geq$ absolute
value of any simple correlation between the two sets). Thus, it
would be advisable to use the union-intersection statistic for the
test of overall independence in a set of p variables. This test
statistic was described in Schuenemeyer and Bargmann (1978). Let
λ_ℓ be the largest and λ_s the smallest characteristic root of the
correlation matrix $R_{\sim}$. Then the union-intersection statistic
for the test of internal independence is the maximum eccentricity,
$r = (\lambda_\ell - \lambda_s)/(\lambda_\ell + \lambda_s)$. It represents the eccentricity of the
two-dimensional ellipse of greatest elongation which can be obtained
by projection of the p-dimensional ellipsoid $x'R^{-1}x = 1$ onto any
two-dimensional subspace through the origin. It is also shown by
Scheunemeyer and Bargmann (1978) that the eccentricity of the two-
dimensional correlation matrix (i.e., not Wishart matrix or sample

covariance matrix) is, in fact, the correlation between the two variables. The maximum eccentricity is the correlation that can be attained between the two linear combinations $u = (e_\ell - e_s)'y$ and $v = (e_\ell + e_s)'y$, where y denotes the p random variables, and e_ℓ and e_s are the eigenvectors associated with the largest and smallest roots of R , respectively. Thus, the maximum eccentricity r is also an upper bound for any canonical correlation which could be obtained by arbitrary decomposition of the p variables in two sets. Many examples were presented in Schuenemeyer (1975) to show that the actual largest canonical correlation is, in practice, only very slightly smaller than the union-intersection statistic of the test of internal independence.

The exact distribution of this union-intersection statistic, under the null hypothesis of internal independence, is under study. Simulation very clearly indicates that for $p = 3$, r^2 (the square of the maximum eccentricity) has the beta distribution with parameters 1 and $(n_e-2)/2$, where n_e denotes the degrees of freedom for error (=n-1, if the sample is from a homogeneous population). For larger values of p , a similarly close correspondence between the distribution based on simulation and the standard characteristic root (Roy, 1957) distribution could not yet be demonstrated.

Unlike the likelihood-ratio statistic, the union-intersection statistic can be defined for singular matrices, also. It is merely necessary to replace the smallest root by the smallest non-zero root of R . Section 2 contains a description of the computational algorithm for the estimation of loadings in factor analysis by minimax eccentricity. Algorithms for other applications are quite similar, and have been described, in outline form, in Schuenemeyer (1975).

2. DESCRIPTION OF THE ALGORITHM

Estimation of factor loadings in Factor Analysis must be done in such a way that the matrix of partial correlations of the observable variables, given the artificial common factors, is as close to the identity matrix as possible. Let there be p variables (attributes or observable attributes), and k common factors (artificial, non-observable variables). In all cases of interest, k is much smaller than p . Let R be the sample correlation matrix of the p observable attributes. The unknown $(p$ by $k)$ matrix of factor loadings is denoted by F . (In the orthogonal representation, this $\underset{\sim}{F}$ can also be regarded as the matrix of sample correlations between the p observable and the k artificial variables.) To approach identity, this $\underset{\sim}{F}$ will be estimated in such a way that the maximum eccentricity of the matrix of partial correlations,

$$\underset{\sim p}{R} = \underset{\sim d}{D}^{-1} (\underset{\sim}{R} - \underset{\sim\sim}{FF'})\underset{\sim d}{D}^{-1} \quad , \tag{1}$$

is a minimum (minimax-eccentricity estimation). In equation (1), $\underset{\sim d}{D}$ is a diagonal matrix with elements $d_i = \sqrt{(1 - h_i^2)}$ along the Principal diagonal, where $h_i^2 = \sum\limits_{\alpha=1}^{k} f_{i\alpha}^2$ (sample communalities, between the ith observable variable and all common factors). The function to be minimized is

$$\phi = (\lambda_\ell - \lambda_s)/(\lambda_\ell + \lambda_s) \quad , \tag{2}$$

where λ_ℓ is the largest and λ_s is the smallest non-zero characteristic root of $\underset{\sim p}{R}$ defined in equation (1). The computer program performs these iterations by the Fletcher-Powell (1963) technique; thus function ϕ and the first derivatives need to be evaluated. Since very good first guesses of $\underset{\sim}{F}$ are easily available (e.g., by the reflected centroid method, Thurstone, 1935), iterations can proceed from such a first guess. The characteristic roots and vectors are obtained by the very fast and very precise Givens-Householder method (see Ralston and Wilf, 1967).

We introduce

$$\underset{\sim}{X} = \underset{\sim d}{D}^{-1} \underset{\sim}{F} \quad \text{and} \quad u_i = (1 + \sum\limits_{\alpha=1}^{k} x_{i\alpha}^2)^{\frac{1}{2}} \quad .$$

Then $\underset{\sim p}{R} = \underset{\sim u}{D}\underset{\sim}{R}\underset{\sim u}{D} - \underset{\sim\sim}{XX'}$, or

$$(\underset{\sim p}{R})_{ij} = u_i u_j r_{ij} - \sum\limits_{\alpha=1}^{k} x_{i\alpha}x_{j\alpha} \quad . \tag{3}$$

Then

$$\partial\phi/\partial X_{i\alpha} = [(\partial\lambda_\ell/\partial X_{i\alpha})(1-\phi) - (\partial\lambda_s/\partial X_{i\alpha})(1+\phi)]/(\lambda_\ell+\lambda_s) \quad , \tag{4}$$

where

$$\partial\lambda_\ell/\partial X_{i\alpha} = 2(X_{i\alpha}e_{i\ell}/u_i)\cdot\sum_{j=1}^{p}e_{j\ell}u_j r_{ij}$$

$$- 2e_{i\ell}\sum_{j=1}^{p}e_{j\ell}X_{j\alpha} \quad , \tag{5}$$

with a similar expression for $\partial\lambda_s/\partial X_{i\alpha}$. Here $e_{i\ell}$ is the $i th$ element of the eigenvector associated with the largest root of $R_{\sim p}$, e_{is} the corresponding element associated with the smallest non-zero root, u_j and $X_{j\alpha}$ are defined in equation (3), and r_{ij} are the elements of the original correlation matrix.

As in many other minimax problems (see, e.g. Bargmann and Baker, 1977), the approach to a solution is indicated by a number of equalities. In the present case, as the process nears a solution, the two largest roots and/or the two smallest non-zero roots of R_p become equal. Because of this, the eigenvectors become nearly indeterminate; the computer program searches for various combinations of eigenvectors associated with a pair of adjacent roots. Even so, the number of iterations required is very large (81 iterations were needed to produce the results reported in Section 3); fortunately, because of the quickness of the Givens-Householder algorithm, CPU time is not excessive (25 CPU sec were needed for the 81 iterations on a Cyber 172, in the 10-variable 2-factor example). Attempts are in progress to modify the program so as to force the two largest and two smallest roots to equality. No satisfactory algorithm has been found thus far.

3.   ILLUSTRATION

The following example, with artificial data, is presented to show that the minimax-eccentricity solution of a singular correlation matrix can recover the solution for the non-singular matrix from which the former was obtained by the addition of two linear combinations. (In the 10 by 10 case, variable 4 is the sum of 1 and 3, and variable 6 is the sum of 1 and 5.) The corresponding matrix for the non-singular model is the matrix obtained from Table 1 by the omission of rows and columns 3 and 6.

Table 2 is the solution obtained for the non-singular matrix (the 8 x 8 matrix embedded in Table 1). Table 3 is the solution for the 10 x 10 singular matrix. To facilitate comparison, the solutions were transformed to a simple structure representation,

TABLE 1:   Correlation matrix.

1	2	3	4	5	6	7	8	9	10
1	.6819	.6413	.9058	.8968	.9738	.8959	.6349	.4643	.4275
	1	.6703	.7463	.6922	.7055	.7156	.8921	.6799	.6943
		1	.9058	.6285	.6519	.6323	.5962	.4596	.4437
			1	.8419	.8973	.8435	.6795	.5099	.4808
				1	.9738	.8978	.6229	.4379	.4366
					1	.9209	.6458	.4632	.4436
						1	.6631	.4239	.4543
							1	.6081	.6390
								1	.4643
									1

TABLE 2:   Factor loadings from non-singular matrix.

Variable	F		h_i^2
	I	II	
(1)	.9431	-.1108	.90171
(2)	.7932	.5988	.98773
(3)	.6739	.2123	.49921
(5)	.9373	-.0796	.88487
(7)	.9439	-.0574	.89424
(8)	.7272	.5212	.80047
(9)	.5183	.4532	.47403
(10)	.5201	.4645	.48626

Roots of the partial correlation matrix:

$\lambda_1 = 1.2284$ $\lambda_2 = 1.2284$ $\lambda_7 = .7989$ $\lambda_8 = .7965$

TABLE 3:   Factor loadings from singular matrix.

Variable	F I	II	h_i^2
(1)	.9683	-.0936	.94637
(2)	.7793	.6065	.97515
(3)	.7360	.2226	.59125
(4)	.8875	.0076	.78771
(5)	.9468	-.0871	.90402
(6)	.9729	-.0786	.95271
(7)	.9438	-.0415	.89248
(8)	.7168	.5440	.80974
(9)	.5122	.5234	.53630
(10)	.4842	.4872	.47181

Roots of the partial correlation matrix:

$$\lambda_1 = 1.9726 \quad \lambda_2 = 1.9726 \quad \lambda_7 = .89139 \quad \lambda_8 = .89128$$

TABLE 4:   Simple structure representations: A = singular (10 variables), B = non-singular (8 variables).

Variable	V_1 A	B	V_2 A	B
(1)	.6989	.6701	-.0174	-.0279
(2)	.0432	.0196	.6658	.6659
(3)	.3075	.2497	.2796	.2704
(4)	.5694		.0772	
(5)	.6800	.6419	-.0126	.0027
(6)	.6904		-.0020	
(7)	.6433	.6286	.0327	.0254
(8)	.0503	.0398	.5985	.5828
(9)	-.0666	-.0359	.5620	.4968
(10)	-.0572	-.0436	.5237	.5082

which is shown in Table 4. Table 3 indicates, in comparison with Table 2, that the communalities of variables (1), (3), and (5) were increased, as expected, since these entered into the construction of the two linear combinations (4) and (6). The computer program regards as essentially non-zero those roots which are greater than 1/p (p = number of variables). Thus, λ_s was λ_8 in both studies. The equality of λ_1 and λ_2 , and of λ_7 and λ_8 is obvious in both studies. The discrepancy in the communalities of variable (9) cannot be explained.

Table 4 shows that, with the exception of variable (9), the non-singular portion (which, in this example was known) has been adequately represented in the solution based on the singular matrix. The transformation which produced the two A columns in Table 4 from Table 3 was

$$T = \begin{bmatrix} .648100 & .078437 \\ -.761555 & .996919 \end{bmatrix} ,$$

and that which produced the two B columns of Table 4 from Table 2 was

$$T = \begin{bmatrix} .618127 & .087448 \\ -.786078 & .996169 \end{bmatrix} .$$

Thus, the correlations between the simple structure factors is .7084 in the ten-variable case, and .7290 in the eight-variable case.

REFERENCES

Bargmann, R. E. and Baker, F. (1977). A minimax approach to component analysis. In *Applications of Statistics*, P. S. Krisnaiah, ed. North Holland, Amsterdam. 55-69.

Chang, J. C. and Bargmann, R. E. (1974). Internal multi-dimensional scaling of categorical variables. Department of Statistics, THEMIS Report 34, University of Georgia, Athens, Georgia.

Fletcher, R. and Powell, M. J. D. (1963). A rapidly convergent descent method for minimization. *The Computer Journal*, 6, 163-168.

Harman, H. H. (1967). *Modern Factor Analysis*, 2nd ed. The University of Chicago Press, Chicago.

Horst, P. (1961). Relations among m sets of measures. *Psychometrika*, 26, 129-149.

Lawley, D. N. (1943). The application of the maximum-likeihood method to factor analysis. *British Journal of Psychology*, 33, 172-175.

Ralston, A. and Wilf, H. S., eds. (1967). *Mathematical Methods for Digital Computers*, Vol. 2. Wiley, New York.

Roy, S. N. (1957). *Some Aspects of Multivariate Analysis*. Wiley, New York.

Schuenemeyer, J. H. (1975). *Maximum eccentricity as a union-intersection test statistic in multivariate analysis*. Ph.D. dissertation, University of Georgia, Athens, Georgia.

Schuenemeyer, J. H. and Bargmann, R. E. (1978). Maximum eccentricity as a union-intersection test statistic in multivariate analysis. *Journal of Multivariate Analysis*, 8, 222-227.

Steel, R. G. D. (1949). *Minimum generalized variance for a group of linear functions*. Ph.D. dissertation, University of Iowa, Ames, Iowa.

Thurstone, L. L. (1935). *The Vectors of Mind*. The University of Chicago Press, Chicago.

[Received July 1978. Revised January 1979]

L. Orloci, C. R. Rao, and W. M. Stiteler, (eds.),
Multivariate Methods in Ecological Work, pp. 11-20. All rights reserved.
Copyright © 1979 by International Co-operative Publishing House, Fairland, Maryland

ON LINGUISTIC APPROACHES TO ECOSYSTEMS AND THEIR CLASSIFICATION

M. B. DALE

CSIRO
Cunningham Laboratory
St. Lucia, QLD., Australia 4067

SUMMARY. This paper discusses the possible use of grammar theory to provide a formal basis for classification of ecasystems. The importance of logical dependency and a natural definition of similarity are discussed. Some thoughts on the inference of grammars and the introduction of semantic constructs are also presented.

KEY WORDS. grammars, linguistics, classification.

1. INTRODUCTION

There has been some considerable debate in the field of pattern recognition concerning the relative merits of *statistical* and *linguistic* approaches to the problem of assigning items to classes. The former makes use of an essentially geometric view, using values of simple properties to provide coordinates for the item in some space. Hopefully items in the same class will lie close together so that assignment can be based on distributional criteria such as likelihood. The difficulties lie first in establishing a suitable set of features to measure, and second in that while behaviorly isomorphic, the statistics are rarely structurally isomorphic with the real system.

In contrast the latter approach is often known as a structural approach, and emphasises the recursive description of objects in

terms of subparts and their relationships. The approach may have
fuzzy or stochastic elements but is essentially deterministic in
outlook. Such descriptions are very similar to parsings in linguistics; thus grammars, that is formal language theories, may
play an important role. The approach suffers, however, in that like
all deterministic models it must approach the real system in
complexity, in effect demanding structural isomorphism. While
inconvenient in many cases of interpolation, where behavioral
isomorphism and hence stochastic approximation are adequate, the
strength of the linguistic approach lies in extrapolation beyond
the available observations.

It is not my intention here to argue in detail the specific
merits of either approach; clearly both have a place. Instead I
would like to examine how the linguistic approach may be useful
in the recognition of classes, and particularly in the classification of vegetation. The use of grammars does throw some interesting light on some darker corners of numerical classification
which have otherwise proved intractable.

2. LOGICAL DEPENDENCY

One of the earliest problems faced in numerical classification
was the existence of logical dependencies, or necessary correlations,
between descriptors (Proctor and Kendrick, 1963). In its simplest
form this is an exclusion statement, exemplified in Williams'
aphorism - some noughts are noughtier than others. A typical
example is the number of hairs on a leaf. You may have some hairs
or no hairs but what happens when you have no leaf?

Obviously some form of weights would be appropriate. Dale
(1968) has discussed the possibility of list-structured descriptors
in general terms while Williams (1969) provided the required
weights for simple cases. However, more complex situations are
computationally intractable. The recommended approach in numerical
classification has been to avoid such situations wherever possible
and where imperative adopt some simple weighting scheme and hope
this was adequate. What is surprising here is that one characteristic of human taxonomies is that they rely heavily on characters with
many logically dependent descriptors. The biological world is
disjointed. Thus numerical methods, which are presently almost all
statistical, seek to avoid using precisely those characteristics
human classifiers most avidly seek! We clearly need some alternative means of classifying which can cope with logical dependency.
Such means operate at the level of definition of similarity, for
the heuristics of sorting are not related to logical dependency.
Fortunately grammars can provide such means, and I shall later
indicate how this is done. First I want to discuss the nature
and variety of grammars.

3. GRAMMARS

A grammar is an n-tuple, in some extreme cases reaching 11-tuples, but more typically a 4-tuple. The four parts are: (1) a set of non-terminal symbols; (2) a set of terminal symbols; (3) a distinguished non-terminal, the start symbol; and (4) a set of productions which specify the conditions under which a specified set of nonterminals can be replaced by a set of terminals and/or other nonterminals.

In linguistic studies the terminal symbols may be words, but in the general case other symbols can be used, such as lines or regions. In linguistics nonterminal symbols include such things as 'subject' and 'predicate,' and the symbols are arranged sequentially. Alternative nonterminal symbols may be appropriate and the relationships between symbols may be complex. All these alternatives are fairly obvious ones, although they may have far reaching consequences. For example, if intersymbol relationships are arbitrary then the production rules will have to specify not only how some nonterminal changes, but also how the relationships are reorganized. In most linguistic studies the rules are applied serially, singly. Alternative rules of application might be to choose between alternatives in some stochastic manner, or to apply together all rules which can be applied, a socalled parallel grammar. Parallel grammars have obvious biological attractions and may be more succint than serial grammars. In Figure 1 some snow-flake curves are shown, for which Figure 2 shows a serial grammar and Figure 3 a parallel grammar.

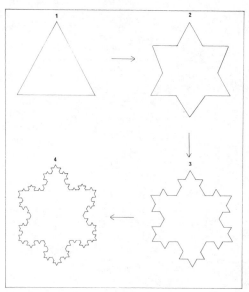

Figure 1 : the first four Snowflake curves.

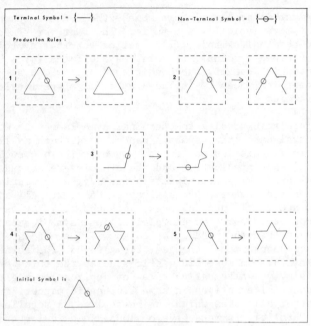

Figure 2 : a serial shape grammar for the Snowflake curve.

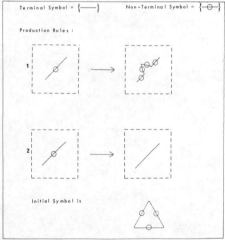

Figure 3 : a parallel shape grammar for the Snowflake curve.

Parallel grammars also seem appropriate in the 3-dimensional systems of Reusch (1976) and Mayoh (1974). However, it is apparently possible to replace most parallel grammars by serial ones, albeit more complex ones. The point of most importance is that the grammar structure reflects a process of generation, either of a terminal sequence, or of some sequence of patterns of nonterminals.

A wide variety of grammars have been suggested including matrix, web, graph, context free, context sensitive, transformational and 2-level grammars. I do not intend to define each and every type, for such a catalog would be rather uninteresting to say the least. As an anonymous Elizabeth air put it: "My love in her attire doth show her wit," and it is to potential uses of grammars I shall now turn.

4.   SERIAL GRAMMARS IN ECOLOGY

There have been two attempts to use serial grammars in ecological studies. Haeffner (1975) used transformational grammars to investigate cooccurrence of species. He assumes the existence of elaborate structured descriptions of every taxon covering autecological information, and also an elaborate environmental description. Essentially he places taxa into the environment and uses his rules to specify the results of this inclusion by adjusting previous taxa and the environment. In discussion, Haeffner admits the essentially descriptive role of his model, which is independent of any actual spatial configuration. My own view is that such an encyclopedic approach demonstrates the essential circulatory and barrenness of the niche concept. The descriptive structures are arbitrary and without biological import, and essentially infinite in scope, which would cause some computational problem.

Bradbury and Loya (1975), in contrast, are concerned to explain the actual spatial distribution of corals, and to learn a grammar for this purpose. It is fairly clear that the grammar learnt was inadequate, and that various modifications are necessary to make the learning process effective. One of the most important features is simply the lack of parallelism, but the method cannot learn bracketed structures either. My own unpublished experiments tend to confirm the impression given by Bradbury and Loya that this is a first, somewhat primitive, attempt which should be followed up.

5.   L-SYSTEMS, SIMILARITY, AND INFERENCE

Parallel grammars have in fact been used in several biological studies, especially in the study of development. The seminal paper here, and a good deal of the subsequent work, is by Lindenmayer (1968)

Lindenmayer, or L-systems are not concerned with the terminal
sequences, which biologically are about always just 'dead,' but
with the successive nonterminal sequences. The work I want to
note here is due to Hogeweg (Hogeweg and Hesper, 1974). By using
a specified grammar (a bracketed, propagating, deterministic, 2-
sided, parallel grammar to be precise), they generated some
sample sequences. These structures were then described using
obvious visual characteristics and submitted to monothetic classification. The resulting groups strongly reflect the rules of the
grammar, although those rules are essentially local in operation.

Thus it seems that a classification can reflect the processes
of generation, so that a natural classification might even be defined as one which does reflect these processes. At least this
joins the phenetic and phylogenetic schools, both of which are seen
to be seeking to infer an appropriate grammar. Equally the taxonomic hierarchy we already have, does presumably reflect our present
hypotheses regarding the grammar of evolution; the higher taxa are
simply reflection of some key production being applied and effectively preventing other productions from being applied later.

Obviously some strong method of inferring the grammar would
be desirable. Coulon and Kayser (1978) suggest that this is not a
standard statistical inference problem, for the concern is with the
rules of the game and not frequently occurring states. Grammar
inference is not a trivial task although various heuristic approaches
have been used (Cook and Rosenfeld, 1976; Fu and Booth, 1975). It
is known that only certain classes of grammars can be sensibly
inferred, and that for some classes the examples used for inference
must have particular properties. Some L-systems are inferrable
as are context-free grammars.

Fu and Lu (1977) have used grammar inference in a novel way to
provide a basis for measurement of similarity. The method is particularly appropriate as it works well with logical dependencies.
Given two items, A and B, we determine the directional similarity
of B to A as follows:

1. Infer a grammar for A.
2. Expand this grammar by a set of error correcting productions.
3. Generate B using this augmented grammar, but noting when
     the error productions are involved, so that n[k] is the
     number of times the kth error production is used.
4. The similarity is then a function of the N[k].

It is important to note that the similarity is determined in
part by the set of error productions incorporated, so that similarity is not a function of the items alone. If the similarity is
simply the sum of the n[k], the resulting value is apparently a
Levenshtein metric. It would be interesting to link this work with

that of Bednarek and Ulam (1977) on Steinhaus metrics. Fu and Lu
have actually used this procedure in classifying hand written
alphabetic character, although there is a considerable computational
cost involved (see also Lu and Fu, 1978).

6. GRAMMARS FOR VEGETATION AND FEATURE SELECTIONS

Besides their uses in general numerical classification, grammars
may have more specific application in vegetation studies. Here the
processes are parallel and it is not clear what symbol sequences
might be used. My own approach treats vegetation as analogous to
a supercooled liquid, with strong local organization but weak
global organization. We can expect, on this approach, to require
two different models of vegetation, one global the other local.

At the global scales, properties comparable to miscibility and
viscosity would be important in temporal studies. For purely
spatial patterns it is the shapes of distribution which are important,
and the texture of the vegetation. These features are the classic
elements of image processing but here, too, grammars have been
used. Pfaltz (1972; see also Rosenfeld, 1976) first introduced
web grammars to deal with graphs of relationships and various
developments have proceeded therefrom, such as tree grammars
(Brayer and Fu, 1977) which are particularly convenient forms.
However, much work would be required to use these formalisms in
vegetation study, especially in identifying appropriate preprocessing
and regionalizing techniques.

At the local scales, the interest lies with the individual
plant and its neighbors. Mayoh (1973) has discussed ring grammars
for situations where the symbol strings are circular, Williams
et al. (1969) have used point clump sampling, where neighbors are
simply spatially close plants, with some success, but use of the
Dirichet Tessalation allows actual neighbors to be identified.
We can clearly seek to infer ring grammars, and classifications
based on such a modified point clump basis would be an interesting
starting place.

There is however one other interesting feature. A grammar still
requires components to make up the symbol strings and there is no
guarantee that our initial labelling of individuals, by species
or life form or anything else, is necessarily appropriate. We
could be looking for a grammar for an inappropriate symbol sequence.
Moreover a classification of individuals in terms of their neighbors,
has the effect of relabelling all individuals with new class labels.
We can clearly repeat the process interating until the labelling
is stable, and only then infer the grammar. Obviously we need
some means of describing the number of labels, or groups, and
perhaps 2-parameter methods will be successful here. However,

this is the only method I know for modifying descriptions during analysis which does not rely on some random generation procedure.

7. SEX, SEMANTICS, AND AFFIXES

While much vegetation description, as indeed much of the English language, is independent of context, this is not true of all features. In the point clump exercise above the actual labelling of the central individual was tacitly ignored. Equally there are interactions which are nonlocal, such as pollination, and these will be constrained to act between particular individuals of opposite sex. There could of course be additional relationships such as those of lianes or epiphytes to their supporting hosts, which are context sensitive.

Now it is well known that context sensitive grammars are not inferable in general. The common solution has been to specify these extra constraints through semantics expressed in English, but with computer languages a recent trend has been to try to formalize these conditions. The idea is to produce a 2-level system within which the semantics are expressable. These 2-level grammars were at first somewhat intractable but recent work by Watt (1974) and Koster (personal communication) shows considerable progress. They utilize extended affix grammars as a means of incorporating semantic information in a relatively transparent manner. What is presently lacking is any information on the inference of such grammars or of interesting subclasses of them, and on equivalent parallel grammars. Progress in these two areas would be extremely interesting biologically.

8. CONCLUSION

The main theme of this paper is really to emphasize that our knowledge of the structure and generating processes for vegetation and ecosystems should be incorporated into models. I believe that grammars have some part to play in this process, although stochastic elements will still play an important role. The above examples do show some interesting clarifications which have not been forthcoming with the statistical approach, and center attention on the local predictivity of interindividual interaction.

Vegetation classification has been a paradise of individuality, eccentricity, heresy, anomalies, hobbies and perhaps humor. I hope that grammars can profitably add some formality.

REFERENCES

Bednarek, A. R. and Ulam, S. M. (1977). *On the theory of relational structures and schemata for parallel computation.* Informal Report LA-6734-MS, Los Alamos Scientific Laboratory, Los Alamos, New Mexico.

Bradbury, R. M. and Loya, Y. (1975). A heuristic analysis of spatial pattern of hermatypic corals at Eilat, Red Sea. *American Midland Naturalist,* 112, 493-507.

Brayer, J. M. and Fu, K. S. (1977). A note on the k-tail method of tree grammar inference. *IEEE Transactions on Systems, Man and Cybernetics,* 7, 293-300.

Cook, C. M. and Rosenfeld, A. (1976). Some experiments in grammar inference. In *Computer Oriented Learning Processes,* J. C. Simon, ed. Nordhoff, Leyden. 157-174.

Coulon, D. and Kayser, D. (1978). Learning criteria and inductive behavior. *Pattern Recognition,* 10, 19-25.

Dale, M. B. (1968). On property structure, numerical taxonomy, and data handling. In *Modern Methods in Plant Taxonomy,* V. H. Heywood, ed. 168-197.

Fu, K. S. and Booth, T. L. (1975). Grammatical Inference: Introduction and Survey Part I and II. *IEEE Transactions on Systems, Man and Cybernetics,* 5, 95-111.

Fu, K. S. and Lu, S. Y. (1977). A clustering procedure for syntactic patterns. *IEEE Transactions on Systems, Man and Cybernetics,* 7, 734-742.

Haeffner, J. W. (1975). Generative Grammars that simulate ecological systems. *Simulation Council Proceedings Series,* 5, 189-211.

Hogeweg, P. and Hesper, B. (1974). A model study of biomorphological description. *Pattern Recognition,* 6, 165-179.

Lindenmayer, A. (1968). Mathematical Models for cellular interaction in development. 1. Filaments with 1-sided inputs. *Journal of Theoretical Biology,* 18, 280-299.

Lu, S. Y. and Fu, K. S. (1978). A sentence to sentence clustering procedure for pattern analysis. *IEEE Transactions on Systems, Man and Cybernetics,* 8, 381-389.

Mayoh, B. (1973). *Mathematical models for cellular organisms.* Technical Report 12, Computer Science Department, University of Aarhus.

Mayoh, B. H. (1974). Multidimensional Lindemayer Organisms. In
    L Systems, G. Rozenberg and A. Salomaa, eds. Lecture Notes in
    Computer Science 15, Springer, Verlag, New York. 302-326.

Pfaltz, J. L. (1972). Web grammars and picture description. *Computer Graphics and Image Processing*, 1, 193-220.

Proctor, J. R. and Kendrick, W. B. (1963). Unequal weighting in
    numerical taxonomy. *Nature*, 197, 716-717.

Reusch, P. J. A. (1976). A common approach to retrieval concepts
    and multidimensional development systems. *Gesellschaft für
    Mathematik und Dataverarbeitungen*, 90, 1-70.

Rosenfeld, A. (1976). Array and Web grammars: an overview. In
    Automata, Languages, Development, A. Lindenmayer and R. Rosenberg,
    eds. North Holland Publishing Co., Amsterdam. 517-529.

Watt, D. A. (1974). *Analysis-oriented two level grammars*. Ph.D.
    thesis, Glasgow University.

Williams, W. T. (1969). The problem of attribute weighting in
    numerical classification. *Taxon*, 18, 369-374.

Williams, W. T., Lance, G. N., Webb, L. T., Tracey, G. T., and
    Connell, J. H. (1969). Studies in the numerical analysis of
    complex rain forest communities. IV. A method for the
    elucidation of small scale patterns. *Journal of Ecology*,
    57, 635-654.

[*Received March* 1979]

L. Orloci, C. R. Rao, and W. M. Stiteler, (eds.),
Multivariate Methods in Ecological Work, pp. 21-53. All rights reserved.
Copyright © 1979 by International Co-operative Publishing House, Fairland, Maryland

ANALYSIS OF ECOLOGICAL FREQUENCY DATA:
CERTAIN CASE STUDIES

D. V. GOKHALE

Department of Statistics
University of California
Riverside, California 92521 USA

SUMMARY. The paper deals with analysis of frequency data from
ecological investigations by using the minimum discrimination
information approach. With the help of several examples it is
shown how the present method is more appropriate and how it often
provides deeper insight into the phenomenon under study, as compared
to the usual methods such as regression analysis, analysis-of-
variance or two-way-chi-square.

KEY WORDS. ecological contingency tables, minimum discrimination
information approach.

1.   INTRODUCTION

     This expository paper is concerned with the analysis of fre-
quency or count data that arise in ecological investigations with
particular reference to problems in fisheries and marine biology.
Methods of analysis presented here offer more valid means of interpreting the data. In many instances, they are more appropriate
than classical methods and provide a deeper insight into the problem.
These statements are illustrated here with the help of numerous
examples taken from recent journals and monographs.

     Basically, the methodology makes use of the principle of
minimum discrimination information (MDI) for model building and
model validation in multinomial experiments. A particular case
of the model building aspect is that of 'log-linear models.'
Application of these models in five dimensional ecological contingency tables is discussed first by Fienberg (1972). The present

paper uses a different parameterization and points out some interesting interpretations not discussed by Fienberg (1972).

Looking at recent issues of many ecological research journals one notices that there are essentially three standard methods of analyzing discrete data: (i) two-way chi-square, (ii) simple or multiple regression, and (iii) analysis of variance. The two-way chi-square often provides some insight into the problem but gives only a pairwise analysis. It completely ignores the effect of other variables that are present and does not permit analysis of three or higher factor interactions. Regression and analysis of variance methods require some or all of the following assumptions for a complete analysis of actual measurements:

(i) random sampling;

(ii) a structure on the population means, (such as

$$E[Y_{ij}] = \mu + \alpha_i + \beta_j);$$

(iii) constancy of variance, $(Var(Y_{ij}) = \sigma^2$ for all i and j);

(iv) normality of underlying distributions.

Without assumption (iv), validity of the F-tests and other tests is open to doubt. As a remedy, some authors use 'distribution-free' procedures. But for these too, assumptions such as continuity of distribution functions are required. Another usual remedy is the use of transformations. But after analyzing the transformed data it often becomes difficult to interpret the results as they relate to the actual measurements. Further, in that process the statistical properties of estimates may become intangibly warped.

The methodology proposed here does away with all the assumptions above *except* (i). On the other hand, there are hardly any applications-oriented statistical techniques which can provide something significant without that assumption. Since the analysis requires minimal assumptions, the conclusions are valid for a wider class of problems. This validity is obtained at the cost of the requirement that the *sample be large*. As the following examples show, there are many ecological studies where paucity of data is not a problem. Also when frequencies are analyzed rather than actual measurements, one immediate advantage is that the underlying variable can be quantitative as well as qualitative.

This paper aims at providing more appropriate substitutes for regression and analysis of variance methods. It also indicates a few other interesting applications of the MDI approach to ecological data. In the following two sections the necessary notation and background is developed. Mathematical details are kept to a

minimum. They can be found in the monographs by Kullback (1959)
or Gokhale and Kullback (1978). The remainder of the paper can be
looked upon as a collection of 'case studies.' Its lay-out is
modular in that each example is self-contained and consists of two
parts. The first part describes the data-sets, the problems under
investigation and the statistical techniques used accompanied by
their partial critique if appropriate. The second part consists
of a reanalysis of the data using the MDI approach, the conclusions
obtained from it and a comparison with earlier analysis.

For other models such as those based on 'logits,' the reader
is referred to Bishop *et al.* (1975), Fienberg (1977), or Gokhale
and Kullback (1978). Also, methods of handling (too many) zero
counts are not discussed here, their nature being rather technical.
They may be found in the monographs mentioned above.

2. NOTATION AND PRELIMINARIES

Consider N observations classified independently into a
certain number (denoted by Ω) of cells according to one or more
characteristics. Each characteristic may have several different
categories (or *levels*) of classification. Data represented in
such grouped forms are called frequency distributions. An important
special case of a frequency distribution is a *contingency table*.
The methodology described in this paper is applicable to all forms
of tabular data for which the table-entries are frequencies. The
table may have one or several dimensions of classification. For
contingency tables the conventional notation '4 × 2 × 5' signifies
that it is a three-way table, the first characteristic has four
categories, the second has two, and the third has five. (In this
instance Ω = 40.) More details are supplied by the format illustrated below which is used throughout this paper. Consider the
data (Table 1) on trapping of mourning doves (Henry, Baskett,
Sadler, and Goforth, 1976). The characteristics and levels are
summarized below the table. A typical frequency of occurrence in
the body of the table is denoted by the symbol x followed by the
cell index in parentheses. Thus x(212) is the number of doves
trapped in 1964 (h = 2), which are adults (i = 1), and in the
field (j = 2); x(212) = 44. General notation for cell frequencies
is x(hij); h = 1,2,···,7, i = 1,2, j = 1,2. The corresponding
observed proportions are denoted by $\hat{\pi}$(hij) with $\hat{\pi}$(hij) = x(hij)/N.
(The definition of $\hat{\pi}$ is different when there are two or more
samples; see Section 3). The underlying probabilities are denoted
by p(hij). Thus p(hij) is the probability that an individual
observation is classified in the cell (hij). Note that the p(hij)
are constant from one individual to the other.

24   D. V. GOKHALE

TABLE 1: *Number of mourning doves trapped in different locations over different years (Henry, Baskett, Sadler, and Goforth, 1976).*

| | Adult (i=1) | | Immature (i=2) | |
Year (h)	Roadside (j=1)	Field (j=2)	Roadside (j=1)	Field (j=2)
1963 (1)	226	50	45	97
1964 (2)	205	44	54	111
1965 (3)	387	58	91	28
1966 (4)	185	83	69	359
1967 (5)	100	221	23	241
1968 (6)	106	152	18	406
1969 (7)	119	78	16	73

Summary of Characteristics and Levels

Characteristic	Index	1	2	$\cdots$	7
Year	h	1963	1964	$\cdots$	1969
Age	i	Adult	Immature		
Location	j	Roadside	Field		

The *dot-notation* is used to denote totals. Thus

$$x(\ldots) = \Sigma_h \Sigma_i \Sigma_j \; x(hij) = N \quad,$$

$$\hat{\pi}(h.j) = \Sigma_i \hat{\pi}(hij) \quad,$$

$$p(.i.) = \Sigma_h \Sigma_k p(hij) \quad.$$

For a two-way table the cell probabilities are denoted by p(hi), for a three-way table by p(hij) etc. It is extremely convenient to use a unified notion p(ω), say, where ω generically denotes an (hi) cell in a two-way table, an (hij) cell in a three-way table tec. This is achieved by arranging the cells of a contingency table in *lexicographic* order. For example, in a 4 × 2 × 3 table, the symbol ω takes values 1,2,$\cdots$,24 and the correspondence between ω and the cell index is given by

ω: 1   2   3   4 $\cdots$ 24

cell index: 111   112   113   121 $\cdots$ 423.

In the process of data analysis the underlying probabilities are usually required to satisfy some *linear constraints*. These constraints are expressed in matrix notation as

$$Cp = \theta \quad , \tag{1}$$

where C is a $(r+1)\times\Omega$ matrix and θ is a $(r+1)\times1$ vector. Both C and θ are assumed to be known. The first row of C consists of all ones and the first element of θ is also one. This incorporates in the equations the *natural* constraint $\Sigma_\omega p(\omega) = 1$.

As a simple example, suppose that in a 2×2 table the underlying probabilities $p(ij)$ are required to satisfy the same marginal totals as the observed proportions. These constraints can be formulated as

$$p(1.) = \hat{\pi}(1.) \quad \text{(This implies } p(2.) = \hat{\pi}(2.)) \quad ,$$
$$p(.1) = \hat{\pi}(.1) \quad \text{(This implies } p(.2) = \hat{\pi}(.2)) \quad , \tag{2}$$

and are expressible as (1) by letting

$$
C =
\begin{array}{cccc}
(11) & (12) & (21) & (22)
\end{array}
$$

$$
C = \begin{bmatrix}
1 & 1 & 1 & 1 \\
1 & 1 & 0 & 0 \\
1 & 0 & 1 & 0
\end{bmatrix}
$$

$$p = (p(11), p(12), p(21), p(22))' \quad .$$

$$\theta = C\hat{\pi}$$

$$= (\hat{\pi}(11), \hat{\pi}(12), \hat{\pi}(21), \hat{\pi}(22))' \quad .$$

It is assumed throughout for the sake of simplicity and without loss of generality that the rows of C are linearly independent, i.e., rank $(C) = r+1$. (Hence in the above 2×2 example the constraints $p(2.) = \hat{\pi}(2.)$ and $p(.2) = \hat{\pi}(.2)$ should *not* be included in addition to (2)). As another example, suppose that in a 2×2 table the underlying probabilities are required to satisfy the constraint $p(12) = p(12)$ (hypothesis of symmetry), then the matrix C and vector θ of (1) become

$$
\begin{array}{cccc}
(11) & (12) & (21) & (22)
\end{array}
$$

$$
C = \begin{bmatrix}
1 & 1 & 1 & 1 \\
0 & 1 & -1 & 0
\end{bmatrix}
$$

$$\theta = (1 \ 0)' \quad .$$

The two examples given above also illustrate two important types of problems that arise in the analysis of multinomial

experiments. In the first example, the underlying probability
distribution is required to have the same marginal totals as the
observed distribution. In other words, the underlying model is
postulated to use that much (and no more) information from the
data as given by the specified (and implied) marginal totals.
More generally, when linear combinations of underlying probabilities
are required to have the same values as given by the same linear
combinations of observed proportions, the problem is called an
Internal Constraints Problem (ICP). ICP can be looked upon as
a *model-building* or smoothing-cum-fitting process. Note that in
ICP, the vector θ of (1) is obtained 'internally', i.e., by
using the observed distribution; in fact $\theta = C \pi$.

In the second example, it was possible to derive the constraint-
equations by using the hypothesis under question, i.e., 'externally'
to the data. Such problems, which arise in *model validation* are
called External Constraints Problems (ECP).

3. THE MDI APPROACH

Let $\bar{\Omega}$ denote the set of elements $\{1,2,3,\cdots,\Omega\}$. For $\omega \epsilon \bar{\Omega}$
if $p(\omega)$ and $\pi(\omega)$ are any two probability distributions defined
on $\bar{\Omega}$, the discrimination information $I(p:\pi)$ in p with respect
to π is given by

$$I(p:\pi) = \Sigma_\omega p(\omega) \ln \frac{p(\omega)}{\pi(\omega)} \ .$$

(It is assumed that $p(\omega) = 0$ whenever $\pi(\omega) = 0$ and $0 \ln 0 = 0$).
The function $I(p:\pi)$ is well known to many ecologists as a measure
of diversity when π is taken as the uniform distribution. $I(p:\pi)$
can be looked upon as a measure of closeness between the distributions
p and π. For example, if $\bar{\Omega} = \{1,2,3,4\}$ and if

$$\pi(\omega) = \tfrac{1}{4} \text{ for } \omega \epsilon \bar{\Omega}$$

$$p_a(1) = 0.3, \quad p_a(2) = 0.3, \quad p_a(3) = 0.2, \quad p_a(4) = 0.2,$$

and $\qquad p_b(1) = 0.4, \quad p_b(2) = 0.3, \quad p_b(3) = 0.2, \quad p_b(4) = 0.1,$

then $I(p_a:\pi) = 0.02014$ and $I(p_b:\pi) = 0.1064$ indicating that the
distribution p_a is closer to π than p_b. In fact $I(p:\pi)$ is
always non-negative and equals zero if and only if $p(\omega) = \pi(\omega)$ for
all $\omega \epsilon \bar{\Omega}$. Suppose that π is a known 'reference' distribution,
not necessarily $\hat{\pi}$, and that p is required to satisfy linear
constraints of the form $Cp = \theta$ given by (1). Consider the problem

of choosing from all probability distributions satisfying (1), the
one which is closest to π in the sense that it minimizes $I(p:\pi)$.
The solution to this problem is known as the minimum discrimination
information theorem (Kullback, 1959) which states that such a p^*,
which minimizes $I(p:\pi)$ subject to (1), exists uniquely and
is given by

$$\ln \frac{p^*(\omega)}{\pi(\omega)} = L + \Sigma_{j=2}^{r+1} C(j,\omega)\tau_j \quad , \quad \omega\epsilon\bar{\Omega} \quad , \tag{3}$$

where parameters L and τ_j, $j=1,\cdots,r+1$, are determined so that
p^* satisfies (1). The corresponding MDI estimate of the cell-
frequency is denoted by $x^*(\omega) = Np^*(\omega)$.

Internal constraints problems. As mentioned earlier, in ICP the
constraints are taken as linear combinations of observed proportions. The objective is to smooth out the data by arriving at a
model which explains the data in terms of as few parameters as
possible. Thus, the reference distribution π may be any distribution which satisfies some constraints of (1) and does not involve
any more parameters than in the ones in the representation (3). It
turns out that the actual choice of such π is immaterial; the
estimates of cell frequencies, test-statistics, and related analysis
remain the same (see Gokhale and Kullback, 1978). The uniform
distribution over $\bar{\Omega}$ is the most obvious choice for π in ICP
since it has a scaling parameter only. On the other hand the observed distribution $\hat{\pi}$ can not be used as π in ICP.

The MDI statistic for ICP is $2I(x:x^*)$ which allows one to
test the goodness-of-fit. It is distributed as a chi-square with
$\Omega-r-1$ degrees of freedom for large sample sizes N. Further, if
(i) x^*_a is a vector of MDI estimates corresponding to a set of
constraints C_a, (ii) x^*_b is such a vector corresponding to constraints C_b, and (iii) the rows of C_a are explicitly or
implicitly (as linear combinations) contained in the rows of C_b,
we have

$$2I(x:x^*_a) = 2I(x^*_b:x^*_a) + 2I(x:x^*_b) \quad . \tag{4}$$

Each term on the right is also an MDI statistic distributed like
a chi-square for large samples. The property (4) is called
analysis of information.

External constraints problems. Here the constraints are provided
by the hypothesis under question and the problem is to determine
whether the departure of $\hat{\pi}$ (if any) from (1) can be attributed to
chance. This is achieved by finding the MDI estimate p^* which,

satisfies (1) and is closest to $\hat{\pi}$ by minimizing $I(p:\hat{\pi})$. Validity of the external hypothesis is tested by the MDI statistic $2I(x^*:x)$, distributed as chi-square with r degrees of freedom in large samples. (Note the difference in the form of the MDI statistics for ICP and ECP and also in their degrees of freedom.)

Analysis of information can also be applied in ECP. If an external hypothesis formulated as $C_2 p = \theta_2$ implies another external hypothesis $C_1 p = \theta_2$, where C_2 is $(r_2+1) \times \Omega$ and C_1 is $(r_1+1) \times \Omega$ with $r_2 > r_1$.

$$2I(x_2^*:x) = 2I(x_2^*:x_1^*) + 2I(x_1^*:x) .$$

The L.H.S. has r_2 degrees of freedom, the first and second terms on the RHS have r_2-r_1 and r_1 degrees of freedom respectively.

Estimate of the covariance matrix. The representation (3) can itself be looked upon as the underlying probability distribution with L and the τ_j being the unknown parameters. Their values calculated from the data are estimates and are hence subject to chance errors. Usually, the covariances of the estimates of tau-parameters are of interest. These covariances are estimated as a matrix denoted by $(S_{22.1})^{-1}$, where

$$S_{22.1} = S_{22} - S_{21} S_{11}^{-1} S_{12}, \text{ with } S \text{ partitioned as}$$

$$S = \begin{bmatrix} S_{11} & S_{12} \\ \hline S_{21} & S_{22} \end{bmatrix}$$

where S_{11} is 1×1, S_{22} is $r \times r$, $S_{12} = S_{21}'$ is $1 \times r$. Now $S = CDC'$ where C is given in (1) and D is a diagonal matrix with $x^*(\omega)$ in the ωth diagonal position.

Outliers. In internal constraints problems it happens on some occasions that the model fits the data very well *except* for one or two 'outlier' cells. These outliers lead to rejection of the model on an overall basis due to large contributions to the goodness-of-fit MDI-statistic $2I(x:x^*)$. For each cell, it is possible to find a lower bound to the contribution of the cell-frequency to $2I(x:x^*)$. These lower bounds are called OUTLIER values and can be used effectively in model search (see Gokhale and Kullback, 1978, p. 64). If there are only one or two cells with large outlier contributions to $2I(x:x^*)$, they often indicate some errors such as in coding or punching or some other feature of the data. Use of outliers is illustrated in one of the examples to follow.

Percentage 'variation' explained. Consider an ICP. From the analysis of information (4) it is clear that the goodness-of-fit MDI statistic $2I(x:x_a^*)$ is decomposed into two components. One, $2I(x:x_b^*)$, is the goodness-of-fit statistic corresponding to model (b) and the other, $2I(x_b^*:x_a^*)$ is a measure of the *effect* of using more information from the data, in terms of the constraints added in C_b. The ratio

$$\frac{2I(x:x_a^*) - 2I(x:x_b^*)}{2I(x:x_a^*)} = \frac{2I(x_b^*:x_a^*)}{2I(x:x_a^*)}$$

can be looked upon as a fraction of the original 'variation' (disparity between the data and model (a)) explained by using model (b) which is more complex (has more parameters) than model (a). A similar interpretation holds for ECP.

The case of several samples. It is easily possible to extend the foregoing methodology to the case of more than one sample. But the notation becomes clumsy and the discussion a little more technical. The reader is referred to Gokhale and Kullback (1978) for the general treatment. However, a specific application is given in one of the examples in Section 4.

Computer programs. The computer programs necessary for application of the MDI approach to frequency data are written in PL-1 with optimizing compiler. They are working at least at the George Washington University Computer Center and at the Computer Center of the University of California, Riverside. Interested readers should contact Department of Statistics, The George Washington University, Washington, D. C. 20052, where tapes of programs can be made available at a nominal charge.

4. EXAMPLES

4.1 Example 1. The data of this example, from Schoener (1970), relate to structural habitats of lizards from Jamaica. This example is included here for the purpose of comparing the analysis with that of Fienberg (1972), who improved the original analysis of Schoener (1970). The data are given in Table 2.

On an *a priori* basis, it is expected that there are associations between perch height and perch diameter and between insolation and time of day. Hence all the analyses are performed by *retaining* these associations from the data. This is achieved by the requirement that the MDI estimates of cell frequencies have the same marginal totals as the observed totals $x(hi...)$ and $x(..jk.)$.

TABLE 2: Counts in structural habitat categories for grahami *and* opalinus *lizards from Whitehouse Jamaica (from Schoener, 1970).*

Cell Index	Observed Count	Cell Index	Observed Count	Cell Index	Observed Count	Cell Index	Observed Count
11111	20	12111	8	21111	13	22111	6
11112	2	12112	3	21112	0	22112	0
11121	8	12121	4	21121	8	22121	0
11122	1	12122	1	21122	0	22122	0
11131	4	12122	5	21131	12	22131	1
11132	4	12131	3	21132	0	22132	1
11211	34	12211	17	21211	31	22211	12
11212	11	12212	15	21212	5	22212	1
11221	69	12221	60	21221	55	22221	21
11222	20	12222	32	21222	4	22222	5
11231	18	12231	8	21231	13	22231	4
11232	10	12232	8	21232	3	22232	4

Summary of Characteristics and Levels

Characteristic	Index	Values		
		1	2	3
Perch height	h	< 5 ft.	$\geq$ 5 ft.	
Perch diameter	i	$\leq$ 2 in.	> 2 in.	
Isolation	j	sun	shade	
Time of day	k	early	midday	late
Species	u	grahami	opalinus	

Analysis of Information

	Source	MDI Statistic	D.F.	P-value
a)	Fitting x(hi...), x(..jk.), x(....u)	$2I(x:x_a^*) = 88.022$	38	0.0000
	Effect	$2I(x_a^*:x_b^*) = 39.517$	2	0.0000
b)	Fitting x(hi...), x(..jk.), x(h...u), x(.i..u)	$2I(x:x_b^*) = 48.505$	36	0.0796

Analysis of Information

	Source	MDI Statistic	D.F.	P-value
a)	Fitting x(hi...), x(..jk.), x(....u)	$2I(x:x_a^*) = 88.022$	38	0.0000
	Effect	$2I(x_a^*:x_c^*) = 15.200$	3	0.0017
c)	Fitting x(hi...), x(..jk.), x(..j.u), x(...ku)	$2I(x:x_c^*) = 72.822$	35	0.0002

If the above associations are absent in the data, x(hi...), for
example, is x(h...) x(.i...)/N and the MDI estimates do take
this fact into account.

The primary question of interest is whether the two species
have different preferences depending upon the levels of the other
four variables. The base model thus corresponds to the null hypothesis that this is not so and the MDI estimates are obtained by
fitting the observed marginals (a) x(hi...), x(..jk.), and
x(....u). (The log-linear representation reflects this by the
absence of any interaction parameters involving the index u.)
The MDI statistic is 88.022 with 38 d.f. which is very highly
significant indicating the presence of species differences of
preference.

The next question may be whether the species have more
differences of preferences with respect to perch related variables
or with respect to temperature-related variables. To investigate
this aspect, two sets of marginals are fitted: (b) x(hi...),
x(..jk.), x(h...u), x(.i..u), and (c) x(hi...), x(..jk),
x(..j.u), x(...ku). The set (b) incorporates in the model pairwise associations between species and perch variables while (c)
does so for the other two variables. The MDI statistic $2I(x:x^*_b)$
is 48.505 with 36 d.f. and a P-value of 0.0796 while the MDI
statistic $2I(x:x^*_c)$ equals 72.822 with 35 d.f. and a P-value of
0.0002. The analysis of information shown in Table 2 reveals
the relative importance of the two types of associations.

It is evident that preference differences between species for
perch variables are markedly more important than those for the
temperature related variables. In fact the model obtained by
fitting the set (b) gives a goodness-of-fit statistic not significant at 5% level. However, looking at the ratio of 'variation'
explained, it is seen to be only 45% for model (b). (It is a meager
17% for model (c).) It is therefore necessary to include in the
model additional pairwise association parameters corresponding
to the species interaction with temperature variables. Fitting the
marginals x(..j.u) and x(...ku) *in addition to* the ones in (b),
separately, shows that it is necessary to include both types of
associations in the final model. (The related analysis of information is not given here.) Thus the final model (d) is arrived
at by fitting marginals x(hi...), x(..jk.), x(h...u), x(.i..u),
x(..j.u), and x(...ku). The goodness-of-fit MDI statistics
$2I(x:x^*_d)$ is 33.303 with 33 d.f. and a P-value of 0.45. Model (d)
accounts for 62% of the 'variation' in the base model (a) and is
therefore considered satisfactory. The estimated cell frequencies
are given in Table 3.

TABLE 3: MDI *estimates of cell-frequencies under Model* (d). *(See Table 2 for notation.)*

Cell Index	Estimate	Cell Index	Estimate	Cell Index	Estimate	Cell Index	Estimate
11111	16.399	12111	10.694	21111	13.881	22111	4.818
11112	2.177	12112	2.957	21112	0.623	22112	0.450
11121	7.012	12121	4.572	21121	5.935	22121	2.060
11122	0.849	12122	1.153	21122	0.243	22122	0.176
11131	8.461	12131	5.517	21131	7.162	22131	2.486
11132	2.235	12132	3.035	21132	0.640	22132	0.462
11211	34.096	12211	22.233	21211	28.860	22211	10.018
11212	10.801	12212	14.667	21212	3.091	22212	2.234
11221	73.565	12221	47.971	21221	62.269	22221	21.614
11222	21.248	12222	28.855	21222	6.081	22222	4.396
11231	14.817	12231	9.662	21231	12.542	22231	4.353
11232	9.340	12232	12.683	21232	2.673	22232	1.932

It is interesting to see that Fienberg (1972) arrives at the same model as (d) by following a different approach to model selection. On the other hand, he recommends as 'acceptable' another model which replaces the pairwise associations between perch diameter and species and time-of-day and species by the pairwise associations between perch diameter and time-of-day. Since associations between perch-diameter and time-of-day may be explained by their individual associations with the common variable species differences, model (d) appears more reasonable.

Once an acceptable model which smoothes out the data but retains all the important features has been found, other relevant questions can be answered with ease. For example, given that the perch height is less than five feet (h = 1), perch diameter is above two inches (i = 2), and late in the day (k = 3), the odds-ratio

$$\frac{x_d^*(12132) \times x_d^*(12231)}{x_d^*(12232) \times x_d^*(12131)} = \frac{3.035 \times 9.662}{12.683 \times 5.517} = 0.2393/0.5710$$

$$= 0.419$$

shows that the odds of finding an *opalinus* lizard in the sun are 41.9% of the corresponding odds for *grahami*.

When the time of the day is early (k = 1), the respective odds ratio is

$$\frac{x_d^*(12112) \times x_d^*(12211)}{x_d^*(12212) \times x_d^*(12111)} = \frac{2.957 \times 22.233}{14.667 \times 10.694} = 0.419$$

as before. This is so since there are no three-factor interactions in the model.

4.2 Example 2. This data set is taken from Gilbert (1977) and deals with cannibalism in the rotifer *Asplanchna*. In one of his experiments (Experiment 2) clone 12Cl campanulate predator females are presented with males from 12Cl and 10C6 clones. An attempt by the predator to capture prey is recorded as response, its absence recorded as no response. Data are given in Table 4.

TABLE 4: *Rotifer cannibalism data from experiment 2, Gilbert, 1977.*

Replicate Number	Clone of Prey	Predator Response Response	No Response	Replicate Number	Clone of Prey	Predator Response Response	No Response
1	12Cl	1	55	5	12Cl	6	63
	10C6	42	9		10C6	53	35
2	12Cl	0	72	6	12Cl	0	44
	10C6	53	8		10C6	43	9
3	12Cl	1	77	7	12Cl	1	70
	10C6	58	12		10C6	48	19
4	12Cl	1	33	8	12Cl	1	63
	10C6	27	3		10C6	40	12

Summary of Characteristics and Levels

Characteristic	Index	Values of Index							
		1	2	3	4	5	6	7	8
Replicate No.	i	1	2	3	4	5	6	7	8
Prey clone type	j	12Cl	10C6						
Predator response	k	Response	No response						

Analysis of Information

Source	MDI Statistic	D.F.	P-value
b) Fitting x(i..), x(.jk)	$2I(x:x_b^*) = 43.594$	21	0.0026
Effect of removing the outlier (5 2 2)	$2I(x_c^*:x_b^*) = 19.864$	1	0.0000
c) Fitting x(i..), x(.jk), x(5 2 2)	$2I(x:x_c^*) = 23.730$	20	0.2570

Gilbert (1977) has analyzed the data separately for each replicate as a set of 2 × 2 tables. Thus there is no way of knowing
whether the response-pattern has changed from replicate to replicate,
which may be possible, for example, due to changes in experimental
conditions. Re-analysis of the data in Table 4 is given below.
It illustrates the use of OUTLIER values.

The purpose of the analysis is to find out whether the predators prefer prey of a clone type other than their own. The null
hypothesis states that this is not the case so that response is
independent of clone type. Also, in a properly planned experiment
the proportion of prey clones should remain the same. Hence the
base model (a) corresponds to fitting the observed marginals
$x(i..)$, $x(.j.)$, and $x(..k)$, i.e., the absence of any association
parameters. The MDI statistic for goodness-of-fit is $2I(x:x^*_a)$ =
717.184 with 22 d.f. which is very highly significant. The next
model (b) includes associations between prey clones and response
by fitting the marginals $x(i..)$ and $x(.jk)$. This corresponds
to saying that there is an association between prey clone type
and predator response, but this association does not depend on
the replicate number. The MDI statistic $2I(x:x^*_b)$ = 43.594 with
21 d.f. with a P-value of 0.0026. The effect of inclusion of the
above associations is very highly significant as shown by $2I(x^*_b:x^*_a)$
= 673.59 with 1 d.f. At this stage, because of the low P-value,
the question whether model (b) is satisfactory remains unanswered.
Looking at the 'explained variation,' model (b) accounts for 94% of
variation from model (a). There is one *outlier*; the observed
cell frequency of 35 in cell (5 2 2) gives an OUTLIER value of
13.81 which is very high. Assuming that some extraneous factors
may have influenced this observation, yet another MDI estimate of
the table is obtained under model (c) by fitting marginals $x(i..)$,
$x(.jk)$, and $x^*_c(5\ 2\ 2) = x(5\ 2\ 2) = 35$. The analysis of information
is shown in Table 4.

The conclusion is that female predators of clone 12Cl strongly
prefer 10C6 clone males as prey and that the significantly high
no response for 10C6 in replicate 5 needs further investigation.
Apart from this fact, there does not seem to be any change in
response-patterns over replicates. On the other hand, if model
(c) gives a poor fit, the next model one may try is the one which
includes replicate-response associations by fitting the marginals
$x(i.k)$.

An academic point to be noted is that $2I(x^*_c:x^*_b)$ = 19.864
is larger than the OUTLIER value 13.81. This is to be expected
since the latter is a lower bound to $2I(x^*_c:x^*_b)$.

4.3 Example 3. The data set relates to recruitment of postlarval penaeid prawns to nursery areas in Moreton Bay, Queensland, Australia (Young and Carpenter, 1977). Four stations are set up at the same altitude to test if a horizontal stratification of the catch has occurred. A three-hour catch is totalled for five separate nights to yield Table 5.

TABLE 5: Total nocturnal catch of pelagic postlarval P. plebejus from horizontally aligned surface stations at Nerang River Bar (Young and Carpenter, 1977).

Prawn Catch at:	Date (1973)				
	May 17	May 20	May 23	May 26	May 30
Station 8	911	1515	497	66	498
Station 9	1041	527	333	30	444
Station 2	3650	6187	307	52	236
Station 4	12110	2559	339	25	191

Summary of Characteristics and Levels

Characteristic	Index	1	2	3	4	5
Station Number	i	8	9	2	4	
Date	j	17	20	23	26	30

Analysis of Information

Source	MDI Statistic	D.F.	P-value
a) Uniform distribution	$2I(x:x_a^*) = 62367.644$	19	0.0000
Effect	$2I(x_b^*:x_a^*) = 14526.222$	3	0.0000
b) Fitting x(i.) - Stations	$2I(x:x_b^*) = 47841.422$	16	0.0000
Effect	$2I(x_c^*:x_a^*) = 38479.605$	4	0.0000
c) Fitting x(.j) - Nights	$2I(x:x_c^*) = 23888.039$	15	0.0000

The authors carry out analysis of variance on the natural logarithms of observed totals since the variances of the separate counts making up each station total are found proportional to the mean counts to the power of 1.8. Stations are not found to be significantly different in their analysis.

The following analysis with actual frequencies provides an alternative to two-way analysis of variances. It also brings out some interesting features when the number of observations is extremely large.

36   D. V. GOKHALE

In order to assess differences between stations and between
nights the reductions in MDI statistics are compared with respect
to the MDI statistic fitting the uniform distribution (no station
differences, no night differences). See Table 5 for the analysis
of information.

First to be noted are the gigantic values of the MDI statistics,
all being extremely highly significant. But this is known to occur,
especially with goodness-of-fit statistics, when the sample size
is large, such as 31518 in this case. Any slight departure in the
estimated probability gets blown up tremendously when multiplied
by the sample-size. Moreover, when the sample size is large the
assumption (i) of Section 1, in particular, the assumption that
probabilities of classification remain constant from individual
to individual, is often violated. Another (somewhat ad hoc) way
of judging the situation is to examine the percentage 'variation'
explained by each of the models. Taking into account station
differences only, results in a reduction of (14526.22 ÷ 62367.644)
× 100 = 23.29% in the 'variation' inherent in the model which
assumes no station - as well as night - differences. Similar
percentage for night - differences alone is 61.70%. When both
station - and night differences are included in the model (but no
interactions) the net reduction is 84.99%. This may be considered
as satisfactory. In any case, station differences are significant
though not to the same extent as differences between catches on
the five nights. Also, a study of the log-odds $\ln[x(ij)/x(i5)]$
for each night, j=1,2,3,4 plotted for each fixed value of station
i=1,2,3,4 is helpful in studying the station-differences and
their interactions. Table 6 gives the above log-odds.

TABLE 6: Values of $\ln[x(ij)/x(i5)]$.

	$\ln \frac{x(1j)}{x(15)}$	$\ln \frac{x(2j)}{x(25)}$	$\ln \frac{x(3j)}{x(35)}$	$\ln \frac{x(4j)}{x(45)}$
j = 1	0.6039	0.8521	2.7387	4.1495
j = 2	1.1126	0.1714	3.2664	2.5951
j = 3	-0.0020	-0.2877	0.2630	0.5737
j = 4	-2.0209	-2.6946	-1.5126	-2.0334
j = 5	0	0	0	0

From the graph (not shown) it becomes apparent that the interactions between stations are mainly attributable to the catch-
patterns on the first two-nights during which Stations 2 and 4
form one group and Stations 8 and 9 form another. It is interesting
to observe that Stations 8 and 9 are over shallow bed while the
other two are over deep bed (Figure 3 in Young and Carpenter, 1977).
In the terms of the log-linear representation (3), the corresponding
model, worth trying, is given by

$$\tau_{11}^{AB} = \tau_{21}^{AB} \ ; \quad \tau_{12}^{AB} = \tau_{22}^{AB} \ ; \quad \tau_{ij}^{AB} = 0 \ , \text{ otherwise.}$$

Under this model the goodness-of-fit MDI statistic $2I(x:x^*)$ equals 5852.34. The OUTLIER values show that Stations 8 and 9 can not be put in one group on the first night and also that Stations (2 and 4) and (8 and 9) have different interactions on the second night. The overall conclusion is that (i) as far as the first two night are concerned only Stations 2 and 4 are probably alike, that too only on the first night and (ii) the Station × Night interactions resemble one another for the remaining three nights.

Various tests of hypotheses about the tau-parameters can be made based upon their estimates and an estimate of their covariance matrix. The latter is provided as output from the computer program. To illustrate such tests, suppose one wants to test the hypothesis that for the first night, stations 8 and 9 are alike. Symbolically,

$$\text{H:} \quad \tau_1^B + \tau_{11}^{AB} = \tau_1^B + \tau_{21}^{AB} \ , \ \text{i.e.,} \quad \text{H:} \quad \tau_{11}^{AB} = \tau_{21}^{AB} \ .$$

From the computer output we have

$$\hat{\tau}_{11}^{AB} - \hat{\tau}_{21}^{AB} = -3.2944 - (-3.0491) = -0.2453 \ .$$

Also $\quad \text{Var}(\hat{\tau}_{11}^{AB} - \hat{\tau}_{21}^{AB}) = \text{Var}(\hat{\tau}_{11}^{AB}) + \text{Var}(\hat{\tau}_{21}^{AB}) - 2\text{Cov}(\hat{\tau}_{11}^{AB}, \hat{\tau}_{21}^{AB})$

which is estimated by

$$(6.1560 + 6.3962 - 2 \times 4.1265) \times 10^{-3} = 4.2992 \times 10^{-3}$$

A large sample test of $\tau_{11}^{AB} = \tau_{21}^{AB}$ is provided by

$$\frac{(\hat{\tau}_{11}^{AB} - \hat{\tau}_{21}^{AB})^2}{\text{Var}(\hat{\tau}_{11}^{AB} - \hat{\tau}_{21}^{AB})} = \frac{(-0.2453)^2}{4.2992 \times 10^{-3}} = 13.99 \ ,$$

distributed as a chi-square with one degree of freedom. This value is highly significant indicating that the station × night interactions for first night and stations 8 and 9 are not the same.

In conclusion it should be reemphasized that since the sample is very large any slight departure from a simpler model is going to be blown up as significant. However, it can be safely stated by looking at the percentage variation explained that there *are*

significant differences between catches of prawns at different
stations and also at different nights. Further for the first two
nights at least, there are significantly different station × night
interactions. This finding calls for further investigation into
the nature of the experiment. It is possible that the depth of
the river bed at different stations and/or a common factor for the
first two nights as compared to the remaining three nights (such
as time or turbulence) affects the behavior of the prawn catch.

4.4 Example 4.   In some situations the data can be analyzed either
as an ICP or as an ECP. The present example is a case in point.
The data, in Table 7 due to Beverton and Bedford (1963), are taken
from Gulland (1969, p. 77) and deal with tagging mortality of
whiting. It is assumed, as usual, that the tagged fish are subject
to constant fishing and natural mortality rates that are the same
as in the natural untagged population. The question of interest
is whether there is any evidence of tagging mortality and if so,
whether it depends more on the condition of the whiting or on
the quality of its scales.

The data are analyzed first as ICP. If there is no tagging
mortality the proportion of tagged fish that return will be the
same over all combinations of conditions and scale qualities. It
is expected that condition and scale quality may be associated,
hence the first set (a) of marginals to be fitted consists of
$x(ij.)$ and $x(..k)$. The goodness-of-fit MDI statistic $2I(x:x_a^*)$
is 53.514 and 5 d.f. is very highly significant giving a strong
evidence of tagging mortality, as seen by the nonconstancy of tag-
return-rates over the six condition × scale quality combinations.
To judge whether tag-return is more associated with condition of
the fish rather than quality of its scales two sets of marginals
(b) $x(ij.)$, $x(i.k)$ and (c) $x(ij.)$, $x(.jk)$ are fitted. The
results of the analysis of information are set out in Table 7.

The important point that emerges is that tag-return is almost
entirely 'explained' by scale quality (which, in turn, may be
associated with sluggishness or otherwise of the whitings). In
fact model (c) corresponds to the statement that given scale
quality, condition and tag-return are independent, i.e.,

$$p(ijk|j \text{ fixed}) = \frac{p(ij.)\ p(.jk)}{p(.j.)} \quad .$$

(For more complex models obtained by marginal fitting, it may not
be possible to express the underlying probabilities by such equivalent multiplicative structures.)

TABLE 7: Recapture of tagged whitings (Beverton and Bedford, 1963).

Condition	Scale quality	Number tagged	Number returned	Number did not return
Lively	Good	60	18	42
Lively	Moderate	98	21	77
Lively	Poor	49	6	43
Sluggish	Good	254	60	194
Sluggish	Moderate	765	159	606
Sluggish	Poor	509	43	466

Summary of Characteristics and Levels

Characteristic	Index	1	2	3
Condition	i	Lively	Sluggish	
Scale quality	j	Good	Moderate	Poor
Tag-return	k	Yes	No	

Analysis of Information

	Source	MDI Statistic	D.F.	P-value
a)	Fitting $x(ij.)$, $x(..k)$	$2I(x:x_a^*) = 53.514$	5	0.0000
	Effect of (ik) interactions	$2I(x_b^*:x_a^*) = 2.518$	1	0.113
b)	Fitting $x(ij.)$, $x(i.k)$	$2I(x:x_b^*) = 50.996$	4	0.0000
	Effect of (jk) interactions	$2I(x_c^*:x_a^*) = 51.736$	2	0.0000
c)	Fitting $x(ij.)$, $x(.jk)$	$2I(x:x_c^*) = 1.778$	3	0.6197

Analysis of Information

	Source	MDI Statistic	D.F.	P-value
0)	Homogeneity of six binomials	$2I(x_0^*:x) = 57.037$	5	0.0000
	Effect of assumed additional homogeneity	$2I(x_0^*:x_1^*) = 55.352$	2	0.0000
1)	Homogeneity of condition	$2I(x_1^*:x) = 1.685$	3	0.6200

The same data can be analyzed as an ECP as follows. The primary question of existence of tagging mortality can be reformulated as homogeneity of six binomial tag-return probabilities over the combinations of levels of condition and scale quality. This is a six-sample ECP, since the hypothesis can be formulated as H_0:

$p(111) = p(121) = p(131) = p(211) = p(221) = p(231)$ and the constraints $p(ij1) + p(ij2) = 1$ for $i=1,2,$ $j=1,2,3$. This hypothesis is expressed in the form $C_0 p = \theta_0$ of (1) by letting

	i	1	1	1	1	1	1	2	2	2	2	2	2
Cell Index	j	1	1	2	2	3	3	1	1	2	2	3	3
	k	1	2	1	2	1	2	1	2	1	2	1	2

$$
C_0 = \begin{bmatrix}
1 & 1 & 0 & 0 & 0 & 0 & 0 & 0 & 0 & 0 & 0 & 0 \\
0 & 0 & 1 & 1 & 0 & 0 & 0 & 0 & 0 & 0 & 0 & 0 \\
0 & 0 & 0 & 0 & 1 & 1 & 0 & 0 & 0 & 0 & 0 & 0 \\
0 & 0 & 0 & 0 & 0 & 0 & 1 & 1 & 0 & 0 & 0 & 0 \\
0 & 0 & 0 & 0 & 0 & 0 & 0 & 0 & 1 & 1 & 0 & 0 \\
0 & 0 & 0 & 0 & 0 & 0 & 0 & 0 & 0 & 0 & 1 & 1 \\
1 & 0 & -1 & 0 & 0 & 0 & 0 & 0 & 0 & 0 & 0 & 0 \\
0 & 0 & 1 & 0 & -1 & 0 & 0 & 0 & 0 & 0 & 0 & 0 \\
0 & 0 & 0 & 0 & 1 & 0 & -1 & 0 & 0 & 0 & 0 & 0 \\
0 & 0 & 0 & 0 & 0 & 0 & 1 & 0 & -1 & 0 & 0 & 0 \\
0 & 0 & 0 & 0 & 0 & 0 & 0 & 0 & 1 & 0 & -1 & 0
\end{bmatrix}
$$

$p = (p(111)\ p(112)\ p(121)\ p(122)\ p(131)\ p(132)$

$\quad p(211)\ p(212)\ p(221)\ p(222)\ p(231)\ p(232))'$,

and $\qquad \theta_0 = (1\ \ 1\ \ 1\ \ 1\ \ 1\ \ 1\ \ 0\ \ 0\ \ 0\ \ 0\ \ 0\ \ 0)'$.

The MDI statistic $2I(x_0^* : x)$ to test validity of this hypothesis equals 59.037 with 5 d.f. This is very highly significant showing the presence of tagging mortality.

If tag-return probabilities were dependent only on the scale quality and not on condition, the corresponding hypothesis is

H_1: $p(111) = p(211),\ p(121) = p(221),$ and $p(131) = p(231).$

This is expressed as $C_1 p = \theta_1$ with the first six rows of C_1 the same as those of C_0 and the last three being

		1	1	1	1	1	1	2	2	2	2	2	2
Cell Index		1	1	2	2	3	3	1	1	2	2	3	3
		1	2	1	2	1	2	1	2	1	2	1	2
		1	0	0	0	0	0	-1	0	0	0	0	0
		0	0	1	0	0	0	0	0	-1	0	0	0
		0	0	0	0	1	0	0	0	0	0	-1	0

$$\theta_1 = (1 \quad 1 \quad 1 \quad 1 \quad 1 \quad 1 \quad 0 \quad 0 \quad 0)' \ .$$

The MDI statistic for testing this hypothesis is $2I(x_1^*:x)$ and equals 1.685 with 3 d.f. and is not significant. Note that the hypothesis of equality of all six tag-return probabilities implies the present (weaker) hypothesis. Hence it is possible to analyze the MDI statistic $2I(x_0^*:x)$ as shown in Table 7. This shows that the additional constraints imposed by C_0 which are not in C_1 (those corresponding to homogeneity over scale quality) are untenable.

The MDI estimates of cell-frequencies under model (c) of ICP and hypothesis H_1 are given in Table 8 for comparison.

TABLE 8: *Comparison of MDI estimates under ICP and ECP -- Tagged whittings data (Table 7). Characteristic 1 is condition (1 = lively, 2 = sluggish); Characteristic 2 is scale quality (1 = good, 2 = moderate, 3 = poor); and Characteristic 3 is tag-return (1 = yes, 2 = no).*

Cell Index	Observed	Estimate under model (c)	Estimate under H_1
111	18	14.904	14.859
112	42	45.095	45.141
121	21	20.440	20.440
122	77	77.560	77.560
131	6	4.303	4.279
132	43	44.697	44.721
211	60	63.096	62.903
212	194	190.904	191.097
221	159	159.560	159.554
222	606	605.440	605.446
231	43	44.697	44.451
232	466	464.303	464.549

Note that under model (c) the MDI estimates of probabilities
of tag-return are 0.24841 for good scales 0.20857 for moderate
scales and 0.087814 for poor scales. Respective estimates under
H_1 are 0.24765, 0.20857 and 0.087331. They are close but not all
equal. This is due to the difference in the rationales of ICP and
ECP. In ICP the estimates are based on a log-linear model utilizing a certain amount of summary information from the data (in
terms of $C\hat{\pi}$) while, in ECP the estimates retain as many features
from the data as possible (being closest to $\hat{\pi}$) at the same time
being consistent with the hypothesis under question. (See Gokhale
and Kullback, 1978.)

Incidentally, under the hypothesis of homogeneity over scale
quality,

H_2: $p(111) = p(121) = p(131)$ and $p(211) = p(221) = p(231)$,

the MDI statistic $2I(x_2^*:x)$ is 56.108 with 4 d.f. which is very
highly significant, rejecting H_2.

4.5 Example 5. Nelson, Calabrese, and MacInnes (1977) have
studied survival of juvenile bay scallops, *Argopecten irradians*,
after 96 hours of exposure to mercury at various salinity and
temperature levels. The purpose is to determine whether salinity
and temperature combine to enhance mercury toxicity to juvenile
bay scallops. The authors have used usual regression methods
to obtain a regression equation of percentage mortality on linear
and quadratic terms in the three factors. They find that survival
is significantly affected by mercury concentration (c) and salinity
(s, s^2), and by the interaction between temperature and salinity
(T×s) and mercury and temperature (c×T). Mercury and salinity, in
that order, are the singly most important variables affecting
scallop survival. The authors have then proceeded to study the
nature of the response surface at 100% and at 50% survival rate.

In what follows, MDI approach is applied to analyze the actual
frequencies treating the problem as ICP. It can be regarded as an
analogue of the standard regression technique in which survival is
the 'dependent' variable and salinity, temperature and mercury are
the 'independent' or 'predictor' variables.

At first the levels of predictor variables are regarded as
qualitative and the two most important single predictors are determined. The question whether interactions of predictor variables
significantly affect survival is answered next. For the sake of
comparison, the relative importance of including in the model

temperature-salinity, temperature-mercury and salinity-mercury
interactions with survival is assessed. The best and the worst
combinations of factor levels, which respectively maximize and
minimize survival-odds, are determined. Table 9 gives the data.

The fact that factor levels are quantitative can be used to
put further structure on the various association-parameters. This
is illustrated in the remainder of the analysis.

Since the sample size is fixed to be 50 for each treatment
combination the sequence of marginal fitting begins with (a)
x(hij.) and x(...k). The model corresponds to the hypothesis
that the treatment temperature, salinity, and mercury have no
predictive capability about survival rate. It is expected that
this model would be extremely poor and the subsequent analysis
would arrive at a model that incorporates all pertinent interactions between the dependent and the predictor variables. The
sequence of marginals fitted is listed in the analysis of information (Table 9). Note that models (b), (c), and (d) include,
turn by turn, the associations between survival and the predictor
variables. Model (e) includes all pairwise interactions. Models
(f), (h), and (i) include, in turn, the three factor interactions
between the dependent and the predictor variables. Model (g)
incorporates the interactions between survival and temperature ×
salinity combinations, and, separately, the interaction between
su-vival and mercury. Model (j) includes over and above interactions in model (f) the ones between temperature, mercury, and
survival. Finally, model (k) includes all three factor interactions.

From the analysis of information of models (a) through (d),
it is clear that (i) each of the three variables is extremely
helpful in the predicting of the survival rate, (ii) mercury is
the single most important predictor $2I(x_d^*:x_a^*) = 786.670$, accounting
for 74% of the 'variation' in model (a), and (iii) the next most
important predictor is salinity though it accounts for only 10%
of the variation in model (a). The least important factor in
itself is temperature accounting for only 1% of the variation.
Model (e), which includes all pairwise associations with the
dependent variable, does not provide a good fit to the data, as
the MDI statistic, $2I(x:x_e^*) = 63.198$ with 28 d.f. and P-value of
0.0002, but it must be observed that it accounts for 94% of the
variation from model (a). Model (e) calls for a closer examination
from the point of view of OUTLIER values to weigh the additional
advantages gained by using models with more complex interaction
structures. An examination of the OUTLIER values shows moderately
large (around 4.0) values for cells corresponding to some temperature × salinity combinations. Note also from the analysis of
information of models (b), (c), and (f) that temperature × salinity

44   D. V. GOKHALE

*TABLE 9: Mortality of juvenile bay scallops at various combinations of temperature-
salinity-mercury levels after 96 hours of exposure. (Nelson et al., 1977).*

Cell Index	Observed Frequency	Cell Index	Observed Frequency	Cell Index	Observed Frequency	Cell Index	Observed Frequency
1111	50	1321	49	2231	45	3141	1
1112	0	1322	1	2232	5	3142	49
1121	44	1331	33	2241	14	3211	48
1122	6	1332	17	2242	36	3212	2
1131	23	1341	13	2311	50	3221	49
1132	27	1342	37	2312	0	3222	1
1141	3	2111	47	2321	50	3231	38
1142	47	2112	3	2322	0	3232	12
1211	49	2121	40	2331	39	3241	21
1212	1	2122	10	2332	11	3242	29
1221	50	2131	20	2341	15	3311	50
1222	0	2132	30	2342	35	3312	0
1231	43	2141	4	3111	43	3321	47
1232	7	2142	46	3112	7	3322	3
1241	10	2211	50	3121	25	3331	35
1242	40	2212	0	3122	25	3332	15
1311	50	2221	50	3131	10	3341	15
1312	0	2222	0	3132	40	3342	35

Summary of Characteristics and Levels

Characteristic	Index	1	2	3	4
Temperature	h	15°C	20°C	25°C	
Salinity	i	15°/∞	20°/∞	25°/∞	
Mercury	j	Control	52 ppb	89 ppb	150 ppb
Survival	k	Alive	Dead		

Analysis of Information

	Source	MDI Statistic	D.F.	P-value
a)	x(hij.), x(...k)	$2I(x:x_a^*) = 1060.539$	35	0.0000
		$2I(x_b^*:x_a^*) = 7.679$	2	0.0215
b)	x(hij.), x(h..k)	$2I(x:x_b^*) = 1052.860$	33	0.0000
		$2I(x_c^*:x_a^*) = 108.863$	2	0.0000
c)	x(hij.), x(.i.k)	$2I(x:x_c^*) = 951.676$	33	0.0000
		$2I(x_d^*:x_a^*) = 786.670$	3	0.0000
d)	x(hij., x(..jk)	$2I(x:x_d^*) = 273.868$	32	0.0000
		$2I(x_e^*:x_a^*) = 997.341$	7	0.0000
e)	x(hij.), x(h..k), x(.i.k.), x(..jk)	$2I(x:x_e^*) = 63.198$	28	0.0002
f)	x(hij.), x(hi.k)	$2I(x:x_f^*) = 931.082$	27	0.0000
g)	x(hij.), x(hi.k), x(..jk)	$2I(x:x_g^*) = 37.682$	24	0.0374
h)	x(hij.), x(h.jk)	$2I(x:x_h^*) = 236.655$	24	0.0000
i)	x(hij.), x(.ijk)	$2I(x:x_i^*) = 62.018$	24	0.0000
j)	x(hij.), x(hi.k), x(h.jk)	$2I(x:x_j^*) = 27.327$	18	0.0731
k)	x(hij.), x(hi.k), x(h.jk), x(.ijk)	$2I(x:x_k^*) = 11.991$	12	0.4464

interactions with survival are significantly important. Thus a model incorporating temperature × salinity × survival and mercury × survival interactions seems appropriate. This is model (g). The goodness-of-fit MDI statistic $2I(x:x_g^*)$ = 37.682   with 24 d.f. with a P-value of 0.0374. Model (g) accounts for 96% of the variation from that of model (a). The few moderately large OUTLIER values occur only at control level of mercury when the survival rate is 100%. With a sample as large as 1800, the fit is good at the 3.75% level of significance. Hence model (g) can be regarded as satisfactory from all angles.

   An important distinction to be noted in model (g) and the regression model used by Nelson *et al.* (1977) concerns the three factor interaction temperature × salinity × mercury. The authors find that "toxicity of mercury at low concentrations was enhanced by high temperatures and low salinity, whereas at high mercury concentrations this effect diminished." Model (g), on the other hand, postulates that scallop survival is significantly affected by the interactive combinations of levels of temperature and salinity and also by levels of mercury. *but three factor interactions survival × temperature × mercury and survival × salinity × mercury need not be included.* Model (k) corresponds to the authors' model. Note that the difference $2I(x_k^*:x_g^*)$ = 25.691 with 12 d.f. which is significant at 5% but not significant at 1%. The effect of adding temperature × mercury interactions with survival to model (g) is seen from $2I(x:x_j^*)$ = 27.327. The differences $2I(x_j^*:x_g^*)$ = 10.355   with 6 d.f. which is not significant at 10%.

   The estimated probabilities of survival and the odds of survival against death are given in Table 10.

   A striking feature of Table 10 is the drastic reduction in odds for survival from control to other mercury levels. For mercury level 4(105 ppb) survival odds are uniformly (over all levels of temperature and salinity) smaller than 1 and can go as low as 174: 1000. For a specific level of mercury one can study the effect of temperature and salinity on the odds from Table 11. Observe that the odds are maximum for each mercury level at salinity level of 20°/∞. As for temperature levels they decrease monotonically for salinity level of 15°/∞, but for other salinity levels, reach a peak at 20°C temperature. The overall conclusion is that among the experimental levels 20°C temperature and 20°/∞ salinity is the optimum combination at *every* mercury level.

TABLE 10:   Estimated values for probability of survival and odds of survival under
model (g).   See Table 9 for data and notation.

Cell Index	Prob.	Odds	Cell Index	Prob.	Odds	Cell Index	Prob.	Odds
1111	.9700	32.37	2111	.9567	22.08	3111	.8567	5.98
1121	.8766	7.10	2121	.8289	4.84	3121	.5674	1.31
1131	.4676	0.88	2131	.3745	0.60	3131	.1396	0.16
1141	.0860	0.09	2141	.0603	0.06	3141	.0171	0.02
1211	.9929	139.2	2211	.9950	197.5	3211	.9941	169.7
1221	.9683	30.54	2221	.9774	43.33	3221	.9738	37.23
1231	.7906	3.78	2231	.8427	5.36	3231	.8215	4.60
1241	.2880	0.40	2241	.3647	0.57	3241	.3303	0.49
1311	.9900	99.41	2311	.9935	153.6	3311	.9909	109.32
1321	.9562	21.81	2321	.9712	33.70	3321	.9600	23.98
1331	.7295	2.70	2331	.8064	4.17	3331	.7478	2.97
1341	.2242	0.29	2341	.3086	0.45	3341	.2411	0.32

TABLE 11:   Survival odds for different levels of mercury (except
control).   See Tables 9 and 10 for details.

Mercury levels ppb	Temperature levels °C	Salinity Levels °/∞		
		15	20	25
52	15	7.10	30.54	21.81
52	20	4.84	43.33	33.70
52	25	1.31	37.23	23.98
89	15	0.88	3.78	2.70
89	20	0.60	5.36	4.17
89	25	0.16	4.60	2.97
150	15	0.09	0.40	0.29
150	20	0.06	0.57	0.45
150	25	0.02	0.49	0.32

4.6 Example 6.   This data set is taken from the monograph by
Andrewartha (1971). The experiment presented here examines the
effect of starvation and dessication on the behavior of two species
of flour beetles, Tribolium confusum and I. castaneum, in a
gradient of moisture. Prior to the experiment one of three samples
for each species is starved and dessicated for eight days, one
sample is starved and dessicated for four days, and the third

sample is freshly fed.   Each sample is placed on a moisture gradient and after 10 minutes the number of beetles on the moist side and on the dry side is counted, yielding Table 12.

TABLE 12:   *Behavior of* Tribolium confusum *and* T. castaneum *in a moisture gradient.*

		Response to moisture gradient	
Species	Days without food	Dry	Wet
T. *confusum*	0	74	24
	4	42	58
	8	24	54
T. *castaneum*	0	70	30
	4	43	57
	8	6	38

Summary of Characteristics and Levels

		Values of Index		
Characteristic	Index	1	2	3
Species	i	T. *confusum*	T. *castaneum*	
Condition	j	0 days	4 days	8 days
Response	k	dry	wet	

Analysis of Information

	Source	MDI Statistic	D.F.	P-value
a)	Fitting x(ij.) and x(..k)	$2I(x:x_a^*) = 85.525$	5	0.0000
	Effect of (jk) interactions	$2I(x_b^*:x_a^*) = 79.986$	2	0.0000
b)	Fitting x(ij.) and x(.jk)	$2I(x:x_b^*) = 5.539$	3	0.1363
	Effect of (ik) interactions/ (jk) interactions	$2I(x_c^*:x_b^*) = 1.828$	1	0.1763
c)	Fitting x(ij.), x(.jk), and x(i.k)	$2I(x:x_c^*) = 3.711$	2	0.1564

Andrewartha analyzes the data using the usual chi-square techniques for contingency tables. He concludes that (i) beetles recognize a gradient in moisture and tend to accumulate on one end or the other, (ii) the beetles response to the gradient depends on their condition, and (iii) the change in behavior with a change in condition is more extreme in *I. castaneum* than in *I. confusum*. The last conclusion is based on the analysis of three 2 × 2 contingency tables, one for each of the three conditions.

The following analysis using MDI statistics verifies Andrewartha's first two conclusions but refutes his last conclusion. We first consider the model that the response of the beetles is independent of the species and condition. The goodness-of-fit MDI statistic $2I(x:x_a^*)$ is highly significant so the hypothesis of independence is rejected. The second model includes an association between condition and response. This model provides a good fit -- the statistic $2I(x:x_b^*)$ is not significant. Also, the statistic $2I(x_b^*:x_a^*)$ is highly significant indicating that there is a strong association between condition and response.

Normally, the model-building would stop here with the non-significant goodness-of-fit MDI statistic $2I(x:x_b^*)$. In order to scrutinize Andrewartha's last conclusion, we consider one more model including associations between condition and response and between species and response. The MDI statistic $2I(x_c^*:x_b^*)$ is not significant, implying that the two species do not differ significantly in their response to the moisture gradient after inclusion of the interaction between condition and response. Thus, the model including only the condition × response interaction and the species × condition constraint (the design of the experiment specified the number of beetles at each species × condition combination) satisfactorily explains the data, resulting in a simpler model than suggested by Andrewartha. The results are summarized in the analysis of information (Table 12).

4.7 Example 7. This problem is discussed in the book *Ecological Studies* by E. B. Ford (1964).

Many species of Lepidoptera rest fully exposed upon tree trunks or rocks and are protected from predators by a cryptic resemblance to their background. Since the middle of the nineteenth century the vegetation surrounding manufacturing districts has been blackened by soot. In these areas the cryptic moths no longer blend in well with their background and some of these species have become darker than they are in unpolluted areas. The mechanism for this change in color is the spread through the population of a previously rare gene responsible for excess melanin production.

One theory explaining the spread of the melanic gene is selective elimination by birds; that is, birds hunting by sight are more likely to capture moths that do not match their background than those moths that are inconspicuous. Thus, in unpolluted areas, cryptic moths would be favored by selection and in polluted areas melanic moths would be favored. This theory was examined through a series of experiments by Kettlewell. One of these experiments is described here and its data analyzed using Minimum Discrimination Information statistics.

Kettlewell uses two forms of the moth *Biston betularia;* the typical form of this species is cryptically colored, and the industrial melanic, *carbonaria,* is all black except for a small white dot. The moths rest fully exposed on tree-trunks and are preyed upon by birds hunting by sight.

One part of the experiment is carried out in Dorset, a rural area where tree trunks are light colored, lichen is abundant and pollution is minimal. Large numbers of both forms of moths are marked with a dot of paint (in order to distinguish them from wild specimens) and released. The moths are then recaptured by mercury-vapor traps and by assembling to caged females. The number of moths recaptured is assumed to be proportional to the number still present in the population.

The second part of the experiment follows the same procedure as above except that the moths are released in Birmingham, an industrial area where the tree trunks are blackened by soot and lichens are absent. The experiment is performed twice at the Birmingham location, once in 1953 and once in 1955.

In Dorset the typical form is well matched to its background, while in Birmingham the *carbonaria* form blends in well with its background. Thus, if selected elimination by birds is occurring, it is expected that more typical forms than *carbonaria* forms would be recaptured in Dorset, and the reverse in Birmingham. A quick inspection of the data in Table 13 shows that this is the case. It is also apparent that the recapture rates for Birmingham 1955 are much higher than the rates for Birmingham 1953; however, the relative recovery of the two forms are similar. Kettlewell suggests that perhaps the arrangement of the traps in 1955 was more efficient. Without statistical justification Kettlewell concludes from the data that typicals and *carbonaria* are subjected to differential predation depending on the environment. From this he concludes that moths that match their background have a better chance of escaping predation, whether cryptic colored forms in rural areas or melanic forms in industrial areas; thus predation is a powerful force in the natural selection operating on the color-pattern of moths.

The data presented by Kettlewell is analyzed here using the Minimum Discrimination Information approach. First, an attempt is made to find a model that satisfactorily explains the data. Models (a) through (d) (see Analysis of Information) fit various one and two-way marginals with model (d) fitting all two-way marginals. The MDI statistic $2I(x:x_d^*)$ is significant so model (d) and all simpler models do not provide a good fit of the data. Examining the computer output it is found that the cells corresponding to the Dorset location are outliers. This means that the Dorset

TABLE 13:   Recovery of cryptic and malanic forms of Biston Betularia.

Location	Form	Recapture		% of releases recaptured
		Recaptured	Not Recaptured	
Dorset 1955	Typical	62	434	12.5%
	carbonaria	30	443	6.3%
Birmingham 1953	Typical	18	119	13.1%
	carbonaria	123	324	27.5%
Birmingham 1955	Typical	16	48	25.0%
	carbonaria	82	72	52.3%

Summary of Characteristics and Levels

Characteristic	Index	Values of Index		
		1	2	3
Location	i	Dorset 1955	Birmingham 1953	Birmingham 1955
Form	j	Typical	carbonaria	
Recapture	k	Recaptured	Not recaptured	

Analysis of Information

Source	MDI Statistic	D.F.	P-value
a) Fitting x(ij.) and x(..k)	$2I(x:x_a^*) = 191.507$	5	0.0000
Effect of (jk) interactions	$2I(x_b^*:x_a^*) = 18.868$	1	<0.0001
b) Fitting x(ij.) and x(.jk)	$2I(x:x_b^*) = 172.639$	4	0.0000
Effect of (ik) interactions	$2I(x_c^*:x_a^*) = 152.382$	2	0.0000
c) Fitting x(ij.) and x(i.k)	$2I(x:x_c^*) = 39.115$	3	0.0000
Effect of (ik)/(jk) interactions	$2I(x_d^*:x_b^*) = 136.827$	2	0.0000
Effect of (jk)/(ik) interactions	$2I(x_d^*:x_c^*) = 3.303$	1	>0.0500
d) Fitting x(ij.), x(i.k), x(.jk)	$2I(x:x_d^*) = 35.812$	2	0.0000
Effect of omitting Dorset/ (jk) interaction	$2I(x_e^*:x_b^*) = 135.890$	2	0.0000
e) Fitting x(ij.), x(.jk), x(111), x(121)	$2I(x:x_e^*) = 36.749$	2	0.0000
Effect of omitting Dorset/ (ik) interaction	$2I(x_f^*:x_c^*) = 10.910$	1	<0.0010
f) Fitting x(ij.), x(i.k), x(111)	$2I(x:x_f^*) = 28.205$	2	0.0000
Effect of omitting Dorset/ (ik) and (jk) interactions	$2I(x_g^*:x_f^*) = 35.294$	1	0.0000
Effect of (ik) interaction/omit Dorset and (jk) interaction	$2I(x_g^*:x_e^*) = 36.231$	1	0.0000
Effect of (jk) interaction/omit Dorset and (ik) interaction	$2I(x_g^*:x_f^*) = 27.687$	1	0.0000
g) Fitting x(ij.), x(i.k), x(.jk), x(111)	$2I(x:x_g^*) = 0.518$	1	>0.1000
h) Fitting x(ij.), x(..k), x(211), x(221), x(311), x(321)	$2I(x:x_h^*) = 10.915$	1	<0.0010

data and the Birmingham data are behaving differently, which should be expected simply because Dorset and Birmingham are radically different environments.

The best approach is to fit the Dorset data and Birmingham data separately. The Dorset data set consists of only four points so there are just two models possible: (i) form and recovery are independent (model (h)), and (ii) the complete model. The statistic $2I(x:x^*_h)$ is significant so form and recovery are not independent for the Dorset data. This leaves us with the complete model:

$$\ln \frac{x^*(jk)}{N/4} = L + \tau_j^F + \tau_k^R + \tau_{jk}^{FR} \quad , \quad N = 969.$$

The parameters for this model are

$$L = 0.6036, \quad \tau_1^F = -0.0205, \quad \tau_1^R = -2.6924, \quad *\tau_{11}^{FR} = 0.7465,$$

with all other τ's = 0.

To find the best model for the Birmingham data, models (b), (c), and (d) are tried again but now the cells for Dorset are completely specified in the model, giving us models (e), (f), and (g). Model (g) is chosen because it is the only model that has a non-significant goodness-of-fit MDI statistic $(2I(x:x^*_g))$. The statistics $2I(x^*_g:x^*_e)$ and $2I(x^*_g:x^*_f)$ show that the location × recovery interaction and the form × recovery interaction are both highly significant. Thus the recovery rates are different for the two years 1953 and 1955, and for the two forms of moth. It is important to note that the statistic $2I(x:x^*_g)$ also indicates that there is no third order interaction for Birmingham occurring between location, form, and recovery; this implies that the ratio of the recovery rates for the two forms are the same for both years. This observation is consistent with Kettlewell's suggestion that the arrangement of traps in 1955 was more efficient than in 1953. The equation for model (g) is

$$\ln \frac{x^*_G(ijk)}{N/12} = L + \tau_i^L + \tau_j^F + \tau_k^R + \tau_{ij}^{LF} + \tau_{ik}^{LR} + \tau_{jk}^{FR} + \tau_{ijk}^{LFR} \quad ,$$

where $N = 1771$; the parameters are

$$L = -0.7177, \quad \tau_1^L = 1.7940, \quad \tau_2^L = 1.4760 \quad ,$$

$$\tau_1^R = 0.0865, \quad \tau_1^F = -0.4638,$$

$$\tau_{11}^{LF} = 0.4433, \quad \tau_{21}^{LF} = -0.5186, \quad \tau_{11}^{LR} = -2.7789 ,$$

$$\tau_{21}^{LR} = -1.0364, \quad *\tau_{11}^{FR} = -1.0503, \quad \tau_{111}^{LFR} = 1.7968 ,$$

with all other τ's = 0.

It is informative to compare the parameters for the Dorset model with the parameters for the Birmingham model, in particular, the form × recovery interaction parameter τ_{11}^{FR} .
This parameter is positive for Dorset and negative for Birmingham, indicating that the typical form has the higher recovery rate in Dorset but the *carbonaria* form has the higher recovery rate in Birmingham. This is statistically sound evidence in support of the theory of selective elimination by birds.

ACKNOWLEDGEMENTS

My special thanks are due to Sue Picquelle for running the computer programs and to Peggy Franklin for her typing the manuscript. Financial assistance from the National Marine Fisheries Service provided for partial support of a graduate student during final preparation of the paper and also for my participation in the Satellite Program at Parma, Italy.

REFERENCES

Andrewartha, H. G. (1971). *Introduction to the Study of Animal Populations*. Chapman and Hall, London.

Beverton, R. J. H. and Bedford, B. D. (1963). The effect of return rate of condition of fish when tagged. *Special Publications of International Commission for the North-West Atlantic Fisheries*, 4, 106-116.

Bishop, Y. M. M., Fienberg, S. E., and Holland, P. W. (1975). *Discrete Multivariate Analysis: Theory and Practice*. The MIT Press, Cambridge, Massachusetts.

Fienberg, S. E. (1972). Analysis of cross-classified data in ecology. Mimeographed lecture notes: *The Advanced Institute of Statistical Ecology* at The Pennsylvania State University, University Park, June-August 1972.

Fienberg, S. E. (1977). *The Analysis of Cross-classified Categorical Data*. The MIT Press, Cambridge, Massachusetts.

Ford, Edmund Briscoe (1964). *Ecological Genetics*. Wiley, New York.

Gilbert, J. J. (1977). Defenses of males against cannibalism in the rotifer *Asplandhna:* size, shape and failure to elicit tactile feeding responses. *Ecology*, 58, 1128-1135.

Gulland, J. A. (1969). *Manual of Methods for Fish Stock Assessment.* Food and Agricultural Organization of the United Nations, Rome.

Gokhale, D. V. and Kullback, S. (1978). *The Information in Contingency Tables.* Marcel Dekker, New York.

Henry, D. L., Baskett, T. S., Sadler, K. C., and Goforth, W. R. (1976). Age and sex selectivity of trapping procedures for mourning doves. *Journal of Wildlife Management*, 40, 1, 122-25.

Kullback, S. (1959). *Information Theory and Statistics*. Wiley, New York.

Nelson, D. A., Calabrese, A., and MacInnes, J. R. (1977). Mercury stress of juvenile bay scallops, Argopecten irradians, under various salinity-temperature regimes. *Marine Biology*, 43, 293-297.

Schoener, T. W. (1970). Nonsynchronous spatial overlap of lizards in patchy habitats. *Ecology*, 51, 408-418.

Young, P. C. and Carpenter, S. M. (1977). Recruitment of postlarval penaeid prawns to nursery areas in Moreton Bay, Queensland. *Australian Journal of Marine and Freshwater Research*, 28, 745-773.

[Received December 1978. Revised April 1979]

L. Orloci, C. R. Rao, and W. M. Stiteler, (eds.),
Multivariate Methods in Ecological Work, pp. 55-63. All rights reserved.
Copyright © 1979 by International Co-operative Publishing House, Fairland, Maryland

EXAMPLES OF DISCRETE MULTIVARIATE METHODS IN ECOLOGY

JOHN A. HENDRICKSON, JR.

Division of Limnology and Ecology
Academy of Natural Sciences
Philadelphia, Pennsylvania 19103 USA

SUMMARY. Discrete multivariate observations are often collected
in ecological studies. The choice of analytic technique can often
affect the ease of interpretation in terms of the specific ecological questions. Some common situations are illustrated.

KEY WORDS. contingency tables, interaction, prediction, ordering,
natural selection.

1. INTRODUCTION

Most real problems involve multivariate observations; at
least in ecology, many of these multivariate observations involve
wholly or partially discrete variables. As little as a decade
ago, virtually no surveys of methods for analyzing discrete multivariate data were available either to the statistician or to the
ecologist.

Perhaps fortunately, we since have been inundated by many
useful compendia of methods for the analysis of discrete multivariate data. A sampling of these are listed among the references.
Over the same time span, computer program packages, often associated with a single set of methods, have proliferated. The
associated terminology, having originated in statistics and in
various subject matter fields, verges on the chaotic. It now
appears probable that another decade will pass before practitioners
are able to come to reasonably complete agreement on which groups
of approaches are equivalent in terms of support to subject-matter
inference for any given case beyond the conventional 2×2 table.

In the interim, it seems useful to consider some examples
which are typical of at least some ecological problems, and to
suggest particular approaches that seem to involve relevant quan-
tifications of those examples. It is my opinion that most ecological studies are conceived with the direct intent to evaluate
the plausibility of particular predictive ecological hypotheses;
hence there is some emphasis in the paper on methods which quantify
such predictions, as well as a few comments on what qualifies
as meaningful prediction.

2. INDUSTRIAL MELANISM IN MOTHS

E. B. Ford (1965) included a review of the evidence for
natural selection operating on normal and melanic forms of the
moth, *Biston betularia*, in Great Britain. In this classic case,
melanic forms were rarely collected prior to the industrial revolu-
tion, while in recent decades, the melanic forms have predominated
in sooty forests to the east of industrialized areas, while the
normal forms are found to predominate in the west, where tree
barks are relatively unsullied by airborne particulates.

In the course of establishing the case that natural selection
was indeed operating through differential survival of the two
forms (and hence differential reproductive contribution to the
next generation), studies were done in which marked moths of both
forms were released at each of two sites (one in the west, the
other in the industrialized area), and their recovery at traps
was compared between the two forms at each site. Table 1 presents
the data from these two release studies.

*TABLE 1: Number of marked and released moths recovered or not
recovered, categorized as to typical or carbonaria phenotype of*
Biston betularia *at two sites (Birmingham and Dorset) in 1955.
(Data from Table 16 of Ford, 1965).*

	Typical	Carbonaria	Total
Birmingham			
Recovered	16	82	98
Not recovered	48	72	120
	64	154	218
Dorset			
Recovered	62	30	92
Not recovered	434	443	877
	496	473	969

If these studies are treated separately, using standard approaches (e.g., Fienberg, 1977), each one has a significant interaction ($x^2 = 13.46$ for Birmingham; $x^2 = 9.98$ for Dorset), while the odds ratios for survival (recovery) of typicals relative to melanics is 0.29 at Birmingham and 2.11 at Dorset, both of which are significantly different from 1.0, indicating that different forms have different relative survivals in the two sites, as would have been predicted.

If these two tables are combined in a $2 \times 2 \times 2$ table, we can in fact reach the same biological conclusion in the process of interpreting the significant three-factor interaction. Table 2 presents maximum likelihood estimates for the model which includes all two-factor interactions, but excludes a three-factor interaction. A goodness of fit test for this model can be evaluated either by $x^2 = 22.6$ or $G^2 = 25.9$, which are equivalent statistics to be compared with a chi-square distribution with no degree of freedom. This goodness of fit test is equivalent to a test for the difference between the two odds ratios, which yields $z^2 = 24.6$, also to be compared to the same chi-square distribution.

All of the above methods are presented in Fienberg (1977). Equivalent inference (if not quite numerically identical statistics) would follow from using Minimum Discrimination Information, which is described by Gokhale (1979).

TABLE 2: Maximum-likelihood estimates for the model with no three-way interaction, but all two-way interactions, for the data of Table 1, after 12 iterations.

	Typical	Carbonaria	Total
Birmingham			
Recovered	29.5	68.5	98
Not recovered	34.5	85.5	120
	64	154	218
Dorset			
Recovered	48.5	43.5	92
Not recovered	447.5	429.5	877
	496	473	969

3.   ARE TURTLES BIGGER AT HIGHER LATITUDES?

Lindsey (1966) followed the suggestion of several earlier authors that the pattern described by Bergmann's rule (developed for individual homeotherms within species) might apply to global trends in the sizes attained by species of poikilotherms within groups. One of his examples was based on non-marine turtles, and he concluded that there is no evidence that larger species of non-marine turtles are more prevalent at higher latitudes. The example seems to be suitable for introducing some of the methods termed prediction analysis (Hildebrand *et al.*, 1977), and thus for motivating the use of prediction analysis in place of generalized tests for interaction when only particular predictions are of interest.

The data are presented in Table 3. Both rows and columns are ordered, with rows going from tropical through warm temperate to cool temperate latitudes, while columns go from small (less than 20 cm) to large (at least 40 cm) turtles as measured by carapace length. Table 4 presents the maximum likelihood estimates on the assumption that cell totals depend only on the row and column totals, and not on interaction. This model is readily rejected for (3-1) (3-1) = 4 degrees of freedom, since $x^2 = 19.30$ ($G^2 = 19.62$), indicating highly significant interaction. This does not, however, evaluate the specific prediction, namely that larger turtles should be found at higher latitudes.

Part of the general idea of prediction analyses is that we should be able to specify certain results which would be inconsistent with our predictions, and if there were many such results, our prediction would be no better than a model of complete independence. If the prediction is phrased as tending from small turtles in the tropics to large turtles in the cool temperate latitudes, we can specify two results which would be inconsistent with the prediction, namely many large turtles in the tropics or many small turtles in the cool temperate latitudes. In other words, our prediction becomes phrased in terms of the cell totals in Table 3 versus the expected cell totals in Table 4; we predict fewer than 9.0 small, cool temperate turtles and fewer than 17.5 large tropical turtles. With perfect prediction, we would observe zero small, cool temperate or large, tropical turtles, and our reduction in proportion of errors would be identical 1. We find,

$$\nabla_p = 1 - \frac{\text{observed error cell totals}}{\text{expected error cell totals}}$$

$$= 1 - \frac{10 + 26}{9.0 + 17.5} = 1 - \frac{36}{26.5} = -0.36,$$

and our reduction in error is negative, meaning that our prediction does even worse than a model assuming no interaction.

TABLE 3: Numbers of species of non-marine turtles found in three latitudinal regions and categorized by maximum recorded adult carapace length (Data recomputed from percentages of marginal totals given in Table 1 of Lindsey, 1966).

	Small	Medium	Large	Total
Tropical	14	37	26	77
Warm temperate	24	18	6	48
Cool temperate	10	16	3	29
	48	71	35	154

TABLE 4: Maximum likelihood estimates for cell totals for the data of Table 3, assuming no interaction:
expected = (row total) (column total)/(grand total)

	Small	Medium	Large	Total
Tropical	24.0	35.5	17.5	77
Warm temperate	15.0	22.1	10.9	48
Cool temperate	9.0	13.4	6.6	29
	48	71	35	154

This should indicate the importance of selecting a method which permits evaluating the specific predictions of the model, rather than a generalized test for interactions.

4. GROWTH FORMS OF PINES NEAR THE TREE-LINE

One characteristic observed on climbing to high altitudes is the gradual disappearance of tall trees and the substitution of stunted growth forms at the tree line. In some tree species, such as *Pinus albicaulis,* growth form is known to be genetically determined, and it can be established that there is gene flow within the population and across the tree line. The stunted growth forms would be expected to do poorly under the shade of taller growth forms, and hence few stunted forms would be expected to reach maturity in the tall forest. On the other hand, tall forms would be subject to damage from wind, snow, and ice if they projected above the surrounding growth of stunted trees above the tree line.

Clausen (1965) censused tracts of forest on south-facing and east-facing slopes of Slate Creek Valley, California, in three

altitude ranges, from somewhat below tree line to just above
tree line, characterizing the growth forms of *Pinus albicaulis*
(among other species) as to whether they were tall (single-trunk
or multi-trunk), stunted (elfinwood), or intermediate, using only
mature, cone-bearing trees so as to include only the reproducing
population in evaluating the role of natural selection in maintaining the tree line.

The data for *Pinus albicaulis* are presented in Table 5 while
Table 6 presents the maximum-likelihood estimates for the model
with no interaction between elevation and growth-form. The x^2
value for the goodness of fit of Table 6 to the data of Table 5
is 1916.2, which is to be compared with a chi-square distribution
with 6 degrees of freedom; obviously there is significant interaction.

If one were to predict zero surviving tall-form trees above
tree line and zero surviving stunted trees (intermediate or elfinwood) in the lowest elevation range, this would be a considerable
improvement over the no interaction model. Our calculation
(following Hildebrand *et al.*, 1977) would yield, comparing with
no interaction expected cell totals,

$$\nabla_p = 1 - \frac{\text{observed error cell totals}}{\text{expected error cell totals}}$$

$$= 1 - \frac{34 + 34 + 0 + 14}{171.3 + 345.4 + 20.2 + 358.5}$$

$$= 1 - \frac{82}{895.4} = 0.91.$$

In other words, a prediction of zero for those four cell totals
is 91% better than a prediction based on no interaction. This
is clearly a dramatic improvement based on an *a priori* prediction
of the results of natural selection.

Prediction analysis should not be confused with certain *a
posteriori* methods (e.g., Goodman and Kruskal, 1959; Colwell, 1974)
which attempt to measure how well one can predict the row (or
column) given the marginal totals and cell totals. One example
of such an index is Goodman and Kruskal's $\lambda_{C|R}$, which predicts
$\lambda_{C|R} = 0.38$ which indicated that one cannot do very well at predicting the column from knowledge of the row relative to predicting
the column without selecting the row.

TABLE 5: Census of growth forms of mature, cone-bearing Pinus
albicaulis *in three altitude ranges (*A = 10,000–10,500
feet, B = 10,500–10,800 *feet,* C = *over* 10,800 *feet). (Cell
totals recomputed from marginal totals in Table 5 of Clausen,*
1965.)

	A	B	C	Total
Single-trunk	107	9	0	116
Multi-trunk	1610	437	14	2061
Intermediate	34	136	124	294
Elfinwood	34	164	395	593
	1785	746	533	3064

*TABLE 6: Maximum-likelihood estimates of cell totals for Table 5,
under the model of no interaction.*

	A	B	C	Total
Single-trunk	67.6	28.2	20.2	116
Multi-trunk	1200.7	501.8	358.5	2061
Intermediate	171.3	71.6	51.1	294
Elfinwood	345.4	144.4	103.2	593

5. DISCUSSION

This presentation, although brief, should serve the purpose of giving a brief worm's eye tour of the potential use of discrete multivariate methods in ecology. Two points seem important to me.

The first point is that in the analysis of contingency tables, the interaction terms are the basis for determining whether or not interpretable significance is likely to be present.

The second point is that *a priori* ecological hypotheses can guide the analysis to the estimation of readily interpretable forms of interaction.

Many methods are now available, often in computerized packages. The statisticsl literature is becoming more accessible to the user through many summary volumes, most of the pages of which are devoted to estimating how much interaction is present at

various levels of stratification of a contingency table; such
methods are particularly useful in exploring the unknown. Often,
however, the ecologist suspects what should be going on and merely
needs guidance in establishing the plausibility of that suspicion.
Hopefully, this paper offers some guidance to effective
cooperation in such cases.

ACKNOWLEDGEMENTS

Many people have offered suggestions or encouragement,
including D. V. Gokhale, G. P. Patil, G. Policello, E. Russek,
D. Hildebrand, W. Stiteler, D. Simberloff. I am solely responsible
for the choice of illustrations and the direction of the polemics.

This writing was done under research funds provided by the
Division of Limnology and Ecology, The Academy of Natural
Sciences of Philadelphia.

REFERENCES

Bishop, Y. M. M., Fienberg, S. E., and Holland, P. W.(1975).
Discrete Multivariate Analysis. M.I.T. Press, Cambridge.

Clausen, J. (1965). Population studies of alpine and subalpine
races of conifers and willows in the California High Sierra
Nevada. *Evolution*, 19, 56-68.

Colwell, R. S. (1974). Predictability, constancy, and contingency
of periodic phenomena. *Ecology*, 55, 1148-1153.

Cox, D. R. (1970). *The Analysis of Binary Data*. Halsted Press,
London.

Fienberg, S. E. (1977). *The Analysis of Cross-Classified Categorical Data*. M.I.T. Press, Cambridge.

Fleiss, J. L. (1973). *Statistical Methods for Rates and Proportions*. Wiley-Interscience, New York.

Ford, E. B. (1965). *Ecological Genetics*. Second edition.
Methuen, London.

Gokhale, D. V. (1979). Analysis of ecological frequency data.
In *Environmental Biomonitoring, Assessment, Prediction, and
Management - Certain Case Studies and Related Quantitative
Issues*. J. Cairns, Jr., G. P. Patil, and W. E. Waters eds.
Satellite Program in Statistical Ecology. International
Cooperative Publishing House, Fairland, Maryland.

Gokhale, D. V. and Kullback, S. (1978). *The Information in Contingency Tables*. Marcel-Dekker, New York.

Goodman, L. A., and Kruskal, W. H. (1959). Measures of association for cross-classifications, II. Further discussions and references. *Journal of the American Statistical Association*, 54, 123-163.

Hildebrand, D. E., Laing, J. C., and Rosenthal, H. (1977). *Prediction Analysis of Cross Classifications*. Wiley-Interscience, New York.

Lindsey, C. C. (1966). Body sizes of poikilotherm vertebrates at different latitudes. *Evolution*, 20, 456-465.

[*Received August 1978. Revised March 1979*]

L. Orloci, C. R. Rao, and W. M. Stiteler, (eds.),
Multivariate Methods in Ecological Work, pp. 65-86. All rights reserved.
Copyright © 1979 by International Co-operative Publishing House, Fairland, Maryland

CLUSTER ANALYSIS OF THE DISTRIBUTION PATTERNS
OF DUTCH CARABID SPECIES (COL.)

R. HENGEVELD   P. HOGEWEG

Department of Geobotany   Bioinformatica
Catholic University   University of Utrecht
Toernooiveld   Padualaan 8
Nijmegen, The Netherlands   Utrecht, The Netherlands

SUMMARY. We describe the degree of similarity of distribution
patterns of carabid species on two scales: that of The Netherlands
and that of Europe. It appears that immigration into a new polder
area is mainly affected by habitat factors rather than by assess-
ibility. Distribution patterns of the beetles within The Netherlands are similar to some degree to the patterns on the European
scale, especially with respect to the amount of dispersion.

KEY WORDS. carabid species, zoogeographical districts, cluster
analysis in biogeography, invasion and colonization in polders.

1. INTRODUCTION

Some years ago, Hengeveld was faced with the problem of
estimating which variables might affect the immigration of carabid
beetles into the newly reclaimed IJsselmeer polders in The Netherlands, and their eventual colonization. In his opinion, a general
evaluation of these variables should not be confined to just a few
species in a restricted area. Instead one should ignore the border
separating ecologists and biogeographers in order to put the problems
and their solutions into proper perspective.

If we want to characterize the group of species which have
invaded and colonized the polder, we have to compare them with a
group of species which did not colonize the area. Which species
to include in the non-colonizing group is a matter requiring *ad hoc*
decisions. We want to include the species of the surrounding area,

but how large should this area be? In this study we have included all species which occur in The Netherlands.

The variables which may determine immigration and colonization can be divided into four categories: 1) the numbers of the species occurring in the surrounding area; 2) their spatial distributions in the surrounding area; 3) the vagility of the species; and 4) their habitat requirements. For evaluating the first two categories of variables we need to decide, apart from which species should be included, also the size of the surrounding area. This is also an *ad hoc* decision. Here two scales for the surrounding area were chosen: The Netherlands and Europe.

Two species may occur in the same area, as defined for a particular grid size, but still differ in the number and type of locations which they occupy within this area. This is true for whatever scale is taken. Thus one more *ad hoc* decision is needed for evaluating distribution patterns, i.e., the size of the grid squares. We chose squares of 100 km^2 for distributions within The Netherlands. For Europe, a much coarser grid was used, dividing Europe into 54 squares each of approximately 250,000 km^2.

In this study we concentrate on the characterization of invading species in terms of the geographical distribution patterns of the species. An analysis in morphological and physiological terms will be given in another paper. Geographical distribution patterns may indicate constraints on invasion and colonization of the polders brought about by factors such as species number, accessibility, and habitat. However the present study may alternatively be viewed in two other ways: 1) As a summary of the information on distribution patterns of individual carabid species given in Turin *et al.* (1977) by describing similar distribution patterns. We want to stress that recognition of similar distribution patterns implies neither interactions among the beetles nor the existence of 'guild structure,' 2) To obtain some insight into relationships, if any, among different distribution patterns on different scales.

2. MATERIAL

The material consists of the presence–absence data for 372 carabid species as given by Turin *et al.* (1977). There, a record was considered as one observation by a collector on a certain day and at a certain site. When two or more such observations had been made, it was still considered as one record. In this study, the frequency of records in a grid square of 10×10 km^2, as calculated over the period from about 1900 to 1975, was reduced to the

mere presence in that square. The original number of records
totaled 51,525 and the presences 17,977.

To estimate the location of the European ranges, we recorded
the presence or absence of each species in the 54 squares of a
rectangular grid laid over the maps of the European distributions
in Turin *et al.* (1977). We did not consider the Russian part of
the distribution ranges, because the indications of the presence
in Russia are less precise than those of other European countries.
When comparing Dutch distribution patterns, we excluded those
species which occupied only one grid square in The Netherlands,
but they were not excluded from the comparison of the European
distributions.

3. METHODS

3.1 Introduction. There are two basically different approaches
to the problem of recognizing common distribution patterns. These
two approaches are called 'supervised' and ' non-supervised.' In
the supervised approach (e.g. discriminant analysis), it is the
user who distinguishes groups of objects on *a priori* grounds, and
the aim of the analysis is to produce characterizations of these
groups in terms of the data under investigation. In the non-
supervised approach (e.g. cluster analysis), the analysis produces
groups which are similar with respect to the data, and which may
afterwards be compared to any *a priori* grouping.

In our case a supervised approach would amount to an attempt
to distinguish between the geographical distributions of the groups
of species which did and did not invade the polder. The non-
supervised approach would amount to generating groups of species
of similar geographical distribution and checking whether these
groups coincide with the groups of species present or absent in
the polder, or, alternatively, to generating groups of locations
(grid squares) which are similar with respect to their carabid
fauna, and checking which of these groups contain locations in
the polders.

The supervised approach is limited to cases where *a priori*
groups can be distinguished and may be hampered by multimodal
distributions of these groups. In our case, the approach is only
useful for distinguishing polder species, but not for the other
problem posed.

The non-supervised approach can give no information about the
polder immigration when, for example, the polder quadrats are not
confined to just a few clusters of species of a similar distribution
pattern in The Netherlands. However, when the groups do coincide,
the non-supervised approach gives results which may generate a

variety of biological hypotheses when additional, although partial, knowledge about the species or grid squares comprising the groups is used.

When using non-supervised methods (cluster analysis in our case) and subsequent characterizations and representations of clusters, the heuristic nature of such methods should be kept in mind. The correct interpretation of the results of such an analysis is crucially dependent on an appreciation of its properties. In particular the following properties of the method should be stressed: 1) The results are of the form: If we want to distinguish groups of similar objects, here species or geographical areas, the generated groups may be useful. The analysis does not solve problems about whether species coincide significantly in their geographical distribution. 2) The validation of the groups generated is not statistical but semantic. A statistical validation is impossible with this type of procedure because, apart from other reasons, we are dealing with the entire population instead of a sample. The semantic validation is based on the interpretation of the groups using knowledge which was not included in the analysis and which is often partial and only added afterwards. 3) The results are crucially dependent on the choice of data set. In our particular case the results depend on (a) the size of the grid squares; (b) the size of the area sampled; and (c) the selection of the species. These are all *ad hoc* choices. Nevertheless, each of these choices may turn any positive association between species into a negative one and *vice versa*.

3.2 Analysis of the Data Sets. All analyses, including the representation of the results, were carried out with the BIOPAT program for biological pattern analysis (Hogeweg and Hesper, 1972).

Cluster analysis. As indicated above, cluster analysis is used in this study in order to group together species which happen to have similar geographical distributions and not because we expect groups of species which coincide significantly in distribution. This is done to see whether a similar geographical distribution on one scale implies anything about the similarity of distribution on another scale, or about the invasion of the species in the polder.

The chosen technique is that of agglomerative clustering, using Ward's criterion of minimal mean square error at each step. This will give a dendogram with clearly discernable clusters even with data sets having a regular distribution. In such data sets the clusters are not really disjunct but they do group similar points together. Since the purpose of this study is the latter rather than the former, the method is appropriate. The method also exhibits a bias towards the formation of equally sized groups. This

bias is seen by us an an advantage rather than a drawback, because
the more frequently occurring patterns are exposed in more detail.

Subdivision of dendrograms. Subdivision of the dendrogram into
clusters was guided by the criterion for optimality of splitting
levels by which the within- and between-clusters ultrametric distances are compared (Hogeweg, 1976). The clustering corresponding
to the highest optimum was not used in all cases since a similar
amount of detail was preferred for the comparison of distributional
patterns in Europe and in The Netherlands.

Characterization of clusters. Cluster analysis produces an extensive definition of classes. An intensive characterization of the
clusters is the next step to be taken.

Clusters of objects were characterized by the means, standard
deviations, and frequencies of each of the characters, i.e., clusters of localities were characterized by the mean, standard deviation, and frequency of each species in the group of samples.
Clusters of species were characterized by the frequency of occurrence of these species in each of the localities, e.g. half of the
species of a certain species-cluster occurs in a particular locality. Moreover, the entire clustering was characterized by the
'importance' of characters for the subdivision into clusters. Chi-
square was taken as the measure of importance (Hogeweg, 1976).

3.3 Representation of the Results. Since we are interested in
geographical distributions, the results are presented in the form
of maps. The mapping of location clusters is straightforward:
for each locality the cluster to which it belongs is indicated
(Figure 2). Species clusters are mapped on the basis of their
occurrence in grid squares. Thus, for each species cluster, a
map is constructed in which the darkness of shading indicates the
frequency of occurrence of the cluster (i.e., the number of species
belonging to the cluster) in each grid square.

4. RESULTS

4.1 Invading versus Non-invading Species. The distribution
patterns of the groups of species which had and had not invaded
the polder at the time of preparing the *Atlas* (Turin *et al.*, 1977)
were compared by looking at the maps of these distribution
patterns. Figure 1 shows that the overall pattern of these two
groups of species is dissimilar. The species which did invade
could have orginated from any locality in The Netherlands; their
frequency of distribution strongly resembles the overall pattern
of the frequency of the species in The Netherlands. The species
which did *not* invade the polders occur mainly just to the west and

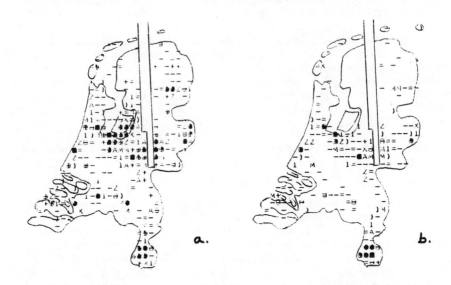

FIG. 1: Density-prints of the carabid species which (a) *did, and* (b) *did not invade the new polder area.*

east of the new land and in Limburg, the southeastern part of the country. From Figure 1 it also appears that the non-invading species may occur in the vicinity of the polders.

4.2 Clustering of Dutch Grid Squares. Comparison of the grid squares according to their faunal composition shows that certain regions within The Netherlands with a similar carabid fauna can be recognized (Figure 2). It also appears that three of these are types (clusters 2, 3, 5) which occur in the polders whereas the others do not.

The relationships among these nine clusters are shown in Figure 3. The clusters were characterized after comparing the map of the clusters with maps of soils and soil usage and also after examining the known habitat preferences of species 'important' for distinguishing the clusters. The characterization of the clusters may be summarized thus: cluster 1, underworked squares; cluster 2, pasture land; cluster 3, arable land; cluster 4, dunes; cluster 5, marshes; cluster 6, diluvial sand; cluster 7, river banks; cluster 8, region of much small-scale variation; cluster 9, loess and limestone. We should bear in mind, however, that there is a large variation in collecting (sampling) intensity over the grid squares (see Figure 5 in Turin *et al.*, 1977). We think that the

FIG. 2: *Distribution of the nine clusters based on the faunal
similarity of the grid squares.*

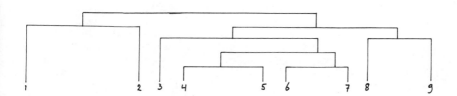

FIG. 3: *Dendrogram diagram showing the nine clusters of grid
squares distinguished. Length of the forks is arbitrary.*

structure of clusters 1 and 8 especially has been biased by this.
Cluster 1 and cluster 2, of grid squares in pasture land, are
grouped together because both the beetles and the collectors share
a dislike for this type of habitat.

The polder squares fit very well in the above scheme of
clusters. Cluster 3 is restricted to the northernmost polders
which are well-drained and have been almost exclusively converted
into arable land. Cluster 5 is restricted to the youngest and
still undrained and swampy polder, while the less common cluster 2
probably represents grassland. As the soil of these polders is
mainly clay, it is understandable that the faunas of clusters 4,
6, and 7 do not occur there.

The coincidences of these clusters with different soil type
and usage in The Netherlands indicate that the faunal composition
of the new land is mainly dependent on edaphic variables. The
close vicinity of areas which another fauna indicates that accessibility plays only a minor or no role at all on this spatial scale.

4.3 Clustering of Dutch Species in The Netherlands. The clustering leads to ten clusters or types of distributions within The
Netherlands. The interrelationships among these clusters are
shown in Figure 4. The complete dendrogram including species names
is available from the authors upon request.

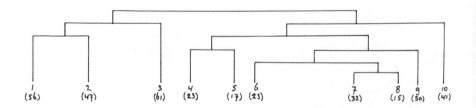

*FIG. 4: Dendrogram diagram showing the ten clusters of species
distributions distinguished within The Netherlands. Length of
the forks is arbitrary. The number of component species of each
cluster is given in parentheses.*

The distribution of these clusters in The Netherlands can be expressed in so-called density-prints which give the percentage of the species of a particular cluster occurring in each grid square. The density-prints in Figure 5 show a marked regional heterogeneity of species over The Netherlands. We can easily distinguish very widespread patterns, as well as very restricted ones, such as those occurring mainly in the western, eastern, southwestern, and southeastern parts of the country. Apparently the beetles do not experience The Netherlands as a homogeneous area.

It can be seen that species which occur frequently in the polders are confined mainly to species clusters 3 and 4. Cluster 3 is distributed over most of The Netherlands, whereas cluster 4 is restricted to the dune area in the western part of the country and around the polders. Similarly cluster 3 has the widest distribution in the polders and the species of cluster 4 occur most frequently along the dikes. Species from the other clusters only occassionally occur in the polders or not at all.

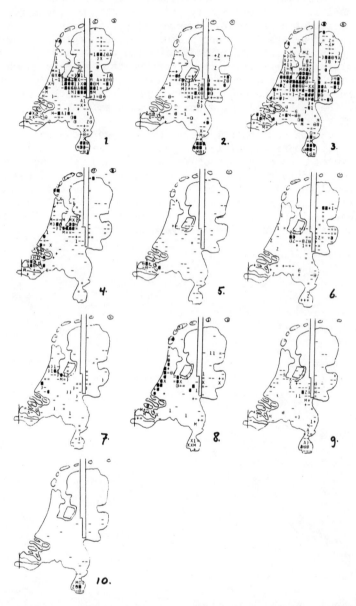

FIG. 5: *Density-prints of the distribution of the ten species clusters within The Netherlands. Clusters are based on the similarity of the distribution patterns of carabid species within The Netherlands.*

4.4 Clustering of the Dutch Species in Europe. When clustering the Dutch species with respect to their European distribution, it should be kept in mind that the restriction to the use of only Dutch species for the analysis instead of all European species, drastically influences the results. In particular, the differences in species distribution will be based on their occurrence at the fringe of the area studied, as all of them occur in The Netherlands and its immediate surroundings. Since the occurrence of a species at the fringe of Europe as well as within The Netherlands implies a wide distribution, the clustering will strongly reflect the amount of dispersion.

Figure 6 shows the interrelationships among the nine clusters recognized on the basis of the European species distribution. As expected, the main division of the dendrogram reflects the amount of dispersion, and lies between clusters 1–4 and clusters 5–9. Figure 7 shows the frequency of occurrence of the species clusters in each European grid square.

The dispersion of the clusters 1–4 resemble each other most: the species of cluster 2 are, however, the most widespread, ranging from North Africa to northern Scandinavia and Iceland. Species of clusters 1 and 2 extend more to the north than those of clusters 3 and 4. Species ranges of cluster 3 are the most restricted of the four, being absent in Scandinavia and North Africa. Clusters 5 and 6 extend mainly to the south, being frequently found in North Africa, and have a much smaller range than the above-mentioned clusters. Clusters 7–9 contain mainly central or western European species, and all show a relatively restricted distribution on the European scale.

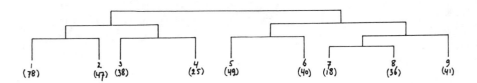

FIG. 6: Dendrogram diagram showing the nine European clusters. The number of species in each cluster is shown in parentheses.

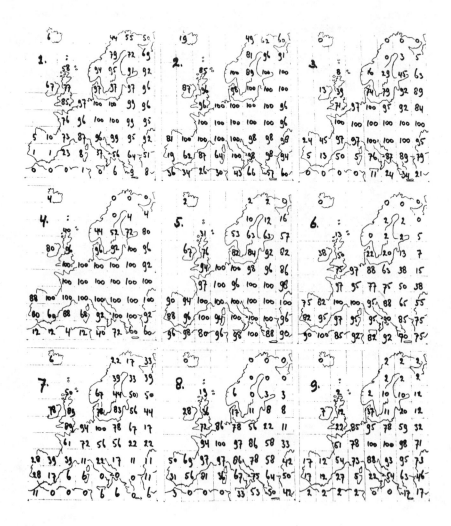

FIG. 7:   Density-prints of the distribution of the nine species
clusters within Europe.   Clusters are based on the similarity of
the distribution patterns of carabid species within Europe.

4.5 Coincidence of Dutch and European Species Clustering. As
mentioned above, changes in the scale of area and grid size may
turn any positive association into a negative one. If, nevertheless, species do show a similar distribution on several scales,
either accessibility is restricted, which is not likely on the
scale of The Netherlands, or similar ecological factors are
operating on a macro- (e.g. European), meso- (e.g. Dutch), and
micro- (e.g. within one grid square) scale. This may be the
case for climatological factors. Alternatively the restriction
of diluvial sand soils to some parts of Europe and also to
certain parts of The Netherlands may result in this phenomenon.

The degree of overlap between the clusterings of species
distributions on a European and Dutch scale is shown in a contingency table (Table 1) representing the frequencies of cooccurrence of species in Dutch (rows) and European (columns)
clusters. The contingency coefficient is $C = \sqrt{0.41}$, with
$x^2 = 2.523$ for 8×9 degrees of freedom, which implies the
probability of independence of the two clusterings (p << 0.1%).

From Table 1 we see that some Dutch clusters are restricted
to a few European clusters, and *vice versa*, but that other Dutch
clusters are more or less evenly distributed over all the European
clusters, and conversely this is true for some European clusters.
Dutch species cluster 3 is an example of the former case, and
shows a wide distribution in The Netherlands which includes the
polder area. Its members are largely confined to a few European
clusters, in particular they occur far more in European cluster 2
than would be expected from the frequencies. As mentioned above,
European cluster 2 shows the widest distribution in Europe.
Conversely, species of Dutch cluster 3 occur far less than expected
in the European clusters of species with a restricted geographical
area, i.e., clusters 6-9.

Also the other Dutch cluster, cluster 4, whose members occur
frequently in the polder, is confined to a few European clusters,
in particular to cluster 7. Dutch cluster 4 is restricted to the
coast line in The Netherlands and European cluster 7 has a restricted,
mainly northern atlantic distribution in Europe. Another striking
example of a coincidence of Dutch and European distribution clusters
is that of Dutch cluster 5, which is restricted to the south-
western part of The Netherlands, and the province of Zeeland, which
strongly overlaps with European cluster 6, with a very restricted,
and also the most southern, distribution in Europe. Examples of
Dutch species clusters which are unrelated to the European ones
are clusters 1, 7, 8, and 9 of which the last three show a
restricted distribution in The Netherlands and occur in all the
European clusters. The habitat requirements of these species
apparently limit them to certain areas on the small scale of The

TABLE 1:   Coincidence of the clusters based on the distribution
patterns within The Netherlands (rows) and within Europe (columns).
The figures in the cells represent the numbers of coinciding
species.

Clusters of Dutch patterns	Clusters of European patterns								
	1	2	3	4	5	6	7	8	9
1	20	12	5	6	6	2	1	3	1
2	11	3	4	7	7	2	1	6	6
3	18	20	3	5	14	0	0	1	0
4	2	1	0	0	5	4	9	0	2
5	1	1	0	1	0	8	0	2	4
6	8	2	5	0	1	2	4	1	0
7	7	4	1	2	2	6	1	4	5
8	1	1	4	1	4	1	1	0	2
9	5	3	11	2	6	6	0	8	9
10	4	0	4	1	4	9	1	9	9

Netherlands, whereas on the larger European scale they show diverse
distributions.

The interdependence of distribution patterns on the two
different scales is represented in another way in Figures 8 and 9.
Here the frequency of species of the Dutch clusters in the European
grid squares (Figure 8) and of the European clusters in the Dutch
grid squares (Figure 9) is plotted. Similar conclusions may be
drawn from this representation to the ones mentioned above.

It should be noted that our data on species distribution
in The Netherlands and Europe are independent; The Netherlands
constitute just one of the European grid squares, and one which
does not contribute just one of the European grid squares, and
one which does not contribute to the differentiation of species
distributions in Europe since all species included in this study
occur in it.

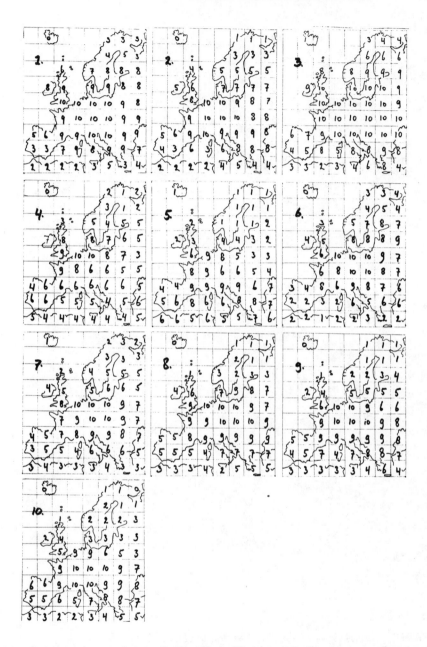

FIG. 8: *Density-prints of the distribution of the ten species clusters within Europe. The clusters are based on the similarity of the distribution patterns of carabid species within the Netherlands.*

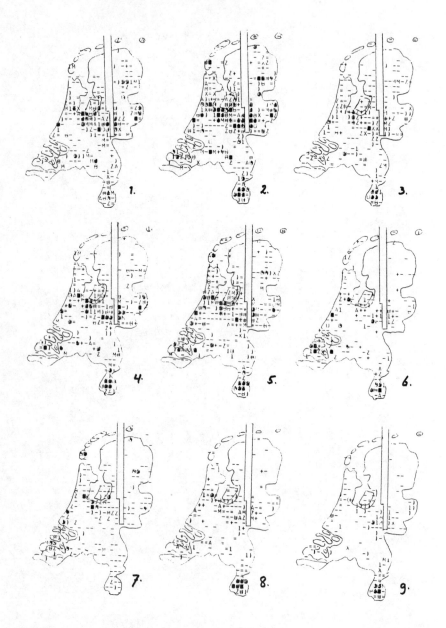

FIG. 9:   *Density-prints of the distribution of the nine species clusters within The Netherlands.   The clusters are based on the similarity of the distribution patterns of carabid species within Europe.*

4.6 Dispersion of Species on Different Scales. The above representation of the relationship between the patterns of distribution
of Dutch carabid species in The Netherlands and in Europe suggests
that the amount of dispersion itself, irrespective of the location
of the distribution, is strongly related on the different scales
used. This is indeed the case, as shown in Figure 10, in which
the number of grid squares in which a species occurs in The
Netherlands and in Europe is plotted. The rank correlation
coefficient, Kendall's tau, of the amount of dispersion at different
scales, is τ = 0.54, which is highly significant from the 365
species for which it applies. More informative, however, is the
distribution of points shown in Figure 10. It appears that a wide
dispersion in The Netherlands invariably coincides with a wide
dispersion in Europe. In fact for a certain dispersion in The
Netherlands, there is a well-defined minimal dispersion in Europe.
However, a limited dispersion in The Netherlands does not imply a
limited dispersion in Europe; almost any degree of dispersion in
Europe is found in species with a limited dispersion in The
Netherlands. This agrees with the results mentioned above in
which the Dutch species clusters showing a restricted dispersion
in The Netherlands are evenly distributed over the European
clusters. This is easily understood in terms of habitat types,
which are confined to particular areas within The Netherlands, but
which occur all over Europe with a minimum frequency of once per
(large) European grid square. If species with a limited dispersion
in The Netherlands and a wide dispersion in Europe were found at
the fringes of their range in The Netherlands, the results of
Hengeveld and Haeck (in preparation) would be consistent with
our results about dispersion. They found a correlation between
the frequency of occurrence and the distance from the center of
the range of the species.

Disregarding constraints on dispersion mechanisms, the strong
relationship between the amount of dispersion in The Netherlands
and a minimal amount of dispersion in Europe, would suggest that
most habitat types occurring in The Netherlands occur anywhere
in Europe although they may be scarce in some places. Therefore,
if a species occurs everywhere in The Netherlands, it can find a
suitable place anywhere in Europe. Alternatively, species occurring
in a large proportion of the grid squares of The Netherlands are
very tolerant species occurring in a large variety of habitats,
and therefore occur in a large part of Europe. Habitat tolerance
clearly increases migration possibilities, having more or less
suitable habitats all along the way. The fact that the species
with wider European distributions are frequently found in the
polders seems to agree with this conjecture.

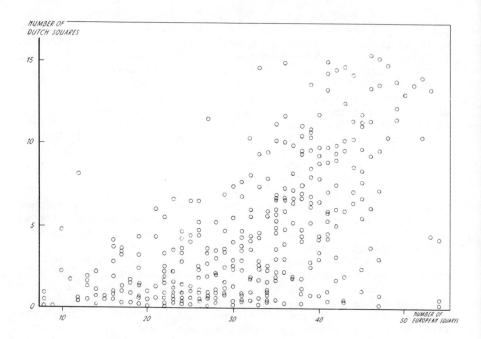

FIG. 10:   *Relationship between the number of grid squares occupied by each species within The Netherlands and within Europe.*

5. DISCUSSION

The present study about dispersion patterns in relation to the migration into newly generated land should be regarded as supplementary to more local studies on the accessibility in relation to wing dimorphism (Haeck, 1971) or seed dispersion mechanisms (Feekes, 1936; Feekes and Bakker, 1954; Van der Toorn *et al.*, 1969; Nip, Haeck, and Hengeveld, in press). These studies on accessibility may be complicated by the heterogeneity of the material from the aspect of habitat requirement. Such heterogeneity could easily be demonstrated by the clustering of the geographical distributions of the species in the area surrounding the new land. As has already been discussed, the choice of scale is a crucial, but arbitrary decision. Considering species which occur in The Netherlands and their distribution patterns within this country, we chose a scale on which dispersion mechanisms seemed to be of little importance, whereas the opposite applies on the European scale which is of an order of magnitude of geographical species ranges. Only by consideration on both scales could we obtain a representative picture showing that particular amounts of dispersion and

TABLE 2: Coincidence of the clusters based on the comparison of Dutch grid squares (columns) and the species patterns (rows) within The Netherlands. The figures within the cells represent the relative numbers of coinciding species.

Clusters of species patterns	Clusters of grid squares								
	1	2	3	4	5	6	7	8	9
1	0.02	0.10	0.16	0.39	0.28	0.44	0.53	0.80	0.62
2	0.01	0.05	0.11	0.16	0.19	0.19	0.43	0.47	0.75
3	0.03	0.18	0.43	0.50	0.76	0.45	0.70	0.86	0.75
4	0.04	0.04	0.12	0.41	0.40	0.07	0.09	0.29	0.08
5	0.01	0.00	0.01	0.12	0.04	0.02	0.02	0.04	0.04
6	0.01	0.02	0.02	0.07	0.06	0.12	0.12	0.36	0.14
7	0.01	0.02	0.05	0.14	0.19	0.04	0.11	0.26	0.12
8	0.02	0.02	0.06	0.60	0.10	0.09	0.15	0.48	0.37
9	0.00	0.01	0.05	0.07	0.08	0.05	0.15	0.21	0.30
10	0.00	0.00	0.01	0.01	0.01	0.04	0.03	0.03	0.51

habitat types were strongly related to the chance of migrating into the polder. If data had been available we would have liked to have looked at the pattern on the still smaller scale of the micro-habitat, rather than on that of the landscape type.

A point of interest arising from this study concerns the nature of biogeographical districts within The Netherlands. Clustering the species together on the basis of their localities and the dispersion of their distribution, and the mapping of the occurrence of these clusters, it can be seen that the localities overlap strongly, owing to large differences in dispersion among the clusters and overlap of clusters with a similar, sometimes restricted dispersion. Moreover, when we cluster grid squares, we see that the clusters coinciding with certain landscape types may be characterized by a combination of species clusters rather than by a single one (Table 2). This indicates a mosaic structure within the grid squares. The overlap of species clusters indicates that the borderlines between the species groups are not sharp and apply for some species and not for others. This holds for whatever level of variation is chosen. Our choice was guided by the optimal splitting level of the dendrogram.

Reading the biogeographical literature, one gets the impression that botanists and biogeographers think otherwise, considering the boundaries between districts or higher order biogeographical units as sharp, even without defining the level of variation on which one looks. Thus boundaries are given an absolute meaning, which in The Nehterlands even led to separate research projects in such boundary areas (cf. Mennema, 1978).

Lastly we want to point out that in this study clustering of geographical distributions is done not for its own sake, but as an aid to explain something else, namely immigration into the polder area. Cluster analysis on geographical data is being carried out more and more, but these studies are confined solely to a description of similar distribution patterns and their explanation in historical (e.g., Holloway and Jardine, 1968) or economic (Ezcurra *et al.*, 1978) terms.

When using cluster analysis as an aid for analyzing an independent process, the necessity of making *ad hoc* choices is less distracting than when using it as an endpoint analysis.

6. CONCLUSIONS

The major conclusions that we can draw from these investigations on the distribution patterns of Dutch Carabid species in relation to immigration into the newly reclaimed land, are:

1. The carabid fauna in the polders is at present roughly as one would expect from the type of habitat, the polder quadrats being included in clusters of mainland quadrats which resemble them with respect to habitat.

2. Most carabid species occurring in the polder quadrats have a very wide distribution, both in The Netherlands and in Europe, suggesting that a wide geographical dispersion may be the outcome of easy migration.

3. Distribution patterns of carabid species on different scales of observation are not independent. There is a clear relationship between observations on different scales, especially with respect to the amount of dispersion. This relationship is asymmetric, however, a wide dispersion in The Netherlands invariably implies a wide dispersion in Europe, and a limited dispersion in Europe implies a limited dispersion in The Netherlands. However, the reverse is not true: a wide dispersion in Europe does not imply a wide dispersion in The Netherlands as well, and a limited dispersion in The Netherlands does not imply a limited dispersion in Europe. We think that this asymmetry is easily explained as an effect of the tolerance ranges of species, but it excludes an explanation in terms of history and arrival.

4. In some cases, not only the amount of dispersion but also the location of occurrences seem to be related in data obtained on different scales. This applies for the species occurring almost exclusively along the coast in The Netherlands (Dutch clusters 4 and 5). These clusters coincide strongly with European clusters 6 and 7, which show respectively, restricted southern and western

distributions in Europe. This suggests that a specific habitat requirement is only realized in that area, for example, that the days of frost are relatively few.

5. Although the patterns found on different scales are not independent of each other, it is evident that they are not the same. The recognition of species groups is therefore not meaningful if the scale of investigation is not taken into account.

REFERENCES

Ezcurra, E., Rapoport, E. H., and Marino, C. R. (1978). The geographical distribution of insect pests. *Journal of Biogeography*, 5, 149-157.

Feekes, W. (1936). De ontwikkeling van de natuurlijke vegetatie in de Wieringermerpolder, de eerste groote droogmakerij van de Zuiderzee, *Nederlands Kruidkundig Archief*, 46, 1-295.

Feekes, W. and Bakker, D. (1954). De ontwikkelig van de natuurlijke vegetatie in de Noordoostpolder. *Van Zee tot Land*, 6, 1-92.

Haeck, J. (1971). The immigration and settlement of carabids in the new Ijsselmeer-polders. In *Dispersal and Dispersal Power of Carabid Beetles*. P. J. den Boer, ed. Miscellaneous Papers of the Landbouwhogeschool. Wageningen, The Netherlands, 8, 33-52.

Hengeveld, R. and Haeck, J. (in preparation). The distribution of abundance.

Hogeweg, P. (1976). Iterative character weighing in numerical taxonomy. *Computers in Biology and Medicine*, 6, 199-211.

Hogeweg, P. and Hesper, B. (1972). BIOPAT, program system for biological pattern analysis. Bioinformatica, University of Utrecht.

Mennema, J. (1978). Floristisch onderzoek naar Van Soests planten-geografische districten van Nederland, *Gorteria*, 9, 142-154.

Nip-van der Voort, J., Hengeveld, R. and Haeck, J. (in press). Immigration rates of plant species in three Dutch polders. *Journal of Biogeography*.

Toorn, J. van der, Donougho, B. and Brandsma, M. (1969). Verspreiding van wegbermplanten in Oostelijk Flevoland. *Gorteria*, 4, 151-160.

Turin, H., Haeck, J. and Hengeveld, R., (1977). Atlas of the carabid beetles of The Netherlands. Verhandelingen van de Koninklijke Akademie van Wetenschappen, Afdeling Natuurkunde, 2e Reeks, 68, 1-228.

[*Received January* 1979. *Revised March* 1979]

L. Orloci, C. R. Rao, and W. M. Stiteler, (eds.),
Multivariate Methods in Ecological Work, pp. 87-100. All rights reserved.
Copyright © 1979 by International Co-operative Publishing House, Fairland, Maryland

SPECIES WEIGHTING: A HEURISTIC APPROACH

R. C. JANCEY

Department of Plant Sciences
University of Western Ontario
London, Ontario, Canada   N6A 5B7

SUMMARY. The rationale of species weighting is discussed and
the available techniques are reviewed. Consideration is given
the dichotomies of purpose and applicability. A heuristic
approach, based on known data set partitions is described.

KEY WORDS. species weights, ranking.

1. INTRODUCTION

The problem of optimality of sampling strategy is of fundamental importance to phytosociologists. Setting aside the actual
positioning of sampling units (relevés) as a separate problem,
the most important question remains: Given a finite availability
of sampling effort, how to strike a balance in its partition
between the number of sampling units, number of species, and the
level of species data recording effort. A number of factors
point to descriptors (species) as the sector where economies
might be made most profitably.

The demonstration of pattern in vegetation depends
ultimately on an adequate number and distribution of sampling
units. In addition, the value of the extra information
generated by the more time consuming methods of data recording
(e.g. percentage cover) is related to sampling intensity (Goodall,
1970). It is also known that species vary in their sensitivity
of response to environmental factors, and hence in their ability
to create pattern, and that most species are correlated in their
environmental responses which creates redundancy of pattern information. Clearly then, the ordering of species on their ability
to reveal pattern should be of concern to phytosociologists.

2. WEIGHTING TECHNIQUES

Perhaps conceptually the simplest, a ranking technique described (but not advocated!) by Orlóci (1978a) is important since
it demonstrates a fundamental difficulty in computation. In a
pilot study composed of N quadrats and P species, let p
represent a subset of the P species. D is an N × N Euclidean
                                         $\sim$
distance matrix based on all P species and D   a similar matrix
                                              $\underset{\sim}{p}$
based on p. Possible distortion in D   due to the reduced
                                      $\underset{\sim}{p}$
species set is measured by a stress function σ(p;N,P), relating
D   to D. A correlation coefficient between the two matrices
$\underset{\sim}{p}$ $\sim$
would be one such measure. A value of σ(p;N,P) is computed
for all combinations of P species, in groups of p, and the
combinations ordered on the basis of values for σ(p;N,P). If
p is not specified, there are

$$\sum_{P} P!/[p!(P-p)!] \quad p=1,\cdots,P-1$$

stress values to compute. Clearly, whatever its virtues from a
phytosociological standpoint, this and similar methods based on
combinatorials are computationally unrealistic and other approaches
must be found.

The majority of ranking techniques use the computationally
simpler approach of summing the comparisons of each of the P
species in turn with the remaining P-1 species. An example of
such an approach is seen in Williams, Dale, and Macnaughton-Smith
(1964) and also Macnaughton-Smith, Williams, Dale and Mockett
(1964):

$$\sum_h \chi_{hi}^2/N \quad h=1,\cdots,P-1, \tag{1}$$

where χ_{hi}^2 is derived from a 2 × 2 contingency table of presence
(absence) data for species h and i in the N quadrats.

In such a technique it will be seen that all P-1 species
are involved in computing a rank for the *ith* species. An alternative philosophy is that once the species of rank 1 has been determined, it should play no further part in the ranking process, since
it is thus unable to influence the rank achieved by other species.
Similarly, the second ranked species must be eliminated after it
has been identified. This alternative is well illustrated by a
technique due to Orlóci (1973).   It employs a P × P matrix of
sums of squares and cross-products S and determines, for each
                                    $\sim$
species, its independent share of the total sum of squares.

Having found the first ranked species, the residual of S is computed in which the contribution of the first ranked species is eliminated. The process is then repeated through subsequent residuals of S until all species have been ranked and the cross-products matrix reduced to zeros.

An alternative to procedures based on the summation of pair-wise comparisons is given by Rohlf (1977). It starts with a sums of squares and cross-products matrix S of P species. An element in the matrix is S_{hi} and an element in the inverse, S^{-1}, of the matrix is $(S^{-1})_{hi}$. The object is to partition S_{hh} into a specific component S_{hhs} and a common component S_{hhc} such that $S_{hh} = S_{hhs} + S_{hhc}$. These components can be described in terms of the multiple correlation of species h with the remaining $P-1$ species:

$$R_h^2 = S_{hhc}/S_{hh} = (S_{hh} - S_{hhs})/S_{hh}. \tag{2}$$

Hence

$$S_{hhc} = S_{hh} R_h^2 \quad \text{and} \quad S_{hhs} = S_{hh}(1-R_h^2),$$

since

$$R_h^2 = 1 - [1/(S_{hh}(S^{-1})_{hh})],$$

$$S_{hhs} = 1/(S^{-1})_{hh}, \quad \text{and} \tag{3}$$

$$S_{hhx} = S_{hh} - 1/(S^{-1})_{hh}. \tag{4}$$

This technique suffers from the disadvantage that the matrix S may prove to be singular. But this can be overcome (Rohlf, 1977; Orloci, 1978a).

The techniques so far described illustrate a major dichotomy in ranking criteria. The method of Orlóci (1973) and equation (3) above, rank species on the basis of maximal unique variance. Equation 4, however, ranks on maximal shared variance. How to decide? The ranking technique should use the same rationale as that of any subsequent operation. For example, in the case of clustering procedures based on some kind of Euclidean distance measure of quadrat relationship, to use species ranked on the basis of methods such as that embodied in equation (4), would

result in the greatest redundancy and least discrimination. In
the case of ordinations, however, co-variation of species is of
major importance. In this context it would seem desirable to
rank species on the basis of techniques represented by criteria
such as (4).

Hill, Bunce, and Shaw (1975) combine species weighting with
a divisive, polythelic technique for classification. The stands
are ordered on the first axis of a reciprocal averaging ordination. They are then divided into two groups at the center of
gravity of the axis. The quantity

$$I_j = \left| (m_1/M_1) - (m_2/M_2) \right|$$

is then computed for each species, where m_1 and m_2 are the
numbers of stands possessing the jth species on the negative and
positive sides of the axis, respectively, and M_1, M_2 are the
total numbers of stands in each group. Thus I_j is the 'indicator value' for the jth species. The five species with the
greatest indicator values are then used to effect the first division of the stands. The process is repeated for successive
division.

With the exception of equation 1, the methods so far described assume that inter-species relationships are linear in form,
except that in the case of multiple correlations, the linear
regression is embedded in a hyperplane. Such techniques, however, clearly place a constraint upon the data. While it is
possible to envision methods based on curvilinear polynomial
regressions, it would seem more desirable to by-pass the problem
altogether. This can be achieved by turning to methods based on
probabilistic models: a method described by Feoli (1973)
incorporates a test of significance based on the t-distribution.
A matrix of quadrat similarities is computed, (Feoli used Sorensen's
(1948) index, but any symmetric measure of similarity would be
appropriate). The quadrats are then divided into two groups;
those containing species h, say H^+, and those from which
the species is absent, H^-. Two similarities are now computed;
$\overline{S}_{H+}$, the average similarities of quadrats in H^+ and $\overline{S}_{H-}$, the
average similarity of quadrats in H^+ with quadrats in H^-. This
gives a weight

$$W_h = 1 - \frac{\overline{S}_{H^-}}{\overline{S}_{H^+}} ,$$

(5)

upon which the species may be ranked. This technique has the advantage that it measures the sharpness of isolation of group H^+ from H^- (but not the other way) in terms of the total species composition. Since a simple test can be applied to the difference of the two means, a cutoff level can be established for the ranked species, thus providing an objective means of determining the size, for any chosen probability level, of the list of species to be finally used.

One of the problems associated with ranking techniques involving species covariances, or similar sums of squares related techniques, is the necessary assumption that covariance is meaningful in terms of the data available. Phytosociological data are frequently expressed in terms of abundance symbols, which may be ordered but not necessarily quantified. Such data are inappropriate for ranking by covariance related techniques, or indeed by any technique requiring data quantification. Fortunately, an alternative exists in the form of a number of Information Theory related measures. Information Theory is concerned with the distributions and joint distributions of data element frequencies. It is thus ideally suited to data expressed in terms of abundance classes. As in the case of the previously described methods, different ranking strategies are available using Mutual Information (Orlóci, 1976), which ranks species on the basis of maximum common information, i.e. maximum redundancy, and Equivocation Information (Orlóci, 1978b) ranking on maximum information describing group structure in the sample (comparable to specific, or unique variance).

3. A SPECIAL CASE

Whilst one of the major uses of mathematical techniques of data analysis for the phytosociologist lies in the elucidation of group structure, or pattern in vegetation, another possibility exists: Suppose some area under study possesses readily identifiable differences in topography, soil type, etc. One may wish to ask the question – do these areas also differ in their vegetation, and if so, which species have the greatest diagnostic value?

The ranking techniques so far reviewed would fail to take advantage of all the information available in this particular case. Since the quadrats are already known to come from one or other of the physically distinguishable areas already mentioned, a great increase in ranking efficiency can be achieved by employing this information in the ranking procedure. It is towards such a situation that this final technique is directed (Jancey, 1978). The only other example known to the author of weighting following partition of the set is one using binary data (Goodall, 1953).

The set of vegetation data which will be used to illustrate the technique is composed of 45 quadrats, for which 73 species were recorded on the Braun–Blanquet cover/abundance scale. The quadrats were located on three distinct river terraces, near Hope, British Columbia. The terraces differed both in soil type and moisture content. The objective was to rank the species on their ability to discriminate between the terraces. In addition, it was proposed to find that least subset of species which would successfully reproduce a group structure corresponding to the three terraces. For a fuller description of the vegetational area, see Fewster and Orlóci (1978).

As has been shown previously, measures of species interaction commonly form the basis of ranking techniques. While such approaches are undoubtedly valuable when the final objective is an ordination, or a classification of an otherwise undifferentiated area of vegetation, in the particular type of problem presently being described, it was considered advantageous to consider each species independently. In such a situation, a very simple measure of discriminatory power exists.

Let x'_{ij} be an element of the data matrix, representing the score for the jth of P species in the ith of N quadrats. The data are first standardized within species,

$$x_{ij} = (x'_{ij} - \overline{x}_j)/S_j ,$$

where S_j is the standard deviation and $\overline{x}_j$ the mean of the jth species. Since these values are deviations from the species mean, the total species sum of squares is

$$T_j = \sum_{i=1}^{N} x_{ij}^2 .$$

The quadrats are divided into E groups (in the example given, E = 3, the number of terraces). The number of quadrats in the kth group is N_k. Within group sums of squares are now computed. If x_{ij} is a standardized datum and $\overline{x}_{kj}$ the mean in the kth group for species j, the pooled within groups sums of squares W_j, is given by

$$W_j = \sum_{i=1}^{E} \sum_{i \epsilon k}^{N_k} (x_{ij} - \overline{x}_{kj})^2 .$$

Hence the between groups sum of squares for species j may be obtained by subtraction, $B_j = T_j - W_j$. The between groups variance is $V_{Bj} = B_j/(E-1)$, and similarly, W_j may be converted to a variance, $V_{Wj} = W_j/(N-E)$. Thus a variance ratio may be formed, $F_j = V_{Bj}/V_{Wj}$.

The computation is then repeated for all P species. Apart from thus ranking the species, a means now exists for determining the size of the subset of species. Since the ranked values are variance ratios, they have an F distribution (under certain restrictive assumptions) with the appropriate degrees of freedom. Hence, a decision on the probability level p determines the size and composition of the species subset.

In a well structured data set, such as the present example, p might well be set to a very small value indeed and still achieve a classification based on vegetation which accurately reflects environmental grouping. Conversely, in other situations where structure is virtually absent, one might find very few species with a variance ratio significant at even the p = .1 level. If these few species were unable to reproduce the physical group structure, it would be appropriate to conclude that a vegetational reflection of the environmental grouping did not exist.

The ranking for this particular data set is illustrated in Table 2d (see Table 1 for the species list). The F values in Table 2 have been rounded to two decimal places, causing *apparent* ties.

The algorithm as so far developed may be further refined: Let us suppose that the three river terraces were not more or less equally distinct, but that two, (A,B) were very similar while one, (C) was very distinct. The algorithm as so far described would give highest ranking to those species which distinguish between C and the other two, (A,B) combined, since it is this distinction which would generate the greatest variance ratios. As a consequence, it would prove very difficult with such a ranking to discriminate between the A and B terraces. A partitioning of the ranking into all pairwise site comparisons overcomes this problem (see Table 2a-c). A number of possibilities now exist for determining the size and composition of the species subset to be used for a substantive study. Assuming that we wish each pairwise group comparison to receive equal weight, one approach would be to take, for any given probability level, that group with the least number of significant F values, and use those significant species plus an equal number of the most highly ranked species from each of the other pairwise comparisons. In the case of species being highly ranked in more than one list, the size of the final species subset would be correspondingly smaller.

TABLE 1:   Cross-reference list of species names and corresponding
numeric codes used in Table 2.

Species	No.	Species	No.
Abies grandis	1	Bromus vulgaris	38
Acer macrophyllum	2	Chimaphila umbellata	39
Betula papyrifera	3	Circaea alpina	40
Pseudotsuga menziesii	4	Clintonia uniflora	41
Thuja plicata	5	Dicentra formosa	42
Tsuga heterophylla	6	Disporum hookeri	43
Acer circinatum	7	Dryopteris assimilis	44
Acer glabrum	8	Galium triflorum	45
Acer macrophyllum	9	Goodyera oblongifolia	46
Alnus rubra	10	Lactuca canadensis	47
Amelanchier alnifolia	11	Lilium columbianum	48
Betula papyrifera	12	Linnaea borealis	49
Cornus nuttallii	13	Claytonia sibirica	50
Corylus cornuta	14	Osmorhiza chilensis	51
Gaultheria shallon	15	Polypodium glycirrhiza	52
Holodiscus discolor	16	Polystichum munitum	53
Lonicera ciliosa	17	Pteridium aquilinum	54
Menziesia ferrunginea	18	Pyrola grandiflora	55
Mahonia aquifolium	19	Cinna latifolia	56
Mahonia nervosa	20	Eurhynchium oreganum	57
Paxistima myrsinites	21	Streptopus amplexifolius	58
Philadelphus lewisii	22	Streptopus roseus	59
Rhamnus purshianus	23	Tolmiea menziesii	60
Ribes bracteosum	24	Trientalis latifolia	61
Rosa gymnocarpa	25	Trillium ovatum	62
Rubus parviflorus	26	Urtica dioica	63
Rubus spectabilis	27	Veronica officinalis	64
Sambucus pubens	28	Viola glabella	65
Spiraea douglasii	29	Dicranum scoparium	66
Symphoricarpos albus	30	Hylocomium splenden	67
Taxus brevifolia	31	Hylocomium splendens	68
Thuja plicata	32	Mnium insigne	69
Tsuga heterophylla	33	Mnium spinulosum	70
Vaccinium membranaceum	34	Rhacomitrium canescens	71
Vaccinium parvifolium	35	Rhytidiadelphus loreus	72
Achlys triphylla	36	Rhytidiopsis robusta	73
Athyrium filix-femina	37		

TABLE 2: Species ranking based on their ability to discriminate between (a) groups 1 and 2, with 1 and 32 df, (b) groups 1 and 3, with 1 and 23 df, and (c) groups 2 and 3, with 1 and 29 df; (d) species ranking based on all quadrats, with 2 and 42 df.

Rank	(a) No.	(a) F Value	(b) No.	(b) F Value	(c) No.	(c) F Value	(d) No.	(d) F Value
1	46	56.85	15	1455.97	29	68.29	42	111.50
2	41	43.01	53	89.97	34	40.58	15	104.80
3	54	33.75	39	55.82	16	14.61	53	73.97
4	53	31.54	29	45.56	49	11.83	21	65.19
5	15	16.85	35	39.15	39	7.42	29	58.16
6	6	9.70	49	35.37	35	3.44	39	51.32
7	21	8.04	21	31.79	15	2.91	49	37.97
8	7	5.30	34	27.82	48	1.87	41	36.83
9	3	5.04	7	20.48	66	1.87	54	36.38
10	42	4.57	46	20.22	71	1.87	34	34.42
11	23	3.28	17	14.68	64	1.87	30	32.70
12	30	2.02	1	12.88	10	1.86	43	31.92
13	26	2.01	16	9.60	32	1.86	35	30.85
14	27	1.75	3	6.47	20	1.76	7	23.17
15	4	1.70	42	3.42	21	1.28	68	22.91
16	68	1.54	43	3.13	36	1.15	46	22.54
17	18	1.45	36	2.35	41	1.13	50	20.67
18	20	1.43	55	2.29	54	1.12	69	20.55
19	17	1.35	25	2.15	17	.95	47	16.55
20	55	1.33	47	2.01	43	.90	20	15.50
21	35	1.29	30	1.97	47	.68	36	14.94
22	39	1.16	68	1.39	5	.62	40	13.30
23	61	1.14	32	1.29	11	.57	27	12.42
24	67	.97	33	1.28	7	.53	16	12.34
25	43	.96	66	1.28	25	.50	51	11.29
26	69	.89	64	1.28	27	.38	17	10.19
27	50	.84	23	1.28	53	.35	5	9.84
28	5	.70	71	1.27	4	.30	2	8.98
29	1	.69	48	1.27	6	.27	6	8.88
30	33	.69	10	1.27	45	.22	61	8.44
31	59	.68	11	.99	31	.18	4	8.40
32	49	.68	69	.90	55	.17	25	7.25
33	22	.65	51	.74	46	.14	57	7.17
34	40	.54	45	.69	72	.13	67	6.95
35	51	.51	61	.67	62	.13	45	5.44
36	31	.49	41	.65	51	.12	9	4.65
37	14	.48	50	.63	8	.11	11	4.59
38	2	.41	67	.53	13	.11	38	3.94
39	57	.32	57	.51	69	.11	28	3.90
40	9	.27	2	.45	57	.08	31	3.81
41	47	.25	40	.41	9	.07	19	3.75

Table 2:   (Continued)

Rank	(a) No.	(a) F Value	(b) No.	(b) F Value	(c) No.	(c) F Value	(d) No.	(d) F Value
42	38	.16	27	.40	68	.05	72	3.25
43	28	.16	72	.39	26	.05	3	3.00
44	19	.15	12	.35	30	.05	63	2.41
45	25	.13	13	.33	18	.04	65	2.31
46	63	.10	14	.32	2	.04	52	2.27
47	65	.10	5	.18	44	.04	58	2.22
48	72	.10	9	.13	23	.03	24	2.13
49	52	.09	38	.12	12	.02	70	2.01
50	24	.09	24	.12	1	.02	26	1.79
51	58	.09	28	.12	73	.02	13	1.73
52	70	.08	65	.11	22	.02	23	1.71
53	45	.08	19	.11	59	.02	55	1.66
54	12	.08	20	.11	65	.02	62	1.62
55	60	.05	62	.08	37	.02	14	1.59
56	56	.05	63	.07	24	.02	32	1.58
57	62	.03	70	.07	70	.02	66	1.58
58	11	.03	52	.07	67	.01	64	1.58
59	13	.03	58	.07	61	.01	71	1.58
60	36	.01	8	.06	3	.01	48	1.57
61	8	.01	60	.03	14	.00	10	1.57
62	37	.01	56	.03	33	.00	18	1.29
63	73	.00	37	.03	56	.00	60	1.11
64	44	.00	73	.03	19	.00	56	1.11
65	34	.00	44	.03	60	.00	8	.77
66	29	.00	31	.03	58	.00	12	.73
67	32	.00	4	.02	40	.00	1	.61
68	66	.00	54	.00	28	.00	59	.60
69	16	.00	59	.00	38	.00	22	.57
70	64	.00	22	.00	42	.00	33	.42
71	48	.00	6	.00	50	.00	44	.42
72	10	.00	18	.00	52	.00	73	.36
73	71	.00	26	.00	63	.00	37	.31

4.   RESULTS

Since this is a methodological presentation, no comments
will be included on the ecological significance of individual
species.   The changes in rank order when different group pairs
are considered are summarized in Table 3.   It will be seen that
in most cases the rankings for species differ greatly when
different group pairs are considered (only *Gaultheria shallon*
being consistently highly ranked).   Despite these differences
in pairwise ranking, however, only *Holodiscus discolor* fails to

TABLE 3: Summary of changes in rank order when the five top ranked species in each group pair are included, also the lowest ranked species (Athyrium filix-femina) for the all quadrats ranking.

Species	Groups:	Ranking 1-2	1-3	2-3	All
Goodyera oblongifolia		1	10	33	16
Clintonia uniflora		2	36	17	8
Pteridium aquilinum		3	68	18	9
Polystichum munitum		4	2	27	3
Gaultheria shallon		5	1	7	2
Chimaphila umbellata		22	3	5	6
Spirea douglasii		66	4	1	5
Vaccinium parvifolium		21	5	6	13
Vaccinium membranaceum		65	8	2	10
Holodiscus discolor		69	13	3	24
Linnaea borealis		32	6	4	7
Athyrium filix-femina		62	63	55	73

reach the top 18 in the ranking based on all quadrats. As might be expected, the species ranked lowest overall: *Athyrium filix-femina* has consistently low rankings in the pairwise comparisons.

In order to demonstrate the effect of reduced species number on group structure, two hierarchical clusterings are shown (Figures 1 and 2). Figure 1 is based on all 73 species, and Figure 2 on those with the 18 greatest F values (corresponding to a species ranking cutoff at $p < .05$). Based on Euclidean distances between all pairs of quadrats, the quadrats were clustered using a sums of squares criterion. It will be seen that the clusterings are remarkably similar, differing only at the lower fusion levels. The clustering achieved clear separation of the terraces, with no incorrectly assigned quadrats, even when only 18 species were used.

5. DISCUSSION AND CONCLUSIONS

In the preceding review of species weighting techniques, it has been seen that a number of dichotomies exist. In general, techniques manipulating sums of squares and cross-products make the assumption of linearity, while those having a probabistic basis, e.g. the Information Theory models, make no such assumption. The decision whether to rank on maximal shared variance as opposed to specific variance has not been resolved. In many cases both alternatives are available as options in the same program. Perhaps more important than the question of which is

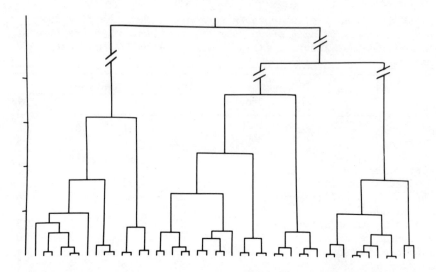

FIG. 1:   *Hierarchical clustering based on all species. Numbers on the horizontal axis identifying quadrats, while numbers on the vertical axis represent sums of squares at group fusion.*

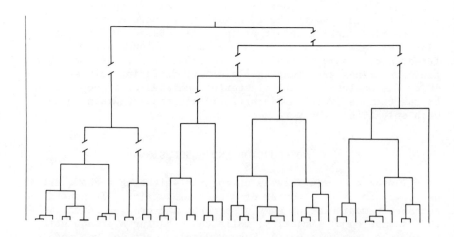

FIG. 2:   *Hierarchical clustering based on the 18 most highly ranked species from Table 2d. Numbers on the horizontal and vertical axes are as in Figure 1.*

fundamentally correct is the need to match the ranking philosophy to that of subsequent data analyses. In any case, methods representing all combinations of these two dichotomies are available, whether the data is binary, multistate, or in the form of qualitative frequencies.

In the case where group membership is pre-determined by external criteria, great power is added to the ranking procedure, though as yet only two such techniques exist: Goodall (1953) for binary data and Jancey (1978) for multi-state data.

A computer program written in BASIC is available from the author.

REFERENCES

Feoli, E. (1973). Un indice che stima il peso dei caratteri per classificazioni monotetiche. *Giornale Botanico Italiano*, 107, 263-268.

Fewster, P. and Orlóci, L. (1978). Stereograms to aid group recognition and trend identification in vegetation data. *Canadian Journal of Botany*, 56, 162-165.

Goodall, D. W. (1953). Objective methods for the classification of vegetation. II. Fidelity and indicator value. *Australian Journal of Botany*, 1, 434-456.

Goodall, D. W. (1970). Statistical plant ecology. *Annual Review of Ecology and Systematics*, 1, 99-124.

Hill, M. O., Bunce, R. G. H., and Shaw, M. W. (1975). Indicator species analysis, a divisive polythelic method of classification, and its application to a survey of nature pinewoods in Scotland. *Journal of Ecology*, 63, 597-613.

Jancey, R. (1978). Species ordering on a variance criterion. *Vegetatio* (in press).

Macnaughton-Smith, P., Williams, W. T., and Mockett, L. G. (1964). Dissimilarity analysis: a new technique of hierarchical subdivision. *Nature*, 202, 1034-1035.

Orlóci, L. (1973). Ranking characters by a dispersion criterion. *Nature*, 244, 371-373.

Orlóci, L. (1976). Ranking species by an information criterion. *Journal of Ecology*, 64, 417-419.

100   R. C. JANCEY

Orlóci, L. (1978a). *Multivariate analysis in vegetation research*, 2nd ed. Junk, The Hague.

Orlóci, L. (1978b). Ranking species based on the components of equivocation information. *Vegetatio*, 37, 123-125.

Rohlf, F. J. (1977). A note on the measurement of redundancy. *Vegetatio*, 34, 63-64.

Sorensen, T. (1948). A method of establishing groups of equal amplitude in plant sociology based on similarity of species content. *Dansk Videnskaberves Selskab, Copenhagan. Biologiske Skrifter*, 5, 1-34.

Williams, W. T., Dale, M. B., and Macnaughton-Smith, P. (1964). An objective method of weighting in similarity analysis. *Nature*, 201, 426.

[*Received July* 1978. *Revised January* 1979]

L. Orloci, C. R. Rao, and W. M. Stiteler, (eds.),
Multivariate Methods in Ecological Work, pp. 101-126. All rights reserved.
Copyright © 1979 by International Co-operative Publishing House, Fairland, Maryland

A COMPARISON OF TWENTY-ONE MEASURES OF SITE DISSIMILARITY

BYRON B. LAMONT

Department of Biology
Western Australian Institute of Technology
Perth, Western Australia 6102

KEN J. GRANT

Department of Mathematics and Computing Studies
Western Australian Institute of Technology
Perth, Western Australia 6102

SUMMARY. Over sixty measures of (dis)similarity proposed in the
literature are described. After dismissing those unsuitable for
ordination, the remainder reduce to 26 dissimilarity formulae
suitable for *binary* data. The profound effect of choice of measure
on the results of data reduction is pointed out and the relative
merits of binary data are noted. It is argued that the best
measure of biotic distance (difference in species composition)
between two sites is one which closely reflects their ecological
distance along an environmental gradient. Attempts to define this
ideal measure founder on ignorance about the environmental tolerances (autecology) of species encountered in a synecological study,
the multiplicity of response curves that are ecologically possible,
and the lack of guidelines from ecological theory. The behavior
of 21 binary measures in response to six hypothetical cases of
extremes in species presence at two sites (S_i, S_j), species in
common (σ), and total number of species in the collection (N)
was examined and compared against the family of acceptable curves.
Few measures satisfied the criteria given here (versatility, environmental sensitivity, and constancy of slope), but, among the
unstandardized measures which disregard conjoint absences, the
simple Absolute $[S_i + S_j - 2\sigma]$ and Euclidean $[(S_i + S_j - 2\sigma)^{\frac{1}{2}}]$
can be strongly recommended. Standardization and recognition of
conjoint absences increase calculation time but have no ecological
advantages.

102   B. B. LAMONT AND K. J. GRANT

KEY WORDS. dissimilarity measure, ordination, environmental
gradient, binary data, ecological distance, standardization, continuum, species clumping.

1. THE PROBLEM

    As a result of recent developments in multidimensional scaling,
it is now possible to find the best distance (ordination) matrix
to represent a given dissimilarity matrix derived from the original
data matrix. It is a suitable time then to return to the more
elementary issue: choice of measure of dissimilarity. Most
ecologists must accept the ordination and classification programs
that are available to them, and being more concerned with interpretation of the results, are satisfied to have at least understood
the derivation of the dissimilarity measure used, let alone to
question it. Goodall (1973) allows too much latitude when he says,
"To a large extent, the choice of index is a matter of *taste* and
of one's particular purpose...." After exploring the latter, we
outline here our procedures and results in improving the object-
ivity of the former.

2. EFFECT OF DISSIMILARITY MEASURES

    The profound influence that choice of dissimilarity measure
may have on distortion of the original matrix and the consequent
ordination results have been demonstrated by Gauch and Whittaker
(1972), Orloci (1973), and references in Goodall (1973). A similar
effect of different manipulations of the data before applying a
dissimilarity measure (in many ways equivalent to using a different
measure of dissimilarity at the outset, Noy-Meir and Whittaker,
1977) can be seen in Austin and Greig-Smith (1968) and Noy-Meir,
Walker, and Williams (1975). In a non-ecological context, Green
and Rao (1969) have gone a step further to illustrate graphically
(PCA) the relationship between eight quantitative measures of dissimilarity for a given set of data. If the data are well spread,
this gives us an idea about the numerical relationship between the
measures, but not which is 'best'.

    As an example from our own work, Table 1 highlights how
definition of the 'best' axes using the 'objective' variance method
of selecting endpoints as in principal axes ordination (van der
Maarel, 1969) and optimized polar ordination (Swan, Dix, and Wehrahn,
1969) is influenced by the binary dissimilarity measure used. (We
are not necessarily advocating these methods of ordination.) The
frequent appearance of sites E, P, U, and V as 'best' endpoints
reflects both similarity between some of the eight measures used
and real trends in the data. If four sites are deliberately set
as endpoints and the length of the range along one axis (in this

TABLE 1: Effect of dissimilarity measure on the composition of the 'best' axes using the variance method of selecting endpoints in polar ordination. Data from 24 sites with a total of 135 species (Lamont, unpub.). See also Figure 1.

Measure	Best axis	Second best axis
Absolute	Sites V,H	Sites U,E
Jaccards	V,P	U,E
Sorensens	U,E	R,P
Orlocis chord	V,P	W,E
Kulczynskis 2nd	U,E	R,P
Product moment	U,E	V,P
Chi-square	W,Q	S,E
Dice-Sorensen	O,W	V,R

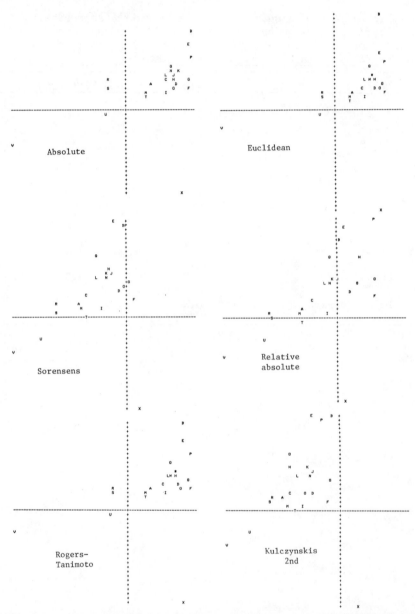

FIG. 1:   Polar ordination of 24 sites with a total of 135 species
(Lamont, unpub.) using twelve binary dissimilarity measures.   For
comparative purposes the endpoints are the same (V, W for X-axis,
B, X for Y-axis) and the distance between the X-endpoints is held
constant.   To save space above W (on the extreme right) has not
beel plotted.   * indicates two coincident sites.

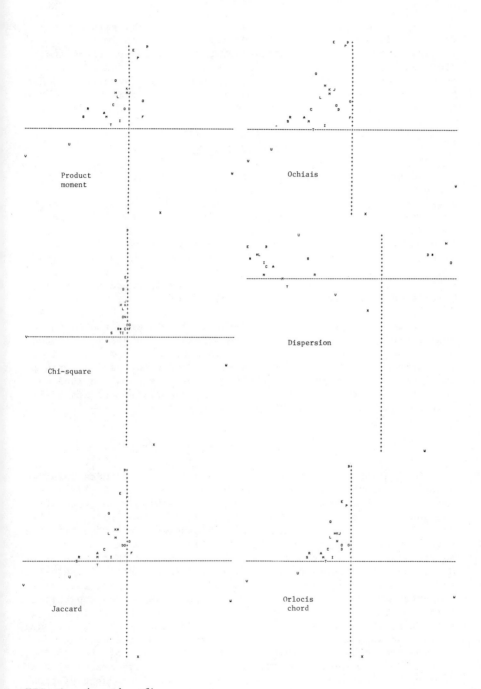

FIG. 1: (continued).

case   X) is equalized for each case, we may observe directly any
variation in site positions due entirely to the dissimilarity
measure (Figure 1). The rankings of the positions of the sites
along each axis do not change much, but the spread and skewness
about the origins and their absolute distances vary markedly. Note
for example the swing in position of B and P relative to the
Y axis and their variable distance apart in the X direction.

3. SCOPE OF THE STUDY

We are not concerned here with sample size, sampling error,
and replication, and their effects on dissimilarity values in
relation to niche overlap and environmental heterogeneity. See,
for example, the measure by Mountford (1962) for comparing communities that fit the log-series distribution model. Neither do we
take up the issue of whether or not to remove rare species, which
of course is related to the above. In addition, we only consider
site relationships (Q analysis) but believe similar procedures
could be followed if species (R) relationships were the major
interest and similar conclusions would be drawn.

In this study, actual or hypothetical data are treated as
transformed into a binary form with presence of a species denoted
by 1, and absence by 0. Had the data been further relativized
(e.g., by dividing by sum of site scores for each species, Orloci,
1967) use of binary measures would no longer be possible, and we
would need to revert to the more complex quantitative forms available for many of the measures.

4. BINARY *VERSUS* QUANTITATIVE MEASURES

Ordination (PCA) of binary and quantitative measures based on
a set of transect data by Smartt, Meacock, and Lambert (1976)
showed the qualitative measure was generally widely separated from
the quantitative measures (no matter now crudely collected or
severely transformed) but nevertheless concluded that "qualitative
data provide best ecological groupings in terms of the major
habitats...." Van der Maarel (1969) has asserted that the use of
binary data is "the most satisfactory general approach in vegetation complexes." There seems to be general agreement that binary
measures may be just as effective in elucidating broad ecological
relationships, provided species variation in a data matrix is high
(Noy-Meir, 1971; Walker, 1974; Bouxin, 1975). This being so, the
further advantages of binary measures are: a) the enormous saving
in time and costs of collecting binary data only, b) the simplicity
of calculations required compared with quantitative measures (these
days more a saving of costs than time), c) the avoidance of problems of observer error, choice and validity of the many indices of

species importance available, d) the automatic overcoming of the
disproportionate effect of widespread, dominant species - perhaps
preferable even to site standardization of quantitative data,
and e) the conceptual ease with which changes in species richness
and commonality (in contrast to species diversity, including as it
does on abundance component) can be followed in relation to environmental changes. It is not our intention to convince the reader
about general adoption of binary measures, however, for we largely
chose to examine them on the basis of point e) above.

5. RANGE OF DISSIMILARITY MEASURES

Table 2 gives details of the binary forms of 51 measures of
dissimilarity. For consistency, the symbols have meanings appropriate for site (Q) comparisons, but they could be given equivalent meanings for species (R) comparisons: S_i = number of
species in site (sample) i, S_j = number of species in site j,
σ = number of species in common between i and j, $S = S_i + S_j$
- σ, $S - \sigma$ = difference in number of species between i and j,
N = total number of species in the data set. Many of the measures
have only ever been used in the quantitative and/or similarity
form and thus needed conversion before their relationship with
other measures could be examined. In addition, measures which
have a constant scaling factor (usually N) can be regarded as
synonymous with equivalent unscaled measures. As an extreme
example it is interesting to note the collapse of twelve measures
to the form $(S - \sigma)$. The measures may be placed into two broad
groups - those disregarding conjoint absence of other species in
the collection (no reference to N except as a scaling factor)
and those taking conjoint absence into account (involving (N - S)
terms). Within each group they may be separated into those that
are standardized (*sensu* Clifford and Stephensen, 1975) and those
that are not. Standardization may be by the arithmetic mean (e.g.,
van der Maarels), the harmonic mean (e.g., McConnaughys) or the
geometric mean (e.g., cosine). Whether standardized or not, the
measures are either transformed on a Euclidean (square root) basis
or, to a lesser extent, contingency (squared) basis, or untrans-
formed (Manhattan-related). Of the 26 distinct measures, 13 in
the first group and 8 in the second were selected for further
analysis. The range of possible values for each of these 21
measures is given in Table 4.

6. MEASURES NOT SUITABLE FOR ORDINATION

The following similarity measures (additional to those in
Table 2) were rejected because they have no upper limit, and hence

108   B. B. LAMONT AND K. J. GRANT

TABLE 2: *The binary Q-analysis form of 51 dissimilarity measures cited in the literature, grouped according to whether or not they are standardized or recognize conjoint absences. * - dissimilarity form of the original similarity measure. ϕ - binary form of the original quantitative measure.*

NAME	FORMULA	SYMBOL (Figs. 3-10)	ORIGIN	MORE RECENT REFERENCE
1. Comparison between two samples, disregarding conjoint absence of other species in the collection.				
a) Not standardized, but may be scaled by N.				
Absolute	$S-\sigma$	a	-	-
Manhattan	$S-\sigma$ $^{\phi}$	a	'Minkowski' in Kruskal (1964)	Noy-Meir & Whittaker (1977)
Simple-matching	$(S-\sigma)/N$ *	a	Sokal & Michener (1958)	Clifford & Stephenson (1975)
Average distance	$(S-\sigma)/N$ $^{*\,\phi}$	a	-	Wishart (1970)
Hamanns	$2(S-\sigma)/N$ *	a	Hamann (1961)	Clifford & Stephenson (1975)
Canberra	$(S-\sigma)/N$ $^{\phi}$	a	Lance & Williams (1967)	Clifford & Stephenson (1975)
Museum	$(S-\sigma)/N$ $^{\phi}$	a	Goldman (1973)	Clarke (1975)
Error sum of squares	$(S-\sigma)/N$ $^{*\,\phi}$	a	-	Wishart (1970)
Mean character difference	$(S-\sigma)/N$ $^{\phi}$	a	Cain & Harrison (1958)	Clifford & Stephenson (1975)
Euclidean squared	$S-\sigma$ $^{\phi}$	a	-	Clifford & Stephenson (1975)
Variance	$(S-\sigma)/4N$ $^{\phi}$	a	-	Clifford & Stephenson (1975)
Information	$2(S-\sigma)\ell n\ 2$ $^{\phi}$	a	Williams *et al.* (1966)	Pielou (1969)
Euclidean	$\sqrt{(S-\sigma)}$ $^{\phi}$	b	'Minkowski' in Kruskal (1964)	Orloci (1975)
Euclidean (scaled)	$\sqrt{(S-\sigma)}/N$ $^{\phi}$	b	-	Wishart (1970)
Pattern difference	$(S_i-\sigma)(S_j-\sigma)/N^2$	c	-	Wishart (1970)
b) Standardized, but not scaled by N.				
Interspecific correlation	$(S-\sigma)/S$ $^{*\,\phi}$	f	Ellenberg (1956)	Goodall (1973)
Jaccards/Coeff. community	$(S-\sigma)/S$ *	f	Jaccard (1901)	Clifford & Stephenson (1975)
% co-occurrence	$(S-\sigma)/S$ *	f	Angrell (1945)	Whittaker (1967)
Similarity ratio	$(S-\sigma)/S$ $^{*\,\phi}$	f	-	Wishart (1970)
Ruzickas	$(S-\sigma)/S$ $^{*\,\phi}$	f	Ruzicka (1958)	Goodall (1973)
Czekanowskis/% similarity	$(S-\sigma)/S+\sigma$ $^{*\,\phi}$	g	Czekanowski (1909)	Clifford & Stephenson (1975)
Sorensens	$(S-\sigma)/S+\sigma$ *	g	Sorensen (1948)	Gauch & Whittaker (1972)
Bray-Curtis (unstandardized)	$(S-\sigma)/S+\sigma$ $^{*\,\phi}$	g	Bray & Curtis (1957)	Gauch & Whittaker (1972)
Van der Maarel	$2(S-\sigma)/S+\sigma$ *	g	Van der Maarel (1969)	-
Pandeyas	$(S-\sigma)/S+\sigma$ $^{*\,\phi}$	g	Pandeya (1961)	Goodall (1973)
Relative total	$S/S+\sigma$	h	Lamont (unpub.)	-
Orlocis chord	$\sqrt{[2\sqrt{S_iS_j}-\sigma)/\sqrt{S_iS_j}]}$ $^{\phi}$	i	Orloci (1967)	Orloci (1975)
Ochiais	$(\sqrt{S_iS_j}-\sigma)/\sqrt{S_iS_j}$ *	j	Ochiai (1957)	Goodall (1973)
Cosine/Angular separation	$(\sqrt{S_iS_j}-\sigma)/\sqrt{S_iS_j}$ $^{*\,\phi}$	j	Gower (1967)	Wishart (1970), Orloci (1975)
Relative absolute 1	$(S-\sigma)/\sqrt{S_iS_j}$	m	Lamont (unpub.)	Clifford & Stephenson (1975)
Relative absolute 2	$(S-\sigma)/S_iS_j$	k	Lamont (unpub.)	Clifford & Stephenson (1975)
Relative euclidean	$\sqrt{(S-\sigma)}/\sqrt{S_iS_j}$	ℓ	Orloci (1967)	Clifford & Stephenson (1975)

TABLE 2: (continued)

NAME	FORMULA	SYMBOL (Figs. 3-10)	ORIGIN	MORE RECENT REFERENCE
Kulczynskis 2nd	$(2S_i S_j - S_i\sigma - S_j\sigma)/S_i S_j$ *	n	Kulczynski (1927)	Clifford & Stephenson (1975)
Gleasons	$(2S_i S_j - S_i\sigma - S_j\sigma)/2S_i S_i$ * φ	n	Gleason (1920)	Orloci (1975)
McDonnaughy	$(2S_i S_j - S_i\sigma - S_j\sigma)/S_i S_j$ *	n	McDonnaughy (1964)	Clifford & Stephenson (1975)
Unnamed 1	$2(S-\sigma)/2S-\sigma$ *	o	Sokal & Sneath (1963)	Wishart (1970)

2. *Comparison between two samples which takes into account conjoint absence of other species in the collection.*
   a) Not standardized, but may be scaled by N.

NAME	FORMULA	SYMBOL (Figs. 3-10)	ORIGIN	MORE RECENT REFERENCE
Dot product	$(N-\sigma)/N$ * φ	d	-	Wishart (1970)
Russell-Rao	$(N-\sigma)/N$ *	d	Russell & Rao (1940)	Clifford & Stephenson (1975)
Dispersion ($S_{max} = S_i$ or S_j)	$\dfrac{(S_{max}-\sigma)(N-S)-(S_i-\sigma)(S_j-\sigma)}{N^2}$ * φ	e	-	Wishart (1970)
Shape difference	$N(S-\sigma)-(S_i-S_j)^2/N^2$ φ	-	-	Wishart (1970)

b) Standardized, but not scaled by N.

NAME	FORMULA	SYMBOL (Figs. 3-10)	ORIGIN	MORE RECENT REFERENCE
Product-moment correlation	$1-\dfrac{N\sigma - S_i S_j}{\sqrt{(S_i S_j (N-S_i)(N-S_j)}}$ * φ	p	-	Wishart (1970)
χ^2 association	$N-\dfrac{[(\sigma(N-S)-(S_i-\sigma)(S_j-\sigma)]^2 N}{S_i S_j (N-S_i)(N-S_j)}$ *	r	-	Clifford & Stephenson (1975)
Mean square contingency	$1-\dfrac{[(\sigma(N-S)-(S_i-\sigma)(S_j-\sigma)]^2}{S_i S_j (N-S_i)(N-S_j)}$ *	r	-	Clifford & Stephenson (1975)
Kendalls rank correlation	$1-\dfrac{(N-S)-(S_i-\sigma)(S_j-\sigma)}{\sqrt{S_i S_j (N-S_i)(N-S_j)}}$ *	q	Looman & Campbell (1960)	Lie & Kelly (1970)
Unnamed 2	$1-\dfrac{\sigma(N-S)}{\sqrt{S_i S_j (N-S_i)(N-S_j)}}$ *	-	Sokal & Sneath (1963)	Wishart (1970)
Rogers - Tanimoto	$2(S-\sigma)/N+S-\sigma$ *	s	Rogers & Tanimoto (1960)	Clifford & Stephenson (1975)
Dice - Sorensen	$(S-\sigma)/2N-S+\sigma$ *	t	-	Sokal & Sneath (1963)
Yule	$\dfrac{2(S_i-\sigma)(S_j-\sigma)}{\sigma(N-S)+(S_i-\sigma)(S_j-\sigma)}$ *	u	Yule (1912)	Clifford & Stephenson (1975)
Forbes	$2S_i S_j - \sigma N/S_i S_j$ *	-	Forbes (1907)	Whittaker (1967)
Coles	$2-\dfrac{\sigma N-(S_i-\sigma)(S_j-\sigma)}{S_i S_j}$ *	-	Cole (1949)	Whittaker (1967)
Segregation coefficient (plotless)	$\dfrac{(S-\sigma)N}{S_i(N-S_j)+S_j(N-S_i)}$ *	-	Pielou (1961)	Goodall (1973)

cannot be converted into dissimilarity measures for ordination
purposes: Kulczynskis first (Clifford and Stephenson, 1975) =
$\sigma/S-\sigma$, Unnamed (Sokal and Sneath, 1963) = $(N-S+\sigma)/S-\sigma$, Fager and
McGowan (Clifford and Stephenson, 1975) = $\sigma/\sqrt{S_i S_j} - 1/2\sqrt{S}_{min}$
(related to Ochiais), and Mountford (1962) = $2\sigma/2S_i S_j - \sigma(S_i + S_j)$.
Measures based on the results of a preliminary discriminant function or PCA (e.g., Mahalanobis) were not used because of the
possible inappropriateness of these methods of multidimensional
scaling and the vast increase in computations required. The dissimilarity measures for differences in shape (Wishart, 1970) =
$N(S-\sigma)-(S_i-S_j)^2/N^2$ and size (*ibid.*) = $(S_i-S_j)^2/N^2$ are clearly
inappropriate for ordination (e.g., as $S_i \to S_j$, $\delta_i \to \delta_j$, independent of σ). Another unnamed measure (Sokal and Sneath, 1963) with
the dissimilarity form $1 - \sigma(S_i + S_j)/4S_i S_j - (N-S)(2N-S_i-S_j)/$
$4(N-S_i)(N-S_j)$ was rejected on grounds of its complexity and apparent
relationship to the other contingency measures used here (Table 2).
The corrected forms of χ^2 (Goodall, 1973) were considered too
complex for the small rewards they offered. The segregation coefficient (Table 2) is based on the identity of the nearest neighbor
and, whatever its virtues, cannot be used with most data collected
by ecologists.

7. SITE DISSIMILARITY *VERSUS* ENVIRONMENTAL DIFFERENCE

Though only binary data are considered here, we are not just
interested in an inventory of the species per site or group of
sites. Species richness and composition are clearly a function of
other ecological factors, and this has been no where better
expressed than by Billings (1970) who states: "vegetation is a
delicate integrator of environmental conditions and can be used as
an indicator of such conditions." Billings points out that there
are even advantages in using vegetation data as monitors of past
conditions, soil properties - especially at depth, and likely
herbivores present. Inventories of animal species in turn should
indicate both the state of plant *and* abiotic components of the
environment. Our aim then is to find that measure of biotic distance (dissimilarity) between two sites which best reflects their
ecological distance.

Whittaker (1967) and Gauch (1973) seem to be among the few to
have looked at this question directly. Figure 2 shows the curve
of Sorensens values that Whittaker obtained for an environmental
gradient based on increase in elevation. Even without knowledge
of the actual data, we can obtain a close fit with Sorensens values
based on a model of a steady drop of species in j and of species

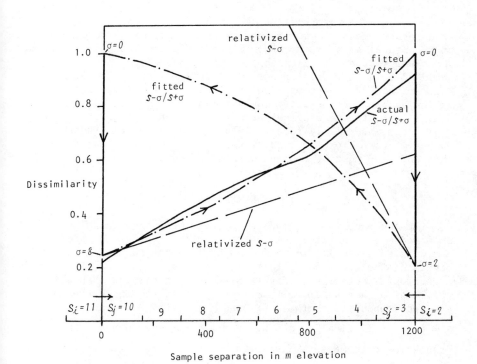

FIG. 2: Change in (S-σ)/(S+σ)(Sorensens dissimilarity measure)
along an elevation gradient (Whittaker, 1967) and its relationship to a model of species change fitted by the present authors
based on a uniform drop in S_j and σ with increase in elevation.

in common (σ) across the gradient. This gives an exponential
curve for which Gauch derived an error function term based on
Gaussian species-abundance curves to linearize the relationship
between the measure and gradient. The same effect (though with a
different slope) could have been obtained simply by using a linear
measure (S - σ) at the outset, as illustrated in Figure 2.

We would be incorrect to conclude however that it is a property
of Sorensens measure to increase exponentially with increase in
environmental separation. For, if we proceeded along the gradient
the other way (steady increase in j, steady drop in σ), there
would be a 'hysteresis' effect, Sorensens this time increasing
logarithmically. Another problem is that steady increase in elevation does not necessarily imply a linear change in total environmental (let alone ecological) conditions - an exponential increase
would seem more likely (Scott, 1974). Because of the indeterminate
number of variables involved, finally we do not know the answer to
this one, and must be content with a 'perceived' ecological gradient.

8. THE APPROACH

It would be instructive to observe the change in all 21
selected measures (Table 2) to the simulated change in species
composition given in Figure 2. Hopefully, the resulting curves
might then be compared against the expected curves associated with
an ecological gradient. Figure 3 (Case 1) shows the change in the
values of each dissimilarity measure with increasing difference in
species composition between i and j, with S_j fixed and S_i =
σ + 2, until finally σ = 0. For comparative purposes, the
values for each measure have been scaled so that the first value
always equals one. Though all measures (except χ^2 for N = 20)
show a monotonic increase with increase in (S - σ), the steepness
and rate of change in their slopes varies markedly.

If we could plot the curve for change in ecological distance
with increase in (S - σ), what properties would it have in
relation to these dissimilarity curves? Figure 2 suggests it may
be linear, but what of its slope? The answer is clearly related
to both *width* of the environmental range of each species and extent
of their *overlap* relative to the sampling interval or size (Figure
4).

It is quite easy to demonstrate that if the species have
moderate-sized environmental ranges which are randomly (provided
S_i and S_j are sufficiently large) or uniformly distributed along
an environmental gradient (continuum concept of species distribution) then the slope will be moderate (Figure 5a, based on Figure

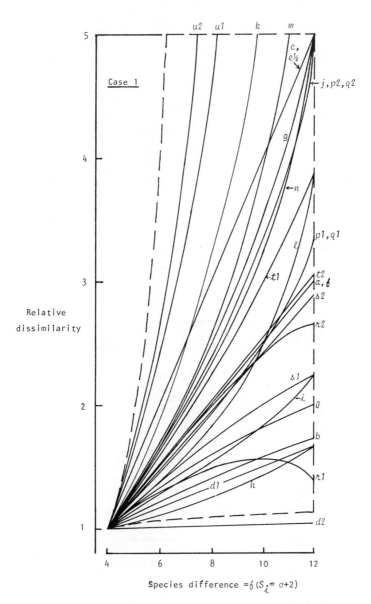

FIG. 3: (Case 1). *The curves for* 21 *binary dissimilarity measures (a-u) associated with increase in* (S-σ), *where* S_j=10 *and* S_i=σ+2, *until finally* σ=0. *The values for each measure have been relativized so that* $(S-σ)_1$=1. *The broken lines bound the (truncated) area of expected curves in terms of possible ecological distance.* 1, *for the case* N = 20; 2, N = 200.

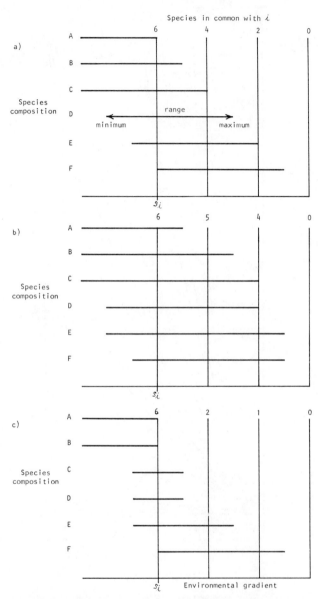

FIG. 4:   *Hypothetical distribution of six species (A-F) about site i with* a) *moderate environmental tolerances, distributed in a continuum;* b) *broad tolerances, distributed in communities;* c) *narrow tolerances, distributed in communities. Note the different rates of drop in species across the environmental gradient (given at top of figures) depend on the pattern of species environmental tolerances.*

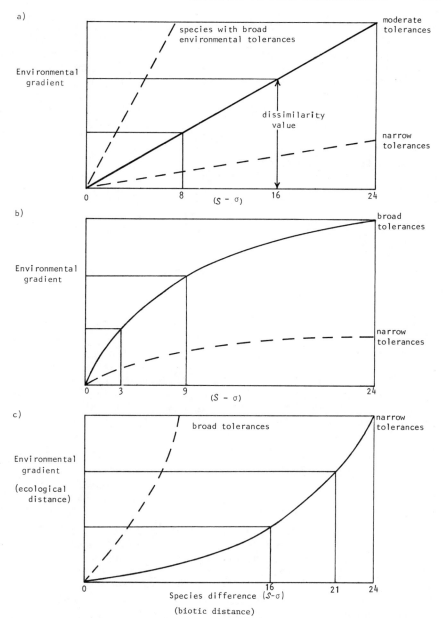

FIG. 5: *Hypothetical change in relationship between difference in species composition between two sites and ecological distances along an environmental gradient (based on Figure 4).* a) *Species distribution continuous;* b) *and* c) *species distribution clumped. For convenience* $S_i = S_j$. *Broken lines indicate other extremes in possible species composition in relation to the gradient.*

4a). If the ranges are predominantly broad, the slope will be
steep; if predominantly narrow, the slope will be slight. If
there is no clumping of species whose ranges are equal, the slope
will remain constant. If there is a tendency for clumping (community concept of species distribution) of predominantly moderate
with some wide ranges, the rate of increase in slope decreases
(logarithmically-related)(Figure 5b, based on Figure 4b). If there
is clumping of moderate with some narrow ranges, the rate of
increase in slope increases (exponentially-related) (Figure 5c,
based on Figure 4c). Uniformly spread ranges varying in width
also lead to an accelerating change in slope. In attempting to
answer this question then, we must know where the tolerance limits
of *every* species lie along the environmental gradient. Since this
question is impossible to answer in the absence of exhaustive autecological studies, to retain flexibility we must accept a family
of possible ecological distance curves. Using Figure 5 as a guide,
and considering some extremes of possible relationships, we can
locate the approximate boundaries of this family as in Figure 3.
Only the dot product measure (N = 200) in this example seems
ecologically unacceptable.

9.   OTHER CASES, INCLUDING $\sigma = 0$

Gauch (1973) states that "...beyond the point which samples
have practically no species in common ... similarity measures of
any sort are ... meaningless." Since it is often the case where
two samples have $\sigma = 0$, we would hope, on the contrary, that at
least some would be meaningful! Figure 6 (Case 2) therefore considers a situation of a steady increase in S_i for a fixed S_j
with $\sigma = 0$. The results are even more variable than in Case 1,
with 13 measures showing nil or a negative slope with increase in
$(S - \sigma)$, which is clearly ecologically unacceptable.

It is possible to conceive of a number of other extreme cases
for plotting purposes, in which one of S_i, S_j, σ, and S are independent, with the others fixed or proportional to it (Table 3). A
further four only (Figures 7-8) were considered as ecologically
realistic. It is not our intention to examine the results for
each case in detail here, but if any are felt to be relevant to
the origins of a set of data, the response of the various measures
will repay careful study, to see which fit in with ecological
theory, however weak.

10.   RATING THE MEASURES

10.1 Versatility. For each of the six cases, each measure was
rated as 1 if it fell within the family of expected curves (i.e.,

TABLE 3: *Values of the four variables in six extreme cases used to examine the relationship between 21 binary dissimilarity measures. For measures sensitive to total species in the collection, N was set at 20 and 200.*

Case	S_i	S_j	σ	S	$S-\sigma$
1	indept.	10	$S_i - 2$	12	$f(S_i = \sigma+2)$
2	indept.	2	0	dept.	$f(S_i)$
3	indept.	2	2	dept.	$f(S_i)$
4	8	8	indept.	dept.	$f(\sigma)$
5	indept.	S_i	0	dept.	$f(S_i = S_j)$
6	indept.	$4S_i$	S_i	$4S_i$	$f(S_i = \sigma = S_j/4)$

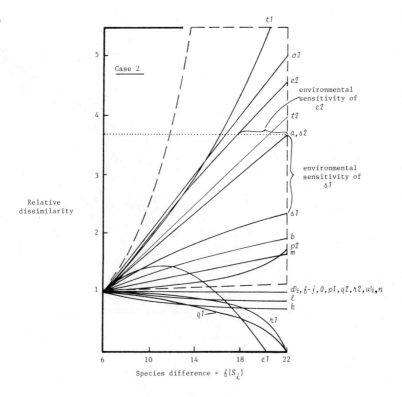

FIG. 6: *(Case 2). The curves for 21 binary dissimilarity measures associated with increase in (S-σ), where S_j = 2, σ = 0.*

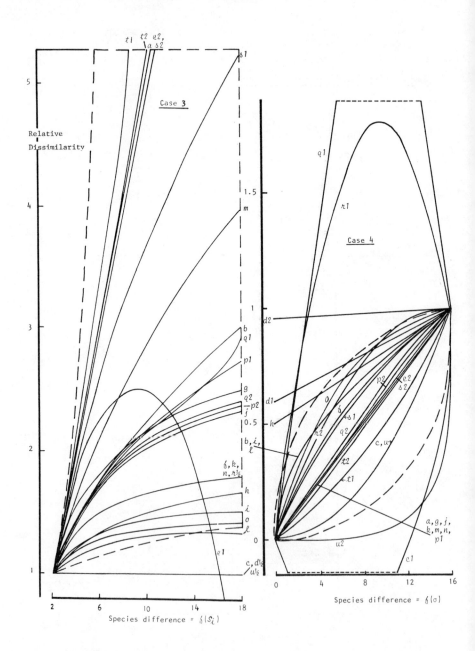

FIG. 7: (Cases 3, 4). The curves for 21 binary dissimilarity measures associated with increase in (S-σ).

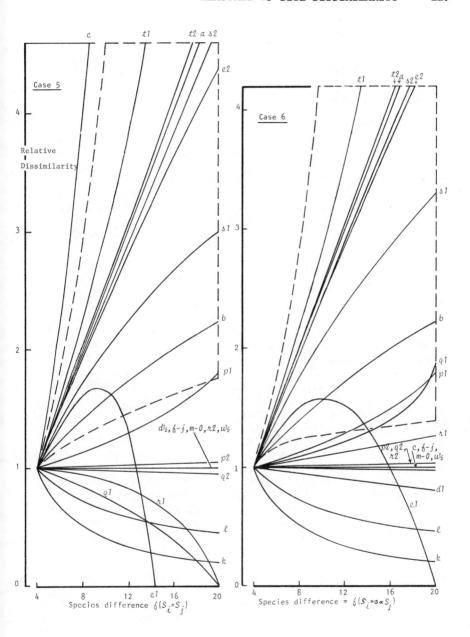

FIG. 8: (Cases 5, 6). The curves for 21 binary dissimilarity measures associated with increase in (S-σ).

within the bounds of the broken lines in Figures 3, 6-8) 0.5 of
it was partly in, and   0   if it was neither. The versatility,
i.e., extent of general applicability of the measure, was then
rated as follows: 5-6, excellent; 3.5-4.5, moderate; 2-3, poor;
and 0-1.5, negligible. Only eight measures were in the first two
categories (Table 4).

10.2 *Environmental Sensitivity*. We are treating organisms as
monitors of their environment. However, if they tolerate a wide
range of environments, especially with binary data, they will not
be good discriminators between environments. Such a case is
assumed by measures which tend to hug the Y-axis in the six situations considered here. On the other hand, if they tolerate only a
narrow range of conditions, there will need to be a large drop in
number of species to register a significant environmental change:
this is assumed by measures which hug the X-axis (recall Figure 5).
In this case many sites may well have insufficient species to provide a large enough (S-σ) to show up their environmental differences. If the actual species tolerances and distributions are known
the appropriate measure can be selected by inspecting the graphs.
But to *impose* a particular relationship on the data without such
knowledge is far from satisfactory. In the absence of such knowledge our preference is clearly for a compromise such that a
moderate change in species composition is associated with a
moderate change in site dissimilarity. In the absence of any
knowledge about what can be regarded as the environmental baseline,
the (S-σ) curve (*a*) may be considered ideal as it assumes a
1:1 relationship between these two variables. For each of the
six cases, each measure was ranked (best = 1) according to the
distance of the intersection point of the curve along one of the
lines drawn parallel to the X and Y axes to their intersection
point at the extreme portion of the (S-σ) curve in each graph
(see Figure 6). Negative slopes were treated as mirror images in
the positive quadrat, and the slope of the longest arm was used
in non-monotonic measures. Environmental sensitivity was rated as
follows: 0-25, high; 26-50, moderate; 51-75, low; and 76-100,
negligible. Only seven measures were in the first two categories.

10.3 *Slope*. The direction and rate of change of slope and its
monotonicity were noted for each measure in the six cases studied
here. Measures were rated highly if the direction of slope remained
the same (but not zero) for the different cases and whose rate of
change was consistent between and within curves. Because of their
elliptic shape with small collections, the dispersion, χ^2 and
Kendall's rank measures have little to commend them. As already
discussed, it was not possible to make a generalization on whether
the ideal curves have a constant, accelerating, or decelerating

TABLE 4: Categorization of 21 binary dissimilarity measures based on their performance in six extreme cases according to their versatility, environmental sensitivity and slope (see text for details).

Symbol	Name	Range of Values	Versatility	Sensitivity	Slope[#] - direction	change[φ]	Overall Suitability
a	Absolute	0...N	excellent	high	+	n	high
b	Euclidean	0...$\sqrt{N}$	excellent	moderate	+	d	high
c	Pattern difference	0..$N^2/4$	poor	low	+, 0	*n*,d	low
d1*	Dot product	0...N	negligible	negligible	+, *0*, -	n	low
d2*	Dot product		negligible	negligible	+, *0*, -	n	low
e1	Dispersion	$-\frac{N^2}{4}...\frac{N^2}{4}$	poor	moderate	+, +/-	n,d/a	low
e2	Dispersion		excellent	high	+	n,d	high
f	Jaccards	0...1	poor	low	+, 0	*n*,d	low
g	Sorensens	0...1	poor	low	+, 0	a,*n*,d	low
h	Relative total	0.5...1	poor	negligible	+, 0	a,*n*,d	low
i	Orlocis chord	0...1	poor	low	+, 0	a,*n*,d	low
j	Ochiais	0...1	poor	low	+, 0	a,*n*,d	low
k	Relative absolute 1	0...$\sqrt{N}$	poor	low	+, -	a,n,*d*	low
l	Relative euclidean	0...$2/\sqrt{N}$	poor	low	+, -	a,d	low
m	Relative absolute 2	0...2	moderate	low	+, 0	a,*n*,d	low
n	Kulczynskis 2nd	0...2	moderate	low	+, 0	a,*n*,d	low
o	Unnamed 1	0...1	poor	negligible	+, 0	c,d	low
p1	Product moment	0...2	moderate	moderate	+, 0	*a*,c,d	moderate
p2	Product moment		poor	low	+	a,c,d	low
q1	Kendalls rank	0...2	poor	moderate	+, +/-, -	*a*,d/a	low
q2	Kendalls rank		poor	low	+, 0, -	a,c,d	low
r1	Chi-square	0...1	poor	low	+, +/-, -	*a*,d/a,d	low
r2	Chi-square		poor	low	+, 0	c,d	low
s1	Rogers-Tanimoto	0...0.5	excellent	moderate	+	d	high
s2	Rogers-Tanimoto		excellent	high	+	*n*,d	high
t1	Dice-Sorensen	0...1	excellent	moderate	+	a	high
t2	Dice-Sorensen		excellent	high	+	a,c	high
u1	Yule	0...2	poor	negligible	+, *0*	a,*n*	low
u2	Yule		poor	negligible	+, *0*	a,*n*	low

*1, number of species in total collection (N) = 20;2, N = 200. [φ] n = nil, a = accelerating, d = decelerating, / = within same curve, [#] script face type indicates most cases in this category.

rate of change in slope. Only four measures maintained a positive
slope between (S-σ) and dissimilarity values for the six cases
considered here.

11. CONCLUSIONS

On the basis of their versatility, environmental sensitivity,
and slope as defined here, among the unstandardized measures which
disregard conjoint absences, the absolute (constant) and euclidean
(decelerating) measures are best (Table 4). None of the ten
standardized measures which disregard conjoint absences meet the
criteria given here, though Kulczynskis 2nd and the simpler relative
absolute 2 may sometimes be acceptable. Among the measures which
recognize conjoint absences but are unstandardized, the dispersion
measure is only suitable for data sets with large N (it is not
monotonic for small N). Of the associated standardized measures,
Rogers-Tanimoto (decelerating) and Dice-Sorensen (accelerating)
are highly suitable, independent of the size of N, while the
product moment (four-fold point) correlation coefficient is of
limited suitability when the collections are small. The reason
for the advantages of the Rogers-Tanimoto and Dice-Sorensen
measures is due to the fact that they approach the absolute
measure, especially as N becomes large (e.g., for 135 species
the three ordinations were almost identical, Figure 1). In view
of its simplicity, compared with these two measures, the absolute
measure then is to be preferred.

It is appreciated that some ecologists would use different
criteria in assessing these dissimilarity measures, while others
might point to undesirable properties of the preferred measures
above not discussed here. We at least hope by this 'first-step'
approach to the problem, to have stimulated interest in its scope.
A greater challenge lies in comparing quantitative dissimilarity
measures.

ACKNOWLEDGMENTS

We acknowledge funds to KJG from the Academic Staff Development Committee and the Environmental Studies Group at W.A.I.T.

REFERENCES

Agrell, I. (1945). The collemboles in nests of warmblooded
    animals with a method for sociological analysis. *Acta
    Universitatis Lundensis, N.F. Afdelring,* 2, 41, 1-19.

Austin, M. P. and Greig-Smith, P. (1968). The application of quantitative methods to vegetation survey II. Some methodological problems of data from rainforests. *Journal of Ecology*, 56, 827–844.

Billings, W. D. (1970). *Plants, Man, and the Ecosystem*, 2nd ed. MacMillan, London.

Bouxin, G. (1975). Ordination of quantitative and qualitative data in a savanna vegetation (Revanda, Central Africa). *Vegetatio*, 30, 197–200.

Bray, J. R. and Curtis, J. T. (1957). An ordination of the upland forest communities of southern Wisconsin. *Ecological Monographs*, 27, 325–349.

Cain, A. J. and Harrison, G. A. (1958). An analysis of the taxonomists judgement of affinity. *Proceedings of the Zoological Society, London*, 131, 85–98.

Clarke, S. S. (1975). The effect of sandmining on coastal heath vegetation in NSW. *Proceedings of the Ecological Society of Australia*, 9, 1–16.

Clifford, H. T. and Stephenson, W. (1975). *An Introduction to Numerical Classification*. Academic Press, New York.

Cole, L. C. (1949). The measurement of interspecific association. *Ecology*, 39, 411–424.

Czekanowski, J. (1909). Zur differential Diagnose der Neandertalgruppe. *Korrespbl. dt Ges. Anthrop*, 40, 44–47.

Ellenberg, H. (1956). Aufgaben and Methoden der Vegetationskunde. In *Einfuhrung in die Phytologie, Vol. 4, Part 1*, H. Walter, ed. Ulmer, Stuttgart.

Forbes, S. A. (1907). On the local distribution of certain Illinois fishes: an essay in statistical ecology. *Bulletin of the Illinois State Laboratory of Natural History*, 7, 273–303.

Gauch, H. G. (1973). The relationship between sample similarity and ecological distance. *Ecology*, 54, 618–622.

Gauch, H. G. and Whittaker, R. H. (1972). Comparison of ordination techniques. *Ecology*, 53, 868–875.

Gleason, H. A. (1920). Some applications of the quadrat method. *Bulletin of the Torrey Botanical Club*, 47, 21–33.

Goldman, B. (1973). *Aspects of the ecology of coral reef fishes of One Tree Island.* Ph.D. thesis, MacQuarie University, New South Wales.

Goodall, D. W. (1973). Sample similarity and species correlation. In *Handbook of Vegetation Science, V. Ordination and Classification of Vegetation,* R. H. Whittaker, ed. W. Junk, The Hague. 106-156.

Gower, J. C. (1967). A comparison of some methods of cluster analysis. *Biometrics,* 23, 623-637.

Green, P. E. and Rao, V. R. (1909). A note on proximity measures and cluster analysis. *Journal of Marketing Research,* 6, 359-364.

Hamann, U. (1961). Merkmalsbestand und Verwandtschaftsbeziehungen den Farinosae. Ein Beitrag zum system der Monokotyledonen. *Willdenowia,* 2, 639-768.

Jaccard, P. (1901). Distribution de la flore alpine dans le Bassin des Dranses et dans quelques regions voisines. *Bulletin de la Societe vaudoise des Sciences Naturelles,* 37, 241-272.

Kruskal, J. B. (1964). Multidimensional scaling by optimizing goodness of fit to a nonmetric hypothesis. *Psychometrika,* 29, 115-129.

Kulczynski, S. (1927). Die Pflanzenassoziationen der Pieninen. *Bulletin International de l'Academic Polonaise des Sciences et des Lettres. Classe des Sciences Mathematiques et Naturelles. Serie B,* (suppl. 2), 57-203.

Lance, G. N. and Williams, W. T. (1967). Mixed-data classificatory programs. I. Agglomerative systems. *Australian Computer Journal,* 1, 15-20.

Lie, U. and Kelly, J. C. (1970). Benthic fauna communities off the coast of Washington and in Puget Sound, identification and distribution of the communities. *Journal of the Fisheries Research Board of Canada,* 27, 621-651.

Looman, J. and Campbell, J. B. (1960). Adaptation of Sorensen's K (1948) for estimating unit affinities in prairie vegetation. *Ecology,* 41, 409-416.

McConnaughy, B. H. (1964). The determination and analysis of plankton communities. *Marine Research in Indonesia,* 1-40.

Mountford, M. D. (1962). An index of similarity and its application to classificatory problems. In *Progress in Zoology*, P. W. Murphy, ed. International Society Soil Science, London.

Noy-Meir, I. (1971). Multivariate analysis of desert vegetation. II. Qualitative/quantitative partition of heterogeneity. *Israel Journal of Botany*, 20, 203-213.

Noy-Meir, I., Walker, D., and Williams, W. T. (1975). Data transformations in ecological ordination. II. On the meaning of data standardization. *Journal of Ecology*, 63, 779-800.

Noy-Meir, I. and Whittaker, R. H. (1977). Continuous multivariate methods in community analysis: some problems and developments. *Vegetatio*, 33, 79-98.

Ochiai, A. (1957). Zoogeographical studies on the soleiod fishes found in Japan and its neighboring regions. *Bulletin of the Japanese Society of Scientific Fisheries*, 22, 526-530.

Orloci, L. (1967). An agglomerative method for classification of plant communities. *Journal of Ecology*, 55, 193-206.

Orloci, L. (1973). Ordination by resemblance matrices. In *Handbook of Vegetation Science, V. Ordination and Classification of Communities*, R. H. Whittaker, ed. W. Junk, The Hague.

Orloci, L. (1975). *Multivariate Analysis in Vegetation Research*. W. Junk, The Hague.

Pandeya, S. A. (1961). On some new concepts in phytosociological studies of grassland. II. Community coefficient (FXC) ICC. *Journal of the Indian Botanical Society*, 40, 267-270.

Pielou, E. C. (1961). Segregation and symmetry in two-species populations as studied by nearest-neighbor relationships. *Journal of Ecology*, 49, 255-269.

Pielou, E. C. (1969). *An Introduction to Mathematical Ecology*. Wiley-Interscience, New York.

Rogers, D. J. and Tanimoto, T. T. (1960). A computer program for classifying plants. *Science*, 132, 1115-1118.

Russell, P. F. and Rao, T. R. (1940). On habitat and association of species of anopheline larvae in south-eastern Madras. *Journal of the Malaria Institute of India*, 3, 153-178.

Ruzicka, M. (1958). Anwendung mathematish-statistischer. Methoden in der Geobotanik (Synthetische Bearbeitung von Aufnahman). *Biologia*, Bratisi, 13, 647-661.

Scott, J. T. (1974). Correlation of vegetation with environment: a test of the continuum and community-type hypothesis. In *Handbook of Vegetation Science, VI. Vegetation and Environment,* B. R. Strain and W. D. Billings, eds. Dr. W. Junk, The Hague.

Smartt, P. F. M., Meacock, S. E., and Lambert, J. M. (1976). Investigations into the properties of quantitative vegetational data II. Further data type comparisons. *Journal of Ecology,* 64, 41-78.

Sokal, R. R. and Michener, C. D. (1958). A statistical method for evaluating systematic relationships. *University of Kansas Science Bulletin,* 38, 1409-1438.

Sokal, R. R. and Sneath, P. H. A. (1963). *Principles of Numerical Taxonomy.* Freeman, New York.

Sorensen, T. (1948). A method of establishing groups of equal amplitude in plant sociology based on similarity of species content. *Kongelige Danske Videnskabernes Selskab, Biologiske Skrifter,* 5, 1-34.

Swan, J. M. A., Dix, R. L., and Wehrahn, C. F. (1969). An ordination technique based on the best possible defined axes and its application to vegetation analysis. *Ecology,* 50, 206-212.

van der Maarel, E. (1969). On the use of ordination models in phytosociology. *Vegetatio,* 19, 21-46.

Walker, B. H. (1974). Some problems arising from the preliminary manipulation of plant ecological data for subsequent numerical analysis. *Journal of South African Botany,* 40, 1-13.

Whittaker, R. H. (1967). Gradient analysis of vegetation. *Biological Review,* 49, 207-264.

Williams, W. T., Lambert, J. M., and Lance, G. N. (1966). Multivariate methods in plant ecology. 5. Similarity analyses and information analyses. *Journal of Ecology,* 54, 427-445.

Wishart, D. (1970). *The treatment of various similarity criteria in relation to CLUSTAN IA.* Computing Laboratory, University of St. Andrews.

Yule, G. U. (1912). On the methods of measuring association between two attributes. *Journal of the Royal Statistical Society,* 75, 579-642.

[*Received August* 1978. *Revised March* 1979]

L. Orloci, C. R. Rao, and W. M. Stiteler, (eds.),
Multivariate Methods in Ecological Work, pp. 127-174. All rights reserved.
Copyright © 1979 by International Co-operative Publishing House, Fairland, Maryland

USE OF DUAL STRUCTURES IN INERTIA ANALYSIS ECOLOGICAL IMPLICATIONS

A. LAUREC, P. CHARDY, P. DE LA SALLE AND
M. RICKAERT

Départment Environment Littoral et Gestion du
Milieu Marin
Centre National pour l'Exploitation des Oceans
Centre Oceanologique de Bretagne
B. P. 337   29273 BREST Cedex FRANCE

SUMMARY. Ecological studies commonly lead to two-dimensional
tables. It is essentially considered in this paper that the
rows are associated with the observations and the columns with
variables derived from a set of taxa. Such tables may be
analyzed using mathematical methods, among which ordination
techniques appear particularly important. The basic purpose
of the study is the ordination of the observations. More
precisely, factorial (inertia) analyses are considered, within
Euclidean structures, with a possible weighting of the rows
and/or the columns of the table. In the first section the
basic properties are summarized. Particular attention is paid
to the definition of the contributions of the variables to the
axes and reciprocally, to the duality between observations and
variables. These points are studied in a general context, then
within the special cases of Principal Component Analysis, and
Analysis of Correspondences (reciprocal averaging).

KEYWORDS. data processing, ordination, factorial analysis,
duality, mathematical ecology, marine ecology, phytoplankton.

1. INTRODUCTION

      In ecology, and especially in the field of marine ecology
with which we are more familiar, the gathering of data often
leads to the construction of two-dimensional tables. Conventionally,

we can associate the rows with the different samples and the columns with the characters studied. We shall adopt this framework in the present study. For a long time, this situation has brought about the necessity to proceed with mathematical techniques of analysis. Factorial analyses (or inertia analyses) are in this context essential techniques.

In the early applications, these techniques were used somewhat thoughtlessly, by simply applying a standard perspective to the field of ecology. This approach led to some exaggerations which were rightly denounced by Frontier (1975). It is advisable to develop a real methodology concerning the use of these techniques. This type of research has been developed for some years by several authors, in relation to a new conception of the techniques used in factorial analysis which can be compared to what French-speaking people call data analysis. Whereas the standard factorialist approach is based on the notion of identification of a linear model accounting for the relations between variables, this new conception is essentially descriptive, its aim being to account for the main facts in a table of figures.

The present study considers a particular framework, where the analyses focus first and foremost on the main features of the structure of a set of observations, on which a set of characters is studied. These characteristics can be associated with the presence or the abundance of different taxa of the fauna or the flora, or even with environmental variables. The example which is dealt with has to do with phytoplanktonic studies, but this does not affect the general reasonings at all. In a general way, in order to achieve the study one has proposed doing, one must primarily associate the characters under consideration with mathematical variables. The point has been discussed (Dessier and Laurec, 1978; Laurec, 1979). One must also combine the pieces of information derived from the different variables thus determined in order to measure the likeness or the unlikeness between observations. The point has been approached by many authors (Chardy *et al.*, 1976; Blanc *et al.*, 1976). In order to make complete the choices defining a factorial analysis, a mass has also to be chosen for the different observations, as well as a point of reference (Chardy *et al.*, 1976). This point of view strictly corresponds to the metric ordination techniques, leading essentially to principal coordinate analysis (Gower, 1966). When the distance between observations can simply be reduced to a Euclidean distance, it is possible to go beyond this point of view. As a matter of fact, a certain number of concepts relative to the variables can be defined: coordinates in respect to factors, contribution of the variables, similar, at least at first sight, to those defined about the observations. However, whereas with the observations these concepts are relatively easy to use, and do not give rise to confusion, when we deal with the

variables this soon leads to awkwardness and misunderstandings
which has been denounced in particular cases by several authors.
The aim of the present article is to throw light on these points.
In the first place, this implies that the mathematical aspects
are correctly understood. This is why we shall first review
some points. Then, in a general way, we shall try to think over
the preoccupations linked with the variables that can be put
forward by the ecologist, in a study centered on the structure
of the samples. This consideration will enable us to suggest a
use of the different mathematical concepts relative to the
variables. In Section 4 an example will be provided
in order to illustrate the suggestions made in Section 3.

2. MATHEMATICAL SUMMARY

2.1 General Framework. We shall consider a two-dimensional
table of real numbers with I rows and J columns. Concretely,
the I rows will correspond to as many observations, the J columns to
variables. We shall not situate our analysis within the most
general framework possible, but we shall be content with the
simplest analysis including variants such as the traditional
principal component analysis and the analysis of correspondences.
For a presentation which is more elaborate mathematically and
more general, the reader is referred to Benzecri *et al.* (1973),
Lebart *et al.* (1977).

We shall suppose that the rows and the columns are simply
defined from Euclidean structures with possible weightings.
Rows and columns are thus provided with masses (positive numbers
or numbers equal to zero), which are written $p\ell_i$ and pc_j
respectively for row i and column j. It shall be called
general analysis with double weighting, extending Lebart and
Fenelon's (1971) expression. So the scalar product between
rows i_1 and i_2 is given by:

$$s\ell_{i_1,i_2} = \sum_j pc_j\, x_{i_1,j} \cdot x_{i_2,j}$$

The distance between rows i_1 and i_2 is given by:

$$d^2\ell_{i_1,i_2} = \sum_j pc_j\, (x_{i_1,j} - x_{i_2,j})^2$$

Finally the norm of row i, written $n\ell_i$ is given by:

$$n\ell_i^{\,2} = \sum_j pc_j \cdot x_{i,j}^2$$

Similarly, the scalar product of columns j_1 and j_2 is given by:

$$sc_{j_1,j_2} = \sum_j p\ell_i \; x_{i,j_1} \; x_{i,j_2}$$

The norms and the distances relative to the columns can be defined in the same way.

Laurec (1979) has worked within this context, and the reader who wishes to go beyond the few following notions is referred to his presentation.

"Materially", row i can be represented by a point A_i in R^J. If the point i is provided with the mass $p\ell_i$ the set of the rows defines in R^J a solid, the inertia of which in relation to the origin is $\sum_i p\ell_i \; OA_i^2$, if OA_i^2 is the square of the norm of row i, that is to say $\sum_j pc_j \; (x_{i,j})^2$. Consequently, the total inertia is:

$$\sum_{i,j} p\ell_i \; pc_j \; (x_{i,j})^2$$

This inertia can be divided into contributions of the different columns. The contribution of column j is $pc_j \sum_i p\ell_i (x_{i,j})^2$.

Moreover it could be possible to specify the solid defined in R^I by t J column points ($x_{i,j}$ is the ith coordinate of the column J) provided with masses pc_j. We would find again the preceding total inertia and its decomposition in relation to either the rows or the columns.

2.2 *Configurations and Contributions Linked with the Observations.*

2.2.1 *Axes of inertia of the solid of the column points.* Let, as above, A_i be the point associated with row i, in R^J. Let us consider a subspace of a given dimension and let H_i be the projection of A_i on this subspace. We shall seek the sub-space the nearest possible to the points A_i, and in order to get a precise signification we shall minimize $\sum_i p\ell_i \; (A_iH_i)^2$, $(A_iH_i)^2$ indicating the square of the norm of the vector linking the point A_i to its projection H_i. Thus it is the distance from A_i to the subspace. We know that the desired subspace, if it has a dimension k, will be generated by the first k axes of inertia of the solid of the I points. These axes of inertia will be provided by the first k eigen vectors of the form of inertia (which takes into account the metric defined on R^J). The rate of inertia assignable to an axis (complement in relation to the total inertia of the inertia in relation to this axis) will be given by the eigen value associated with this axis. The different axes of inertia will be assigned a unit vector in R^J. If we take the set of the axes of inertia, we can define an orthonormal set, defining a subspace in R^J containing all the points (or more precisely all those having a non zero weight).

The coordinate of point i on the kth axis is written $fo_{i,k}$. The axes k_1 and k_2 define a plane on which we shall be able to project the I rows, thus defining a plot of observations. If the importance of an axis is quantified by the inertia that it explains, we shall be able in a more general way to define the inertia explained by a subspace generated by axes of inertia as being the sum of the inertias explained by the different axes. It will be interesting to relate these inertias to the total inertia of the cluster in order to reduce everything to percentages of explained inertia.

These percentages of inertia will globally give the importance of an axis, of a plane or more generally of a subspace. For a particular observation, we shall also be able to quantify the importance of a subspace. If, as above, we write A_i and H_i for the point and its projection on a subspace, the importance of this subspace for the observation will be measured by $(OH_i)^2/(OA_i)^2$. This will be a number less than or equal to 1, equal if the point A_i belongs to the subspace. This ratio will be called the relative contribution from the subspace to observation i. This notion is commonly used in the analysis of correspondences and is easily generalized.

Now, if we consider a particular subspace, the inertia that it explains is not different from $\sum_i p\ell_i (OH_i)^2$, keeping the previous notations. We can see that it is the sum of the terms associated with the different observations. Thus it will be possible to consider the fraction assignable to the ith observation. This will be the relative contribution from observation i to the subspace. This concept, which is reciprocal to the preceding one to a certain extent, is equally common in the analysis of correspondences and can be generalized as easily.

*2.2.2.   Optimal representation and metric analysis of the
proximities.* We shall assume that in R^J, the origin is located
at the centroid of the observation points. This amounts to a
mean equal to zero for every column.

$$\sum_i p\ell_i \, x_{i,j} = 0$$

The distance between lines i_1 and i_2 is written $d\ell_{i_1, i_2}$.

The matrix of the distances between the observations
characterizes the structure of the set of the observations.
We can try to account for this network of distances in a simple
way, by making projections on a subspace of a small dimension,
and trying at the same time not to distort the structures as
far as possible. In this respect, we must try to get the
distances between projections to be as near as possible (globally)
to the original distances. So let $(H_{i_1} H_{i_2})^2$ be the
square of the distance between the projections associated with
observations i_1 and i_2. We shall try to get the subspace of
a given dimension which enables us to minimize

$$\sum_{i_1 i_2} p\ell_{i_1} \, p\ell_{i_2} \, (d\ell_{i_1, i_2} - (H_{i_1} H_{i_2})^2)$$

(It is easy to verify that the quantity given in parentheses is
always positive). In this way we enter the context of metric
ordination, with a point of view which can be generalized to
non-metric attempts, when for instance we try to keep pre-orders
instead of distances (Kruskal, 1964a,b). An illustration in
ecology of this nonmetric approach is found in Prentice (1977)
and Fasham (1977).

Once the condition of centering is made sure, it can be
demonstrated that the required subspace will simply be provided
by the first k axes of inertia of the solid of the I points
observations, in relation to the centroid of the cluster. When
the origin is not originally situated at the centroid of the cluster
it can be brought back to it by centering the columns. It is
easy to verify that this does not affect the distances between
the observations at all.

The property that has just been presented is quite fundamental: it provides that the projections on the first axes of
inertia give so-called optimal representations. If the purpose
is first and foremost to study the structure of a set of samples,
one way of proceeding then appears, with three phases:

- a certain number of mathematical variables is defined from the characters under consideration.

- these variables are combined so as to define a distance measuring the dissimilarity between the samples.

- the structure of the set of samples can then be reduced to a matrix of mutual distances; we must try to account for this structure by a subspace of a small dimension keeping at best the distances. This will provide an optimal summary of the structure.

2.3. Configurations and Contributions Linked with the Variables.

2.3.1. Definition of the factors.

In the preceding paragraph axes of inertia have been defined, also called factorial axes. Variables can be associated with these axes. The variables are defined on the set of the observations and are called factors. This definition can be obtained in different ways. We shall simply adopt the easiest.

On axis k, the coordinate of observation i is $fo_{i,k}$. By considering the set of observations, we actually define a variable. Thus we have as many variables as factorial axes. These variables, which are defined by their value on the I observations can be classed as elements of R^I. It is shown that they are orthogonal in relation to the previous Euclidean structure. It is customary to norm them, so as to define an orthonormal set. In order to do this, for the observation i, the factor k will be assigned the value $fo_{i,k}/\sqrt{\lambda_k}$, λ_k, being the inertia explained by the axis k, and not the value $fo_{i,k}$.

The J columns of the table which is analysed are also variables defined on the I observations. They can be decomposed over the orthonormal set which groups the factors. Thus, just as in R^J we make a change of basis and project the observations on the new axes, in R^I a new basis is defined to which the variables are related. The relationship between the two aspects goes far beyond this, for if we had considered in R^I the cluster of the J variable points, in order to get its inertia axes, this would have lead us directly to the basis which has just been defined in an indirect way. Besides, the percentages of inertia thus obtained would have been the same as those of the analysis of the cluster of the observations.

2.3.2. Definition of the plots of variables and the contributions.
The projection of the J columns on the factors gives coordinates.
Considering the factor k associated with the factorial axis k,
the coordinate of the variable in relation to this factor, commonly
called factor loading, will be written $fv_{j,k}$. These coordinates
will enable us to build plots of variables, similar to the plots
of observations and often called dual. As previously, it will be
possible to define relative contributions from the variables to
the factors, and from the factors to the variables.

In R^I the norm of column j is $\sum_i p\ell_i (x_{i,j})^2$. The
norm of its projection on a subspace, necessarily smaller, is
all the largest as this variable is well "summarized" by the
subspace. Considering the subspace associated with factor k,
the norm (to the square) of the projection of variable j is
$fv_{j,k}^2$. The relative contribution from factor k to variable j
will be defined as $(fv_{j,k})^2/(\sum_i p\ell_i (x_{i,j})^2)$. Reciprocally, if
we consider the whole inertia linked with a factor, this is equal
to $\sum_j pc_j (fv_{j,k})^2$. The importance of variable j_o
will be given by the inertia fraction which is ascribed to it
and which is called relative contribution from the variable j_o
to the factor (or to the axis) k: $pc_{j_o} (fv_{j_o,k})^2/(\sum_j pc_j (fv_{j,k})^2)$.
Once again relative contributions from the axes to the variables
and from the variables to the axes are commonly used in the
analysis of correspondences, but they can be generalized easily.

2.3.3. Duality and transition formula. While focusing the
analysis on the cluster of the observations, in fact we carry out
a simultaneous analysis of the inertia of the variables. As was
pointed out at the beginning, the two clusters have the same inertia,
and their respective inertia axes in R^J and R^I are associated
systematically and explain the same inertia for the analogous
axes. It is easy to go from one to the other. In fact a more
elaborate mathematical presentation, as that given by Caillez and
Pages (1976), based on the notion of duality, perfectly explains
the symmetry which has been observed. There are not two analyses
but a single one, in which we can assign perfectly symmetrical
roles to the rows and to the columns. However two elements have
to be considered which will moderate the mathematician's enthusiasm
at such a harmony. On the one hand, the use of the property of
optimal representation can give rise to a few problems, on the

other hand, and practically in a much more important way, the
problem initially set does not necessarily confer symmetrical
parts to the rows and to the columns.

If we consider a study based on the structure of the set of
rows, it is possible, while using the distance between the rows
given by:

$$d^2\ell_{i_1,i_2} = \sum_j pc_j \left(x_{i_1,j} - x_{i_2,j}\right)^2$$

to have the property of optimal representation. In this respect,
in the first place, the columns will have to be centered. This
preliminary transformation does not alter the distances between
the rows. However it affects the distances between the columns.
Besides, generally, the centering of the columns does not make
sure that the rows are centered as well. Moving the origin in
R^J to the centroid of the cluster of the observations does not
make sure, except in particular cases, that it was also successful
in R^I regarding the cluster of the variable points. This is
why very often even when we have the property of optimal representation as regards the observations, this is not the case as
regards the variables. In that case, except when the centering
of the rows is equivalent to that of the columns, the symmetry
is not maintained between the rows and the columns.

Lastly and above all, as was pointed out, frequently the
ecological problem under consideration does not assign symmetrical
roles to the rows and the columns. Our study is placed within a
precise context, in which the observations are associated with
the rows and the variables with the columns. As a general rule,
at this level the roles of the rows and the roles of the columns
are different. We have supposed that the study was based on the
structure of the set of the observations. In order to study
these structures as best as we can, the variables will be defined
adequately, a weighting of the variables will possibly be used, so
as to get a distance which appears as a correct measure of the
dissimilarity between observations. The possible weighting of
the observations will be defined once again in order to make
easier the study of these observations, as well as a centering
of the columns which will bring in the property of optimal
representation; doing which we shall define a distance induced
between the variables. It can happen, as shown by the example
which shall be dealt with, that several variables are associated
with the same character. So the distance between variables won't
give a measure of the dissimilarity between characters directly.
The distance between variables will not necessarily be a correct
measure of the dissimilarity between the variables. Lastly the

property of optimal representation will not necessarily be at our disposal. All this implies that whereas a study based on the structure of the observations will give plots of observations which in a way will permit an appreciation of the structure of the observations at best, generally it will not be possible to give the same meaning to the plots of variables. The major mistakes are found at this level. However we should not infer from this that the plots of variables are of no interest. They are generally very interesting, although the nature of the interest has to be known. This is what we propose to demonstrate in the second section and to illustrate in the third one.

In order to grasp the usefulness of the plots of variables, two types of formulas will be extremely valuable. The first ones are the so-called transition formulas. These formulas again are common in the analysis of correspondences but can be generalized easily. They permit us to go from the fo to the fv or reciprocally. As a matter of fact, it is demonstrated that if, as above, λ_k is the inertia explained by the factor k:

$$fo_{i,k} = \frac{1}{\sqrt{\lambda_k}} \sum_j pc_j \; x_{i,j} \; fv_{j,k} \quad \text{and}$$

$$fv_{j,k} = \frac{1}{\sqrt{\lambda_k}} \sum_i p\ell_i \; x_{i,j} \; fo_{i,k} \; .$$

The other formula, (reconstitution or reconstruction formula) is:

$$x_{i,j} = \sum_k \frac{1}{\sqrt{\lambda_k}} \; fo_{i,k} \; fv_{j,k} \quad \text{if} \quad p\ell_i \; pc_j > 0 \; .$$

2.3.4. Interpretation of the formulas in particular cases. The general framework which has been studied leads to particular variants through the choice of the masses assigned to the rows and columns. This can be complicated by possible preliminary transformations. By making use of these transformations and of the weighting, we are led to common techniques.

- Principal component analysis (PCA).

Let us take an analysis entering the preceding general framework, in which the variables (columns) are centered and standardized and the observations have each the same mass 1/I and the variables the same mass 1. This leads us to a traditional

principal component analysis, based on the correlation matrix
between variables. The matrix which has to be diagonalized in
order to obtain the factorial axes and the factors is not different
from this correlation matrix. So we have to do with the most
traditional kind of technique, with a new presentation. First, it
should be noticed that, owing to the fact that the variables are
centered, we shall indeed have the property of optimal representation at the level of the observations. This time the scalar
products between the variables are not different from the
correlation coefficients between variables. Besides, the $fv_{j,k}$
are also coefficients of correlation: $fv_{j,k}$ is the coefficient
of correlation between the variable j and the normed factor k.
Therefore, the plots of the variables present this first interpretation.

As regards the contributions, the relations to the contributions are particularly simple, since no weighting of the
variables will have to be taken into account. From this, we
deduce easily that the relative contribution from an axis (or a
subspace) to a variable is not different from their correlation
coefficient (possibly multiple) to the square. Besides, it is
simply equal to $(fv_{j,k})^2$ for the variable j and the factor k.

Now if the variables are centered, but not necessarily
standardized, again we have a principal component analysis,
this time about the variance-covariance matrix between variables.
The main difference with the preceding case comes from the fact
that the $fv_{j,k}$ are no longer correlations but simple covariances.
This directly affects their interpretation, and alters the relationship between factor loadings and relative contributions from
the axes to the variables.

At last it must be pointed out that with PCA, the total
or explained inertias are also variances or sums of variances.

- Analyses of correspondences.

Let us consider a table $X(I,J)$, all the terms of which are
positive or equal to zero. Let $x_{i.}$ (respectively $x_{.j}$) be the
sum of the terms of row i (respectively column j). We
shall make a preliminary transformation of the table in two
phases: $x_{i,j}$ is transformed into $x_{i,j}/ (x_{i.} x_{.j})$, then the
rows and the columns are assigned the masses $x_{i.}$ and $x_{.j}$
respectively before the centering of the columns. It should be
noted that this time, this is equivalent to the centering of the
rows. Then an analysis is made within the general framework which

has been defined previously, in which the columns and the rows keep the weights $x_{.j}$ and $x_{i.}$.

Then it is verified that between the rows as well as between the columns this analysis implies the use of the so-called χ^2 distance.

The double centering means that this time, we shall have the property of optimal representation between rows as well as between columns. On the other hand, the $fv_{j,k}$ will not generally have a special significance. On the contrary, the transition formulas are simplified and take a particular significance. As a matter of fact they become (Benzecri *et al.*, 1973):

$$\text{(a)} \quad fo_{i,k} = \frac{1}{\sqrt{\lambda_k}} \sum_j (x_{i,j}/x_{i.}) \; fv_{j,k} \quad \text{and}$$

$$\text{(b)} \quad fv_{j,k} = \frac{1}{\sqrt{\lambda_k}} \sum_i (x_{i,j}/x_{.j}) \; fo_{i,k}$$

This may lead to the so-called centroid principle: to within about $\sqrt{\lambda_k}$, the formula (b) shows that if the plots of observations and variables were superimposed, the variable j would appear at the centroid of the I observations, with the weight $x_{i,j}/x_{.j}$ for the *i*th. This shows that if a group of observations is thrown off centre, in the plots of variables, it will draw the variables having high values for these observations towards a similar situation. However we should bear in mind the nuance: to within about $\sqrt{\lambda_k}$. For the centroid principle to have a direct effect in the way mentioned, we shall have to multiply the fv by these coefficients as a preliminary step.

2.3.5. Limitations of the notion of optimal representation. Utility of another point of view. It is mathematically fascinating if one can obtain a theory which almost assigns symmetrical roles to the rows and the columns of the table which is analysed. However, as previously mentioned we must bear in mind that when observations and variables are treated, the two together have not generally to be studied in the same perspective. This is particularly important as regards the concept of optimal representation. Let us consider a principal component analysis, about the variance-covariance or correlation (between variables) matrix. The centering of the variables gives the plots of observations the property of optimal representation, which is fruitful at this level. On the other hand, we do not generally have this property on behalf of the plots of variables. Often, the different

variables have coordinates of the same sign, in relation with
the first factor. So, the first axis does not make a clear
distinction between variables, and we are getting very far from
the property of optimal representation. We should not regret
it systematically. It is interesting to know that the dominant
feature of the structure of the variables is a likeness rather
than an opposition. This will be particularly valuable in the
comparison of different studies, in which the dominance of the
likeness in comparison with the oppositions can be more or less
pronounced, or can even totally disappear. If the analyses are
carried out from the centroid of the variables, we shall lose
this possibility. This weakening is regrettable; it must be
related to the fact that from one study to another, and even
when the variables do not change, their centroid is not a fixed
point. So the analysis is not carried out from the same point
of reference from one study to another. So whatever the interest
of the concept of optimal representation, with the meaning
previously mentioned, one should not be the slave of it. This is
true at the level of the observations and above all, practically,
at the level of the variables. What we called the optimal
representation, is not necessarily the most interesting plotting
in the analysis of the structure of the variables.

To a certain extent, this can lead to a revaluation of the
more traditional point of view based on the interpretation of a
variance-covariance or correlation (between variables) matrix.
In this respect, the point of view developed by Brillinger (1975)
seems to us the most interesting. Although it is limited to
principal component analysis, it seems to us that it could be
generalized to the case when the variables are provided with
unequal masses without major difficulty. Considering a random
vector, not necessarily centered, Brillinger tries to define the
space of a given dimension that can best account for the set of the
random variables making up the vector.

In order to do this, he considers the residuals corresponding
to the complement to the projection of each variable on the space
of a given dimension. Mathematically, for the random vector $\underset{\sim}{X}$, of
size J, the vector $\underset{\sim}{B}$, and the matrix $\underset{\sim}{A}$, are defined so that
the residuals $\underset{\sim}{X} - (\underset{\sim}{A}\,\underset{\sim}{X} + \underset{\sim}{B})$ are the smallest possible, in the
sense which we are going to specify. So he takes an interest in
the variance-covariance matrix of these residuals. Globally he
minimizes all the eigen values of this matrix. It is shown that
doing this, the trace, the determinant, and all the diagonal terms
are minimized. We get a better result than the traditional
one: we have a more global minimization. Moreover it is not a
priori absolutely necessary for the variables to be centered.
The spaces can be defined from any point. One of the results of
the theorems presented by Brillinger (1975) is to demonstrate that

it is actually necessary to center the variables in order to reach
an optimum. That being the case, it is possible to find again all
the formulas and traditional concepts of principal component
analysis. This presentation is interesting for two reasons: on
the one hand it provides an important justification for the
traditional analysis, which is better than that usually put
forward, on the other hand it proves the interest of centering
the variables. It should be noticed that Brillinger does not
necessarily make use of a hypothesis of normality and statistical
independence of the samples, concerning the foundation of the
method. This hypothesis will be essential only in the case of the
application of the traditional processes of inference. Similarly,
the linearity is not strictly required; except perhaps for some
interpretations. Consequently, although Brillinger's presentation
belongs to traditional principal component studies, it permits us
to avoid its most serious drawbacks.

If we have observations and variables at our disposal, we
can refer to the concept of optimal representation for the observations, to the point of view which has been set forth for the
variables. In practice, a convergence appears since in both cases
the variables have to be centered. This makes the plots of
variables of a principal component analysis more valuable, even
when the analysis is based on the structure of the set of the
observations. Therefore the previous remarks made about the non-
optimality of the plots of variables have to be seriously altered:
the criterion of optimality itself has a relative value.

3.   INTERPRETATION OF THE PLOTS AND CONTRIBUTIONS LINKED
WITH THE VARIABLES

Principal coordinate analysis (Gower, 1966) is nothing
more than a metric ordination technique. As soon as we use simple
Euclidean structures, possibly weighted, we can go beyond this
stage. In fact, it is possible to come close to a traditional
factorialist approach, to a certain extent. We shall adopt this
point of view. The fact that we can have contributions linked
with the variables and plots of variables makes it possible
for us to go beyond simple ordinations. If we associate a factor
with the coordinates of the observations on the axes, we can split
the original variable in relation to these factors. This attempt
is no longer the aim of the analysis; nevertheless it can lead
us to useful interpretations afterwards. This will be done by
means of the contributions linked with the variables and the plots
of variables.

3.1. Preoccupations Going Beyond the Framework of a Mere Ordination.
When the aim is to study the structure of a set of observations,
the inertia analyses from the centroid of these observations are
justified by the property of optimal representation. It is also
possible to have plots of variables and contributions linked with
the variables, at least when we use weighted simple Euclidean
structures. The interpretation to be given to the plots of
variables gives rise to great ambiguity, even to wrong ideas
(Jolicoeur and Mosimann, 1960; Gabriel, 1971). In order to make
the situation clearer, we must come back to the primary preoccupations relative to the link between the variables and the
structures of the set of observations.

- The first preoccupation has to do with the role of the
  variables in the setting up of the structures.

- The second preoccupation is relative to the way in which
  the structures which are given prominence to, on the set
  of the observations, explain the variations of the variables.

- The third one is the search for privileged associations
  between one or several observations and one or several
  variables.

- Lastly, one can be tempted to study the proximities between
  variables. This last point has no longer to do with the
  analysis of the structure of the observations. However, one
  can be tempted to try to do two things at once. This may
  be dangerous, it is advisable to know if, and when, it is
  possible to do so.

In order to attain these aims, we fundamentally have plots
of variables, contributions from the axes to the variables and
reciprocally.

*3.2. Role of the Variables in the Definition of the Axes. Expla-
nation of the variables by the Factors.*

3.2.1. Use of the contributions. In order to measure globally the
role of a variable in the definition of the axes, we shall be
interested in the relative contributions from the variables to
the axes. This fits exactly our concern in quantifying the
responsibility of the different variables. In this way so-called
structuring characters will be defined. When we shall use a
logical coding and when a character will give rise to several
variables, it will be useful to add the contributions of the
different associated logical variables, in order to define the
contribution of the character.

In order to quantify the role of an axis or a subspace in the variations of a variable, this time we shall require the concept of relative contribution from the axes to the variables. Thus the variables highly depending on the structures brought out by the analysis will be brought into prominence. It will be wise to restrict the comments on the variables to those well explained by these structures.

3.2.2 Possible use of the plot of the variables. These plots can back up the examination of the contributions in the analysis of the role of the variables in the definition of the structures, or in the analysis of the "explanation" of the variables by the structures.

When the variables are assigned unit masses, the plots of variables visualize their contributions to the axes: the absolute contribution from a variable to an axis is given by the square of its coordinate on this axis. Therefore the plots of variables permit the direct approach of the part of the variables in the definition of the axes, and their examination thus complements that of the tables of contributions from the variables to the axes. This is particularly true in principal component analysis. When the variables are assigned different weights, this direct approach is no longer possible; the remoteness of a variable on an axis does not necessarily imply a major contribution since it can have a small weight. This can be especially true in the case of analysis of correspondences. The visualization of the contributions is nevertheless an advantage which we abandon regretfully. This is why we should readily advise, even if this is not usual, to make graphic displays relative to the variables, in which the coordinate of variable j for factor k would be $\sqrt{pc_j} \ fv_{j,k}$, pc_j always being the mass of column j, and not $fv_{j,k}$ as it was defined in Section 2. It is easy to verify that these graphic displays will permit a visualization of the contributions from the variables to the axes.

When several variables are associated with the same character, for instance with a species, in order to appreciate the global influence of the character, we should, as was said, sum up the contributions of the different variables. It is not inconceivable to represent these added contributions graphically. But then we are getting far from the context of the use of the common plots of variables.

It is also possible that the plots of variables visualize the relative contribution from the axes to the variables. This will be the case when the norm of the variables is 1. This occurs particularly in principal component analyses based on the matrix of

correlation between taxa. The relative contribution from an axis
to a variable is then given by the square of its coordinate in
relation to the corresponding factor. Similarly, on a factorial
plane, the square of the segment relating the origin to the projection of the variable point on the plane gives the relative
contribution from the plane to the variable. This can be grasped
more easily if we draw a circle of radius 1, centered on the
origin, called correlations circle. The mentioned contribution
is in fact less than or equal to 1. It is equal to 1 when the
variable which is studied is fully explained by the factorial
plane.

The visualization of the relative contribution from the
axes and the planes to the variables is made real by the plots of
variables only when the variables are normalized beforehand. This
is not the case for many traditional variants (analysis of
correspondences, principal component analysis on the matrix of
covariance of the variables and not on the correlation matrix).
Then it could be useful to transfer on graphic displays relative
to the variables the points of coordination $fv_{j,k}/(no_j)$, if
no_j designates the norm of the variable j, and not the points of
coordinates $fv_{j,k}$ in relation to the axis k.

*3.3. Privileged Associations Between One or Several Observations
and One or Several Variables.* In this respect, it shall be
considered the simultaneous use of the plots of observations and
variables. It is then tempting to associate variables with
observations having similar positions in homologous planes, and
particularly when the corresponding points are isolated and thrown
off center. Let us suppose such a situation. When the isolated
observations are called O (this will be the term retained), the
homologous variables are called of same V. In order to give a
concrete content to the idea of association, two possibilities
(which are compatible) can be referred to:

- the variables V take higher values for the observations
O than for the other observations. POINT OF VIEW A.

- for the observations O, the variables V take higher
values than the other variables. POINT OF VIEW B.

3.3.1. Transition and reconstruction formulas. Transition
formulas, which are of particular interest within the context of
the analysis of correspondences, are also useful in the general
case.

$$fo_{i,k} = \frac{1}{\sqrt{\lambda_k}} \sum_j pc_j \, x_{i,j} \, fv_{j,k}$$

$$fv_{j,k} = \frac{1}{\sqrt{\lambda_k}} \sum_i p\ell_i \, x_{i,j} \, fo_{i,k}$$

If $x_{i,jo}$ is zero except when $i = io$,

$$fv_{jo,k} = \frac{1}{\sqrt{\lambda_k}} \, p\ell_{io} \, x_{io,jo} \, fo_{io,k}$$

If $x_{io,jo}$ is positive, providing that we multiply the $fv_{j,k}$ by $\frac{1}{\sqrt{\lambda_k}}$ in the dual configurations the observation point i_o and the variable point j_o appear in the same direction. If $x_{io,jo}$ is negative, there will be an inversion. Afterwards, we shall mainly consider the case when $x_{io,jo}$ is positive.[1] When the $x_{i,jo}$ are not equal to zero and when in absolute value $p\ell_{io} \, x_{io,jo}$ is highly dominating, the same phenomenon is found, yet blurred. This also occurs if a set of observations, situated in the same area (in terms of sector in relation to the origin), provides the main part of the contributions to the variable jo.

More precisely, if a variable j is equal to zero except for the two observations i_1 and i_2:

$$fv_{j,k} = \frac{1}{\sqrt{\lambda_k}} \, (p\ell_{i_1} \, x_{i_1,j} \, fo_{i_1,k} + p\ell_{i_2} \, x_{i_2,j} \, fo_{i_2,k})$$

When $x_{i_1,j}$ and $x_{i_2,j}$ are positive, we see that, to within about $\frac{1}{\sqrt{\lambda_k}}$, in a factorial plane, the variable j appears in the area defined by the axes going through the observations i_1 and i_2 in the dual plane. When more than two observations are involved, the second is defined by the set of observations, in a similar way.

Similarly, we can consider the case when only $x_{io,jo}$ is not equal to zero among the $x_{io,j}$. In that case, we must multiply the $fo_{i,k}$ by $\sqrt{\lambda_k}$. We can get near this ideal case when a variable, or a group of variables situated in the same sector, are dominating beside an observation.

The preceding lines show that high values $x_{io,jo}$, either among the $x_{io,j}$, or among the $x_{i,jo}$, draw the points io (observations) and jo (variables) towards similar sectors, to within about $\sqrt{\lambda_k}$. This also occurs when there are sets of dominating points in the same sectors.

Conversely, the observation of similar situations can suggest the hypothesis that some variables have high values for some observations. This is only a hypothesis which has always to be verified in the basic data, but the interest of which can still be justified considering the so-called reconstruction formula. (Benzecri et $al.$, 1973; Laurec, 1979).

$$x_{i,j} = \sum_k \frac{1}{\sqrt{\lambda_k}} \; fo_{i,k} \; fv_{j,k}$$

This proves that high values simultaneously for $fo_{i,ko}$ and $fv_{j,ko}$ contribute to the inflation of the value of $x_{i,j}$.

$x_{io,jo}$ is more likely to be high among the $x_{i,jo}$, if $fo_{io,k}$ is big among the $fo_{i,k}$, among the $x_{io,j}$ if $fv_{jo,k}$ is big among the $fv_{j,k}$. This relative importance of a $fv_{j,k}$ or $fo_{i,k}$ can be measured in terms of relative contribution to the variables or to the observations. If the variables (respectively the observations) are standardized, this will be an advantage for the point of view A (respectively B) in the use of the plots of variables.

It should be noticed that the case is all the more favorable as we deal with points thrown off center and receiving a more important relative contribution from the first axes. Lastly, the formulas can become more easily usable, if we correct them with the factors $\sqrt{\lambda_k}$. This is not always realized. In term of orientation in the factorial planes in relation to the origin, this is all the more important as the axes present very different inertias.

3.3.2. Analysis of correspondences. In this case, the transition
formulas lead to the centroid principle that can be related
to the points of view A and B of the paragraph 3.3.

If we multiply the $fv_{j,k}$ by $\sqrt{\lambda_k}$, or if we divide the
$fo_{i,k}$, by $\sqrt{\lambda_k}$, the variable j appears in a simultaneous
representation of the observations and variables, at the center
of gravity of the observations, with the weight $x_{i,j}/x_{.j}$ for
the *i*th. The effect of this is to bring a variable closer to the
observations for which it takes particularly high values. This
has to do with the point of view A.

Conversely, if we multiply the $fo_{i,k}$ by $\sqrt{\lambda_k}$, or if we
divide the $fv_{j,k}$ by $\sqrt{\lambda_k}$, we can make use of the barycentric
principle in the other direction. The observation i appears
at the center of gravity of the variables, with the weight
$x_{i,j}/x_i$. for the *j*th   This brings the variables which present
high values on observation i closer to this observation
(point of view B).

In practice, positions which are close to each other do not
constitute a proof in themselves, but generally an indication.
The centroid principle is useful mainly for the isolated and
thrown off center points: we can hardly deduce an association
from the proximity, in the middle of the clusters of an observation and a variable. Lastly, it should be interesting to note
that rigorously it is suitable, for a simultaneous representation,
to multiply the $fo_{i,k}$, or the $fv_{j,k}$ by $\sqrt{\lambda_k}$, depending on
whether we follow the point of view A or B.

3.3.3. Principal component analysis (PCA). Let us consider a
PCA based on the matrix of correlation between variables. The
general remarks made about the transition and reconstruction
formulas are entirely pertinent. However, a particular interpretation
can be given. Let us consider homologous observations and
variables, called respectively O and V, and eccentric in
relation to a factor k.

For the variables, it is also supposed that the corresponding
factors loadings are high, bearing in mind that they are also
correlations. They are supposed to be positive, but the reasonings
can be easily transposed. An eccentric location for the observations
O on an axis corresponds to high values of the corresponding
factor for these observations. This factor is highly correlated
to the variables V since the factor loadings are high. So, we

can expect that these variables, as the factors, will take higher
values for the observations 0 than for the others. This
corresponds to the point of view A. We can also expect that the
variables V will take higher values than the other variables
for the observations 0 (point of view B). This is less
obvious; however we can use as an argument the fact that if
other variables had taken higher values than the variables V
on the observations 0, this should have increased their
correlation with the factor k, and consequently given a high
factor loading which does not appear. Of course all these
reasonings based on correlations have not the value of proofs
for a given observation, and as was pointed out in the section
3.3.1, it is wise to compare the hypothesis suggested by the dual
configurations with the basic data. Nevertheless, the closer the
loadings are to 1 in absolute value, the more the hypothesis
is likely to be good.

The preceding reasoning applys when we have an isolated
factor. The examination of the factorial plane defined by the
factorial axes k_1 and k_2, makes it possible to consider
isolated observations in any direction. Just as the factors are
associated with the factorial axes, it is possible to define a
variable V linked with this particular direction. Now let us
consider the homologous direction in the dual plane. We should
expect that the projection of a variable point on this line gives
the correlation with the variable V. This is not strictly true.
The line to be considered in the plane of the variables does not
have the same orientation as the plane of the observations. If
the latter has for direction cosine "a" and "b" for the plane of
the variables, the direction defined by the vector $(a\sqrt{\lambda}_{k_1}, \quad b\sqrt{\lambda}_{k_2})$
has to be considered. There we find again the role of the factors
$\sqrt{\lambda}_k$.

Finally if we mentioned the PCA on the correlation matrices,
it is not inconceivable to extend the reasonings to analyses
based on the variance -covariance matrices. However, the
interpretations are more difficult, and very often less fruitful.
The loadings are no longer correlation coefficients, but simple
covariances. In the preceding case, of the $fv_{j,k}$, the more
important fact was the direct link with the contributions from
the axes to the variables, these being standardized. This is
particularly important in the point of view A, and is lost in
the covariance analysis.

3.3.4. Binary variables. Considering a variable V and a group
of observations 0, it is then possible to link the points of
view A and B with conditional probabilities. Let us consider as

an example the logical variable associated with the presence of
a species. We can then take an interest in:

- the rate of presence of the species for the observations
  0. This frequency can be usefully related to the global
  rate of occurrence of the species, (point of view A).

- the frequency of the observations 0 among those in
  which the species is present. This can be related to
  the proportion of observations 0 to the set of the
  observations, (point of view B).

The examination of the dual configurations can help to
focus attention on the associations which could deserve a complementary examination, and particularly the calculation of these
frequencies. If we carry out an analysis of correspondences, the
use of centroid principles can be valuable. Other analyses,
resting on indexes of similarity can be reduced to general analysis
with double weighting in the studies in presence-absence. This is
notably the case when one uses the point correlation coefficient, or
the Jaccard index (Roux and Roux, 1973; Laurec, 1979). For each
case we must come back to the significance of the $fv_{j,k}$. As
regards the Ochiai's index, Blanc *et al.* (1976) point out that
the plots of variables account very badly for the concept of
characteristic species, corresponding to the point of view B. This
is also true for the other mentioned indexes. They refer to
another concept, that of structuring species, defined by Verneaux
(1973). This corresponds to the idea that some species can be
responsible for the structures, that is to say for the inertia
assignable to the first axes, more than others. This corresponds
to the notion of contribution from the variables to the axes and
subspaces. When, as for the mentioned indexes, the variables have
the same weight, this is relatively close to the point of view A.
However, some difficulty arises when the same species has given
rise to several logical variables. We shall add the influences
of the different variables in order to actually define the
structuring species. The common plots of variables are not longer
so directly useful. We are brought back to the problem mentioned
about the contributions from the species and variables to the axes
and planes.

3.4. Analysis of the Relations Between Variables. One can be
tempted to use the plots of variables in order to account for the
proximities between variables. Yet the precise aim of such an
interpretation has to be specified. We do not adopt the context
of the traditional factorialist approach, the aim of which is to
identify both hidden factors and the multilinear model accounting
for their action. Three points can be taken into account.

3.4.1. Optimal representation of the variables. Considering the
analysis of the set of the variables, it is possible to adopt the
same process as that adopted for the observations. It is thus
possible to consider the distances between variables inferred by
the choice of the masses for the observations, and possible preliminary transformations. The plots of variables will then be
considered as "summaries", the aim of which is to account for the
network of relative proximities defined by the distance matrix.

In order to appreciate the relevance of the plots of
variables, from this point of view, two criteria will be examined:

- Do the chosen coding and distance account for the links
  between the variables adequately?

- Is the origin situated or not on the centroid of the
  cluster of the variable points in the inertia
  analysis (what determines whether or not we have the
  property of optimal representation)?

The answers to these two questions are not necessarily
favorable. For each problem, the two criteria will have to be
examined. If we leave aside the particular case of the analysis
of correspondences, it will be rather infrequent to obtain a
positive answer for both. If we take a principal component
analysis, for instance, based on the matrix of correlation between
taxa, it is possible that the coefficient of correlation is a
correct measure of the links between the variables. This is not
necessarily true, especially because of the problems of non-
linear relations. Generally, if the origin is situated on the
center of gravity of the cluster of the observations, this will
not be the same for the variables. The plots of variables are
not optimal representations, unless we center the observations;
doing which we would alter the distances between the observations.
This shows the danger of having one's finger in more than one
pie, when studying simultaneously the structure of the observations
and that of the variables. If we come to principal component
analyses based on the matrix of covariance between variables, then
the situation becomes trickier: at the start we use a structure
which measures the links between the variables inadequately.

*3.4.2. Giving prominence to the indirect links, in reference to
the factors.* When we study the links between the variables, we
may often take an interest, not in a direct link, but in similarities in the association with a set of other variables. (Two
taxa can be considered as similar not on the basis of parallel
abundance variations, but from a common answer to a set of abiotic
factors: temperature and salinity for instance). In this per-

spective, we may wish to put forward by means of graphic representations the fact that two variables are associated with the factors
sorted out by the analysis of all the sample elements in a more or
less similar way. If we define in this way a measure of linkage
between analysed variables and a factor, considering each pair of
factors, it will be possible to make graphic representations of
the variables, using these measures of linkage as coordinates on
the axes associated with the factors. In this perspective, if we
want to use the plots of variables we shall simply have to ask
ourselves whether or not the $fv_{j,k}$, as they were defined
previously, are correct measures of the links between the variable
j and the factor k.

The case will be particularly favorable if the variables
have the same norm, equal to 1; we are brought back to a visualization of the relative contributions from the axes to the variables,
apart from the fact that the $fv_{j,k}$ involve a sign which is
absent from the contributions. So the remarks which have been
made in this respect are still relevant. This is particularly
interesting in principal component analyses based on the matrix
of correlations between variables. Since $fv_{j,k}$ then represents
the correlation between the variable j and the factor k, two
variables will be near to each other on the plots, if they
present correlations similar to the factors. If the variables
are not standardized, the $fv_{j,k}$ are no longer correlations.
As was suggested about the contributions from the axes to the
variables we could set up the graphic displays, using the $fv_{j,k}/$
no_j, if no_j is the norm of the jth variable.

3.4.3. "Representation" of the scalar products between variables.
We specified in 2.3.5 that the concept of optimal representation,
as it had been defined, was not always the only suitable one.
This was particularly underlined for the representations of the
variables. If we start from a matrix of scalar products between
variables, and no longer from a distance matrix, the problem is
somewhat different, even when we can associate the scalar products
with a distance, by the relation:

$$dc^2_{j_1,j_2} = sc_{j_1,j_1} + sc_{i_2,j_2} - 2\ sc_{j_1,j_2}\ .$$

Yet there is not a strict equivalence between the information
contained in a matrix of scalar products between variables, and
the information of the associated distances. The distance matrix
is intrinsic, independent from the origin. It is not the same
with the scalar products.

As was emphasized in section 2.3.5, it can be important to maintain the part of original information contained in the matrix of scalar products. So we shall try at best to keep the matrix by a projection. We do not have such an elegant result as with the concept of optimal representation associated with the distance. Nevertheless, if we consider the principal component analysis, the point of view which we borrowed from Brillinger (1975) and which is an extension of traditional studies, makes it possible to progress towards the keeping of the matrix of scalar products (variance-covariance, or correlations).

Therefore, when we want to obtain a correct representation of the relationships between variables, if at the start these relationships are quantified by a scalar product matrix, another path appears than that leading to optimal representation through the distances. This one is complete only within the framework of principal component analyses, but as was pointed out in section 2.3.5, the generalization is probably not very delicate. With this new path, we must simply question the importance of the scalar products as a measure of the links between variables. In this perspective, the principal component analysis of the correlations will be more fruitful than other types of analysis. As to the analysis of correspondences, this last point of view, which has not been developed as far as we know, does not in fact present any originality compared with the first one.

If we sum up globally the points relative to the relations between variables, three paths appear in which the plots of variables can be useful:

- If we take an interest in the global, direct, relations between variables, quantified by distances, we shall follow the path leading to the concept of optimal representation. It requires, in order to be effective, that the distance which is used be an adequate measure of the dissimilarities between variables, and that the analysis be carried out of the center of gravity of the variables (observation lines centered). If the second condition is not fulfilled, the plots of variables are not uninteresting, but they lose their optimality.

- The second point of view leads to look for links between variables, but not intrinsic ones, since the interest is now focussed on eventual similar links of the variables with the factors brought out by the analysis. The question will then be about the values of the $fv_{j,k}$ as measures of the links between variables and factors.

- The last one is close to the first since intrinsic relations
are studied, but they are now measured by scalar products.
In order to estimate the value of the plots of variables as
summaries of the links between them, one will first consider
the adequacy of the scalar products as measures of links.

4. ILLUSTRATION

4.1. Data and Method.

4.1.1. Data. The data are borrowed from Rickaert's work
(1979), on the annual cycle of phytoplankton in the area of Paluel
(East Channel). A series of 153 samples were collected with a
Metzikoff-type valve bottle (a pair of 5 litre bottles), from
September 1976 to August 1977. These samples can be grouped
by cruise. Evaluations of abundance have been carried out on
38 taxa (cf. Table 1).

TABLE 1: List of taxa studied.

1. *Actinoptychus senarius*	20. *Minuscula bipes*
2. *Asterionella japonica*	21. *Navicula* sp.
3. *Biddulphia alternans*	22. *Navicula pelagica*
4. *Biddulphia aurita*	23. *Nitzschia closterium*
5. *Cerataulina pelagica*	24. *Nitzschia delicatissima*
6. *Chaetoceros curvisetum*	25. *Nitzschia* sp.
7. *Chaetoceros perpusillum*	26. *Paralia sulcata*
8. *Chaetoceros sociale*	27. *Rhizosolenia delicatula*
9. *Chaetoceros* sp.	28. *Rhizosolenia fragilissima*
10. *Coscinodiscus*	29. *Rhizosolenia setigera*
11. *Coscinodiscus radiatus*	30. *Rhizosolenia shrubsolei*
12. *Dictyocha speculum*	31. *Rhizosolenia stolterfothii*
13. *Dinoflagelles*	32. *Skeletonema costatum*
14. *Ditylum brightwellii*	33. *Thalassionema nitzschioides*
15. *Euglene*	34. *Thalassiosira* sp.
16. *Gymnodinium*	35. *Thalassiosira decipiens*
17. *Gymnodinium lohmani*	36. *Thalassiosira gravida*
18. *Lauderia borealis*	37. *Thalassiosira levanderi*
19. *Melosira* sp.	38. *Thalassiosira rotula*

4.1.2. Setting up of the tables to be analysed. According to
Rey's terminology (1976), we must associate each taxon with a
numerical describer, corresponding to a mathematical variable. A
consideration on the definition of such describers can be found in
Dessier and Laurec (1978) or Laurec (1979). In the present case,

two processes have been used. The first one corresponds to
quantitative variables. For each taxon, we consider its number
x, submitted to a transformation log(x+1). The second one leads
to logical variables, since for each specified taxon we have two
logical variables associated with the presence and the absence
respectively. According to the options which has been retained,
we shall get tables including 38 or 76 columns. We still have
to define the rows of the tables which will be associated with
the set of samples. Previous analyses have proved that the
variations within the same cruise are small in comparison with
variations between cruises. Thus the study has been based on
the definition of the seasonal cycle. The various samples
associated with each cruise have been grouped. Except for
the bloom periods, the variations between the different cruises
of a given month are low. Consequently the cruises have been
grouped by month, except for those of March and June
which remain separated. This leads to a set of 14 rows corresponding
to what has been termed as observations in the preceding pages.
In a more concrete way, the mentioned groupings have been defined
as follows:

- When we use quantitative variables for a row, the means
  of the log (x+1) obtained for the individual samples have
  been retained. Thus we are lead to a table with 14 rows
  and 38 columns;

- When we make use of the logical variables, for each row
  we have summed up the value of the variables for the
  different samples. Thus if we consider January, for each
  species we keep the number of samples in which it was
  present and the number of those in which it was missing.
  Therefore we are lead to a contingency table including
  14 rows and 76 columns.

The procedures followed for the definition of the analysed
tables are in the absolute very important. The reader is
referred to Dessier and Laurec (1978), or Laurec (1979). In the
case which is treated, the only aim is the illustration of the
use of the concepts associated with the variables. The question
of the optimality of the processes which are used is not essential.
This is still true for the subsequent processing of the tables
thus constituted by the factorial analyses.

4.1.3. Factorial analyses. Different methods will be used all of
them having to do with general analysis with double weighting,
with possible preliminary transformations of the analysed tables.
The table corresponding to the quantitative variables was submitted
to two principal component analyses, based respectively on the
matrix of correlations between variables and on the covariance

matrix. The contingency table was treated through an analysis
of correspondences. So, in fact three analyses have been carried
out to illustrate section 3.

4.2. PCA of the Correlations. Figure 1 illustrates the structure
obtained by the projection of the observations on the first two
axes, which explain 28.2 and 20% of the total variance respectively.
This structure is entirely dominated by the existence of the bloom
periods: March and June. Axis I corresponds to the isolation
of the month of June, and axis II to that of March. The corresponding plot of variables appears in Figure 2.

4.2.1. Contribution from the axes to the variables. Figure 2
directly illustrates the contribution from the variables to the
axes. Thus the species "responsible" for the isolation of the
bloom periods are brought out: *Chaetoceros sociale* and *Chaetoceros
curvisetum*, *Thalassiosira gravida* in March, *Cerataulina pelagica*
and *Gymnodinium sp* in June. *Rhizosolenia shrubsolei* and
Actinoptychus senarius also provide important
contributions to the axes. Here and now, we notice that for the
second species the contribution is in a way "negative." Globally,
the cluster of species is rather homogeneous: there is not a small
number of taxa alone responsible for the main part of the inertia
linked with axes I and II.

The relative contributions from the axes to the variables
can also be studied. This is made easier by the drawing of the
correlations circle. Some species are well explained by the
factors I and II. This is the case for the previously mentioned
taxa, which contrast with *Nitzschia closterium* for instance.
This is a species whose variations go beyond a simple answer to the
bloom phenomena.

*4.2.2. Privileged associations between observations and
variables.* The situation is particularly simple insofar as
axes I and II single out precise observations. Axis I on Figure
1 singles out the bloom of June, and as regards the variables, it
locates a set of species in the same sector. Thus *Cerataulina pelagica* appears as linked to the bloom of June (correlation with
factor I = 0.89). It can even be noticed that this is closer to
June 1.

Since this point is not strictly on axis I, the direction
to be considered at the level of the variables is not strictly
that of June-1 at the level of the observations. The latter
is written out as a dotted line and the corrected direction
appears as a continuous line. It is the same for the points

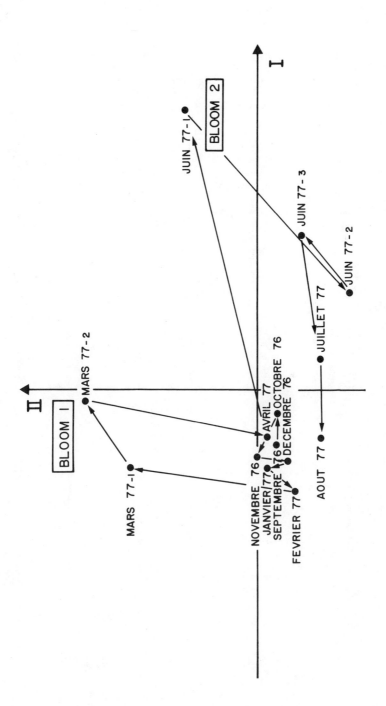

FIG. 1: Annual cycle of phytoplankton at Paluel. Principal component analysis, correlation of variables. Plot of variables obtained in the plane I – II.

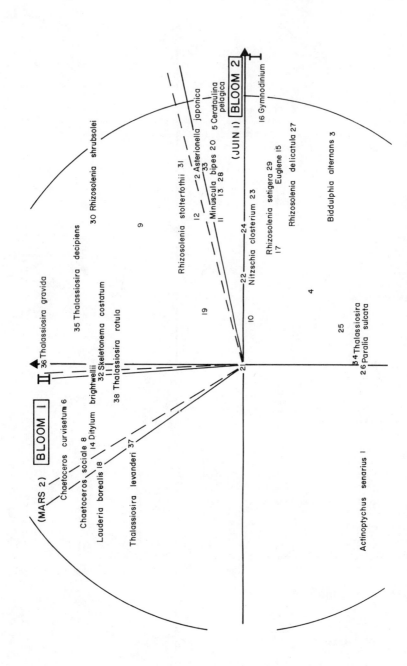

FIG. 2: Annual cycle of phytoplankton at Paluel. PCA of the correlations. Plots of variables obtained in the plane I – II.

March-1 and March-2. The gaps are not very important, especially for the correlations. They would be more important if the difference of inertias explained by the two axes was more pronounced, considering for instance the axes I and III. Even in the present case, the gaps can apply at the level of details. In this respect, the unbroken lines associated with March-1 and 2 define the area where all the species pledged to the month of March will appear.

We also notice the position of a species such as *Rhizosolenia shrubsolei* in the middle of the first quadrant, suggesting either an appearance during the blooms of both June and March, or a remarkable absence during the period determined by the third quadrant, in particular August 1977. The examination of the basic data shows that the first hypothesis is the right one (the species is plentiful in March 77-2 and June 77-1). *Actinoptychus senarius* poses the opposite problem because of the position in the third quadrant. Once again examination of the data is imperative and shows that this species is plentiful essentially from September 76 to February 77.

4.2.3. Proximities and links between variables. In the first point of view (optimal representation) the distance between the variables is interesting in spite of the drawbacks inherent to the problems of non-linearity. The distance is defined for the variables j_1, j_2 by

$$dc^2_{j_1,j_2} = 2 \ (1-C_{j_1,j_2})$$

if C_{j_1,j_2} is the coefficient of correlation between the two variables. On the other hand, the plots of variables are not optimal representations. In order to get such representations, while keeping the same definition of the distances, before the analysis we must center the rows again so as to shift the origin of the centre of gravity of the variables. This leads to a double centering analysis, the interest of which has been underlined by Orloci (1967). The results are recorded on Figure 3 which displays a notably different structure from that of Figure 2. The structure defined on Figure 3 permits a finer distinction. In fact, the main distinctions (oppositions of species linked with the blooms 1 and 2) are maintained, with nuances which appear only on the third axis in the simple PCA. This is particularly true for the set of the species qualified as "out of bloom" as well as for the species common to the two blooms.

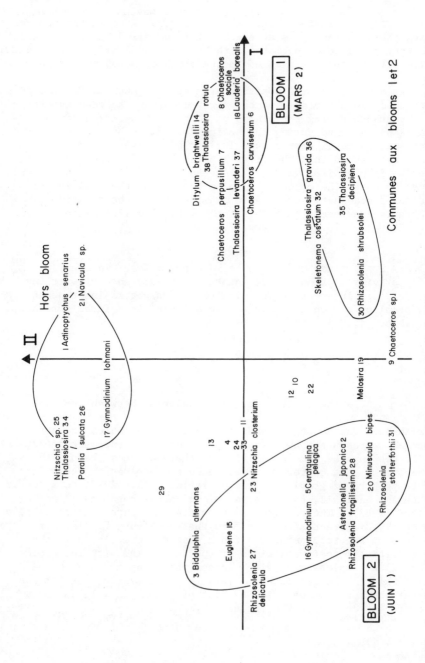

FIG. 3: Annual cycle of phytoplankton at Paluel. PCA of the correlations of variables with recentering. Plots of variables in the plane I – II.

The second point of view brings out relationships between variables which are not direct links between species but which express similarities of links with the axes. This is particularly interesting in the example which is studied, in which axes I and II are quite obviously associated with identified blooms. So Figure 2 shows that the species *Rhizosolenia stolterfothii* and *Asterionella japonica* fit the bloom phenomena of March and June in a similar way.

The third point of view, corresponds to the expression by a plane graphic display not of a distance matrix but of a scalar product matrix. The use of Figure 2 is justified by this point of view. This is particularly true insofar as the mentioned scalar products are directly interpretable since they are correlation coefficients. Consequently, whereas Figure 3 has some advantages over Figure 2, Figure 2 has nonetheless specific advantages. The orthogonality of the species linked with the blooms I and II suggests different preferences rather than a mutual exclusion. This interpretation is possible only in reference to the scalar products. In a way, Figure 3 permits a better discrimination of the species, and Figure 2 a better perception of the nature of these relations. This is because the first structure is based on distances, and the second one on correlations.

4.3 PCA of the covariances. The first two axes explain 30.9 and 26.5% respectively of the total inertia. The observations are visualized on Figure 4. On this diagram the blooms are found again with distinctions due to the fact that the non-standardized species play more or less important parts according to their variance. Figure 5 constitutes the corresponding plot of variables.

4.3.1. Contributions associated with the variables. Figure 5 makes it possible to visualize the contributions from the species to axes I and II, that is to say to the bloom of March and June (especially June-1). These contributions appear much more heterogeneous than in the preceding analysis. This is logical insofar as the global contribution from a species to the total variance is given by its variance. The species being no longer standardized, these global contributions are much more heterogeneous and this is found back at the level of the contributions to sub-spaces. Thus the dominating influence of species such as *Chaetoceros sociale, Chaetoceros curvisetum, Thalassiosira decipiens* or *Rhizosolenia delicatula* appears. As a matter of fact these species have high variances. They are also the ones which have the highest abundances, which is a common fact (Pielou, 1969) corresponding to what Chardy *et al.* (1976) term as effect of abundance.

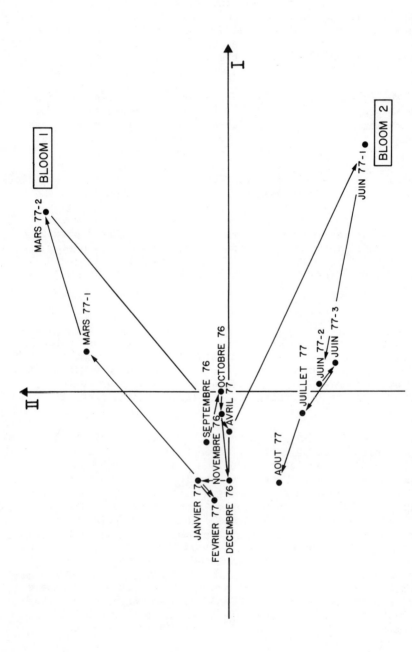

FIG. 4: Annual cycle of phytoplankton at Paluel. PCA of the covariances. Plots of observations in the plane I - II.

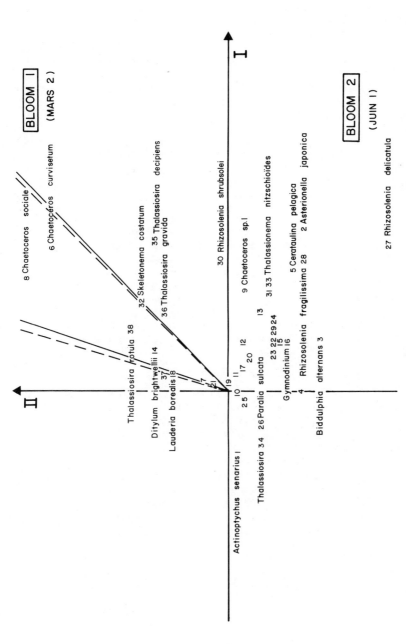

FIG. 5: Annual cycle of phytoplankton at Paluel. PCA of the covariances. Plots of variables in the plane I – II.

Figure 5 does not make it possible to perceive the relative contributions from the axes to the variables. Therefore the "fv" can be divided by the norm of the variables in order to get a new structure (Figure 6). This time, as for the preceding analysis, the coordinates are correlations, whence the appearance of the circle of the correlations. Thus the species *Ditylum brightwellii* having a low contribution because of its low variance, appears as well pledged to the March bloom, what Figure 5 cannot show. This type of figure is interesting inasmuch as a visual representation is more pleasant than a table of contributions.

4.3.2. Privileged associations between observations and variables. If we stick to Figure 5, it is no longer possible to rely on the fact that the coordinates are correlations. There is still the possibility to compare the direction where the observations and the variables appear, with the subsequent precautions to be taken when dealing with the "$\sqrt{\lambda_k}$". The graphic convention of the rows as continuous lines or as dotted lines has been taken over. Thus the lines associated with March 2 and March 1 appear on Figure 5.

The deductions are made less probable since we can read on the graphic display neither the contributions from plans to the observations nor that from the plans to the variables. The latter possibility was given by Figure 2 in the preceding analysis by comparing with the correlations circle. It is found back in Figure 6; in fact this figure is interpreted mainly as Figure 2.

4.3.3. Proximities between variables. Although we do not deny any interest that may well appear in such or such a particular case, the case is globally unfavorable, whatever the point of view which is considered. With the first point of view, we immediately come up against the fact that the distances between variables do not appear as a correct measure of their dissimilarity. With the second point of view, the interpretation is less fruitful since the factor loadings are no longer correlations. In the third perspective, we may wonder whether the covariances, which are scalar products, deserve an analysis in depth, since generally they are not correct measures of the links.

Figure 6 can be more interesting, especially for the second point of view since the coordinates are not correlations. This reinforces the interest of such a figure, even if it is not commonly used.

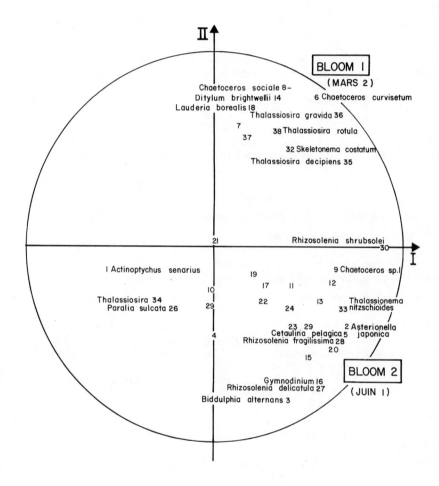

FIG. 6: Annual cycle of phytoplankton at Paluel. PCA of the co-variances. Representation of the "plots of variables" in the plane I - II.

3.4. Analysis of Correspondences. This analysis is realized not
from the average of the numbers, but from the frequency of appearance of each species for each cruise. In order to use the
analysis of correspondences in the best conditions of its application, for a given cruise, each species is defined by two variables:
rate of presence (+) and rate of absence (-).

The plot of observation in the plane I - II is represented in
Figure 7. This structure reveals a weakening of the blooms in
comparison with the preceding plot. This is due to the transformation of the basic data: the bloom, an essentially quantitative
phenomenon does not appear at the qualitative level as a particular
period. The plots of variables of Figure 8 corresponds to a
traditional dual structure in which each species is represented
twice: a variable associated with the presence (+) and a
variable associated with an absence (-).

4.4.1. Contributions linked with the variables. Figure 8 does
not represent the contributions from the variables to the axes,
the assigned masses being different. Figure 9, in accordance
with the suggestions of Section 3 corresponds to these contributions, the "$fv_{j,k}$" being multiplied by $\sqrt{x}_{.j}$. The differences
are not important. They would be more pronounced if the $x_{.j}$ were
more heterogeneous than they are in the example dealt with.

In order to get a visualization of the relative contributions
from the axes to the variables, the $fv_{j,k}$ have been divided by
the norm of the variables (no_j). This leads to Figure 10.

This structure makes it possible to draw the circle of
radius 1, homologous to the correlations circle of PCA. The
graphic display let appear the species for which the variations
of the rate of occurrence are well explained by the plane of the
axes I and II. Thus *Navicula pelagica* is a well-explained
species in comparison with *Rhizosolenia stolterfothii*.

4.4.2. Privileged relations between observations and variables.
It is always possible and useful to look for associated directions
in the analogous planes, especially in Figures 7 and 8. The
sector corresponding to the "Winter-Spring" period has been
transferred to Figure 8. This sector is double, since both
presence and absence can appear. Besides this suggests that the
corresponding period is primarily marked by absences. Conversely,
Spring is primarily identified by the presence of certain taxa
(*Gymnodinium*, *Cerataulina pelagica*, *Coscinodiscus radiatus*).

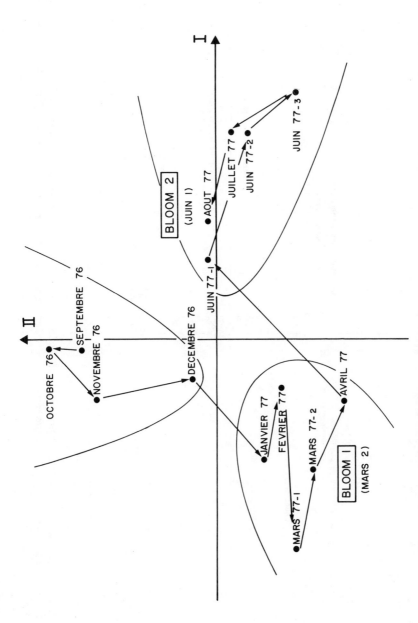

FIG. 7: Annual cycle of phytoplankton at Paluel. Analyses of correspondences. Plots of observations in the plane I – II.

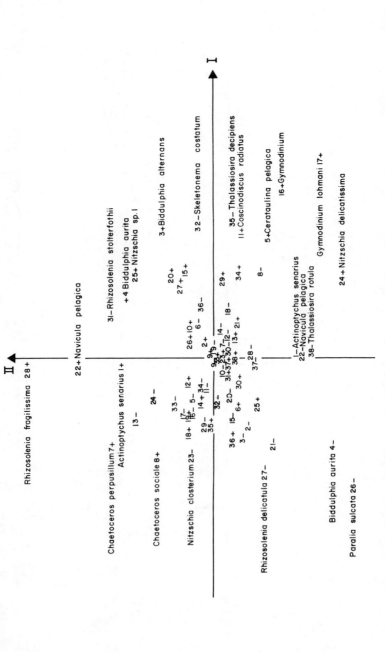

FIG. 8:   Annual phytoplankton cycle at Paluel.   Analysis of correspondences.   Plots of variables in the plane I - II.   Each species is represented by the variables: rate of presence (+) and rate of absence (-).

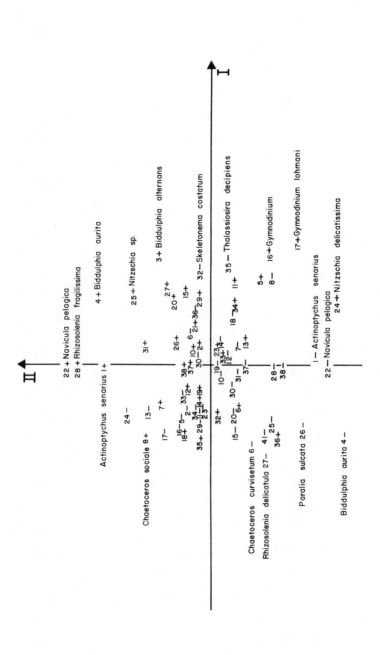

FIG. 9: Annual cycle of phytoplankton at Paluel. Analysis of correspondences. Representation of the relative contributions from the variables to the axes in the plane I - II.

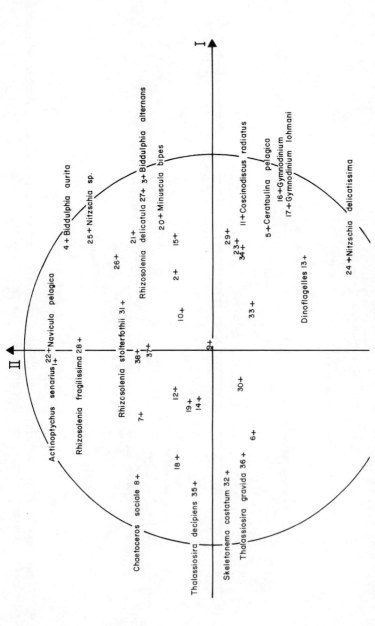

FIG. 10: *Annual cycle of phytoplankton at Paluel. Analysis of correspondences. Representation of the relative contributions from the axes to the variables in the plane I – II. The variables (+) and (-) of a same species being symmetrical, only the (+) are represented on the plane I-II.*

The analysis is more delicate if we make use of the centroid principle. In order for this principle to play in the way corresponding to the point of view A (paragraph 3.3, the variables distinguish between the observations), it is convenient to multiply the $fv_{j,k}$ by $\sqrt{\lambda_k}$ or to divide the $fo_{i,k}$ by $\sqrt{\lambda_k}$. For the sake of simplicity, the second perspective has been chosen. It provides Figure 11. On this diagram, only the variables thrown off center have been represented. The observations are globally much more remote from the center than the variables (consequences of the centroid principle). It cannot happen that a variable is merged with an observation, since there is no species strictly confined to one line of the table. On the other hand, the few variables which are very close to certain periods allow for a rate of presence equal to 1, only for cruises of the given period. This is particularly the case of the variable associated with the species I7 (*Gymnodinium lohmani*) which displays a preference for the Summer period. Similarly, the species 28 (*Rhizosolenia fragilissima*) appears as very directly associated with the samples of September to November 76.

Figure 11 shows that the degree of association is high, insofar as some variables are found in remarkably remote positions. Within the same kind of idea, and we are getting closer to the process followed in the PCA, it can be verified on Figure 10 that these taxa are well "explained" by the plane of the axes I and II.

In order to follow the point of view B (Section 3.3:   the observations distinguish between the variables), the plots of observations and variables can be superposed after the $fo_{i,k}$ have been multiplied by $\sqrt{\lambda_k}$. This time the variables are found at the most distant positions from the origin. The example dealt with does not provide an interpretable diagram. The observations are precisely too much grouped towards the center for the plots to be exploited graphically. In fact, there is no observation which is sufficiently marked out by a variable. The centroid principle takes into account too many variables having very different positions.

4.4.3 Proximities between varibles. The plots of variables constituting optimal representations, and the χ^2 distance providing an obvious interest in the measuring of the links between variables, this analysis meets in a very favorable way the requirements mentioned in Section 3.4.1 (proximities between variables). Figure 8 makes clear the proximity between the variables associated respectively with *Gymnodinium lohmani* or *Nitzschia delicatissima*. It is the same for the taxa *Rhizosolenia stolterfothii, Biddulphia*

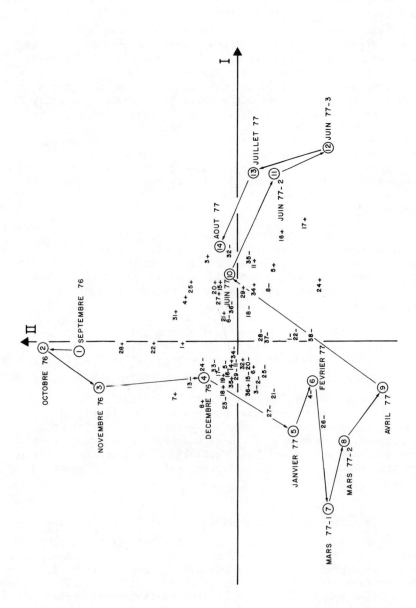

FIG. 11: *Annual cycle of phytoplankton at Paluel. Analysis of correspondences. Superposition of the variables and the observations requiring the barycentric principle.*

aurita and *Nitzschia sp.* The only problem concerning the example
dealt with comes from the fact that two variables being associated
with each taxon, we can perceive the proximities between variables
and not between taxa.

The second point of view, developed in Section 3.4.2
corresponds to bringing out indirect links in reference to the
factors. The direct interpretation is not possible as in the
PCA on the correlations, since the $fv_{j,k}$ are not correlations.
On the other hand, we could take an interest in Figure 10 within
the same perspective. The fact remains that a correlation is not
necessarily the best measure of a link between a quantitative
variable (a factor) and a logical variable.

5.   CONCLUSION

Basically ordination techniques imply that from the start
we take an interest in either the links between the observations,
or the links between the variables. Within the context of general
analyses with double weighting, the analyses can provide important
elements of information relative to the observations as well as to
the variables. Even when the study is based on the structure of the
observations, the general analyses with double weighting gives
extremely useful information about the variables. In most cases,
it is illusory to try to use a similar process for the interpretation of the observations and the variables in one single analysis.
Generally this constitutes an excessive, or even dangerous,
endeavour. Conversely, one should not neglect the interest of
the contributions linked with the variables and of plots of
variables in an analysis based on the links between observations.
Even then, it should be known exactly what they can precisely
tell, as regards precise questions that an ecologist is lead
to ask himself. This depends on the variants which are particular
to each factorial analysis, and particularly on the fact that the
variables have or have not the same norms and weights, and on the
fact that the observations are centered or not. If one is well
aware of the ecological aim which is pursued and of what the
different options defined mathematically can provide, important
progress is possible. This constitutes one of the main advantages
stemming from the general analysis with double weighting.

This appears as very important when the use of other ordination techniques is considered. Nonmetric techniques have obvious
advantages, because they are operative from mere pre-orders, and
are bound to be developed. On the other hand, they do not permit,
in an ordination of the observations, a detailed analysis of the
associated variables. This constitutes a major drawback. If we
refer to principal coordinate analysis, it is not possible to
study the variables delicately (absence of duality). These analyses

are not different from metric ordinations, justified by the concept
of optimal representation. The general analyses with double
weighting are more than mere metric ordinations. This is why as
regards the variables they have obvious advantages. The framework
is more restrictive, but the structures are richer mathematically.

ACKNOWLEDGEMENTS

This work would not have been seen through without the help
of those who gathered and studied the samples, and without the
fruitful collaboration of Madame Jézéquel and Mademoiselle
Derrien.

REFERENCES

Benzecri, J. P., ed. (1973). *L'Analyse des Données*. Tome I,
La Taxinomie. Tome II, *L'Analyse des Correspondances*
Dunod, Paris.

Blanc, F., Chardy, P., Laurec, A. and Reys, J. P. (1976). Choix
des métriques qualitatives en analyse d'inertie. Implications
en écologie marine benthique. *Marine Biology*, 35, 49-67.

Brillinger, D. R. (1975). *Time Series: Data Analysis and Theory*.
Holt, Rinehart & Winston, New York.

Caillez, F. and Pages, J. P. (1976). *Introduction à l'Analyse
des Données*. Smash, Paris.

Chardy, P., Glemarec, M. and Laurec, A. (1976). Application of
inertia method to Benthic Marine Ecology: practical implications of the basic options. *Estuarine Coastal Marine
Science*, 4, 170-205.

Dessier, A. and Laurec, A. (1978). Le cycle annuel du zooplancton
à Pointe-Noire (RP Congo). Description mathématique.
Oceanologica Acta 1(3), 285-304.

Fasham, M. J. R. (1977). A comparison of nonmetric multidimensional
scaling, principal components and reciprocal averaging for
the ordination of simulated coenoclines and coenoplanes.
Ecology, 58, 551-561.

Frontier, S. (1975). L'analyse factorielle est-elle heuristique
en ecologie du plancton? *Cah. ORSTOM. Sér. Océanogr.* XII,
77-81.

Gabriel, K. R. (1971). The biplot graphic display of matrices
with application to principal component analysis. *Biometrika*,
53, 453-467.

Gower, J. C. (1966). Some distance properties of latent roots
and vectors methods used in multivariate analysis. *Biometrika*,
53, 325-338.

Jolicoeur, P. and Mosimann, J. (1960). Size and variation in
the painted turtle. A principal component analysis.
Growth, 24, 339-354.

Kruskal, J. B. (1964a). Multidimensional scaling by optimizing
goodness of fit to a nonmetric hypothesis. *Psychometrika*,
29, 1-27.

Kruskal, J. B. (1964b). Nonmetric multidimensional scaling: a
numerical method. *Psychometrika*, 29, 115-129.

Laurec, A. (1979). *Analyse des données et modèles prévisionnels
en écologie marine*. Thèse d'Etat, Université d'Aix-Marseille.

Lebart, L. and Fenelon, J. P. (1971). *Statistique et Informatique
Appliquée*. Dunod, Paris.

Lebart, L., Morineau, A., and Tabard, N. (1977). *Techniques de
la Description Statistique*. Dunod, Paris.

Orloci, L. (1967). Data centering: a review and evaluation,
with reference to component analysis. *Systematic Zoology*,
16, 208-212.

Pielou, E. C. (1969). *An Introduction to Mathematical Ecology*.
Wiley, New York.

Prentice, I. C. (1977). Non-metric ordination methods in ecology.
Journal of Ecology, 65, 85-94.

Reys, J. P. (1976). Les peuplements benthiques (zoobenthos) de la
région de Marseille (France):   aspects méthodologiques de la
délimitation des peuplements par les méthodes mathematiques.
Marine Biology, 13, 123-134.

Roux, G. and Roux, M. (1973). A propos de quelques méthodes de
classification en Phytosociologie. In *L'Analyse des Données*,
J. P. Benzecri, ed. Tome I, p. 361-374.

Rickaert, M. (1979). Etude écologique du site de Paluel:
Phytoplancton. In *Rapport EDF*.

Verneaux, J. (1973). Cours d'eau de Franche-Comté (Massif du Jura) Recherches écologiques sur le réseau hydrographique du Doubs. Essai de biotypologie. *Ann. Sci. Univ. Besançon (Zool)*, 9, 1-257.

FOOTNOTE

(1) On a positive table, after centering the columns, some properties are kept, although the "positivity" is not kept in the transformed table. Let $\underset{\sim}{Y}$ be the original table, $\underset{\sim}{X}$ the table transformed, ym_j the mean of the column j in the original table.

$$fv_{j,k} = \frac{1}{\sqrt{\lambda_k}} \sum_i p_i\, x_{i,j}\, fo_{i,k} \;,$$

an ordinary transition formula. Let us replace $x_{i,j}$ by $y_{i,j} - ym_j$:

$$fv_{j,k} = \frac{1}{\sqrt{\lambda_k}} \sum_i p\ell_i\, x_{i,j}\, fo_{i,k}$$

$$= \frac{1}{\sqrt{\lambda_k}} \sum_i p\ell_i\, (y_{i,j} - ym_j)\, fo_{i,k}$$

$$= \frac{1}{\sqrt{\lambda_k}} \sum_i p\ell_i\, y_{i,j}\, fo_{i,k} - ym_j \sum_i p\ell_i\, fo_{i,k}$$

$\sum_i p\ell_i\, fo_{i,k} = 0$ since the transformed columns are centered.

Thus the transition formula can also be written:

$$fv_{j,k} = \frac{1}{\sqrt{\lambda_k}} \sum_i p\ell_i\, y_{i,j}\, fo_{i,k}$$

and we can refer to the fact that the $y_{i,j}$ are positive. This property is used in the analysis of correspondences, in the studies in the presence-absence (Blanc *et al.*, 1976). It can also be useful in the PCA. In ecological applications, and especially in ours, in fact, we very often start with a positive table, even if it is necessary to center later on.

[*Received June* 1979.]

L. Orloci, C. R. Rao, and W. M. Stiteler, (eds.),
Multivariate Methods in Ecological Work, pp. 175-189. All rights reserved.
Copyright © 1979 by International Co-operative Publishing House, Fairland, Maryland

SIZE AND SHAPE VARIABLES

JAMES E. MOSIMANN AND JAMES D. MALLEY

Division of Computer Research and Technology
National Institutes of Health
Bethesda, Maryland 20014 USA

SUMMARY. The concepts of shape and size are considered and
approaches to quantify these are discussed. The first approach
uses the components of squared distance. The second relies
directly on size and shape variables.

KEY WORDS. size variable, shape vector, Mahalanobis D^2, coefficient
of racial likeness, Penrose size and shape.

1. INTRODUCTION

The terms 'growth', 'size', and 'shape' occur commonly in the
biometric literature. As soon as a precise meaning is sought for
one of these terms one of two things is apt to happen. Either the
process of precision results in a definition devoid of the original
richness of the concept, or, a general definition results which
is so porous as to have little utility. The very richness of a
concept means that there are a number of precise, interesting, and
quite different possible meanings and these need to be distinguished.

Nearly thirty years ago, P. Weiss as quoted in Zuckerman
1950, p. 433) speaking of the term 'growth' said it well: "It is
not even," he says," a scientific term with defined and constant
meaning, but a popular label that varies with the accidental
traditions, predilections, and purposes of the individual or
school using it. It has come to connote all and any of these;
reproduction, increase in dimensions, linear increase, gain in
weight, gain in organic mass, cell multiplication, mitosis, cell
migration, protein synthesis, and perhaps more."

Weiss' remarks about the term 'growth' apply equally well today, and equivalent remarks can be made for the terms 'size' and 'shape'. In this paper we shall consider various implicit and explicit definitions of 'shape' and 'size' as they have occurred in the literature.

The qualitative utility of separate definitions of size and shape variables is strikingly illustrated by Gould's (1977) precise formulation of long-used morphogenetic concepts like 'neoteny', 'hypermorphosis', 'acceleration', etc. The quantitative utility of separate definitions is reflected in the works of various authors since Penrose (1954) (for example, see Jolicoeur and Mosimann, 1960; Rao, 1964; Mosimann, 1970; and Spielman, 1973). Two different approaches are found in these papers. The first approach is to partition a squared distance between two vectors into size and shape components after Penrose and Spielman. The second approach, after Jolicoeur and Mosimann, is to define directly size and shape variables applicable to a single vector of measurements. We discuss these directly defined variables, as well as show what directly defined variables are implicit in the distance decompositions.

The notion of geometric similarity of two individuals may or may not underly the definition of shape used in either approach. Our discussion will be based on the following important premise: two individuals which are geometrically similar with respect to a given set of measurements have the same 'shape' with respect to those measurements.

Our purpose here is not to discuss problems inherent in defining a suitable set of measurements to begin with (cf. Blum, 1973; Blum and Nagel, 1977; Bookstein, 1977). We take the vector of measurements as given. Even so, as we shall see, many problems remain.

2. SIZE AND SHAPE ARISING FROM PARTITIONS OF DISTANCE

In this section we discuss the partitioning of Euclidean distance into size and shape components, as done by Penrose (1954) for K. Pearson's (1926) coefficient of racial likeness, and by Spielman (1973) for Mahalanobis' D^2 (Rao, 1973). For the moment, like Gower (1972), we are not concerned with statistical notions, and we examine the geometrical properties of the distance partitions used by these authors.

We begin by establishing some notation. We consider k-dimensional Euclidean space, E^k, and let $\underset{\sim}{x}$ and $\underset{\sim}{y}$ ($\neq 0$) be two vectors of this space. We denote their difference $\underset{\sim}{x}-\underset{\sim}{y}$ by $\underset{\sim}{d}$. Their coordinates are written, respectively, $\underset{\sim}{x}' = (x_1,\cdots,x_k)$, $\underset{\sim}{y}' = (y_1,\cdots,y_k)$, and $\underset{\sim}{d}' = (d_1,\cdots,d_k)$. The distances of $\underset{\sim}{x}$ and $\underset{\sim}{y}$ from the origin are, respectively, a, b. The distance of $\underset{\sim}{x}$ from $\underset{\sim}{y}$ is c. Therefore, $\underset{\sim}{x}'\underset{\sim}{x} = a^2$, $\underset{\sim}{y}'\underset{\sim}{y} = b^2$, and $\underset{\sim}{d}'\underset{\sim}{d} = c^2$.

2.1 Penrose's Partition of Pearson's Coefficient of Racial Likeness. We now consider Penrose's partition of the squared distance of two vectors into size and shape components. The squared distance from $\underset{\sim}{x}$ to $\underset{\sim}{y}$ is

$$c^2 = \underset{\sim}{d}'\underset{\sim}{d} = \Sigma_1^k d_i^2.$$

If we let $\bar{d} = \Sigma_1^k d_i/k$, and note that the 'variance' of the d_i's is

$$\Sigma(d_i-\bar{d})^2/k = (\Sigma d_i^2/k) - \bar{d}^2 = c^2/k - \bar{d}^2,$$

we obtain, with Penrose, his partition (see Figure 1)

$$c^2/k = \bar{d}^2 + \Sigma(d_i - \bar{d})^2/k = \text{'size'} + \text{'shape'}.$$

Some remarks are in order. The difference vector of $\underset{\sim}{x}$ and $\underset{\sim}{y}$ will have a zero size component if and only if the totals of their respective coordinates are equal, $\Sigma x_i = \Sigma y_i$. This leads us to define two vectors as having the *same Penrose size* if $\Sigma x_i = \Sigma y_i$, and it is then natural to call Σx_i and Σy_i the Penrose sizes of $\underset{\sim}{x}$ and $\underset{\sim}{y}$ respectively. Then two vectors are of the same size t if and only if each lies in the plane which is both orthogonal to the equiangular line through the origin, and passes through the point $(t,0,\cdots,0)$. Similarly we define two vectors $\underset{\sim}{x}$ and $\underset{\sim}{y}$ as having the *same Penrose shape* if their difference vector has shape component zero; that is $\Sigma(d_i - \bar{d})^2 = 0$. Equivalently then $\underset{\sim}{d}' = \Delta(1,\cdots,1)$ for some Δ, so $x_i - y_i = \Delta$ $(i=1,\cdots,k)$. To have the same shape as $\underset{\sim}{x}$, the vector $\underset{\sim}{y}$ must lie on the equiangular ray originating from $\underset{\sim}{x}$.

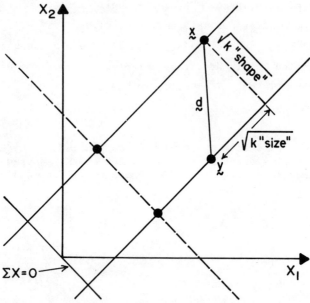

FIG. 1: *Geometry underlying Penrose's decomposition of squared distance into 'size' and 'shape' components.*

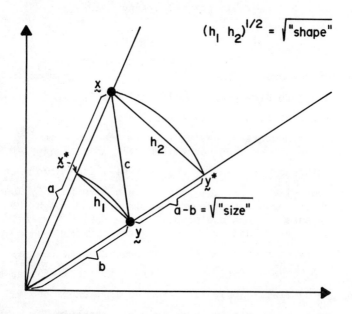

FIG. 2: *Geometry underlying Spielman's decomposition of squared distance into 'size' and 'shape' components.*

Two different vectors $\underset{\sim}{x}$, $\underset{\sim}{y}$ with the same Penrose shape cannot lie on the same ray from the origin, except when both lie on the equiangular ray through the origin. Finally, let $\underset{\sim}{x}$ and $\underset{\sim}{y}$ be of the same size $\Sigma\, x_i = \Sigma\, y_i$. Then the distance from $\underset{\sim}{x}$ to $\underset{\sim}{y}$ is all shape distance, and so

$$c^2 = \text{'shape'}.$$

Now define vectors $\underset{\sim}{u}$, $\underset{\sim}{v}$ as follows. Take $\alpha > 1$, and let $\underset{\sim}{u} = \alpha\, \underset{\sim}{x}$, $\underset{\sim}{v} = \alpha\, \underset{\sim}{y}$. Then $\underset{\sim}{u}$ and $\underset{\sim}{v}$ are on the same rays from the origin, respectively, as $\underset{\sim}{x}$ and $\underset{\sim}{y}$ but at a greater distance $\alpha\, a$ and $\alpha\, b$, respectively, than $\underset{\sim}{x}$ and $\underset{\sim}{y}$. Both $\underset{\sim}{u}$ and $\underset{\sim}{v}$ have the same size since $\Sigma\, u_i = \alpha\, \Sigma\, x_i = \alpha\, \Sigma\, y_i = \Sigma\, v_i$. The distance from $\underset{\sim}{u}$ to $\underset{\sim}{v}$ is also all shape distance, and so

$$c^2 = \text{'shape'}.$$

With increasing α, $\underset{\sim}{u}$, and $\underset{\sim}{v}$ move further out their respective $\underset{\sim}{x}$ and $\underset{\sim}{y}$ rays. The 'shape' component increases without bound, and although the proportions remain invariant, i.e., $u_i / \Sigma\, u_i = x_i / \Sigma\, x_i$, the shape component reflects the increasing divergence of the rays.

2.2 *Spielman's Partition of Mahalanobis'* D^2. We now consider Spielman's partition of the squared distance of two vectors in the D^2-discriminant space. We use the notation of the previous section and let $\underset{\sim}{x}$ and $\underset{\sim}{y}$ be vectors in this discriminant space so that c^2 is D^2, and a, the length of $\underset{\sim}{x}$. Spielman began with

$$c^2 = a^2 + b^2 - 2\, ab\, \cos\theta,$$

where θ is the angle between $\underset{\sim}{x}$ and $\underset{\sim}{y}$. Spielman takes $\underset{\sim}{x}$ and $\underset{\sim}{y}$ to be of the same size if they are of the same length (a=b). Now suppose, without loss, that $\underset{\sim}{x}$ is further from the origin than $\underset{\sim}{y}$ (a>b). Then the difference in size is (a-b), and the squared difference is $(a-b)^2$. The equation displayed above can be written,

$$c^2 = (a-b)^2 + 2\ ab - 2\ ab\ \cos\ \theta = (a-b)^2 + 2\ ab\ (1-\cos\ \theta)$$

$$= \text{'size'} + \text{'shape'}.$$

Spielman gives his shape component a geometric interpretation. Recall $a > b$. Let x^* be the vector of length b on the same ray as x, and let h_1 be the distance from x^* to y. Then

$$h_1^2 = 2\ b^2 - 2\ b^2\ \cos\ \theta = 2\ b^2(1-\cos\ \theta).$$

Correspondingly, let y^* be the vector of length a on the same ray as y, and let h_2 be the distance from y^* to x. Then

$$h_2^2 = 2\ a^2(1-\cos\ \theta),\quad \text{and}\quad h_1\ h_2 = 2\ ab\ (1-\cos\ \theta) = \text{'shape'}.$$

This is the square of the geometric mean of h_1 and h_2; that is of the two chords in Figure 2.

Two non zero vectors x and y have a zero Spielman shape component if and only if they lie on the same ray through the origin. Equivalently y has the same shape as x if $y = h\ x$ for some $h > 0$.

Finally, consider x and y with the same Spielman size $(a=b)$ so that their squared distance is entirely due to shape: $c^2 - 2\ a^2\ (1-\cos\ \theta)$. As before let $u = \alpha\ x$, $v = \alpha\ y$, for $\alpha > 1$. Then u and v have the same size $(\alpha a = \alpha b)$ and their distance is entirely due to shape. Since u and x are on the same ray, as also are v and y, then θ for u, v is the same as for x, y. Hence the squared distance for u, v is $\alpha^2\ c^2 = 2\ \alpha^2\ a^2(1-\cos\ \theta)$. As was the case with Penrose shape, the further out along the two rays we go, the greater the difference due to shape becomes.

3.   DIRECT DEFINITIONS OF SIZE AND SHAPE VARIABLES

In this section, we discuss direct definitions of size and shape variables. In one case the first principal component of a covariance matrix is used as a size variable (Blackith and Reyment, 1971, chapter 12; Rao, 1964; Jolicoeur and Mosimann, 1960). In the second instance, size and shape variables are defined directly from considerations of geometric similarity, Mosimann (1970).

3.1 The Principal Component Approach. We first consider 'size' and 'shape' variables arising from principal component analysis. Let $\underset{\sim}{X}$ be a k-dimensional random vector with finite covariance matrix Σ. Let $\underset{\sim}{U}$ be the orthonormal matrix of eigenvectors of Σ, $\underset{\sim}{U}'$ $\underset{\sim}{\Sigma}$ $\underset{\sim}{U}$ = diag (λ_i) = Λ. Let $\lambda_i > 0$ (i=1,$\cdots$,k). If some eigenvector of Σ has all positive direction cosines, then call this eigenvector $\underset{\sim}{a}'$ = $(a_1,\cdots,a_k)$, $a_i > 0$ (i=1,$\cdots$,k). Then define the 'size' of $\underset{\sim}{x} \in E^k$ to be $w_1 = \Sigma_1^k a_i x_i$. (Note there is no necessary reason to say that the λ_i associated with $\underset{\sim}{a}$ is the largest of the λ's.) Call the other (k-1) eigenvectors of Σ

$$\underset{\sim}{b}'_j = (b_{j1},\cdots,b_{jk}) \quad (j=2,\cdots,k).$$

These eigenvectors define (k-1) 'shape' variables

$$w_j = \Sigma_1^k b_{ji} x_i \quad (j=2,\cdots,k).$$

Jointly, they characterize the 'shape' of $\underset{\sim}{x}$. In terms of random variables we have $\underset{\sim}{W} = \underset{\sim}{U}' \underset{\sim}{X}$, where W_1 is size, $W_2,\cdots,W_k$ span a 'shape' space, the W_j's are mutually uncorrelated, and their defining coefficients are mutually orthogonal.

The geometric flavor of this approach is the same as that arising in Penrose's size and shape. Instead of the size of a vector $\underset{\sim}{x}$ being Σx_i as for Penrose, it is now $\Sigma a_i x_i$, and two vectors $\underset{\sim}{x}_1,\underset{\sim}{x}_2$ have the same *PC (Principal Component) size* if $\Sigma a_i x_{i1} = \Sigma a_i x_{i2}$. Also $\underset{\sim}{x}_2$ has the same *PC shape* as $\underset{\sim}{x}_1$ if $\underset{\sim}{x}_2$ lies on the line through $\underset{\sim}{x}_1$ perpendicular to the plane $\Sigma a_i x_i = 0$. Two vectors with the same PC shape cannot lie on the same ray from the origin, except in the case where both are on the ray whose direction cosines are $(a_1,\cdots,a_k)$.

Before proceeding we merely remark that canonical axes in discriminant analyses could be used in a directly parallel way by choosing one axis to be a 'size' axis (between populations), with the other canonical axes orthogonal to size being 'shape' axes. The essential geometric flavor would be the same as above. (See for example, Rees, 1969.)

3.2 Size and Shape Variables Based on Geometric Similarity. We
next consider definitions of size and shape variables developed by
Mosimann (1970). He considered only positive k-dimensional vectors, that is, vectors with every coordinate positive. (Denote
the set of such vectors by P^k.) Then two vectors $\underset{\sim}{x}_1, \underset{\sim}{x}_2$ of P^k
have the same shape if they are both on the same ray of P^k. A
size variable is generally defined to be any homogeneous function
of degree 1 from P^k to P^1, the positive real numbers. In
other words for every positive $\underset{\sim}{x}$,

$$G(\underset{\sim}{ax}) = aG(\underset{\sim}{x}),$$

for arbitrary $a > 0$. With this definition the following functions
from P^k are size variables (cf. Figure 3):

$$\Sigma\ x_i,\ (\Sigma\ x_i^2)^{1/2},\ (\Pi x_i)^{1/k},\ x_k,\ \text{Max}(x_i),\ x_1^2/x_2,$$

since they take only positive values and for each such G, $G(\underset{\sim}{ax})$
$= aG(\underset{\sim}{x})$.

For a particular size variable G, Mosimann defined a *shape
vector* as a function Z from P^k to P^k where

$$Z(\underset{\sim}{x}) = \underset{\sim}{x}/G(\underset{\sim}{x}),$$

for all $\underset{\sim}{x}$. The familiar direction cosines, $\underset{\sim}{x}/\sqrt{(\Sigma\ x_i^2)}$; proportions, $\underset{\sim}{x}/\Sigma\ x_i$; and ratios, $\underset{\sim}{x}/x_k$; are all examples of shape
vectors. Two vectors $\underset{\sim}{x}_1, \underset{\sim}{x}_2$ have equal shape vector $Z_G(\underset{\sim}{x}_1) =$
$Z_G(\underset{\sim}{x}_2)$ if and only if they both lie on the same ray; that is, if
and only if, $\underset{\sim}{x}_1 = a\ \underset{\sim}{x}_2$ for some $a > 0$.

A size variable can alternatively be defined by arbitrarily
choosing a point along each ray and assigning the value 1 to its
length. This specifies the 'unit sphere' for that size variable.
The size variable then is defined everywhere in P^k by the homogeneity property $G(\underset{\sim}{ax}) = aG(\underset{\sim}{x})$.

Mosimann's definition of shape variables reflects a concern
similar to Spielman's, whose notion of shape is related to that of
ray. Spielman's size, $\sqrt{(\Sigma\ x_i^2)}$, is one of Mosimann's directly
defined size variables. Yet it is not possible to associate a

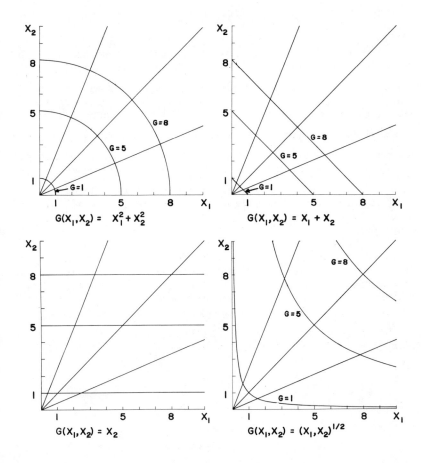

FIG. 3: Examples of size variables following Mosimann (1970).

single vector of shape with an individual ray using Spielman's decomposition of distance, since different shape components result for the same two rays, depending on distance along the ray. (Spielman discusses this problem with respect to his own data set (1973, p. 699).)

4.   GEOMETRIC SIMILARITY

Consider an individual organism on which k measurements have been taken. These are all positive and in the same physical

dimension (say lengths) expressed in the same units (say, mm.).
We write these as a vector $\underset{\sim}{x}' = (x_1, \cdots, x_k)$.

We have seen several serviceable definitions of the shape of
$\underset{\sim}{x}$. With respect to any definition of shape we believe it important and desirable that two organisms which are geometrically
similar, with respect to k measurements, should be recognized
as having the same shape. Thus if $\underset{\sim}{x_1}$ is the measurement vector
of one individual, and $\underset{\sim}{x_2}$ that of a second individual, then the
proportions among the measurements of the first, are the same as
those of the second if and only if $\underset{\sim}{x_1} = a\,\underset{\sim}{x_2}$, for some a > 0.
The two organisms are then geometrically similar with respect to
the k measurements. They lie on the same ray. Not all of the
shape variables discussed are derived from the notion of geometric
similarity.

A word of caution is required immediately. The geometric
similarity of two vectors need not be invariant under even simple
transformation of the data. At this point to situate ourselves
concretely we consider N turtles each with measurement vector
$\underset{\sim}{x_j}$ (j=1,$\cdots$,N), where the measurements (mm.) are length,
width, height, square root of cross sectional area,$\cdots$. Let the
population mean of the N turtles be $\underset{\sim}{\mu}$. Define the centered
variable

$$\underset{\sim}{z_j} = \underset{\sim}{x_j} - \underset{\sim}{\mu} \quad (j=1,\cdots,N).$$

Proportionality of the two $\underset{\sim}{z}$ vectors, $\underset{\sim}{z_i} = a\,\underset{\sim}{z_j}$ for some
a > 0, does not imply geometric similarity of turtles i and j.
Various situations are illustrated in Figure 4.

Another simple transformation of $\underset{\sim}{x}$ we should consider is
unequal scale changes. If we scale each measurement by its sample
standard deviation letting $w_i = x_i/s_i$ (i=1,$\cdots$,k), then statements of geometric similarity are not altered. More generally, no
nonsingular linear transformation on $\underset{\sim}{x}$ disrupts statements of
geometric similarity. Let $\underset{\sim}{A}$ be a k×k, nonsingular matrix; and
consider $\underset{\sim}{w} = \underset{\sim}{A}'\,\underset{\sim}{x}$. Turtles one and two are geometrically similar
if and only if $\underset{\sim}{x_1} = a\,\underset{\sim}{x_2}$, a > 0. This implies that $\underset{\sim}{w_1} = a\,\underset{\sim}{w_2}$
since $\underset{\sim}{w_1} = \underset{\sim}{A}\,\underset{\sim}{x_1} = a\,\underset{\sim}{A}\,\underset{\sim}{x_2} = a\,\underset{\sim}{w_2}$. Since $\underset{\sim}{A}^{-1}$ exists the converse
also follows and geometric similarity with respect to $\underset{\sim}{w}$ is
equivalent to that with respect to $\underset{\sim}{x}$.

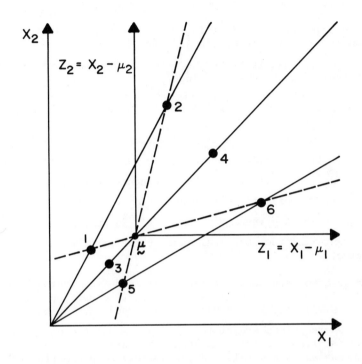

FIG. 4:   *The effect of centering at mean* $\underset{\sim}{\mu}$. *Turtles* 1 *and* 2
are geometrically similar with respect to (X_1, X_2) *as are turtles*
3 *and* 4 *and also turtles* 5 *and* 6. *Turtles* 1 *and* 6 *are*
not geometrically similar but lie on the same ray in the centered
space so $\underset{\sim}{z}_6 = a\, \underset{\sim}{z}_1$ *for some* $a \neq 0$. *Finally, turtles* 3 *and*
4 *are similar, and lie on the ray through* $\underset{\sim}{\mu}$. *Hence* $\underset{\sim}{x}_3 = b\, \underset{\sim}{x}_4$,
$b > 0$, *and* $\underset{\sim}{z}_3 = a\, \underset{\sim}{z}_4$ *for some* $a \neq 0$. *(Note this centering is*
at the mean $\underset{\sim}{\mu}$ *which is not the sample mean of these six points.)*

One consequence of this has relevance to D^2 analysis, where
we consider the shape $\underset{\sim}{z}$ obtained by rotation of the axes of $\underset{\sim}{x}$
with the orthonormal eigenvectors $\underset{\sim}{U}$ of a (common) within covariance matrix $\underset{\sim}{\Sigma}$ of full rank, with subsequent standardization by
$\Lambda^{-1/2}$, the inverse of the diagonal matrix of the square root of

the eigenvalues of $\underset{\sim}{\Sigma}$. Let $\underset{\sim}{z}_j = \underset{\sim}{\Lambda}^{-1/2} \underset{\sim}{U}' \underset{\sim}{x}_j$ (j=1,2). Then $\underset{\sim}{z}_1 = a \underset{\sim}{z}_2$ if and only if the two turtles are geometrically similar with respect to the measurements of $\underset{\sim}{x}$.

Of course centering in this D^2 space would disrupt the relation of rays with geometric similarity. Further let $\underset{\sim}{B}$, possibly not of full rank, be the between covariance matrix for the original measurements $\underset{\sim}{x}$. Then

$$\underset{\sim}{C} = \underset{\sim}{\Lambda}^{-1/2} \underset{\sim}{U}' \underset{\sim}{B} \underset{\sim}{U} \underset{\sim}{\Lambda}^{-1/2}$$

is the between covariance matrix of the measurements $\underset{\sim}{z}$ in the D^2 space. If we further rotate the $\underset{\sim}{z}$ axes with orthonormal eigenvectors, $\underset{\sim}{V}'$, of $\underset{\sim}{C}$ to obtain canonical axes, $\underset{\sim}{w} = \underset{\sim}{V}' \underset{\sim}{z}$, then geometric similarity with respect to $\underset{\sim}{w}$ is the same as that with respect to $\underset{\sim}{x}$ since $\underset{\sim}{V}' = \underset{\sim}{\Lambda}^{-1/2} \underset{\sim}{U}'$ is invertible. Again however centering the canonical axes at the grand mean would be disruptive.

As we have seen shape variables derived from Penrose and Principal Component analysis have no direct connection with geometric similarity in the measurement space $\underset{\sim}{x}$. In the special instance where points are closely grouped about the equiangular ray, and far from the origin so that rays are nearly parallel, then Penrose shape can relate approximately to geometric similarity. Similarly, comparable remarks depending on the orientation of the size axis can be made for Principal Component shape. However, it should be stressed that essentially Penrose and Principal Component shape in the $\underset{\sim}{x}$ space are unrelated to geometric similarity.

We now consider a logarithmic transformation of the measurement space $\underset{\sim}{x}$. Let $y_i = \log x_i$ (i=1,$\cdots$,k), with the vector of log measurements denoted by $\underset{\sim}{y}' = (y_1,\cdots,y_k)$. Mosimann, Malley, Cheever, Clark (1978) and Mosimann and James (in press) have used analyses in the log space to relate to hypotheses relating Mosimann's size and shape variables in the original $\underset{\sim}{x}$ measurement space. Penrose's size variable, applied to $\underset{\sim}{y}$, is $\Sigma_1^k y_i$ which divided by k given the log of the geometric mean size variable $(\Pi x_i)^{1/k}$ in the $\underset{\sim}{x}$ space. His shape in this space is directly related to geometric similarity; since two vectors $\underset{\sim}{y}_1$, $\underset{\sim}{y}_2$ are on the same line parallel to the equiangular line if and only if $\underset{\sim}{x}_1 = a \underset{\sim}{x}_2$. That is $\underset{\sim}{x}_1 = a \underset{\sim}{x}_2$, implies $\underset{\sim}{y}_1 = \underset{\sim}{y}_2 + \underset{\sim}{c}$, where $\underset{\sim}{c}' = (\log a)(1,\cdots,1)$, and conversely. Penrose's partition of distance in the log space is related to geometric similarity in the $\underset{\sim}{x}$ space.

The connection of principal component analysis of the covariance matrix of logs with size and shape has been noted since Jolicoeur (1963) and Mosimann (1970) discussed the issue. Here it suffices to say that while a PC size variable $\Sigma\, a_i\, y_i$ can always be converted to a log size variable by taking $\Sigma\, a_i = 1$, the PC shape space of the log variables cannot represent the log of a Mosimann shape vector unless $a_i = 1/k$ $(i=1,\cdots,k)$.

5. CONCLUSIONS AND SUGGESTIONS

To the researcher whose concept of shape is closely associated with that of geometric similarity, we offer the following suggestions. Consider a vector of positive measurements $(x_1,\cdots,x_k)$ all of the same physical dimensions expressed in the same units.

Given an interest in partitioning distance into size and shape components:

(1) If the interest is to partition distances in the space of the original measurements (or the D^2 space derived from them), use Spielman's partition, not that of Penrose. However, be aware that Spielman's shape component can reflect size as well as shape. Also be aware that, if the partitioning is done in the original space, that a specific size variable has been chosen; namely, the length $\sqrt{(\Sigma\, x_i^{\,2})}$. (If the partitioning is done in the D^2 space the size variable in that space is $\sqrt{(\Sigma\, z_i^{\,2})}$ which may be difficult to interpret for the original measurements.)

(2) If the interest is to partition distance in the space of the log measurements, use Penrose's partition, not Spielman's. However, be aware that the choice of the size variable in the original measurements is then the geometric mean, $(\Pi\, x_i)^{1/k}$.

Given an interest in the direct definition and study of size and shape variables:

(3) Do not use the principal component definition in the space of the original measurements, but use Mosimann's definitions which allow a wide range of size variables. Statistically, wherever it is feasible to assume a multivariate lognormal distribution for the original measurements, do so. Then carry out exact normal statistical procedures on the logs of the measurements as described by Mosimann and James (in press) and Mosimann, Malley, Cheever, and Clark (1978).

REFERENCES

Blackith, R. E. and Reyment, R. A. (1971). *Multivariate Morphometrics*. Academic Press, New York.

Blum, H. (1973). Biological shape and visual science (Part I). *Journal of Theoretical Biology*, 38, 205-287.

Blum, H. and Nagel, R. (1977). Shape description using weighted symmetric axis features. In *Proceedings IEEE Computer Society Conference in Pattern Recognition and Image Processing: RPI*. Troy, New York.

Bookstein, F. L. (1977). The study of shape transformation after D'Arcy Thompson. *Mathematical Biosciences*, 34, 177-219.

Gould, S. J. (1977). *Ontogeny and Phylogeny*. The Belknap Press of Harvard University Press, Cambridge, Massachusetts.

Gower, J. C. (1972). Measures of taxonomic distance and their analysis. In *The Assessment of Population Affinities in Man*, J. S. Weiner and J. Huizinga, eds. Oxford University Press, London.

Jolicoeur, P. (1963). The multivariate generalization of the allometry equation. *Biometrics*, 19, 197-499.

Jolicoeur, P. and Mosimann, J. E. (1960). Size and shape variation in the painted turtle. A principal component analysis. *Growth*, 24, 339-354.

Mosimann, J. E. (1970). Size allometry: Size and shape variables with characterizations of the lognormal and generalized gamma distributions. *Journal of the American Statistical Association*, 65, 930-945.

Mosimann, J. E. and James, F. C. (in press). New statistical methods for allometry with application to Florida red-winged blackbirds. *Evolution*.

Mosimann, J. E., Malley, J. D., Cheever, A. W., and Clark, C. B. (1978). Size and shape analysis of schistosome egg-counts in Egyptian autopsy data. *Biometrics*, 34, 341-356.

Pearson, K. (1926). On the coefficient of racial likeness. *Biometrika*, 18, 105-117.

Penrose, L. S. (1954). Distance, size, and shape. *Annals of Eugenics*, 18, 337-343.

Rao, C. R. (1964). The use and interpretation of principal component analysis in applied research. *Sanhkyā, Series A*, 26, 329-358.

Rao, C. R. (1973). *Linear Statistical Inference and its Applications*. Wiley, New York.

Rees, J. W. (1969). Morphologic variation in the mandible of the white-tailed deer (*Odocoileus virginianus*): A study of populational skeletal variation by principal component and canonical analysis. *Journal of Morphology*, 128, 113-130.

Spielman, R. S. (1973). Do the natives all look alike? Size and shape components of anthropometric differences among Yanomama Indian villages. *American Naturalist*, 107, 694-708.

Zuckerman, S. (1950). The pattern of change in size and shape. (In a discussion on the measurement of growth and form.) *Proceedings of the Royal Society B, Biological Sciences*, 137, 433-443.

[*Received July* 1978. *Revised January* 1979]

L. Orloci, C. R. Rao, and W. M. Stiteler, (eds.),
Multivariate Methods in Ecological Work, pp. 191-202. All rights reserved.
Copyright © 1979 by International Co-operative Publishing House, Fairland, Maryland

NON-LINEAR DATA STRUCTURES AND THEIR DESCRIPTION

LÁSZLÓ ORLÓCI

Department of Plant Sciences
University of Western Ontario
London, Ontario, Canada N6A 5B7

SUMMARY. This paper makes the point that the linear methods of data analysis can have no more than limited relevance to ecologists who traditionally deal with systems characterized by non-linearity.

KEY WORDS. vegetation, data, structure, linear, non-linear, method selection.

1. INTRODUCTION

The most commonly used methods of statistical analysis and inference routinely assume that the variate correlations are linear. Component analysis is a case in point. It assumes linear variate correlations and it relies on a linear condition for independence. But component analysis is not the only one of this kind. The multivariate analysis of variance, the different factor analysis methods, discriminant analysis, and the ordinations to which Seal (1963) referred as canonical analysis are all of the same kind in that they all rely on the linear measure $\underset{\sim}{S} = \underset{\sim\sim}{AA}'$ when they define variate relationships. The assumption of linear correlations cannot be made lightly in ecology, however, knowing that such kind of correlations are conditional on linear response and that this type of response is rarely found in ecological systems.

It is well known that much of the ecologically interesting information in the data is tied up with non-linear correlations.

192   L. ORLÓCI

And yet little has been done in the past to develop statistical
methods which can efficiently analyze non-linear correlations.
Even in the more recent past, based on the published record, it
appears that little productive time is spent by the most qualified of the Art to expand the theoretical base which underlies
non-linear statistical analysis and inference.

This is not to say that the field of non-linear multivariate
analysis is completely barren. But the efforts are in the main
heuristic and the methods developed are deterministic rather than
statistical. Ideally, a new statistical methodology should be
developed. This methodology should be able to do to non-linear
correlations what the conventional methods of statistics can do
to linear correlations. Such a methodology is not yet in sight.

In this paper, I will make the point that the linear methods
of data analysis have only limited relevance to ecologists and
that ideally the methods should not assume that the variate
correlations are of a specific kind. I shall be descriptive and
I shall present a personal view. The main topics of discussion
include: variate response, variate correlations, and method
selection. I shall begin with the description of a typical set
of vegetation data.

2. VEGETATION DATA

The information available about species performance, community
composition and ecological significance is carried by the data:

$$\begin{bmatrix} X_{11} & \cdots & X_{1n} \\ . & \cdots & . \\ X_{p1} & \cdots & X_{pn} \end{bmatrix} .$$

A representative row vector in such a set is the n-dimensional
descriptor of the response of a given species. The column
vectors are p-dimensional quadrat descriptors. The latter are
referred to by vegetation scientists as the relevés.

The physical properties which are the carriers of information
in the data exist in two basically different forms: (a) variation
within and covariation (correlation) among the species; (b) variation within and similarity among the relevés. Should the species
be uncorrelated, or should by chance the relevés be completely
dissimilar, no one should contemplate applying the methods of
multivariate analysis.

3. GENERAL CONCEPTS

Species respond to environmental influences by changing levels of performance. The manner of the response determines the type of correlations to be expected. That species response determines the species correlations has long been recognized (van Groenewoud, 1965). The relationship is such that the linear response determines linear correlations and the non-linear response brings about non-linear correlations.

To facilitate an orderly presentation, the following terms have to be defined: response space; sample space, and data structure.

Response space means the trajectory of the response (Y) as a function of the changing environmental influence (X). Figure 1a gives a simple example. In this two species respond, in a perfectly linear manner, to a given environmental influence. Both influence and response are continuous.

Sample space means a set of points and a given resemblance function. It is often described as a metric space (Figure 1b) in which a point cloud represents the joint distribution of responses. However, the sample space need not be pictured as such.

The manner in which the data points are arranged in sample space determines the *data structure*. In the example of Figure 1b the points form a single, straight cluster with increasing density toward the center.

A distinction should be made between continuous and disjoint sample spaces. It is convenient to consider the condition of continuity satisfied for the environmental influence X if its states $X_1, X_2, X_3, \cdots$ are distributed according to some continuous,

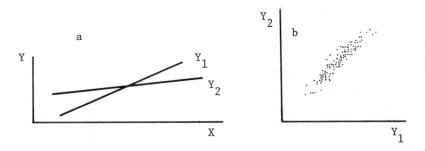

FIG. 1: Linear (a) response space (b) sample space.

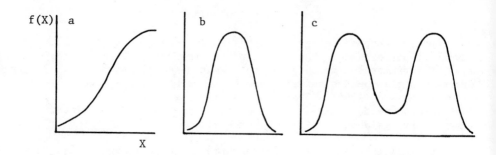

FIG. 2: Continuous distributions: a. *monotone,* b. *unimodal,* c. *bimodal.*

monotone or unimodal, distribution function F(X)   (Figure 2a,b).
The probability density is f(X) at state X. If the species
respond to a continuous influence, continuous response graphs
are expected. The example in Figure 1a pictures the linear case.
The sample space has a single, straight cluster of points (Figure
1b). Exceptions can be conceived in which the environmental
influence is continuous but the response is not. But for this to
happen would be unusual and it would probably indicate interference with the process of response by competition.

    The environmental influence is said to be discontinuous if
its distribution is strongly bimodal or multimodal (Figure 2c).
This means that specific states of the influential factor are
missing or infrequent. When this happens the response graphs
are disjoint (Figure 3a), and typically, the sample space has
several density phases (Figure 3b).

    A distinction has to be made between linear and non-linear
cases. If the condition of linearity is satisfied the graphs of
species response are straight lines (Figure 1a) and the clusters
of points in sample space are also straight (Figure 1b). Should

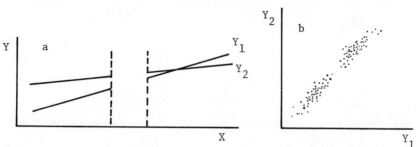

FIG. 3: Disjoint (a) response space (b) sample space.

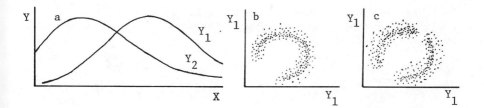

FIG. 4: (a) *Continuous, curved response.* (b) *Continuous, curved sample space.* (c) *Disjoint, curved sample space.*

the response be non-linear (Figure 4a), the clusters would also be curved (Figure 4b,c).

4. CLASSIFICATION OF DATA STRUCTURES

The data structure is expected to be linear and continuous when species respond in a completely linear manner to a continuous environmental influence:

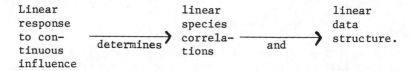

Since the correlations are linear, the covariance matrix is an appropriate descriptor of the data structure. Group mean vectors may be required to complete the descriptions in specific problems.

The data structure is expected to be linear and disjoint if the species respond in a linear manner to a discontinuous environmental influence:

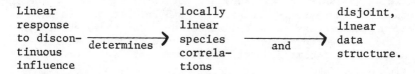

The description of a disjoint, linear data structure is best accomplished based on mean vectors and covariance matrices specific to the different clusters, and a distance matrix whose

elements define the relative placement of the tips of the mean
vectors (cluster centroids).

When the species respond in a non-linear manner to a
continuous environmental influence, the data structure is expected
to be curved and continuous:

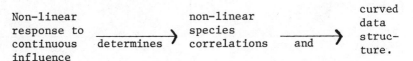

The   descriptions in this case may be based on different methods:

a. Metric co-ordinates or a distance matrix. From inspection
    of co-ordinates or distances the form of the non-linearity
    may however not readily appear.

b. Lines fitted to trace unidirectional trends to serve as axes
    of a new, curved reference system. These are revealing, but
    they may be difficult to obtain.

When the response is discontinuous and curved, a disjoint,
non-linear data structure is expected:

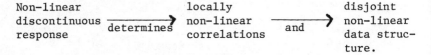

The description of such a data structure is difficult since it
requires isolation of the clusters, measurement of the spatial
relationships of the cluster centroids and the construction of
a non-linear reference system within the clusters.

5.   RECOGNITION OF DATA STRUCTURES

This should precede method selection. The data structure
may be very complex and difficult to identify. Test procedures
are unavailable now and they are not to be expected in the near
future. The need is however real and data structures have to be
identified to assist method selection. How to go about doing
this? The evidence bearing on the type of data structure in the
sample should come from different sources:

(a) The conditions in the survey site should be considered. If
     environmental extremes are found, such as for instance the
     presence of highly productive and very poor soils, and
     intermediates, a non-linear, continuous data structure is
     indicated. If only extremes are found but no intermediates,

the data structure is likely to be disjoint. Representative
(random) sampling is assumed.

(b) Scatter diagrams should be inspected for clues. Such clues
can however be misleading since the full dimensionality of
the samples space may exceed the dimensionality of the
scatter diagram.

(c) Exploratory analyses should be performed with lines and
surfaces fitted to the data to reveal linear or curved
intrinsic trends.

6. DATA STRUCTURE AND METHOD SELECTION

The data structure influences method selection. The contents
in the following table highlight this point:

Data structure	Family of method indicated
Continuous linear	Ordinations based on straight lines for axes to detect trends, e.g., component analysis.
Continuous curved	Ordinations based on curved axes: a. The functional form of response is specified: e.g., Gaussian ordination (Johnson, 1973; Gauch *et al.*, 1974; Ihm and van Groenewoud, 1975; Orlóci, 1978). b. The functional form of the response is not specified: e.g., Kruskal's method (Kruskal and Carmone, 1971; Orlóci, 1978), continuity analysis (Shepard & Carroll, 1966; Sammon, 1969; Noy-Meir, 1974), polynomial ordination (McDonald, 1962, 1967; Phillips, 1978).
Disjoint linear or curved	Cluster analysis to isolate density phases; ordination based on straight lines or curves as axes to detect trends.

7. LINEARIZATION OF DATA STRUCTURES

Different transformations and other manipulations were proposed from time to time to linearize the structure of ecological
data. Van Groenewoud (1965) has sought the solution in restricting the environmental variation of the sample. Others (e.g.

Noy-Meir and Whittaker, 1977) found this a rather unworkable
proposition. Swan (1970) suggested the replacement of zeros
with specified non-zero values while Feoli Chiapella and Feoli
(1977) used log transformation and new quadrat descriptors.
Van der Maarel (1972) proposed the use of variables whose response
to the environmental influence is known to be linear or at least
monotone. Others (e.g., Noy-Meir, 1971; Beals, 1973; Orlóci,
1978) explored the potential of transformations incorporated in
the resemblance functions.

While the suggested manipulations help in the cases of
moderate non-linearity, none are very helpful when the data
structure is strongly curved. The solution of the problem indeed
rests with the more recently developed methods of ordination which
unfold the data structure on suitably chosen curved axes.

8. SELECTED NON-LINEAR METHODS

Ideally, descriptions are sought which are efficient.
Consider a spiral imbedded in a space of 3 dimensions. Any three
perpendicular axes will give a complete description. Yet none
will be efficient since three axes are used where the trend itself
is unidirectional.

Methods have been proposed which extract axes which represent,
in some specified way, efficient descriptors of curved data structures. Some of these methods assume the type of species response
while others make no assumptions. The methods represent a few
distinct cases. Examples are given below:

Case 1. Species response is Gaussian. Its average magnitude is

$$Y_{ij} = e^{-(X_{ij}-a_i)^2/2}$$

in a quadrat j where the level of the environmental influence
associated with gradient i is X_{ij}. The average response to
the influence of gradient i is maximal at level a_i.

Based on the Gaussian assumption the quantity

$$S_{ijk} = e^{-(X_{ij}-X_{ik})^2/4}$$

(cf. Orlóci, 1978) is an admissible measure of similarity
between quadrat j and k, and

$$d(j,k|i) = [2(1-S_{ijk})]^{1/2}$$

$$= \left\{ \frac{1}{\sqrt{\pi}} \int_{-\infty}^{\infty} \left(e^{-(X-a_i)^2/2} - e^{-[X+|X_{ij}-X_{ik}|-a_i]^2/2} \right)^2 dX \right\}^{1/2}$$

is the chord distance of the two quadrats. The overall distance, with respect to t separate gradients, is

$$d(j,k) = \frac{1}{t} \left(\sum_{i=1}^{t} d^2(j,k|i) \right)^{1/2}.$$

What has so far been given accords with what Gauch (1973) has described. But how to imbed $d(j,k)$ in an ordination algorithm? We note that the X_{ij} are the unknowns. Let the ith gradient be identified as the ith ordination axis of the Kurskal and Carmone (1972) method. Sets of ordination co-ordinates can be found, given by $X_{i1}, \cdots, X_{in}$ for n quadrats on each of $i = 1, \cdots, t$ axes, such that the divergence of the $d(j,k)$ values and the observed compositional distances $\delta(j,k)$ of the quadrats is minimized. The algorithm can be simply stated:

(a) Specify t the number of ordination axes to be extracted.

(b) Select t sets of n arbitrary numbers. These will serve as first-order approximations of the unknown X_{ij}.

(c) Compute $d(j,k)$ values and determine their divergence, σ, from the observed $\delta(j,k)$ values.

(d) Change the X_{ij} values a little to reduce σ.

(e) Iterate through steps 3 and 4 until σ is stabilized or a desired degree of precision is reached. The definition of σ is arbitrary. One possibility is $\sigma = 1-\rho^2$ where ρ measures the correlation of the $d(j,k)$ and $\delta(j,k)$ values.

The Gaussian assumption underlies at least three of the recently proposed algorithms. Two of these (Johnson, 1973; Gauch, Chase, and Whittaker, 1974) are iterative and the third (Ihm and van Groenewoud, 1975) is based on eigenanalysis.

Case 2. The functional form of the curvature of the data structure is known, but no assumptions are made about species response. A curved line S is fitted. The ordination co-ordinate X_j of quadrat j is the distance of the projection of j on S measured from a given origin. An algorithm has been described by Phillips (1978).

Case 3. No assumptions are made for species response or for the curvature of the data structure. The algorithms proposed are iterative and they seek descriptions on the basis of co-ordinate sets which yield ordination distances for quadrats which are maximally continuous with their compositional distances. Different continuity measures were used by different authors within similarly formulated algorithms (e.g., Shepard and Carroll, 1966; Sammon, 1969).

Case 4. When the data structure is disjoint and curved, the algorithm that yields the description of the curvature must first be able to detect the clusters. One such algorithm, proposed by Sneath (1965), finds groups and then it fits curves to the groups.

9. DISCUSSION

It is clear that the problem of describing non-linear data structures may entail a cluster analysis to isolate groups and an ordination to unfold the data structure on axes. The axes may be curved such as in the Phillips (1978) algorithm, or they may be straight-line such as in the McDonald (1967) algorithm. Whether curved or straight-line the axes are lines fitted to the data which satisfy the stated conditions of optimality.

Some of the methods may not help to linearize the data structure to the extent required to render the analysis of the sample possible based on linear methods. Better results can be expected when the analysis is based on ordination. Some of the methods assume that the exact functional form of the species response is known (Johnson, 1973; Gauch, Chase and Whittaker, 1974; Ihm and van Gorenewoud, 1975). Some may incorporate assumptions about the observed data structure and the ordination configuration of the sample points (e.g., McDonald, 1967; Kruskal and Carmone, 1972) or the type of curvature in the data structure (Phillips, 1978). Most algorithms are iterative and they rely on stress functions to monitor success at each step in the iterations.

It seems quite appropriate to suggest that the fewer the assumptions the more it is likely that the algorithm will appeal to ecologists. Encouraging results can be reported with the method of continuity analysis (cf. Noy-Neir, 1974) and the Kruskal and Carmone scaling technique (Austin, 1976) neither of which require assumptions about the species response.

Whereas the need for a statistical methodology that can efficiently handle non-linear data structures has been recognized for some time, the efforts, on the whole, were indeed heuristic and the methods produced rarely incorporated the stochastic aspects. The Gauch, Chase, and Whittaker (1974) algorithm is an example of the deterministic approach and the Johnson (1973) algorithm for the statistical approach in which elements of a probabilistic model are recognizable. It seems though that since the multivariate normal assumption is not applicable where a non-linear data structure is known to exist, the incorporation of the notion of probability in the algorithms will be difficult. Axiomatic developments will also be difficult, for in the case of each sample a different underlying distribution may have to be assumed. In the face of such difficulties, the heuristic approach, utilizing stochastic experiments and stressing local relevance, must be appealing.

REFERENCES

Beals, E. W. (1973). Ordination: Mathematical elegance and ecological naivete. *Journal of Ecology*, 61, 23-35.

Feoli Chiapella, L. and Feoli, L. (1977). A numerical phytosociological study of Majella's summits - a multivariate approach. *Vegetatio*, 34, 21-39.

Gauch, H. G. (1973). The relationship between sample similarity and ecological distance. *Ecology*, 54, 618-622.

Gauch, H. G., Chase, G. B., and Whittaker, R. H. (1974). Ordination of vegetation samples by Gaussian species distributions. *Ecology*, 55, 1382-1390.

Groenewoud, H. van (1965). Ordination and classification of Swiss and Canadian coniferous forests by various biometric and other methods. *Ber. geobot. Inst. ETH. Stiftg. Rübel, Zürich*, 36, 28-102.

Ihm, P. and van Groenewoud, H. (1975). A multivariate ordering of vegetation data based on Gaussian type gradient response curves. *Journal of Ecology*, 63, 767-777.

Johnson, R. (1973). *A study of some multivariate methods for the analysis of botanical data.* Ph.D. thesis, Utah State University, Logan, Utah.

Kruskal, J. B. (1964a). Multidimensional scaling by optimizing goodness of fit to a nonmetric hypothesis. *Psychometrika*, 29, 1-27.

Kruskal, J. B. (1964b) Nonmetric multidimensional scaling: a numerical method. *Psychometrika*, 29, 115-129.

202   L. ORLÓCI

Kruskal, J. B. and Carmone, F. (1971). How to use the M-D-SCAL (Version 5M) and other useful information. (Mimeographed.) Bell Telephone Laboratories, Murray Hill, New Jersey, and University of Waterloo, Waterloo, Ontario, Canada.

Maarel, E. van der. (1972). Ordination of plant communities on the basis of their plant genus, family, and order relationships. In *Grundfragen und Methoden in der Pflanzensoziologie*, E. van der Maarel and R. Tüxen, eds. W. Junk, The Hague. 183–206.

McDonald, R. P. (1962). A general approach to nonlinear factor analysis. *Psychometrika*, 27, 397–415.

McDonald, R. P. (1967). Numerical methods for polynomial models in nonlinear factor analysis. *Psychometrika*, 32, 77–112.

Noy-Meir, I. (1971). Multivariate anlaysis of the semi-arid vegetation in southeastern Australia: Nodal ordination by component analysis. In *Quantifying Ecology*, N. A. Nix, ed. *Proceedings of the Ecological Society of Australia*, 6, 159–193.

Noy-Meir, I. (1974). Cantenation: Quantitative methods for the definition of coenoclines. *Vegetatio*, 29, 89–99.

Noy-Meir, I. and Whittaker, R. H. (1978). Recent developments in continuous multivariate techniques. In *Ordination of Plant Communities*, R. H. Whittaker, ed. W. Junk, The Hague. 337–378.

Orlóci, L. (1978). *Multivariate Analysis in Vegetation Research*, 2nd ed. W. Junk, The Hague.

Phillips, D. L. (1978). Non-linear ordination: Field and computer simulation testing of a new method. *Vegetatio*, 37, 43–51.

Sammon, J. W. (1969). A nonlinear mapping for data structure analysis. *IEEE Transactions on Computers*, Vol. C-18, 401–409.

Shepard, R. N. and Carroll, J. D. (1966). Parametric representation of nonlinear data structures. In *Multivariate Analysis*, P. R. Krishnaiah, ed. Academic Press, London, 561–592.

[*Received July* 1978. *Revised February* 1979]

L. Orloci, C. R. Rao, and W. M. Stiteler, (eds.),
Multivariate Methods in Ecological Work, pp. 203-209. All rights reserved.
Copyright © 1979 by International Co-operative Publishing House, Fairland, Maryland

GENERALIZED STRATEGY FOR HOMOGENEITY-OPTIMIZING HIERARCHICAL
CLASSIFICATORY METHODS

JÁNOS PODANI

Research Institute for Botany
Hungarian Academy of Sciences
2163 Vacratot, Hungary

SUMMARY. A new scheme, similar to the route-optimizing strategy
of Lance and Williams (1966), is proposed for homogeneity-
optimizing sorting procedures. Cluster homogeneity is defined
in three ways and three algorithms compatible with the scheme
are briefly discussed.

KEY WORDS. classification, sorting, strategies, clusters,
homogeneity

1. INTRODUCTION

Some cluster analytical procedures used frequently in
mathematical ecology and numerical taxonomy start with a
resemblance matrix between entities and do not require the
initial data. These methods have many computational advantages.
The well-known hierarchical and agglomerative algorithms of this
type have been called 'combinatorial' by Lance and Williams
(1967) and reviewed by Cormack (1971). A basic problem of these
strategies is the definition of inter-cluster similarity, distance,
or dissimilarity. Lance and Williams (1966) gave a recurrence
formula to compute the dissimilarity between group z_h and group
z_{ij} obtained by the fusion of groups z_i and z_j :

$$d_{h(ij)} = \alpha_i d_{hi} + \alpha_j D_{hj} + \beta d_{ij} + \gamma |d_{hi} - d_{hj}| . \qquad (1)$$

The values of parameters α, β, and γ are determined by the

nature of the strategy used (see Cormack, 1971). Algorithms
satisfying relation (1), however, optimize the route by which
the clusters are formed, such as in single linkage and complete
linkage, the sum of squares agglomeration and the centroid
sorting method. Internal structure or some kind of homogeneity
of clusters is taken into consideration by the sum of squares
agglomeration method only (see Ward, 1963; Orloci, 1967). This
strategy, however, minimizes the increase of the within-group
sum of squares so that the homogeneity of the new clusters
is not necessarily optimal. Contrary to these route-optimizing
strategies, it may be desired to optimize the homogeneity of the
new clusters.

In the classification of plant and animal individuals,
communities or other entities, the primary aim is to produce
groupings whose homogeneity is as high as possible. Hierarchies
obtained by even exact and well-defined route-optimizing
procedures may be of secondary significance. It is reasonable
to make a further distinction among the classificatory
techniques. The family of hierarchical and agglomerative
methods can be divided into two groups: the route-optimizing
(called r-hierarchical) and the homogeneity-optimizing (called
h-hierarchical) procedures. It will be shown that some of
the h-hierarchical strategies are 'combinatorial.' This rather
ambiguous term is used and accepted in this paper for lack of
a better terminology.

The concept of homogeneity may of course be defined in a
number of different ways. I shall use three definitions to
illustrate my general classification scheme for h-hierarchical
and combinatorial procedures. It is worth mentioning that most
fruitful information-theroretical definitions are not
compatible with any combinatorial model.

2.   A NEW GENERAL SCHEME AND ITS APPLICATION

2.1 The Basic Equation. Let us assume that in the course of
the computations we have already three clusters denoted by
z_h , z_i , and z_j with the number of elements respectively
n_h , n_i , and n_j . Let w_h , w_i , and w_j denote the
homogeneity or heterogeneity of the clusters and let w_{hi}
denote the homogeneity of group z_{hi} obtained by the fusion of
z_h and z_i (see Sections 2.2, 2.3, 2.4 for definitions of
cluster homogeneity). Thus we have the following semi-matrix,

$$
\underset{\sim}{W} = \begin{bmatrix} w_h & w_{hi} & w_{hj} \\ & w_i & w_{ij} \\ & & w_j \end{bmatrix} \quad ,
$$

and vector

$$
\underset{\sim}{N} = [n_h, \ n_i, \ n_j] \quad .
$$

In the case of homogeneity, let w_{ij} be the greatest value in the upper triangular portion of $\underset{\sim}{W}$. Then we amalgamate groups z_i and z_j to form a new group z_{ij}. After this we can compute w_{hij} from the pre-existing homogeneity measures in $\underset{\sim}{W}$ and values of $\underset{\sim}{N}$ using the following formula,

$$
w_{hij} = \alpha_i w_{hi} + \alpha_j w_{hj} + \beta w_{ij} + \gamma_h w_h
$$

$$
+ \ \gamma_i w_i + \gamma_j w_j \quad . \tag{2}
$$

If heterogeneity measures are given, the same relation holds. The values of the parameters for three h-hierarchical strategies may be found in Table 1.

Computations by the h-hierarchical and combinatorial clustering methods are based on the values of inter-entity matrix $\underset{\sim}{W}$, after calculation of which the original data need not be retained in the memory of the computer. Application of equation (2) differs from that of equation (1) since the values of the diagonal of $\underset{\sim}{W}$ are of importance. Strategies compatible with equation (2) are given below.

2.2 *Optimization of Dispersion within New Clusters.* This strategy is the h-hierarchical version of the sum of squares agglomeration method (Ward, 1963; Orloci, 1967; Wishart, 1969). The sum of squared distance from the centroid within cluster z_h is the measure of z_h's heterogeneity and is denoted by q_h. This quantity can be calculated from the distances between entities,

TABLE 1: *Parameters for three homogeneity-optimizing strategies.* $(n. = n_h + n_i + n_j)$

Name	α_i	α_j	β	γ_h	γ_i	γ_j
Edge-density	$\dfrac{(n_h+n_i)(n_h+n_i-1)}{n.^2 - n.}$	$\dfrac{(n_h+n_j)(n_h+n_j-1)}{n.^2 - n.}$	$\dfrac{(n_i+n_j)(n_i+n_j-1)}{n.^2 - n.}$	$\dfrac{n_h^2-n_h}{n.^2-n.}$	$\dfrac{n_i^2-n_i}{n.^2-n.}$	$\dfrac{n_j^2-n_j}{n.^2-n.}$
Dispersion	$\dfrac{n_h+n_i}{n.}$	$\dfrac{n_h+n_j}{n.}$	$\dfrac{n_i+n_j}{n.}$	$\dfrac{n_h}{n.}$	$\dfrac{n_i}{n.}$	$\dfrac{n_i}{n.}$
Average Dispersion	$\dfrac{(n_h+n_i)^2}{n.^2}$	$\dfrac{(n_h+n_j)^2}{n.^2}$	$\dfrac{(n_i+n_j)^2}{n.^2}$	$\dfrac{n_h^2}{n.^2}$	$\dfrac{n_i^2}{n.^2}$	$\dfrac{n_j^2}{n.^2}$

$$q_h = \frac{\sum\limits_{i=1}^{n_h} \sum\limits_{j=1}^{n_h} d_{ij}^2}{2n_h} , \qquad (3)$$

where d_{ij} denotes the distance between entities e_i and e_j. The analysis starts with matrix $\underset{\sim}{Q} \equiv \{q_{ij}\}$, in which

$$q_{ij} = d_{ij}^2/2 . \qquad (4)$$

2.3 Optimization of Average Dispersion within New Clusters.
This is an improved version of the previous procedure. Let us assume that $q_h = q_i$ such that $n_h > n_i$. Cluster z_h is obviously more compact, therefore less heterogeneous than cluster z_i, thus measuring cluster heterogeneity by the average dispersion seems to be reasonable.

An element of the starting matrix is

$$q_{ij} = d_{ij}^2/4 . \qquad (5)$$

Distance between entities may be defined by the Euclidean distance, such as the chord distance (Orloci, 1967), or other standardized measures in both dispersion-minimizing strategies. These methods are equally applicable to binary and quantitative data.

2.4 Optimization of Edge Density in Subgraphs Representing New Clusters. This strategy (Podani, 1978) is based on graph theoretical considerations and is applicable to binary data only. Let $\underset{\sim}{E} \equiv \{e_i\}$ be the set of entities to be classified, $\underset{\sim}{A} \equiv \{a_k\}$ be the set of attributes describing e_i, and m be the number of attributes. Let, further, $\underset{\sim}{R} \equiv \{r_k\}$ be the set of symmetric relations between entities such that relation $e_i r_k e_j$ holds if e_i agrees with e_j with respect to attribute a_k (joint presence or joint absence). Thus we have an undirected graph $\underset{\sim}{G}$ in which vertex g_i represents e_i and the edges symbolize the existing relations between entities. In this way the maximum number of edges connecting any two vertices is m.

The homogeneity of cluster z_h represented by subgraph

$G_{\sim h}$ may be measured by the edge-density of $G_{\sim h}$. This quantity can be calculated according to ψ_h ,

$$\psi_h = 2 \; \frac{\text{number of edges in } G_{\sim h}}{m \, n_h \, (n_h - 1)} \tag{6}$$

The edge-density of $G_{\sim h}$ may also be determined using the following formula:

$$\psi_h = 1 + \frac{2n_h}{m(n_h - 1)} \; \sum_k \hat{p}_k (\hat{p}_k - 1) \; , \tag{7}$$

where $\hat{p}_k$ is the estimated probability or relative frequency of the presence of attribute a_k in z_h , $0 \le \psi_h \le 1$. If $\psi_h = 1$ then the homogeneity of z_h is maximal. The minimum value of ψ_h is, however, greatly affected by n_h such that min $\psi_h = 0$ if and only if $n_h = 2$. If $n_h > 2$, there will be necessary joint presences and absences in z_h , therefore the edge-density of $G_{\sim h}$ must be greater than zero. The possible minimum of ψ_h can be calculated using the following formulae:

$$\min \psi_h = \frac{\frac{n_h}{2} - 1}{n_h - 1} \tag{8}$$

for even values of n_h , and

$$\min \psi_h = \frac{\frac{n_h}{2} - 1 + \frac{1}{2n_h}}{n_h - 1} \tag{9}$$

for odd values of n_h . This property may or may not be considered in the construction of a sorting algorithm but the strategy is combinatorial in the latter case.

The cluster analysis starts with a similarity matrix $S_{\sim}$

computed based on the coefficient of Sokal and Michener (1958) given by

$$S_{ij} = (a+d)/(a+b+c+d) ,\qquad(10)$$

where the symbols are those regularly used in 2×2 contingency tables. Index (10) is the special case of expression (6) for $n_h = 2$.

ACKNOWLEDGEMENTS

The author is grateful to P. Juhász-Nagy and Z. Szőcs for their helpful suggestions.

REFERENCES

Cormack, R. M. (1971). A review of classification. *Journal of the Royal Statistical Society, Series A*, 134, 321-353.

Lance, G. N. and Williams, W. T. (1966). A generalized sorting strategy for computer classifications. *Nature*, 212, 218.

Lance, G. N. and Williams, W. T. (1967). A general theory of classificatory sorting strategies. I. Hierarchical systems. *Computer Journal*, 9, 373-380.

Orloci, L. (1967). An agglomerative method for classification of plant communities. *Journal of Ecology*, 55, 193-205.

Podani, J. (1978). *Hierarchical classificatory methods for the analysis of binary ecological data*. Ph.D. thesis, Eötvös University, Budapest.

Sokal, R. R. and Michener, C. D. (1958). A statistical method for evaluating systematic relationships. *University of Kansas Science Bulletin*, 38, 1409-1438.

Ward, J. H. (1963). Hierarchical grouping to optimize an objective function. *Journal of the American Statistical Association*, 58, 236-244.

Wishart, D. (1969). An algorithm for hierarchical classifications. *Biometrics*, 25, 165-170.

[*Received June* 1978. *Revised January* 1979]

L. Orloci, C. R. Rao, and W. M. Stiteler, (eds.),
Multivariate Methods in Ecological Work, pp. 211-235. All rights reserved.
Copyright © 1979 by International Co-operative Publishing House, Fairland, Maryland

MULTIVARIATE ANALYSIS IN STATISTICAL PALEOECOLOGY

R. A. REYMENT

Uppsala University
Box 558
S-751 22 Uppsala, Sweden

SUMMARY. Many methods used in modern statistical ecology can be
readily adapted for paleoecological work. Other problems are
peculiar to paleoecology, particularly the analyses of the post-
mortem orientation of fossils. The time-component of paleontology
provides statistical paleoecology with its particular mark, and it
is certainly the main source of problems. The technique of
canonical correlation has been found to be of great value in the
analysis of time-ordered observations on sets of variables. The
'scores' for individual partitioned observational vectors may be
used for producing an *ecolog*. An alternative ecolog may be made
from the first principal coordinates of an association matrix of
Pythagorean distances. The short normal electrical resistivity
log of borehole analysis proves to be a useful indicator of environmental fluctuations such as may be recorded in the electrically
measurable properties of the sediment. In the case-study presented
here, the morphometric variational pattern of the organism studied
(*biolog*) closely follows the oscillations of the short normal log.
New statistical results for the analysis of square asymmetric
matrices have been applied to sedimentary environments in the
Mississippi Delta.

KEY WORDS. paleoecology, multivariate, canonical correlation,
ecology, biology.

1. INTRODUCTION

The fundamental differences between the study of the ecologic
relationships of living organisms and those of the geologic past

have long been recognized and, perhaps, no better formulated than
in the writings of Johannes Walther of more than three fourths of
a century ago. We may think that ecologic problems are difficult
to represent in mathematical terms, but the difficulty is compounded
for fossils as we have not only the biological interactions to
consider but also the manifold of geological agencies, reworking,
diagenesis, transport, and *post-mortem* deformation.

For most purposes, a statistical paleoecologic analysis will
have to be considered at two main levels: (i) The most adequate
and relevant of the extant models or methods of statistical ecology
applicable to the problem; (ii) Special methods of geostatistics
designed to give the approximate analysis obtained under (i) more
substance and geologic meaning. There can be no simple approach
of the 'problem-solution' kind as, by its very nature, paleoecology
is a very complex subject.

In Reyment (1971), I was mainly concerned with fitting paleoecologic problems into the existing framework of statistical
ecology. Obviously, this is an approach which must be variably
successful, depending on how near a particular problem approximates
to the type situation, and how slight the geologic effects are.

2.   ADAPTATION OF ECOLOGIC MODELS

In keeping with accepted terminology in ecology, one speaks,
analogously, of paleoautecology and paleosynecology. The quantitative aspects of these concepts in paleoecology are discussed in
Reyment (1971). Here, I shall only briefly review some of the main
features of the topics treated in that text.

2.1 Orientation Analysis. An important area of statistical
paleoecology is that of the quantitative study of the orientation
of fossils. Two main problems belong here: (1) the orientation
of fossils *in situ*, and (2) the orientations of transported fossils.
The analysis of the former may yield significant information on the
mode of life of the organism or organisms involved. The latter
can provide us with details concerning the current systems pre-
vailing during the time at which the biological flotsam and jetsam
was stranded.

As regards statistical theory for these analyses, we are in a
strong position today, thanks to the developments of the mathematical
analysis of geomagnetism. The entire field of directional statistics can be taken over without modification (cf. Mardia, 1972).

Neoecology does not have the same interest in analysis of
directions, although I can conceive of several situations in which

this type of analysis could be put to more expert use than has
hitherto been done. The most rewarding area of research in
paleoecology has been that of the analysis of the dispersal of
cephalopod shells, after the death of the organism. The essentially
trivariate nature of the majority of orientational problems in
paleoecology remains to be fully exploited, and almost all studies
made are based on the circular distribution.

2.2 Population Dynamics. Population dynamics is one of the main
field of activity of the neo-ecologist. A very large part of the
books on statistical ecology by E. C. Pielou is concerned with
this aspect of the subject. For understandable reasons, population
dynamics cannot be given the same prominence in statistical paleoecology.

Nonetheless, in favorable situations, it is possible to
develop an approximate analysis using a classical population dynamic
approach. Micropaleontology (the study of shell-bearing fossil
microorganisms) offers opportunities as complete growth sequences of
ostracods may occur in sediment that has not been reworked. In
fact, the preparation of a life table (Reyment, 1971, p. 112) for
ostracods of a micropaleontologic sample may be used with great
effect as a means of judging whether a deposit has been reworked
or not, a secondary outcome of the study. For example, life tables
have been prepared for fossil pelecypods, Pleistocene bears, and
other vertebrates, although not always strictly correctly and I
am in some doubt about the validity of the approximate approach
sometimes used.

For more than one species, the number of analyzable paleoecologic situations is rare, being limited, for all practical
purposes, to the predator-prey relationship and semi-quantitative
inferences on competition between species (cf. Reyment, 1971, Figure
24). The predation relationship can only be given adequate statistical study in paleontology for cases where the predator has
left an observable trace on the shell of the prey. The best
example of this is provided by drilling gastropods (cf. Reyment,
1971, p. 130-150). For marine invertebrate paleontologists, at
least, predation by drills happens to be of considerable significance
and it is, therefore, a rewarding subject for paleocological research.
Fossil ostracods, pelecypods, and gastropods are commonly found to
have been drilled by naticids, less often by muricids, and inasmuch as the first-mentioned group is a common component of borehole samples, sufficient material can usually be obtained to permit
a satisfactory statistical study.

It should be noted here that the analysis of a predator-prey
relationship in paleoecology is much of a gamble in that the observed prey and predator frequencies cannot, with certainty, be

claimed to represent the actual maximums attained by them. Not
only are the sampling fluctuations dependent on factors outside the
normal limits of statistics, but there is the added vexation of
the unknown extent of migration as well as post-mortem transport
of the drilled shells. A statistical analysis must therefore be
preceded by a detailed qualitative study of the material.

2.3 Spatial Paleoecology. Within certain limits, it is possible
to carry out useful spatial paleoecologic studies (Reyment, 1971,
Chapter 6). The confines for such studies are, of necessity,
narrow, and may verge on paleobiogeography. I have had occasion
to discuss morphometric variations in Paleocene ostracods occurring
throughout the Early Paleocene epicontinental transgression across
West and North Africa (Reyment & Reyment, in press). The morphometric differences identified could be related to the possible
existence of a climatic gradient. This example is certainly not
referable to the main concept of spatial ecology (cf. Pielou, 1974).
Only sessile organisms, such as corals and byrozoans, are liable to
leave sufficiently good traces of their erstwhile spatial relationships to permit a usual type of spatial analysis such as developed
by Matérn (1960), although Pleistocene plant studies would appear
to offer certain possibilities.

2.4 Ecological Diversity. Ecological diversity for fossil species
can often be analyzed with a fair degree of accuracy and there are,
perhaps, more examples of this category of statistical paleoecologic
analysis in the literature than of any other. Most often these are
of the form of semi-quantitative comparisons of faunal lists, sometimes involving percentages. Considerable use has also been made
of indices of which many variants have been proposed (Reyment, 1971,
p. 160 ff).

One of the preferred tools for analyzing ecological diversity
is the Shannon-Weaver Index, which uses relative abundances (Pielou,
1974, p. 290) which is gaining some vogue of late in statistical
paleoecology, thanks to its desirable properties. Diversity is a
popular subject among statistical ecologists and it is therefore
appropriate to note some of the paleoecologic work.

Some examples of applications are given below. Birks and Deacon
(1973) used lists of species of recent fossil vascular plants in
twelve geographic regions in Britain. Using four similarity indices
(Jaccard's, Dice's, Simpson's, and the Braun-Blanquet index), converted to dissimilarity coefficients, and non-metric scaling, two-
dimensional dispositions of points representing the region were
found for each time interval. A marked north-south floristic gradient was demonstrated. Cheetham and Hazel (1969) studied the per-
formances of various similarity (23) coefficients in analyzing

microfossil associations. Henderson and Heron (1977), in their
work on a probabilistic method of paleobiogeographic analysis,
concluded that diversity studies in paleontology tend to suffer
from defects. This paper also contains references to recent work
on population diversity in paleontology.

Examples of further studies of diversity in paleontology are
those of Hagel (1970), Kaesler (1969), Kaesler and Mulvany (1976).
Measuring the relative abundance of microorganisms has been considered by Forester (1977), who has used the Poisson parameter λt
to produce an abundance coefficient.

2.5 *Analysis of Species Frequencies.* Seventeen species of Early
Paleocene ostracods were analyzed by Jöreskog's maximum likelihood
model of factor analysis (Jöreskog *et al.*, 1976). These data were
originally analyzed in Reyment (1963). Here, it was found that the
relative frequencies of different species in samples may be interpreted in terms of the major environmental factors to which the
organisms react, to wit, five unspecified major environmental
components. It was also concluded here that although the factor
analysis of fossil species associations can seldom be expected
to disclose whether a species is stenohaline or euryhaline, steno-
thermal or eurythermal, it can indicate whether it is stenooic or
euryoic. A more detailed analysis of the material was given in
Reyment (1966), in connection with which it was thought possible
to identify one factor as bathyal, extrapolating from our knowledge
of the depth distribution of living ostracods.

The Shannon-Weaver index, already mentioned, was introduced
into Geology by Pelto (1954) and modified by Miller and Kahn
(1962) as a means of studying multi-species systems (in part,
analogous to the examples of Pielou (1977)). Pelto's (1954)
suggestion was to use the function

$$H = -\sum_i p_i \log_e p_i$$

for studying multicomponent systems. Here, p_i is the percentage
of the *ith* component, and $\sum_i p_i = 100\%$. Pelto made use of the
concept of relative entropy, Hr , which is defined as the ratio
of the actual entropy to the maximum entropy, Hm , for the number
of components under consideration:

$$100\ Hr = \frac{-100 \sum_{i=1}^{N} p_i \log_e p_i}{Hm}\ .$$

Here, p_i is the proportion of the *ith* component in an N-component system and Hm is

$$Hm = - \Sigma \frac{1}{N} \log_e \frac{1}{N} = \log_e N \ .$$

In the paleoecologic application devised by Miller and Kahn (1962), it is required that the species be divided into 'biofacies.' For my study of the Early Paleocene ostracods, I accepted the seven factors as representing paleobiofacies based on the 17 most abundant species. It was found that the entropy approach yields valuable additional information, particularly for the identification of environmental components that may have been overlooked in the earlier analysis. Thus, in addition to the bathyal component found in the factor analysis, calcareous and pelitic components were isolated (Reyment, 1966, p. 48).

2.6 Secular Fluctuations in the Abundance of Species. The study of secular fluctuations in the relative frequencies of species is one that has a specific paleoecologic flavor. A classical, early study is that of Chaney (1924), re-analyzed in Reyment (1971, p. 175 ff). Chaney was concerned with attempting to identify shifts in relative abundances of plants in the Bridge Creek flora of Late Oligocene age in Oregon, U.S.A.

3. ANALYSIS OF CHRONOLOGICAL VARIATION

The study of chronologic variation in microfossils provides the main substance of this paper. Examples are Reyment (1966, p. 90 ff.), and Reyment (1971, Chapter 4).

3.1 The Species × Levels Matrix. The simplest sequential representation of chronological variations in a set of species can be produced in terms of the categories + (= an increase in average size), - (= a decrease in average size, 0 (= no change). The species by stratigraphic levels matrix of these observations is a useful indicator for picking out a sustained ecologic trend in a multi-component set of observation, i.e., a common mode of reaction to the totality of environmental fluctuations. An example is given in Table 1. This representation shows that in the earlier levels of the sequence, most species follow the same pattern of variation, presumably ecologically controlled. Further aspects of the interpretation of this material are given in Reyment (1966, p. 90-91).

TABLE 1: Size-directional changes in a set of time series of seven species of Nigerian Paleocene ostracods (based on fluctuations in the length of the carapace).

Species	Direction of change upwards in borehole								
Cytherella sylvesterbradleyi	–	0	–	0	0	–	+	0	+
Ovocytheridea pulchra	–	+	+	0	–	0	0	–	–
Leguminocythereis lagaghiroboensis	–	+	+	0	–	–	–	–	–
Trachyleberis teiskotensis	0	+	+	0	+	–	+	+	+
Buntonia beninensis	–	0	+	+	0	–	+	+	0
Buntonia bopaensis	0	+	+	0	–	–	–	0	0
Buntonia livida	0	+	+	0	–	–	–	0	0

3.2 Canonical Correlations. I shall now consider an example in which the correlations between frequencies of organisms, on the one hand, and geochemical components of the host sediment, on the other, are used to produce what can be referred to as an *ecolog*, that is, a log in which diagnostic chemical elements are related to fluctuations in the frequencies of species over time. The example is taken from Reyment (1976).

Samples from twenty six levels in a Nigerian borehole in sediments of Late Campanian (Cretaceous) age were analyzed with respect to the 14 elements Si, Fe, Mg, Ca, Na, K, Ti, P, Mn, V, Mo, Sr, Pb, and Zn. The frequencies of the foraminifers *Afrobolivina afra* Reyment, *Gabonella elongata* de Klasz & Meijer, and *Valvulineria* sp. were recorded for these levels. The ostracods, being relatively rare, were pooled for the purposes of the analysis.

The aim of the study was to facilitate the graphical expression of a difficult paleoecologic and biostratigraphic problem. In one direction, it was thought to be of interest to show how all variables considered in the one connexion vary over time. In another direction interest was concentrated on tracing temporal covariation in frequencies and geochemical variables.

The method of canonical correlations was used for studying the relationships between sets. Canonical correlations have been little used in ecology owing to certain problems of interpretation, not the least of which is that a high canonical correlation is not necessarily associated with the greatest part of the information in the material. A biologic example of the application of canonical correlation to an ecologic problem is given by Reyment (1975) for ostracods in the Niger Delta. In this study, pH, Eh, bathymetry, phosphorus, and sulfur formed the predictor set; the response set was composed of total organic substance, Σ CaCO$_3$ and the total frequencies of ostracods. The most significant results of the analysis are (1) a significant canonical correlation with ΣS weighted against ostracods (a thanatocoenetic relationship) and (2) the distribution of the ostracod species is controlled by depth in a negative association with phosphorus. Canonical correlation analysis has an added useful side, to wit, the graphical presentation of the transformed partitioned observational vectors. In the present case, it was found that the samples rich in ostracods form a well defined cluster. The example reviewed now is, however, more complex, as direct observations on known ecologic components could not be obtained. Si is significantly positively correlated with Fe, Mn, and V, and negatively and significantly with Mg, Ca, Na, P, and Sr. The variable Fe is significantly positively correlate with Mn, V, and Mo, and significantly negatively with Na, P, and Sr, while Na is significantly positively correlated with K, P, and Sr. Further significant correlations are as follows: Ti is positively correlated with V, Mo, Pb, and Zn; and P is positively correlated with Sr. Mn is positively correlated with V and Mo, is positively correlated with Pb and Zn, and negatively with Sr.

For the microfossils, the following significant relationships between sets occur. There is a negative correlation between ostracod frequencies and Zn, while *Afrobolivina afra* is not correlated significantly with any of the geochemical variables. The frequencies for the *Valvulineria* are correlated positively with Mn and Fe and negatively with Ca, while *Gabonella elongata* is positively correlated with Fe and Mo, and negatively with Ca and Sr.

In the following, the vector variable z_1 contains the chemovariables and the vector variable z_2 , the frequencies of the micro fossils. The roots of the determinantal equation for the two sets (the R_{ij} are submatrices of the correlation matrix R).

$$|R_{22}^{-1}R_{21}R_{11}^{-1}R_{12} - \lambda_j I| = 0 \qquad (1)$$

are $\lambda_1 = 0.849$, $\lambda_2 = 0.752$, $\lambda_3 = 0.414$, and $\lambda_4 = 0.182$. The first two of these roots are statistically significant. These roots are the squares of the canonical correlations, to wit, $R_{c1} = 0.922$ and $R_{c2} = -.867$ which are the maximum correlations between two linear functions of the two sets of variables.

The structure coefficients for two canonical factors for all 18 variables are given in Table 2. The main steps involved are as follows (extracted from Cooley and Lohnes, 1971). Having found the roots of equation (1), the vector d is obtained from

TABLE 2: Structure coefficients *(Cooley & Lohnes, 1971) for two canonical factors of the geochemical and species-frequencies data. (After Reyment, 1976.)*

	Geochemical set		Set of species frequencies		
Variable	Factor 1	Factor 3	Variable	Factor 1	Factor 2
Si	0.31	−0.22	ostracods	−0.39	−0.66
Fe	0.36	−0.23	*Afrobolivina*	0.12	0.14
Mg	−0.26	−0.08	*Valvulineria*	0.67	−0.74
Ca	−0.38	0.24	*Gabonella*	0.62	−0.20
Na	−0.08	0.31			
K	0.21	0.23			
Ti	0.12	0.39	Factor	0.217	0.197
P	0.05	0.37	redundancy		
Mn	0.49	−0.21	Total	0.565	
V	0.26	0.33	redundancy		
Mo	0.67	0.23			
Sr	−0.27	0.24			
Pb	0.31	0.67			
Zn	0.51	0.36			
Factor redundancy	0.102	0.078			
Total redundancy	0.218				

$$R_{22}^{-1}R_{21}R_{11}^{-1}R_{12} - \lambda_j I) = 0 \quad , \tag{2}$$

with the constraint that $d_j' R_{22} d_j = 1$. The d_j are weights for the jth canonical factor of z_2 . The corresponding weights for the jth canonical factor of z_i are obtained from the relationship

$$c_j = \frac{(R_{11}^{-1}R_{12}d_j)}{\sqrt{\lambda_j}} \quad .$$

These steps are the usual ones of canonical correlation analysis. The expansion of the method into a redundancy analysis (Cooley and Lohnes, 1971) is done by finding the variance extracted by the canonical variables, $s_1' s_1 / p_1$, where p_1 denotes the number of variables in vector z_1 (here, this comprises the 14 chemical elements) and $s_2' s_2 / p_2$, where p_2 denotes the number of variables in vector z_2 (in the present example, this is 4). We have also that $s_1 = R_{11} c$ and $s_2 = R_{22} d$. The redundancy of set 1 given the variables of set 2 (set 1 contains the chemo-variables, set 2 contains the frequencies of the organisms) is defined by

$$R_{d_x} = s_1' s_1 R_{c1}^2 / p_1 \quad ,$$

where R_{c1} denotes the canonical correlation for the first pair of canonical variates and the subscript x labels the canonical factor x. The reverse relationship is expressed by the formula

$$R_{d_y} = s_2' s_2 R_{c1}^2 / p_2 \quad .$$

The first canonical factors are $x_1 = c_1' z_1$ and $y_1 = d_1' z_2$. Likewise, the second canonical factors are $x_2 = c_2' z_1$ and $y_2 = d_2' z_2$.

The first canonical factor (Table 2) for the left set of variables comprises significant loadings for most of them. Only Na, Ti, and P are so low as to indicate non-significant correlation. The first canonical factor for the right set contains significant loadings for all frequencies except that of *Afrobolivina afra*. The left hand canonical variate is positively correlated with Si, Fe, K, Mn, V, Mo, Pb, and Zn, and negatively correlated with Mg, Ca, and Sr. This canonical variate seems to be explainable as a dipolar relationship between sediment richer in carbonates and clastic sedimentary components. The right hand canonical variate is positively correlated with the frequencies of *Valvulineria* sp. and *Gabonella elongata*, and negatively with ostracods.

3.3 Ecolog from the Canonical Correlations. The plot of the scores obtained by substituting the partitioned mean vectors into the first pair of linear relationships can be used to produce a paleoecologic log in which the fluctuations in the frequencies of the organisms are weighted against variations in the chemical components of the host sediment. The log for the 26 levels analyzed here is shown in Figure 1. As to be expected from the rather high corresponding canonical correlation, the two curves follow the same general trends, although there are numerous deviations in the middle and

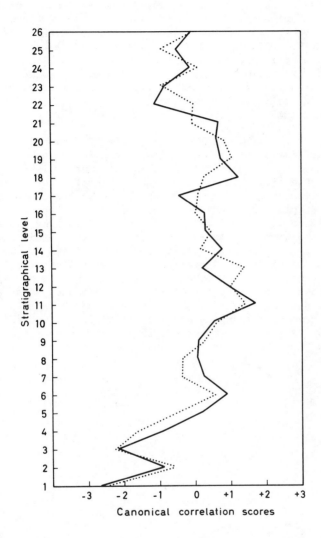

FIG. 1: *Fluctuations in the frequencies of micro-organisms weighted against variations in chemical components of the host sediment used to produce a canonical correlations ecolog.*

upper thirds of the plots. These deviations are small but might
mark periods during which the chemical influences were over-
printed by other environmental factors. The lower third of the
figure might be an indication of a phase in development during
which the chemical components of the environment dominated, such
as arises during periods of pronouncedly chemical sedimentation,
as in the formation of a marl (a sediment composed of varying
proportions of calcium carbonate and clay).

3.4 Ecolog by Principal Coordinates. Using Pythagorean distances
between individuals, all 18 variables of the foregoing analysis were
collected into a single principal coordinates analysis (Gower, 1966).
The plot of the first set of coordinates against location in the
borehole, illustrated in Figure 2, shows the existence of trend in
the points. This could indicate that there was a largely unidirectional ecologic trend in the paleo-environment over the time
covered by the samples. An interesting property of this ecolog
is that the youngest samples appear to be in a state of ecologic
equilibrium (levels 17-26 inclusive), while the older samples may
reflect an ecologically perturbed system. The system could have been
in the process of becoming stabilized in some manner or other, not
necessarily optimal for the proliferation of benthic microorganisms.
In fact, the youngest samples are characterized by the predominance
of *Afrobolivina afra* over the other three categories.

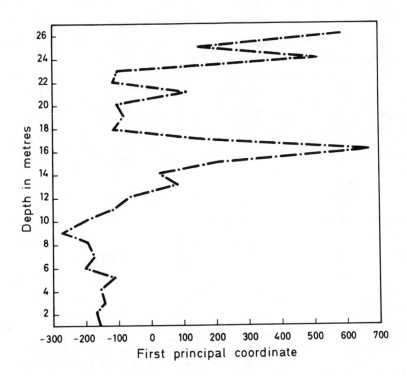

FIG. 2: Ecolog for the same data as used in Figure 1 constructed
from the first set of principal coordinates.

3.5 Stabilized Canonical Variates in Paleoecologic Analysis.
Canonical variate analysis of living and fossil organisms, based on
morphological characters, can be greatly distorted, from the aspect
of the biologic interpretation of the coefficients of the eigenvectors
forming the canonical variates, through the inclusion of redundant
within-group directions. Instability is associated with the smallest
eigenvalues, particularly if these do not greatly differ from zero.
In a study of borehole samples of Afrobolivina afra from Nigeria,
stability of the canonical variate coefficients was attained by
removal of a near-redundant direction of within-group variation.
This leads to improved interpretability of the morphometric relationships in this species (Campbell and Reyment, 1978).

The characters measured on Afrobolivina afra are: (1) = length
of the test, (2) = width, (3) = width of final chamber, (4) = height

of final chamber, (5) = height of second last chamber, (6) = diameter of proloculus, (7) = breadth, (8) = width of aperture, (9) = location of aperture on second last chamber. The main computational steps are set out below.

The within-groups sums of squares and cross products matrix W on n_W degrees of freedom, and the between-groups sums of squares and cross products matrix B are computed in the usual manner of canonical variate analysis, together with the matrix of sample means. It is then recommended that the matrix W be standardized to correlation form, with similar scaling for B. The standardization is obtained by pre- or post-multiplying by the inverse of diagonal matrix S , the diagonal elements of which are the square roots of the diagonal elements of W . Consequently,

$$W^* = S^{-1}WS^{-1} \quad , \text{ and, } \quad B^* = S^{-1}BS^{-1} \quad .$$

The eigenvalues e_i and eigenvectors u_1 of W^* are then computed. The corresponding orthogonalized variables are the principal components. With $E = \mathrm{diag}(e_1, \cdots, e_v)$ and $U = (u_1, \cdots, u_v)$,

$$W^* = UEU^T \quad .$$

(N. B. T denotes matrix and vector transpose in this section.) Usually, the eigenvectors are now scaled by the square root of the eigenvalue; this is a transformation for producing within-groups sphericity. Shrunken estimators are formed by adding shrinking constants k_i to the eigenvalue e_i before scaling the eigenvectors. Write $K = \mathrm{diag}(k_1, \cdots, k_v)$ and define $U^* = U(E + K)^{-\frac{1}{2}} = U^*(k_1, \cdots, k_v)$. Next form the between-groups matrix in the within-groups principal component space, that is,

$$G_{(k_1, \cdots, k_v)} = U^{*T}_{(k_1, \cdots, k_v)} B^* U^*_{(k_1, \cdots, k_v)}$$

and set d_i equal to the ith diagonal element of G . The ith diagonal element d_i is the between-groups sums of squares for the ith principal component.

An eigen-analysis of the matrix $G_{(0, \cdots, 0)}$ yields the usual canonical roots of f and canonical vectors for the principal components, a^u . The usual canonical vectors c^u are given by

$$c^u = U^*_{(0, \cdots, 0)} a^u \ .$$

Generalized shrunken (or generalized ridge-) estimators are determined directly from the eigenvectors a^s of $G_{(k_1, \cdots, k_v)}$, with $c^s = U^*_{(k_1, \cdots, k_v)} a^s$. A generalized-inverse solution results when $k_i = 0$ for $i \leq r$ and $k_i = \infty$ for $i > r$. This gives $a_i^{GI} = a_i^u$ for $i \leq r$ and $a_i^{GI} = 0$ for $i > r$. The generalized inverse solution results from forming $G_{(0, \cdots, 0, \infty, \cdots, \infty)} = U_r^{*T} B U_r^*$, where U_r^* corresponds to the first r columns of $U^*_{(0, \cdots, 0)}$. The generalized canonical vectors $c^{GI} = c^s_{(0, \cdots, 0, \infty, \cdots, \infty)}$ are given by $c^{GI} = U_r^* a^{GI}$, where a^{GI} , of length r , corresponds to the first r elements of a^u . In practice, it is found that marked instability is associated with a small value of e_v and a correspondingly small diagonal element d_v of G . A generalized inverse solution with r=v-1 frequently provides stable estimates and is usually conceptually simpler than using shrinking constants.

An easy rule to use is to examine the contribution of d_v to the total group separation, trace $(W^{-1}B)$; the latter is merely trace $(G_{(0, \cdots, 0)})$ or $\sum_{i=1}^{v} d_i$. In situations where one or two canonical variates describe much of the between-groups variation, it may be better to examine the relative magnitudes of the first one or two canonical roots derived from $G_{(0, \cdots, 0)}$ and $G_{(0, \cdots, 0, \infty)}$ rather than a composite measure. Either way, if $d_v/\Sigma d_i$, or the corresponding ratio of canonical roots, is small (say less than 0.05) then little loss of discrimination will result from excluding the smallest eigenvalue-eigenvector combination ($k_v = \infty$) or, equivalently, from eliminating the last principal component.

The eigenvalues and eigenvectors for all nine variables are listed in Table 3. The smallest eigenvalue accounts for only 1.8 per cent of the variation within groups. The eigenvector corresponding to the smallest eigenvalue (hereinafter referred to as the smallest eigenvector) reflects a contrast between variables 2 and

TABLE 3: Eigenvalues and eigenvectors of the within-groups correlation matrix W^* for all nine variables; between-groups sums of squares for each principal component.

Eigenvalues (e_i)	1	2	3	4	5	6	7	8	9
	4.31	1.25	1.00	0.69	0.48	0.43	0.38	0.29	0.16

Eigenvectors	v1	v2	v3	v4	v5	v6	v7	v8	v9
u_1	0.35	0.43	0.42	0.37	0.37	0.15	0.35	0.21	0.23
u_2	0.33	0.09	0.20	0.19	0.21	-0.30	-0.31	-0.55	-0.53
u_3	0.28	-0.15	-0.03	-0.10	-0.07	-0.85	0.23	0.31	0.09
u_4	-0.08	-0.06	0.03	0.09	0.05	0.13	0.01	0.64	-0.74
u_5	-0.44	-0.05	-0.18	0.44	0.51	-0.25	-0.42	0.17	0.24
u_6	-0.01	-0.10	0.03	0.74	-0.66	-0.02	0.01	-0.03	0.03
u_7	0.27	0.26	0.24	-0.20	-0.28	0.03	-0.73	0.33	0.20
u_8	0.64	-0.48	-0.44	0.14	0.19	0.29	-0.11	0.08	0.10
u_9	-0.10	-0.69	0.70	-0.05	0.08	0.06	-0.04	0.00	0.10

	1	2	3	4	5	6	7	8	9
diag $\{G_{(0,\cdots,0)}\}$	1.01	0.25	0.23	0.27	0.26	0.07	1.57	0.18	0.21

trace $\{G_{(0,\cdots,0)}\} = 4.05$.

3, to wit, the width of the test, and the width of the last chamber. These loadings are quite large and, it may be suspected that if the corresponding between-groups sum of squares is small, as is the case in our example, instability in the corresponding canonical variate coefficients may result.

The between-groups sums of squares for all principal components shows that 39 per cent of the between-groups variation is associated with the seventh principal component and 17 per cent with the first principal component. The variation for the seventh principal component results from a contrast between variable 7 and most of the other variables; the first principal component is a 'size component' (cf. Blackith and Reyment, 1971).

The canonical variate analysis can be carried out in terms of the principal components (the coefficients for the original variables are found by projecting back to the space of the original variables). The coefficents for the first canonical variate (a_i^u in Table 4) highlight the contribution from the seventh principal component. The coefficients for the original variables are determined explicitly from the principal component canonical vector; any inflation in these latter coefficients results in inflated coefficients for those among the original variables contributing to the eigenvector from which the principal component is derived. Note that the first principal component contributes most to the second canonical variate (see a_2^u in Table 4).

The first canonical variate amounts to 56 percent of the between-groups variation. The first two canonical variates account for 75 per cent of the variation between groups. The coefficients for the standardized original variables for the first two canonical variates are shown in Table 4, namely, c_1^u and c_2^u.

The effect of shrinking the contribution of the smallest eigenvector (and associated eigenvalue), namely, the ninth principal component, is shown in Table 4 (here, $k_9 = \infty$ implies the elimination of the ninth principal component from the analysis). The two sets of coefficients for the original variables (c_1^u) and $k_9 = \infty$ $(c^{GI}_{i(0, \ldots, \infty)})$ are similar, except for variables 2 and 3. The decrease in the magnitude of the coefficient for the second variable and the change in sign of the coefficient for variable 3 are apparent. The sum of the coefficients for these two variables is relatively stable and the canonical roots are little affected by the elimination of the smallest eigenvector.

TABLE 4: Standardized canonical vectors for nine variables, including shrunken estimates.

	v1	v2	v3	v4	v5	v6	v7	v8	v9	Canonical roots f
a_1^u	0.54	-0.11	-0.01	0.11	0.06	-0.07	0.79	-0.18	-0.13	
a_2^u	-0.64	-0.32	0.14	-0.13	-0.44	-0.11	0.39	0.05	-0.32	
c_1^u	0.00	-0.59	-0.09	0.43	0.44	0.07	0.99	-0.51	-0.10	2.28
c_2^u	0.65	0.81	-0.27	-0.44	-0.40	0.15	0.04	0.28	0.25	0.76
$c_1^{GI}(0,\cdots,\infty)$	0.04	-0.37	-0.31	0.43	0.41	0.06	1.01	-0.51	-0.12	2.25
$c_2^{GI}(0,\cdots,\infty)$	0.58	0.29	0.31	-0.53	-0.36	0.25	-0.03	0.37	0.32	0.70
$c_1^{GI}(0,\infty,\infty)$	-0.17	-0.21	-0.17	0.32	0.42	-0.04	1.06	-0.53	-0.16	2.17
$c_2^{GI}(0,\infty,\infty)$	0.53	0.32	0.35	-0.40	-0.50	0.22	-0.03	0.36	0.33	0.68

A plot of the shrunken estimates for all nine variables of the canonical variate means for the first canonical variate against location indicates that there is a general drift over time in the morphology of the species, manifested here as a trend to the right. As shown further on this shift seems to be due to a long-term environmental effect. The full analysis is given in Campbell and Reyment, 1978).

3.6 Measuring the Paleoenvironment. Benthonic organisms normally react morphologically to one or more environmental factors. A well studied factor is that of salinity, which may profoundly influence the size and shape of shell-secreting organisms. For example, ostracods are well known to adjust morphologically, often quite markedly, to relatively slight shifts in salinity.

Attempts at producing a graphical display of biologic variation in micro-organisms have run into the difficulty of interpreting morphometric oscillations from the aspect of extracting genuine morphological changes in an organism from random variation, and spurious shifts in averages resulting from varying mixtures of growth stages in the samples.

An important concept for using micro-organisms to establish a variational pattern for, say, a borehole sequence is that average changes in the dimensions of a benthonic organism mirror shifts in gross ecological conditions during the life of the organism. As shown by Burnaby (1966) and Reyment and Banfield (1976), this concept may require considerable modification, depending on the animal concerned. For ostracods, for example, the application of this idea is relatively straightforward, because individuals of the advanced growth stages can be recognized without much difficulty. This is not the same as saying that any organism with comparable dimensions will have been exposed to the same ecologic changes at a given site, as is sometimes thought.

The situation for continuously growing organisms, such as foraminifers, is less simple as any given sample will be confounded by growth effects. This means, that a sample will not be statistically homogeneous, being a mixture of growth stages and growth-inhibited morphologies, even though it may be biologically homogeneous.

In order to compensate for such variations, one will wish to find canonical variates among several populations confounded by growth effects (and size differences), representable as size gradients. With k size-difference components and v variates, the effects to be eliminated may be represented by a v × k matrix K , the *rth* column of which consists of elements proportionate to the direction cosines of the *rth* component. The idempotent symmetric matrix

$$Q = I - K(K'K)^{-1}K'$$

projects every sample value onto the space orthogonal to K where
they are free from growth differences. What we are in effect saying
here is that in the plane of the growth gradients, a specimen may
be situated anywhere, depending on the size it had attained at
fossilization. If we confine ourselves to a three-dimensional
case, this space can be subdivided into component spaces, Q , of
one dimension, the space in which growth differences do not occur,
and M , two-dimensional, in the space containing the growth
gradients. Any individual in M is also specified in Q , with
all its intrinsic properties, but with the important difference
that it will be located at a point in Q , but on a line in M.
Matrix K may comprise any number of size-growth effects but in
the present case, it is a vector.

When there are p populations, all with the same K ,
canonical variates can be obtained by solving the matrix equation

$$Q(G'G - W)Q = 0$$

where G is the $p \times v$ matrix of sample means and W is the
pooled within-populations dispersion matrix. (G'G replaces the
between-populations dispersion matrix of canonical variate analysis.)

Gower (1976) has considered several ways of going about the
estimation of K , both by internal as well as external estimation
by various multivariate statistical methods. It is also possible
to attempt direct estimations of the growth vectors. In the
present example for *Afrobolivina afra*, internal estimation by
principal component analysis gave a very satisfactory solution.
The reason for this is that most of the size variation resides in
the length variate of *Afrobolivina*, a result of the manifestation
of size and shape variation deriving from the morphologic differences
associated with the life cycle of this genus. The first principal
component does extract variation due to relative size differences as
in *Afrobolivina*, the great range of variation in length is directly
reflected in the variance of this variate. K can thus be estimated as the first eigenvector of W , the calculations being
made on the logarithmically transformed observations.

The solution of equation (3) requires the computation of a
generalized inverse for QWQ,

$$C = Q(QWQ)^{-1}Q \ .$$

The required solution is

$$(CG'G - \lambda I) = 0 \ .$$

The generalized Mahanolobis' distances between the means of groups
i   and   j   when projected onto the Q-space are given by

$$D_Q^2 = (g_i - g_j)C(g_i - g_j)' \ .$$

The generalized distances are used for finding the coordinates
of the group means in the Q-space by principal coordinates (Gower,
1966).

The data used in illustrating this method were obtained from
10 samples (borehole sampling levels) of *Afrobolivina afra* Reyment.
The same 9 characters as before were used for the analysis of the
211 specimens.

The first principal coordinate with one growth vector extracted
for the 10 samples was used to construct a size-standardized
chronocline (growth-invariant curve) by plotting the values of
the coordinates for the samples against their stratigraphical
positions (Figure 3). If matrix   P   contains the eigenvectors of
GCG' = T , with   E   being the corresponding eigenvalues, then   T
is a $p \times p$ matrix,   P   is $p \times v$ and   E   is a diagonal $v \times v$
matrix. Scaling the vectors so that   P'P = E   ensures that the
rows of   P   are the coordinates of points referred to principal
axes and   P'P = I.   The squared distances between the ith and
jth points are $t_{ii} + t_{jj} - 2 t_{ij}$.

The plot of the first coordinate against stratigraphical
position for the 10 samples is shown above the heading 'biolog'
in Figure 3. The short normal resistivity log is also shown in
this figure. The curves display the same trend, in the statistical
sense. It will be seen that the variational curve, or biolog, is
remarkably conformable with the course of the electric log. For
readers unfamiliar with procedures of borehole logging, a useful
reference is Lynch (1964).

It is important to bear in mind when comparing the biolog with
physical logs that the latter represent records of continuous
observations, whereas the biolog derives from discrete sampling
points. Direct comparisons can, therefore, not be made, although
directions of trend can be compared at points on the physical
curves that correspond with the levels at which the sediment
samples were taken. Perhaps the best way of contrasting these
logs is by means of directions of oscillations at the sampling
points, as suggested by Table 1 for the ostracod data. Residual
variation expressed by the growth-reduced discriminant coordinates
seems most probably to derive largely from the response of the
foraminiferal organism to environmental factors. Physical logs
of the sediments encountered in a borehole reflect some properties
of the ecologic background in which these were deposited.

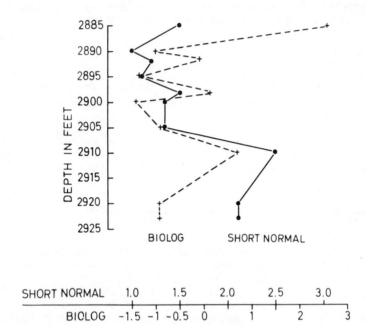

FIG. 3: *Biolog formed from the growth-reduced principal co-ordinates of* 10 *samples of* **Afrobolivina afra.** *The right hand curve is the short normal electrical resistivity log (cf. Lynch, 1964).*

4. ANALYSIS OF ECOLOGIC MAPS

I wish to make brief mention of some work, which although peripheral to the topic of this paper, may interest phytosociologists. Reyment and Banfield (in press) considered the quantitative inter-pretation of spatial relationships between sedimentary environments in the Mississippi Delta.

There are many problems in ecology and paleoecology where the relationship between two 'specimens' i and j , is not symmetric. If a_{ij} denotes a measure of dissimilarity calculated from measure-ments on i and j , $a_{ij} \neq a_{ji}$. One example is counts on the number of times plant species i is the nearest unlike neighbor of plant species j . There is a known geographic relationship between the sites at which i and j occur; the significance lies in the asymmetry of the observations.

A square non-symmetric matrix A, the rows and columns of which are classified by the same specimens, cannot be treated by the same methods as used in the standard multivariate analysis of square symmetric matrices. Gower (1977) has developed a canonical analysis for square non-symmetric matrices whereby the differences between a_{ij} and a_{ji} can be represented graphically.

The example studied by Reyment and Banfield (in press), gives a meaningful interpretation of the inter-relationships between six deltaic depositional environments.

REFERENCES

Birks, H. J. B, and Deacon, J. (1973). A numerical analysis of the past and present flora of the British Isles. *New Phytol.*, 72, 877-902.

Blackith, R. E. and Reyment, R. A. (1971). *Multivariate Morphometrics.* Academic Press, London.

Burnaby, T. P. (1966). Growth-invariant discriminant functions. *Biometrics*, 22, 96-110.

Campbell, N. C. and Reyment, R. A. (1978). Discriminant analysis of a Cretaceous foraminifer using shrunken estimators. *Mathematical Geology*, 10, 347-359.

Cheetham, A. H. and Hazel, J. E. (1969). Binary (presence-absence) similarity coefficient. *Journal of Paleontology*, 43, 1130-1136.

Forester, R. M. (1977). Abundance coefficients, a new method for measuring microorganism relative abundance. *Mathematical Geology*, 9, 619-633.

Gower, J. C. (1966). A Q-technique for the calculation of canonical variates. *Biometrika*, 53, 588-589.

Gower, J. C. (1976). Growth-free canonical variates and generalized inverses. *Bulletin of the Geological Institutions of the University of Uppsala NS*, 7, 1-10.

Gower, J. C. (1977). The analysis of asymmetry and orthogonality. In *Recent Developments in Statistics*, J. Barra *et al.*, eds. North Holland Publishing Company, Amsterdam

Hazel, J. E. (1970). Binary coefficients and clustering in biostratigraphy. *Bulletin of the Geological Society of America*, 81, 3237-3252.

Henderson, R. A. and Beron, M. L. (1977). A probabilistic method of paleobiogeographic analysis. *Lethaia*, 10, 1-15.

Jöreskog, K. G., Klovan, J. E., and Reyment, R. A. (1976). *Geological Factor Analysis*. Elsevier, Amsterdam.

Kaesler, R. (1969). Aspects of quantitative distributional paleoecology. In *Computer Applications in the Earth Sciences*. Plenum Press. 99-119.

Kaesler, R. L. and Mulvany, P. S. (1976). Fortran IV program to compute diversity indices from information theory. *Computers and Geosciences*, 509-514.

Lynch, E. J. (1964). *Formulation Evaluation*. Harper & Row, New York.

Mardia, K. (1972). *Statistics of Directional Data*. Academic Press, London.

Matérn, B. (1960). Spatial variation. *Meddelamden fran Statens Skogsforskningsinsttitut*, 49, 1-144.

Miller, R. L. and Kahn, J. S. (1962). *Statistical Analysis in the Geological Sciences*. Wiley, New York.

Pelto, C. R. (1954). Mapping of multicomponent systems. *Journal of Geology*, 62, 501-511.

Pielou, E. C. (1974). *Population and Community Ecology*. Gordon and Breach, New York.

Pielou, E. C. (1977). *Mathematical Ecology*. Wiley, New York.

Reyment, R. A. (1963). Multivariate analytical treatment of quantitative species associations: an example from palaeoecology. *Journal of Animal Ecology*, 32, 535-547.

Reyment, R. A. (1966). Studies on Nigerian Upper Cretaceous and Lower Tertiary Ostracoda. III, Stratigraphical, palaeoecological and biometrial conclusions. *Stockholm Contribution in Geology*, 14, 1-151.

Reyment, R. A. (1971). *Introduction to Quantitative Paleoecology*. Elsevier, Amsterdam.

Reyment, R. A. (1975). Canonical correlation analysis of hemicytherinid and trachyleberinid ostracods in the Niger Delta. *Bulletin of American Paleontology*, 65(282), 141-145. Ithaca, New York.

Reyment, R. A. (1976). Chemical components of the environment and Late Campanian microfossil frequencies. *Geologiska Föreningens, Stockholm Förhandlingar*, 98, 322-328.

Reyment, R. A. (1978). Graphical display of growth-free variation in the Cretaceous benthonic foraminifer *Afrobolivina afra*. *Palaeogeography, Palaeoclimatology, Palaeoecology*, 25, 267-276.

Reyment, R. A. and Banfield, C. (1976). Growth-free canonical variates applied to fossil foraminifers. *Bulletin of the Geological Institutions of the University of Uppsala*, 7, 11-21.

Reyment, R. A. and Banfield, C. F. (in press). Analysis of asymmetric relationships in geological data. In *Future Trends in Geomathematics*, R. G. Craig and M. L. Labovitz, eds. University of Pennsylvania Press.

Reyment, R. A. and Reyment, E. R. (in press). The Paleocene trans-Saharan transgression and its ostracod fauna. *Proceedings of the 2nd Conference on Geology of Libya*. (Tripoli, 1978).

[*Received June* 1978. *Revised March* 1979]

L. Orloci, C. R. Rao, and W. M. Stiteler, (eds.),
Multivariate Methods in Ecological Work, pp. 237-251. All rights reserved.
Copyright © 1979 by International Co-operative Publishing House, Fairland, Maryland

CORRELATION AND SUGGESTIONS OF CAUSALITY: SPURIOUS CORRELATION*

E. L. SCOTT

Department of Statistics
University of California, Berkeley
Berkeley, California   94720   USA

SUMMARY. Spurious correlation is a common statistical pitfall;
several examples are provided. The first is the classical
example, "Do storks bring babies?", and the second relates the
number of deaths to the level of pollution in the case where the
population at risk varies from one locality to another. The
consequences of applying spurious correlation are derived and contrasted with the use of partial correlation.

KEY WORDS. correlation, spurious correlation, causality, partial
correlation.

The subject of my talk is, I hope, familiar to all of you.
Yet, I think that it merits taking your time here simply because
spurious correlation is such a common statistical pitfall.
Indeed it is a powerful method for discovering statistical regularities that are false in the sense that they are only manufac-
tured by the analysis and thus a kind of statistics that is
frequently a lie although, presumably, an unintentional lie---
"a little knowledge is a dangerous thing." I shall phrase the
problem very much as it was stated (see Neyman, 1972) at a recent
conference (in which many of the papers employed the methods of
spurious correlation).

---

*Based on a talk given by the author at the First Advanced Study
Institute on Statistical Ecology in the United States held at The
Pennsylvania State University during 1972. The paper was pre-
pared with the partial support of the National Institutes of Health.

Consider the situation in which the phenomenon of interest
is not directly controllable. This phenomenon manifests itself
in some variable Y that is of public concern. Now Y is
observed, perhaps in different localities in a given year or over
a sequence of years in the same locality, and it is found that
the currently observed values of Y are unacceptably high (or
low). Also it is suspected that Y may be connected somehow
with another variable X that is subject to at least partial
control. In one example, which I will refer to, Y is the number
of deaths from a particular disease in a specified section of the
population and X is the level of a pollutant. In another
example, a classical example (in Statistics classical means more
than a quarter century in age), Y is the number of babies born
in a county and X is the number of storks. There are many
other examples: number of crimes with number of bars, cost of
operating a railroad with both number of passengers and amount of
freight (so that X is a vector), number of deaths with number
of hospital admissions.

In these situations, public action is contemplated, perhaps
legislation, to require a change in the values of X with the
hope that this change in X would result in the wished-for change
in Y, at least on the average. How can the statistician advise
whether changes in X will cause changes in Y? A well designed
experiment will provide information. However, experiments are
not possible in the situations we are considering; we must fall
back on an observational study. Such a study will not provide
the causal relation between X and Y but it can tell us how
the average values of Y in those units where X is large differ
from those where X is small. This information may be valuable
in deciding on the contemplated public action.

We want to find observational units that are identical in
all respects except for the values of X and Y. In practice, we
cannot do this; the observational units will vary not only in X
and Y but in many other respects. There will be some variables
$Z_1, Z_2, \cdots, Z_s$ whose variation is likely to influence X or Y
or both. For example, when X is the level of a pollutant and
Y is the number of deaths among a specified group then the
interfering variables Z may be the numbers exposed to risk or
may be some other pollutants than the one under study. Clearly,
we need to examine the estimated changes in Y that may result
from the contemplated changes in X when the values of Z are
held constant; we need to study the conditional regression $Y(x|z)$
of Y on X with all the Z maintaining some fixed values
symbolized by the letter z. When the situation is favorable,

that is, the regression is linear and there are only a few of the
interfering variables   Z, this conditional regression analysis is
not difficult. Still, there is always the temptation to take
a shortcut--to construct an index to replace the several measures
of qualitatively different pollutants with a single measure for
all pollutants combined, or to correct the numbers of deaths for
the fact that the number exposed to risk varies from one unit to
another as do the numbers belonging to different socio-economic
groups. Most frequently, such shortcuts lead one to spurious
correlations.

In the usual case, for each observational unit, the analyst
computes the supposedly corrected value V, say, of X and/or
the supposedly corrected value W of Y which these variables
X and Y would have had if all the Z were constant, at some
typical values, rather than varying. Thus, V and W are
certain functions $V = f_1(X,Z)$ and $W = f_2(Y,Z)$ of the directly
observable X, Y, and Z. The trouble comes when the correlation
between V and W is taken to represent the conditional
correlation between X and Y with the influence of the interfering Z eliminated.

Emphatically, no matter how convincing the correcting
functions f_1 and f_2 may be, THE CORRELATION OF V AND
W NEED NOT INDICATE in any way THE PARTIAL CORRELATION OF X
AND Y WHEN THE INTERFERING VARIABLES Z ARE FIXED. Let us
illustrate this difficulty in one of the simplest but most common
cases. In order to eliminate the effect of the variation in the
number of Z exposed to risk, we might change to a death rate
and to a pollution density by dividing by the values of Z, so that

$$V = f_1(X,Z) = X/Z$$

$$W = f_2(Y,Z) = Y/Z.$$

The essential point is that both of the functions depend on a
common argument Z   so that the fluctuations of Z will create
simultaneous effects on both V and W. Any abnormal kind of
increase in Z will diminish both V and W since Z is in
the denominator of each. Any decrease in Z will increase both
V and W will be positively correlated. Let us carry out the
computations to support this intuitive argument.

Consider the correlation coefficient R between the variables
V and W when X and Y are independent. To simplify the com-
putations, we assume linear relations with the interfering variable
Z and so forth, as follows. Specifically we assume that:

(i) Given Z, the variables X and Y are independent.

(ii) X and Z are correlated and the regression of X on Z is linear, say,

$$E[X|Z] = A_0 + A_1 Z.$$

Moreover, the conditional variance of X given Z, say $\sigma^2_{X|Z}$, is independent of Z.

(iii) Y and Z are correlated and the regression of Y on Z is linear, say,

$$E[Y|Z] = B_0 + B_1 Z.$$

The conditional variance of Y given Z, say $\sigma^2_{Y|Z}$, is independent of Z.

(iv) The expectation and the variance of the reciprocal of Z both exist. We shall denote them be $E[1/Z]$ and $\sigma^2_{1/Z}$, respectively.

According to the usual definition, the correlation coefficient

$$R = \frac{E[VW] - E[V]E[W]}{[Var(V)\ Var(W)]^{1/2}}.$$

Thus, in order to compute R, we need to compute the indicated expectations and variances. Straightforward computations give

$$E[V] = E[\tfrac{1}{Z} E(X|Z)] = E[\tfrac{1}{Z}(A_0 + A_1 Z)] = A_0 E[1/Z] + A_1.$$

$$E[V^2] = E[\tfrac{1}{Z^2} E(X^2|Z)] = [\sigma^2_{1/Z} + E^2(1/Z)](\sigma^2_{X|Z} + A_0^2)$$
$$+ 2 A_0 A_1 E[1/Z] + A_1^2.$$

Combining these results,

$$Var(V) = [\sigma^2_{1/Z} + E^2(1/Z)]\ \sigma^2_{X|Z} + A_0^2\ \sigma^2_{1/Z}.$$

Similarly,

$$E[W] = B_0 E[1/Z] + B_1,$$
$$Var(W) = [\sigma^2_{1/Z} + E^2(1/Z)]\sigma^2_{Y|Z} + B_0^2\ \sigma^2_{1/Z},$$

and

$$E[VW] = A_0 B_0 [\sigma^2_{1/Z} + E^2 (1/Z)] + (A_0 B_1 + A_1 B_0) E [1/Z] + A_1 B_1.$$

Thus, the correlation coefficient R between V and W is

$$R = \frac{A_0 B_0 \sigma^2_{1/Z}}{F_1 F_2}$$

where $F_1 = [\text{Var } (V)]^{1/2}$,

$F_2 = [\text{Var } (W)]^{1/2}$.

We notice that our intuitive considerations are only partly correct. Under the assumptions for which R was computed, the correlation R between V and W is zero if and only if $A_0 = 0$ or $B_0 = 0$ or both are zero. If neither is zero, the correlation R is positive whenever A_0 and B_0 have the same sign, and negative whenever they have different signs. We recall that A_0 and B_0 are the intercepts of the regression lines of X on Z and of Y on Z, respectively.

The preceding analysis was made on very simplifying assumptions. However, there is no difficulty in performing it in much more general situations. The difficulty of spurious correlation has been known for some time. The first record of publication appears to be by Karl Pearson (1897) who gives several examples, computes the spurious contribution to the correlation to be 0.45 in some common examples in biology and in economics, in particular, some indices obtained for shrimp organs by Weldon who divided his measures by the total body length or the carapace length of the animal. But notice that there is nothing spurious in the correlation between V and W. Whenever $A_0 \neq 0$ and $B_0 \neq 0$, the correlation between these two is very real, although not very interesting. Thus, the term spurious correlation may seem to miss the point. The real point of discussion is that the reason why we undertook the computation of the quotients V and W was not to study the correlation between these two variables, but the correlation between the variables X and Y. Unfortunately, the correlation between the quotients does not represent the correlation between the variables of interest. It is the method of study that is faulty and the adjective spurious should be applied to this method of studying the correlation between factors of primary interest. The proper method of study is to

compute the partial correlation between X and Y with the
influence of Z eliminated.

Let me give now two numerical examples which have been
simplified, really rounded off, so that the ideas are not obscured
by arithmetic. The first example is the classical example, due
to Neyman (1952), of observations collected to study the question:
Do storks bring babies? The observationsl data are quite comprehensive and refer to different counties. Table 1 shows the raw
data for the 54 counties arranged according to population. Tabulated are the number of women of child-bearing age (given in
units of 10,000), the number of storks in the county, and finally
the number of babies born in the county during a specified period
of time. In the notation we have been using, Y is the number
of babies born, X is the number of storks, and Z is the number
of women of childbearing age.

The problem referred to by Neyman was the biological problem:
Do storks bring babies? These observations might also be used
to provide advice on the question: Assuming that the public
policy is to reduce (increase) the number of babies born, how
effective would it be to reduce (increase) the number of storks,
on the average? In either case we want to study the correlation
between Y, the number of babies born, and X, the number of
storks. We might consider a direct comparison of the numbers X
and Y: Does Y tend to be large whenever X is large and small
when X is small? However, such a comparison would not be
convincing because the counties vary in size, and larger counties
may be expected to have more women, more babies and also more
storks. The variation in the size of the counties appears as a
disturbing factor obscuring the true relationship between the
quantities X and Y.

In order to eliminate the interfering influence of the size
of the county, a friend of Neyman's had the brilliant inspiration
of comparing not the actual numbers of births and the actual numbers of storks but the 'birth rates' on the one hand and the density of storks per 10,000 women on the other. Thus, he obtained
the quantities V and W as follows

$$V = X/Z \quad \text{and} \quad W = Y/Z$$

and then he tried to compare the two quotients V and W.

Naturally, we cannot expect to find absolute regularity with
observations from biology. In particular, we do not expect that
every increase in the quotient V will always be accompanied by
a proportional increase in W. There must be fluctuations and we
can expect to find counties with a large density of storks and a
small birth rate, and *vice versa*. The best we can hope for as

TABLE 1: Do storks bring babies? - Raw data.

County no.	Women in 10,000s	Storks	Babies born	County no.	Women in 10,000s	Storks	Babies born
1	1	2	10	28	4	6	25
2	1	2	15	29	4	6	30
3	1	2	20	30	4	6	35
4	1	3	10	31	4	7	25
5	1	3	15	32	4	7	30
6	1	3	20	33	4	7	35
7	1	4	10	34	4	8	25
8	1	4	15	35	4	8	30
9	1	4	20	36	4	8	35
10	2	4	15	37	5	7	30
11	2	4	20	38	5	7	35
12	2	4	25	39	5	7	40
13	2	5	15	40	5	8	30
14	2	5	20	41	5	8	35
15	2	5	25	42	5	8	40
16	2	6	15	43	5	9	30
17	2	6	20	44	5	9	35
18	2	6	25	45	5	9	40
19	3	5	20	46	6	8	35
20	3	5	25	47	6	8	40
21	3	5	30	48	6	8	45
22	3	6	20	49	6	9	35
23	3	6	25	50	6	9	40
24	3	6	30	51	6	9	45
25	3	7	20	52	6	10	35
26	3	7	25	53	6	10	40
27	3	7	30	54	6	10	45

regularity is that when all 54 counties are classified according to the density of storks and then the birth rate is averaged for each category, the averages will show a variation parallel to the variation in the density of storks. To put this idea professionally: the believer in the proficiency of storks has to be satisfied if he finds a positive correlation between the birth rate and the density of storks.

This was the attitude of Neyman's friend. He computed Table 2 from Table 1. Among the 54 counties studied, there were three which had on the average 1.33 storks per 10,000 women of child-bearing age. For these counties the average birth rate was 6.67. Also, there were three counties with 1.40 as the density of storks and these counties had an average birth rate of 7.00, and so on. As indicated in Table 2, the birth rate is subject to fluctuations but it steadily increases with an increase in the density of storks. The increase becomes even more marked if we make larger categories by dividing the counties into only three classes according to the density of storks: densities below 1.7, densities between 1.7 and 2.1, densities above 2.1. The corresponding class averages, shown in the last column of Table 2, increase decisively.

From Table 2 Neyman's friend concluded that, although there is no evidence of storks actually bringing babies in this table, there is overwhelming evidence that somehow they influence the birth rate! Some of you may be skeptical; you may suspect that the original data of Table 1 were intentionally falsified to produce the astounding results in Table 2. Let me assure you that these suspicions are unfounded. Indeed the transition from Table 1 to Table 2 is just a numerical example of the spurious correlation we have been discussing. Further, the observations in Table 1 are carefully classified so that it is easy to make a complete analysis without performing any arithmetic.

We notice that all the 54 counties fall into six different groups, each with nine counties. It happens that the nine counties forming a group have the same number of women, 10,000 in the first group, 20,000 in the second, etc. We notice next that each group of nine counties falls into three subgroups of three counties each, and that the subgroups have been ordered according to the number of storks. In the first group of counties there are three with 2 storks, three with 3 storks, and three with 4 storks. The same kind of change is repeated in the other groups of counties. Presumably, the counties in the second group are larger than those in the first; certainly, they have more women and the number of storks varies from 4 to 6. Turning to Y, the number of babies born, we see that there are fluctuations. With the same number $X = 2$ of storks in the three counties of the first group, the numbers of babies born are 10, 15, and 20. In the next

TABLE 2:   Do storks bring babies? - Analytical presentation.

Density of storks per 10,000 women	Number of counties	Average birth rate	Class average
1.33	3	6.67	7.12
1.40	3	7.00	
1.50	6	7.08	
1.60	3	7.00	
1.67	6	7.50	
1.75	3	7.50	9.22
1.80	3	7.00	
2.00	12	10.21	
2.33	3	8.33	11.67
2.50	3	10.00	
3.00	6	12.50	
4.00	3	15.00	

subgroup, where there are 3 storks, our experience with Table 2 might lead us to expect that the number of babies born, though fluctuating, will be larger than in the first subgroup where there were only 2 storks. However, Table 1 shows no increase in the number of babies born as long as the number of women remains constant. This is true in every group. So long as we consider a group of counties with the *same number of women,* an increase in the number of storks has no effect whatsoever on the number of babies born.

In more technical terms, the conditional distribution of the number of babies born, given the number of women, is independent of the number of storks. In the symbols we have been using, the distribution of Y, given the value of Z, is independent of X. The conclusion from Table 1 is that increasing (decreasing) the number of storks will not increase (decrease) the number of babies born once the number of women is fixed. Although this finding appears to be contrary to the intuition of Neyman's friend, it coincides with my intuition and yours. That is, apart from a rather unusual regularity, the figures in Table 1 do not involve anything unexpected.

Table 1 shows also that, given the number of women, the birth rate is independent of the number of storks. Then, how can we explain the unexpected regularities of Table 2? The answer is again that here we have X and Y, given Z, are independent

but V = X/Z and W = Y/Z are correlated due to the common Z
in their denominator. The correlation between V and W is
real but the attempt to use it to infer correlation between X
and Y is spurious.

You might say that a false proof like the one Neyman's friend
constructed in Table 2 will do no harm; we already know that
storks do *not* bring babies. But what if we changed the title to:
Do bars cause crime? Now the column headings will be density of
bars per 10,000 persons instead of density of storks per 10,000
women, and crime rate instead of birth rate. Whehter or not the
assertion is true, the argument of the kind shown in Table 2 is
faulty. Although the phenomenon of spurious correlation has been
known for a long time, the general public and many practical statisticians continue to be misled by it.

Let me give another numerical example, provided by Traxler
(see Neyman, 1972) which is more topical and yet simpler than the
storks-babies. It illustrates, with exaggerated precision, what
can come from a shortcut in the case where, given the interfering
Z, the variables X and Y are independent. Here Z denotes
the number of individuals exposed to the risk of death, X
represents the level of a harmless pollutant, and Y is the
number of deaths. In this simple case, only the variable Y will
be corrected for the variation of Z.

The necessary observations are given in Table 3. The localities are classified into six panels, each with the same number
Z of exposed to risk. In the first panel Z = 5, in the second
Z = 6, and so forth up to Z = 10 in the last panel, all measured
in the same units, say 100.

Within each panel the value of Z is constant, but there is
variation in X, the level of the pollutant. In some arbitrary
units, X has three values in each panel: 1, 2, 3 in the first
panel, 2, 3, 4 in the second, etc. This change in the value of
X as we proceed from one panel to the next corresponds to
Traxler's idea that an increase in Z, the number exposed to risk,
may reflect an increase in the total population which results in
an increase in the pollution. This need not be the case and you
may find it interesting to consider the case where the relation
between X and Z is the contrary of that considered by Traxler.

The last column in each panel lists Y, the number of deaths.
Notice that the *same* triplet of values of Y corresponds to all
three values of X. In the first panel this triplet is always
1070, 1100, and 1130 deaths. In the second panel the number of
deaths is always 1270, 1300, and 1330, and so forth. Table 3
shows many facets of regularity not to be expected in any real
study. In a real study, it would not be surprising to find that

TABLE 3: *Six groups of localities with pollutant N, no. of deaths Y, and no. at risk Z.*

Locality no.	No. at risk Z	Pollutant level X	No. of deaths Y	Locality no.	No. at risk Z	Pollutant level X	No. of deaths Y	Locality no.	No. at risk Z	Pollutant level X	No. of deaths Y
1	5	1	1070	19	7	3	1465	37	9	5	1860
2			1100	20			1500	38			1900
3			1130	21			1535	39			1940
4		2	1070	22		4	1465	40		6	1860
5			1100	23			1500	41			1900
6			1130	24			1535	42			1940
7		3	1070	25		5	1465	43		7	1860
8			1100	26			1500	44			1900
9			1130	27			1535	45			1940
10	6	2	1270	28	8	4	1665	46	10	6	2060
11			1300	29			1700	47			2100
12			1330	30			1735	48			2140
13		3	1270	31		5	1665	49		7	2060
14			1300	32			1700	50			2100
15			1330	33			1735	51			2140
16		4	1270	34		6	1665	52		8	2060
17			1300	35			1700	53			2100
18			1330	36			1735	54			2140

every locality has a different value of Z. Then there would be
a problem of how to deal with the variation of Z. The method
that a practicing statistician is likely to use will be spurious.
His reasoning might be: My objective is to study how the frequency of deaths can be expected to change if the level of pollutant is changed. A direct way to study the problem is to compute,
for each locality, the death rate $W = Y/Z$ and to classify all
the localities according to the level X of the pollutant. Then
compute the mean death rate, say $\bar{W}(X)$, corresponding to each
value of X. The same sort of ideas could be used if Table 3
were even more extensive, including less populated and more populated localities, all with a similar pattern of the relation
between Z and X and the lack of relation between Y and X,
given Z.

Table 4 illustrates what the analysis will lead to. This
table is arranged so as to simplify the calculation of $\bar{W}(X)$ and
to clarify somewhat the root of the difficulty. The simplifica-
tion is based on the fact that all of the 54 localities can be
divided into triplets, each triplet characterized by its combination of values of X and Z. Thus, there is no point in calculating the death rate W separately for each locality. It is
enough to calculate the death rate per triplet, then to classify
the triplets by their corresponding (X, Z) and lastly average
$W(X, Z)$ over the values of Z to obtain the desired $\bar{W}(X)$. All
this is done four times in Table 4, each time corresponding to a
different set of observational data according to what population
sizes are available: Population sizes 4-9, then population sizes
5-10 (which corresponds to Table 3), population sizes 6-11, and
lastly population sizes 7-12. Each time the eight lines in the
table correspond to the eight values of X and the six columns
to the six values of Z. The last column gives the average death
rate for increasing values of X.

If any real study exhibited the correspondence between the
average death rates $\bar{W}(X)$ and the level X of the pollutant such
as that shown in Part II of Table 4, the interpretation would be
somewhat as follows:

(i) The pollutant studied does influence the death rates.

(ii) The pollutant studied is beneficial: an increase in
the level of the pollutant decreases the death rate, especially
when the level X is low.

(iii) For these reasons, we should recommend the consideration of adopting a public policy to increase the level of pollution, perhaps by spraying the countryside, at least in those
localities where the current level of the pollutant is low.

TABLE 4: *Dependence of the appraent effect of a totally ineffective pollutant on the range of variation in population sizes.*

I. Population size 4-9

Pollution level X	Mean death rates W̄(X,Z) in triplets of localities crossclassified according to X and Z = no. at risk						Mean death rate W(X) in localities with pollution X
	Z=4	5	6	7	8	9	
1	275.0						275.0
2	275.0	260.0					267.5
3	275.0	260.0	250.0				261.7
4		260.0	250.0	242.9			251.0
5			250.0	242.9	237.5		243.5
6				242.9	237.5	233.3	237.9
7					237.5	233.3	235.4
8						233.3	233.3

II. Population size 5-10.

X	Z=5	6	7	8	9	10	W̄(X)
1	220.0						220.0
2	220.0	216.7					218.3
3	220.0	216.7	214.3				217.0
4		216.7	214.3	212.5			214.5
5			214.3	212.5	211.1		212.6
6				212.5	211.1	210.0	211.2
7					211.1	210.0	210.6
8						210.0	210.0

III. Population size 6-11.

X	Z=6	7	8	9	10	11	W̄(X)
1	183.3						183.3
2	183.3	185.7					184.5
3	183.3	185.7	187.5				185.5
4		185.7	187.5	188.9			187.4
5			187.5	188.9	190.0		188.8
6				188.9	190.0	190.0	189.9
7					190.0	190.0	190.0
8						190.0	190.0

IV. Population size 7-12.

X	Z=7	8	9	10	11	12	W̄(X)
1	157.1						157.1
2	157.1	162.5					159.8
3	157.1	162.5	166.7				162.1
4		162.5	166.7	170.0			166.4
5			166.7	170.0	172.7		169.8
6				170.0	172.7	175.0	172.6
7					172.7	175.0	173.9
8						175.0	175.0

What would be the result of adopting (iii)? Returning to
Table 3, we see that each of the six panels refers to three
different levels of the pollutant: 1, 2, 3 or 4, 5, 6, etc.
We note that if, through spraying or otherwise, the two lower
levels of the pollutant are replaced by the highest, the effect
on the numbers of deaths would be exactly nothing. The data
imply that the number of deaths would be unchanged. The beneficial conclusion suggested by Table 4 is not inherent in the data.
Rather, it is an artifact produced by dealing not with the triplets (X,Y,Z) as given directly by the observations, but with the
values of X and W = Y/Z computed for each locality. Here is
spurious correlation.

Comparing the four parts of Table 4, constructed by considering four overlapping populations sizes, reveals a surprising
phenomenon. Whether the apparent (spurious) effect of increasing
pollution is beneficial or harmful depends, seems to depend, on
the populations size considered! And yet Table 4 was constructed
to try to eliminate the effect of the interfering variable Z,
the population size. Although such changes in the range of Z
should be of little consequence, their effect on the appearance
of Table 4 appears dramatic, completely altering the conclusions
to be drawn from the table. The conslusions are indeed spurious.

The general principle that Traxler's example is intended to
suggest can be formulated heuristically. The object of the empirical study is to estimate the effect on a variable Y of an intentional change in the level X of another variable (or other
variables). The term effect refers not to any single unit of
observation (locality) but to a population of such units. The
information available for the study consists of values of not
only X and Y but also of some other variables $Z_1, Z_2, \cdots, Z_s$
which are suspected of being involved somehow in the mechanisms
that connects X and Y. The safe method of studying the effect
on Y of an intentional change in X, while the values of the
Z are left to vary as they will, is through an investigation of
the *joint* variabilty of all the s+2 variables involved (X, Y,
$Z_1, \cdots, Z_s$). It is this simultaneous variation that characterizes
the complex mechanism involved. We are interested in a single
detail: What will happen to the values of Y (the number of
deaths) if the values of X are modified in a specified manner?
Admittedly, the direct investigation of the variability of
(X, Y, $Z_1, \cdots, Z_s$) is cumbersome so that the tendency to reduce
the number of interfering (often called nuisance) variables Z is
understandable. However, any such reduction is equivalent to
introducing into the mechanism studied some elements that are
extraneous to it. Traxler's example illustrates the pernicious

effect of replacing the triplet (X, Y, Z) by the pair
(X, W = Y/Z) which looks like a natural thing to do. And this is
a common error.

There are other statistical pitfalls common in applied statistics: The confusion between crude rates and net rates; the
introduction of bias and other difficulties by employing transformed (or transgenerated) variables rather than the variables in
which you are interested (see Neyman and Scott, 1960); the incom-
prehensibility introduced by failure to randomize (for example,
by allowing self-selection, see Yerushalmy, 1972); the use of
curve-fitting in place of a solid model footed in theory, to
mention some that are very troublesome.

REFERENCES

Neyman, J. (1952). *Lectures and Conferences on Mathematical
Statistics and Probability*, 2nd edition, Graduate School,
U. S. Department of Agriculture, Washington, D.C.

Neyman, J. (1972). Epilogue of the Health-Pollution Conference.
*Proceedings Sixth Berkeley Symposium on Mathematical Statis-
tics and Probability, Vol. 6.* University of California Press,
Berkeley and Los Angeles. 575-587.

Neyman, J. and Scott, E. L. (1960). Correction for bias introduced
by a transformation of variables. *Annals of Mathematical
Statistics,* 31, 643-655.

Pearson, Karl (1897). Mathematical contributions to the theory
of evolution--On a form of spurious correlation which may
arise when indices are used in the measurement of organs.
Proceedings, Royal Society of London, 60, 489-498.

Yerushalmy, J. (1972). Self-selection--a major problem in observational studies. *Proceedings Sixth Berkeley Symposium on
Mathematical Statistics and Probability, Vol. 4.* University
of California Press, Berkeley and Los Angeles. 329-342.

[Received July 1972]

L. Orloci, C. R. Rao, and W. M. Stiteler, (eds.),
Multivariate Methods in Ecological Work, pp. 253-262. All rights reserved.
Copyright © 1979 by International Co-operative Publishing House, Fairland, Maryland

CONFIDENCE INTERVALS FOR SIMILARITY MEASURES USING THE TWO SAMPLE JACKKNIFE

WOOLLCOTT SMITH, DAVID KRAVITZ
J. FREDERICK GRASSLE

Woods Hole Oceanographic Institution
Woods Hole, Massachusetts 02543 USA

SUMMARY. Similarity between two communities is usually estimated from relatively small samples taken from the two communities. In this paper we assume that we have a random sample of individuals from both communities. We can then apply an approximate method known as the two sample jackknife (Miller, 1974) to reduce the bias and obtain estimates of the sampling variance.

KEY WORDS. communities, similarity measures, confidence interval, jackknife.

1. INTRODUCTION

Measures of community similarity play an important role in the study of community structure and change over time. Similarity between two communities, say A and B , is usually estimated from relatively small samples taken from the two communities. Most discussions of similarity indices overlook this statistical problem. In this paper we assume that we have a random sample of individuals from both communities A and B . We can then apply an approximate method known as the two sample jackknife (Miller, 1974) to obtain relatively unbiased estimates of the true similarity between the communities as well as approximate estimates of the sampling variance. The method depends on forming a set of estimates from the original estimator by successively removing and replacing individuals from the two random samples.

This approach is not applicable to all ecological problems for a number of reasons. First, because of patchiness, individuals sampled from a community by a quadrat or other method often are not a

random sample of the entire community. Second, estimates of the
sampling variance are unnecessary if the samples from a community
are large (i.e., thousands of individuals). In addition, the
methods we will discuss, although theoretically straightforward,
require computing capabilities that until fairly recently were
unavailable or too costly.

Recent work in experimental ecology has stressed the concept
of the microcosm experiment. That is, the study of ecological
processes through the use of relatively small enclosed experimental
ecosystems. For the statistician these experimental ecosystems have
both advantages and disadvantages. An advantage is that it is
relatively easy to obtain samples of individuals that are nearly
random samples of the entire community. The major problem is
that because the experimental ecosystem is small the number of
individuals sampled from the system must be limited so that the
investigator does not become the chief predator of the system that
he is observing. The methods developed in this paper enable us to
take advantage of the random sampling properties of the microcosm
experiments to find the approximately unbiased estimates of similarity and also to obtain the sampling variance, even when the
sample size is small.

Of course our work will also apply to field situations where
random samples of individuals can be drawn from the population.
For instance, in deep sea benthic communities many species are
nearly randomly distributed (Jumars, 1975). In such cases, these
techniques give at least a lower bound on the size of the confidence interval for similarity estimates.

In the following sections we first review the properties of
the normalized expected species shared (NESS) family of similarity
measures. Grassle and Smith (1976) have shown that its estimator
is approximately unbiased and thus it is a likely candidate for
jackknifing. In Section 3 we develop the two sample jackknife
estimator, and in Section 4 we apply this technique to study the
effect of low levels of Number #2 fuel oil on the benthic community
in large fiberglass tanks at the Marine Ecosystem Research Laboratory (MERL) at the University of Rhode Island.

2. PROPERTIES OF THE NESS FAMILY OF SIMILARITY MEASURES

In this section we define and review some of the properties
of the generalized Morisita Index or normalized expected species
shared (NESS) proposed by Grassle and Smith, 1976). Let π denote
the community vector of length K , where π_j is the proportion
of species j in the community. The expected number of species
in common in two multinomial samples of size m from communities
π' and π'' is then

$$\text{ESS}(\underset{\sim}{\pi}',\underset{\sim}{\pi}'';m) = \sum_{j=1}^{K} [1-(1-\pi'_j)^m] [1-(1-\pi''_j)^m] \quad . \qquad (1)$$

A similarity measure should take values between 0 and 1, with similarity equal to one if and only if $\underset{\sim}{\pi}' = \underset{\sim}{\pi}''$. To form equation (1) one can normalize by dividing by the expected number of species shared in two random samples of size m from the same community,

$$\text{NESS}(\underset{\sim}{\pi}',\underset{\sim}{\pi}'';m)$$

$$= 2 \times \text{ESS}(\underset{\sim}{\pi}',\underset{\sim}{\pi}'';m)/[\text{ESS}(\underset{\sim}{\pi}',\underset{\sim}{\pi}';m) + \text{ESS}(\underset{\sim}{\pi}'',\underset{\sim}{\pi}'';m)] \quad . \quad (2)$$

NESS weighs each species by the probability that it will appear in a sample of size m from the community. Thus the larger m the greater the weight given to the rarer species. For $m = 1$ NESS is the Morisita Index (Morisita, 1959).

Let us suppose random samples are drawn from communities $\underset{\sim}{\pi}'$ and $\underset{\sim}{\pi}''$. Let the vector $\underset{\sim}{N}$ denote the random sample where N_j is the number of individuals of species j drawn, and let N denote the total number of individuals in the sample. The estimator for NESS we used is formed by taking the ratio of the minimum variance unbiased estimators of the numerator and denominator of equation (2), Smith and Grassle (1977):

$$\widehat{\text{NESS}}(\underset{\sim}{N}',\underset{\sim}{N}'';m) = \frac{2\ \widehat{\text{ESS}}(\underset{\sim}{N}',\underset{\sim}{N}'';m)}{\widehat{\text{NESS}}(\underset{\sim}{N}';m) + \widehat{\text{ESS}}(\underset{\sim}{N}'';m)} \quad , \qquad (3)$$

where $\widehat{\text{ESS}}(\underset{\sim}{N}',\underset{\sim}{N}'';m)$ is the expected number of species in two sub-samples of size m drawn from samples $\underset{\sim}{N}'$ and $\underset{\sim}{N}''$.

$$\widehat{\text{ESS}}(\underset{\sim}{N}',\underset{\sim}{N}'';m) = \sum_{i=1}^{K} \left[1 - \frac{\dbinom{N'-N'_i}{m}}{\dbinom{N'}{m}} \right] \left[1 - \frac{\dbinom{N''-N''_i}{m}}{\dbinom{N''}{m}} \right]$$

and $\widehat{\text{ESS}}(N;m)$ is the expected number of species in common in two disjoint subsamples drawn from the same sample,

$$\widehat{\text{ESS}}(\underset{\sim}{N};m) = \sum_{i=1}^{K} \left[1 - 2\ \frac{\dbinom{N-N_i}{m}}{\dbinom{N}{m}} + \frac{\dbinom{N-N_i}{2m}}{\dbinom{N}{2m}} \right] \quad .$$

Since $\widehat{\text{NESS}}$ is the ratio of two unbiased estimators, it is natural to assume that $\widehat{\text{NESS}}$ is nearly unbiased; both Morisita (1959) and Grassle and Smith (1976) have confirmed this using simple simulation techniques. However, unlike the case of the expected species diversity index we can not obtain exact unbiased estimators or estimate the sampling variance. The jackknife technique outlined in the next section gives us a practical way of overcoming these difficulties.

3.   TWO SAMPLE JACKKNIFE FOR NESS

In this section we outline the use of the two sample jackknife to reduce the bias of the NESS estimator and to find confidence interval estimates for the NESS similarities. A general review of the jackknife or holdout method introduced by Quenouille (1956) and Tukey (1958) is given by Miller (1974). We outline here only those parts of the theory particular to the similarity measures between communities.

We assume the expected value of $\widehat{\text{NESS}}$ has the form:

$$E[\widehat{\text{NESS}}(N',N'';m)] = \text{NESS}(\underset{\sim}{\pi}',\underset{\sim}{\pi}'';m) + a/N' + b/N'' + c/N'N''$$
$$+ 0(N'^{-2},N''^{-2}) \quad , \tag{4}$$

where N' and N'' are the total number of individuals in sample $\underset{\sim}{N}'$ and $\underset{\sim}{N}''$ respectively. Now let $\widehat{\text{NESS}}'_{-i}$ be the estimated similarity with the ith individual removed from sample $\underset{\sim}{N}'$

$$\widehat{\text{NESS}}'_i = (N'-\tfrac{1}{2})\widehat{\text{NESS}} - (N'-1)\,\widehat{\text{NESS}}'_{-i} \quad . \tag{5}$$

Noting that $\widehat{\text{NESS}}'_{-i}$ is calculated from a sample of $N'-1$ individuals and applying equation (4) we have

$$E[\widehat{\text{NESS}}'_i] \doteq \tfrac{1}{2}\,\text{NESS}(\underset{\sim}{\pi}',\underset{\sim}{\pi}'';m) - \tfrac{1}{2}\,a/N'' + \tfrac{1}{2}\,b/N'' - \tfrac{1}{2}\,c/N'N'' \quad .$$

Similarly, removing a single individual from the random sample N'' and defining $\widehat{\text{NESS}}''_i$ as in (5), we have

$$E[\widehat{\text{NESS}}''_i] \doteq \tfrac{1}{2}\text{NESS}(\underset{\sim}{\pi}',\underset{\sim}{\pi}'';m) + \tfrac{1}{2}\,a/N' - \tfrac{1}{2}\,b/N'' - \tfrac{1}{2}\,c/N'N'' \quad . \tag{6}$$

Let $\overline{\text{NESS}'}$ denote the sample mean for $\widehat{\text{NESS}}'_i$

$$\overline{\text{NESS}'} = \frac{1}{N}\sum_{i=1}^{N}\widehat{\text{NESS}}'_i$$

$$= \frac{1}{N} \sum_{j=1}^{k} N_j \ \widehat{NESS}_j^{\,\prime} \ , \tag{7}$$

where $\widehat{NESS}_j^{\,\prime}$ refers to the result of equation (5) when an individual of species j is removed. We define $\overline{NESS''}$ in the same way. The sample variance of $\widehat{NESS}_i$ is

$$\sigma'^2 = \frac{1}{N-1} \sum_{i=1}^{N} (NESS_i^{\,\prime} - \overline{NESS'})^2$$

$$\tag{8}$$

$$= \frac{1}{N-1} \sum_{j=1}^{k} N_j (\widehat{NESS}_j^{\,\prime} - \overline{NESS'})^2 \ .$$

We define σ''^2 in the same way.

Our two sample jackknife estimator is then

$$\overline{NESS} = \overline{NESS'} + \overline{NESS''}$$

and

$$E[\overline{NESS}] = NESS(\underset{\sim}{\pi}',\underset{\sim}{\pi}'';m) - \frac{c}{N_1 N_2} + O(N'^{-2}, N''^{-2}). \tag{9}$$

Thus the two sample jackknife has removed from the bias the terms of order $1/N$.

If we consider each $NESS_i$ as an independent observation we can compute the approximate standard deviation of $\overline{NESS}$,

$$\sigma_N = \left[\frac{1}{N'} \ \sigma'^2 + \frac{1}{N''} \ \sigma''^2 \right]^{\frac{1}{2}} \ . \tag{10}$$

Equations (10) and (8) are used to form approximate estimates of confidence intervals for the community similarities. Computational efficiency can be increased considerably by storing as intermediate results the hypergeometric probabilities. A Fortran program for this algorithm is available from the first author.

4. EFFECT OF #2 FUEL OIL ON EXPERIMENTAL BENTHIC COMMUNITIES

The methods developed in this paper are particularly applicable to small experimental eocsystems. In this section we describe such an experimental ecosystem and apply the jackknife similarity estimates to the study of the effect of #2 fuel oil on the benthic community. The experiment described was conducted at the Marine Ecosystem Research Laboratory at the University of Rhode Island from the fall of 1976 through the summer of 1977. Nine experimental tanks were used. Each was 5.5 meters high and 1.8 meters in diameter and contained about one ton of sediment. Sea water from Narraganset Bay, Rhode Island was pumped into each tank so that the average residency time of water in the tank was about 30 days. A more detailed description of the system is given in Pilson *et al.* (1977).

In this experiment three tanks were designated as controls, three tanks had no water flow through them and three tanks were treated with #2 fuel oil. In the oiled tanks the concentration in the water column was maintained between 100 to 250 parts per billion, ppb. In this example we will use data from the three oiled tanks and two control tanks. The third control tank was removed from the analysis because three large predatory crabs, two *Libinia emarginata* and one *Ovalipes ocellatus*, appeared in the tank, substantially altering the characteristics of the community. In future experiments these large benthic predators will be removed by trapping or kept at a fixed level in all tanks.

Each tank was sampled once a month. Ten small (4.155 cm^2) cores were taken from the tank. This provided our random sample of individuals from the tank community. The effects of patchiness are minimized by taking a relatively large number of small samples and by the homogeneous nature of the sediment within the tanks.

Community similarity measures can be used to address questions about how well the experimental ecosystems replicate and to assess the difference in community structure between the oiled and control tanks. To do this we used the jackknifed NESS similarity measure at ten individuals, m = 10. One control community (tank 5) was compared to the other control community (tank 8) and the three oiled communities (tanks 2, 7, 9) for each monthly sampling period. Figure 1 displays the five NESS confidence interval estimates for each month. It gives an easily interpretable picture of the change in community structure over time. During the first phase of the experiment all tanks replicated well, with no differences between oiled and control tanks. Since the densities were relatively large the number of individuals sampled in the 10 cores was also large, resulting in rather narrow confidence interval estimates. In the later months, densities in the experimental tanks dropped sharply, Figure 2, and the community structure changed as evidenced in the decrease in similarity between the control and oiled tanks. Since densities decreased, fewer individuals were sampled resulting in wider confidence interval estimates. In summary, both in community structure (Figure 1) and in absolute densities (Figure 2), the oiled tanks differ from the controls.

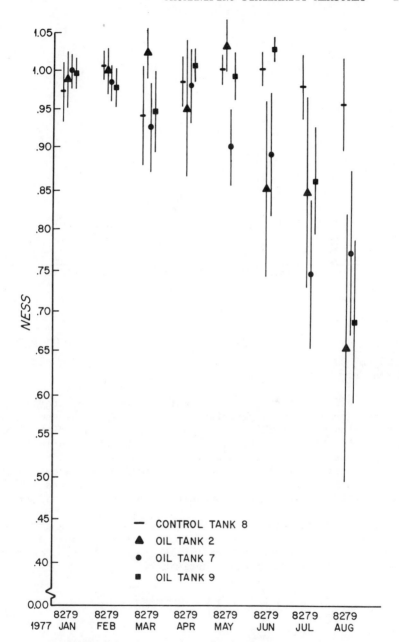

FIG. 1: Control tank 5 compared with control tank 8 and three oiled tanks, using the JNESS estimator to estimate NESS similarity at m = 10 individuals. Vertical lines indicate approximate 95% confidence intervals.

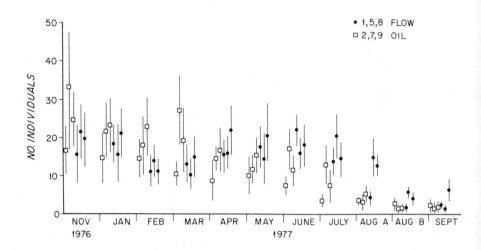

FIG. 2: Mean and 95% confidence interval of individuals per 4.155 cm^2 *sample within each of the flow (control) and oiled tanks. (Taken from Grassle et al. 1978).*

The effects observed on the benthic communities are associated with an accumulation of petroleum hydrocarbons in the sediment. Although the water column concentration was maintained at about 190 ppb, by the end of the 20*th* week the concentration in the surface ooze was 200 ppm, an increase of three orders of magnitude. The concentration in the top one cm of sediment was about 15 ppm. Both these concentrations are in the range where effects have been demonstrated on the benthic communities, Blumer and Sass (1972).

To demonstrate how small sample bias and sampling error can effect results we estimate the similarity between tanks as in Figure 1, using percent similarity rather than JNESS. For small samples the percent similarity estimator,

$$\widehat{PS} = \sum_{i=1}^{k} \min\left(\frac{N_i'}{N'}, \frac{N_i''}{N''}\right) ,$$

will on the average under estimate the true percent similarity between communities,

$$PS = \sum_{i=1}^{k} \min\left(\pi_i', \pi_i''\right)$$

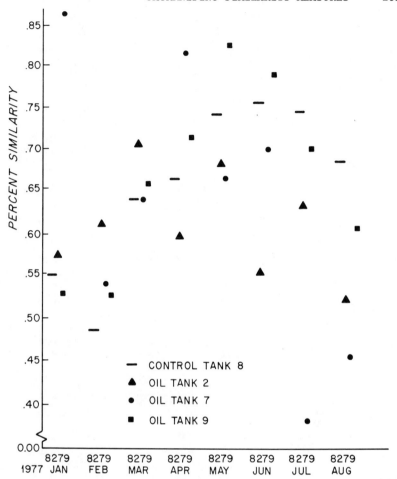

FIG. 3: Control tank 5 compared with control tank 8 and three oiled tanks using standard maximum likelihood estimator for percent similarity.

(Morisita, 1959). There are several points to note in Figure 3. First percent similarity consistantly underestimates the similarity between communities. This gives a false impression of replicability of the five communities. The difference in the July and August controlled and oiled tanks is not as apparent when we use the percent similarity estimator. Presumably this results from the sample size dependent bias and the fact that the percent similarity is highly dependent on the dominant species in the community.

262   W. SMITH, D. KRAVITZ, AND J. F. GRASSLE

info">This work was supported by NOAA Sea Grant 04-8-M01-149 and by The Environmental Protection Agency, Grant Number R803902030 through a subcontract from the University of Rhode Island.

REFERENCES

Blumer, M. and Sass, J. (1972). Oil pollution: persistance and degradation of spilled fuel. *Science*, 176, 1120-1122.

Grassle, J. F. and Smith, W. K. (1976). A similarity measure sensitive to the contribution of rare species and its use in investigation of variation in marine benthic communities. *Oecologia*, 25, 13-22.

Grassle, J. F., Grassle, J. P., Brown-Leger, L. S., Maciolek, N. J., and Lanyon - Duncan, C. H. (1978). Benthic communities in experimental ecosystems and the effects of petroleum hydrocarbons. 1978 Report to the Marine Ecosystems Research Laboratory, Kingston, Rhode Island.

Jumars, P. A. (1975). Environmental grain and polychaete species' diversity in a bathyal benthic community. *Marine Biology*, 30, 253-266.

Miller, R. G. (1974). The jackknife - a review. *Biometrika*, 61, 1-15.

Morisita, M. (1959). Measuring of interspecific association and similarily between communities. *Kyushu University Faculty of Science Memoirs. Series E, Biology*, 3, 65-80.

Pilson, M. E. Q. Vargo, G. A., Gearing, P., and Gearing, J. N. (1977). The Marine Ecosystems Research Laboratory: a facility for the investigation of effects and fates of pollutants. Proceedings of the Second National Conference on the Interagency Energy/Environment Research and Development Program.

Quenouille, M. H. (1956). Notes on bias in estimation. *Biometrika*, 43, 353-360.

Smith, W. K. and Grassle, J. F. (1977). A diversity index and its sampling properties. *Biometrics*, 33, 283-292.

Tukey, J. W. (1958). Bias and confidence in not-quite large samples (abstract). *Annals of Mathematical Statistics*, 29, 614.

[*Received June* 1977. *Revised February* 1979]

L. Orloci, C. R. Rao, and W. M. Stiteler, (eds.),
Multivariate Methods in Ecological Work, pp. 263-278. All rights reserved.
Copyright © 1979 by International Co-operative Publishing House, Fairland, Maryland

ANALYSIS OF NICHE OVERLAP

R. KIRK STEINHORST

Institute of Statistics
Texas A&M University*

SUMMARY. Measures of niche overlap are reviewed. Terms are defined and the underlying theory is considered. The scalar approach of paired comparisons and the multivariate approach of multiple comparisons are contrasted. An example is given.

KEY WORDS. niche, overlap, measures, scalar, multivariate.

1. MEASURES OF NICHE OVERLAP

Overlap measures have been used for a variety of purposes, from assessing potential competition between species in a community to measuring the degree of similarity of two communities. Morisita (1959) defines the similarity of two samples from two (possibly) different communities as

$$C_\lambda = \frac{2 \sum_{i=1}^{S} x_i y_i}{(\lambda_X + \lambda_Y) x . y .} ,$$

where S is the number of species encountered, x_i is the number of individuals of species i in sample X , $x . = \Sigma x_i$, and

---

*Present address: Department of Agricultural Economics and Applied Statistics, University of Idaho, Moscow, Idaho 83843, USA.

$\lambda_X = \Sigma[x_i(x_i - 1)]/[x.(x. - 1)]$; y_i , $y.$, and λ_Y are defined analogously for sample Y . This index is 0 when no species occur in common, and equals

$$\frac{x. - 1}{x.} \cdot \frac{\Sigma x_i^2}{\Sigma x_i^2 - x.}$$

when $x_i = y_i$, for all i . Horn (1966) modified Morisita's index by assuming sampling with replacement. Furthermore, assuming $x. = y.$, he obtains

$$\hat{C}_\lambda = \frac{2\Sigma x_i y_i}{\Sigma x_i^2 + \Sigma y_i^2} .$$

In this form x_i and y_i could just as easily be proportions as numbers and thus the measure can be extended to include overlaps based on weight or cover or other measures in addition to number of individuals. Horn also defines an information based measure,

$$R_o = \frac{\Sigma(x_i + y_i) \log(x_i + y_i) - \Sigma x_i \log x_i - \Sigma y_i \log y_i}{(x. + y.) \log(x. + y.) - x. \log x. - y. \log y.} .$$

When proportions are used one sets $x. = y. = 1$.

In fact, there are numerous measures in addition to these. Kulcyznski's index (see Oosting, 1956) gives the per cent of resources shared by two organisms,

$$K = \frac{\sum\limits_{i=1}^{S} 2 \min(x_i, y_i)}{\sum\limits_{i=1}^{S} (x_i + y_i)} .$$

Hubbard and Hansen (1976) use rank correlation in addition to Kulcyznski's index to gauge dietary similarity. Wiens (1974) defines an areal index of interspecific overlap for birds as

$$IO = \left(\sum_{i=2}^{S} \frac{[i(a_i/a.)]}{S} - M \right) /(1 - M) ,$$

where a_i is the area shared by i species, $a. = \Sigma a_i$, and $M = \min(a_i/a.)$. The index varies from $1/S$ when there is no common areas shared to 1 when there is complete overlap of territories. Diggle (1976) characterizes an individual's competitive

influence as a circle of radius y in a two dimensional space.
He then defines overlap as

$$
q(y_i y_j) = \begin{cases} 0 & \text{if } y_j < r_{ij} - y_i \ , \\ (y_i + y_j - r_{ij})/2y_i & \text{if } r_{ij} - y_i \leq y_j \leq r_{ij} + y_i \ , \\ 1 & \text{if } y_j > r_{ij} + y_i \ , \end{cases}
$$

where r_{ij} is the distance between circle centers. Cormack (1977)
summarizes work in this area. Measures of gamma diversity (Whittaker, 1972) have also been used as simarity measures (Hummon, 1974).

A simple proof that Horn's modification to Morisita's index,
$\hat{C}_\lambda$, lies in the unit interval can be obtained as follows: Let
$\underset{\sim}{x}' = (x_1, x_2, \cdots x_s)$, $\underset{\sim}{y}' = (y_1, y_2, \cdots, y_s)$, and denote the
Enclidean norm by $|| \ ||$. We have

$$
0 \leq ||\underset{\sim}{x} - \underset{\sim}{y}||^2 = (\underset{\sim}{x} - \underset{\sim}{y})'(\underset{\sim}{x} - \underset{\sim}{y}) = \underset{\sim}{x}'\underset{\sim}{x} - 2\underset{\sim}{x}'\underset{\sim}{y} + \underset{\sim}{y}'\underset{\sim}{y} \ ,
$$

or

$$
2\underset{\sim}{x}'\underset{\sim}{y} \leq \underset{\sim}{x}'\underset{\sim}{x} + \underset{\sim}{y}'\underset{\sim}{y} = ||\underset{\sim}{x}||^2 + ||\underset{\sim}{y}||^2 \ .
$$

Now

$$
\hat{C}_\lambda = \frac{2\underset{\sim}{x}'\underset{\sim}{y}}{||\underset{\sim}{x}||^2 + ||\underset{\sim}{y}||^2} \ ,
$$

and, since $x'y$ is nonnegative, $0 \leq \hat{C}_\lambda \leq 1$.

A whole host of overlap measures can be defined from similar
vector inequalities. We might have the triangle inequality index,

$$
0 \leq \frac{||\underset{\sim}{x} + \underset{\sim}{y}||}{||\underset{\sim}{x}|| \quad ||\underset{\sim}{y}||} \leq 1 \ ,
$$

the Cauchy-Schwarz index,

$$
0 \leq \frac{|\underset{\sim}{x}'\underset{\sim}{y}|}{||\underset{\sim}{x}||^{\frac{1}{2}}||\underset{\sim}{y}||^{\frac{1}{2}}} \leq 1 \ ,
$$

or, in more generality, the Hölder inequality index (if $||\underset{\sim}{x}||$ =
$(\Sigma \ x_i|^p)^{1/p}$, $1 \leq p < \infty$) ,

$$0 \leq \frac{|\underset{\sim}{x}'\underset{\sim}{y}|}{||\underset{\sim}{x}||^p \; ||\underset{\sim}{y}||^q} \quad , \quad \frac{1}{p} + \frac{1}{q} = 1 \quad .$$

Clearly, with such a variety of indices available there needs to be some way to analyze the basic underlying theory involved and a rationale developed for choosing an index. A major survey of the fundamental concepts in this area appears in Whittaker and Levin (1975). A paper by Whittaker, Levin, and Root (1973) clarifies the niche-related definitions. The *fundamental niche* of Hutchinson (1958), defined as an n-coordinate hypervolume describing the exis-tence space of an organism in terms of n environmental variables, is termed a *niche* if the dimensions are restricted to intracommunity variables and gradients. If intercommunity dimensions are used the hypervolume describes an organism's *habitat*. *Ecotope* is reserved for an amalgam hypervolume encompassing both niche and habitat characteristics. This formulation parallels the definition of α, β, and γ diversity given by Whittaker (1972). At different scales, niches and habitats form a continuum of hypervolumes. For example, if species differentiation occurs along a food size gradient, one would theoretically have Figure 1. The overlap between species 5 and 6, then, is the shaded area of intersection. One might develop an overlap measure using components of variance by considering vari-ability among and within species. Alternatively, the probability that a randomly selected organism from each species will compete for a given resource might be used as an indication of overlap. In an appendix, Orians and Horn (1969) interpret Morisita's index cal-culated between and within habitats as a 'probability of overlap' measure, for niches and habitats, respectively.

In n dimensions Figure 1 generalizes to the population cloud of Bray and Curtis (1957). For three resources, equi-probability hyperellipsoids might appear as Figure 2. Overlap could be measured as the volume of intersection of γ-probability tolerance ellipsoids. The amount of overlap is a function of the centers of the ellipsoids, $\underset{\sim}{\mu}_1$ and $\underset{\sim}{\mu}_2$, and the orientations and volumes determined by the covariance structures, $\underset{\sim}{\Sigma}_1$ and $\underset{\sim}{\Sigma}_2$. If $\underset{\sim}{\Sigma}_1 = \underset{\sim}{\Sigma}_2$, the distance between ellipsoids is measured by Mahalanobis' distance, $(\underset{\sim}{\mu}_1 - \underset{\sim}{\mu}_2)'$ $\underset{\sim}{\Sigma}^{-1}(\underset{\sim}{\mu}_1 - \underset{\sim}{\mu}_2)$. Overlap, then, is a multivariate concept which depends not only on the mean amounts of a given resource utilized but also on the covariance structure as well. The overlap measures appearing in the literature ignore the covariances. For example, in Horn's R_0 no account is made for the fact that if an organism uses p_1 proportion of resource 1, then it is likely to use a simi-lar amount of a positively related resource but a different amount of a negatively related resource.

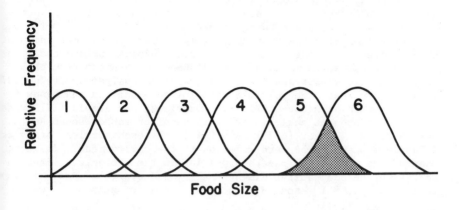

FIG. 1:   Species differentiation by size of prey.

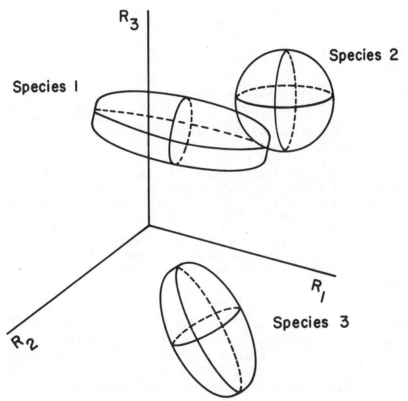

FIG. 2:   Niche relations in 3 dimensions.

There are various examples of multivariate analyses of niche and habitat relations. In studying the bird communities of Northern Georgia, Shugart and Patten (1972) rely on Hotelling's T^2 and the related Mahalanobis distance. They primarily consider pairs of bird species. In attempting to integrate information about all the species, they use the Mahalanobis' distance of a niche from the origin and two scalar measures in a graphical analysis. Green (197 1974) argues that "the multiple discriminant model is analogous to the Hutchinsonian niche model····." Dudzinski and Arnold (1973) use principal component analysis (PCA) in a comparison of sheep and cattle diets. Werger (1973) uses PCA in phytosociological study to illustrate plant niches and relation to the environment. In fact, many of the ordination papers appearing in the ecological literatur address niche and habitat dimensionality and overlap. Poole (1974) tests $H_o: \Sigma_1 = \Sigma_2$ and $H_o: \mu_1 = \mu_2$ in a comparison of two communities consisting of six species of *Drosophila* (fruit flies). He rejects equality of covariance structure and uses factor analyse to investigate the differing patterns of Σ_1 and Σ_2 .

In the sections below a comparison of the univariate scalar or overlap index approach and the multivariate approach are compared using data on diets of cattle, horses, and deer in northwestern Colorado, U.S.A.

2. SCALAR VS. MULTIVARIATE ANALYSIS

2.1: Diet Data. Hubbard and Hansen (1976) report the results of a study of cattle (*Bos taurus*), wild horses (*Equus caballus*), and mul deer (*Odocoileus hemionus*) diets collected in the Piceance Basin, Colorado. Fecal samples were collected according to a factorial de sign (animal species and altitude) in randomized blocks (five dist ridges). A single composited sample for each animal-altitude-spec combination was subjected to microscopic analysis to obtain diet percentages. Using 20 species which represented at least 2% of on of the 45 diets, Hubbard and Hansen analyzed the diets using a MAN suggested by the author followed by classical scalar analysis usin, diversity (H') and overlap (K) indices. In addition, they followe a suggestion of the author to compare diets in terms of rank corre lations.

2.2 Classical Analysis. There are $\binom{9}{2} = 36$ overlap indices to be computed and interpreted if the usual pairwise indices are used. Table 1 gives Morisita's and Horn's values along with a confidence interval on the true overlap obtained by inverting a permutation test for specified overlap (Garratt and Steinhorst, 1976). One hundred permutations were generated to simulate the null distribut of the test statistic. With the exception of the comparison of deer diets at high and medium altitudes, Horn's R_0 is

TABLE 1: Observed overlaps and 90% confidence intervals based on 20 diet species (S = 20). Horn's values below the diagonal and Morisita's above.

	High Cattle	High Horses	High Deer	Medium Cattle	Medium Horses	Medium Deer	Low Cattle	Low Horses	Low Deer
Hi Cattle		.954 (.53,1)	.033 (0,1)	.610 (.35,1)	.592 (.35,1)	.020 (0,.29)	.491 (0,1)	.299 (0,1)	.007 (0,.21)
Hi Horses	.934 (.63,1)		.162 (0,1)	.712 (.43,1)	.710 (.44,1)	.154 (0,1)	.603 (.38,1)	.392 (0,1)	.041 (0,.25)
Hi Deer	.119 (0,.35)	.274 (0,.49)		.025 (0,.26)	.022 (0,.25)	.988 (.51,1)	.018 (0,.27)	.009 (0,.22)	.275 (0,1)
Med Cattle	.790 (.54,1)	.834 (.58,1)	.114 (0,.35)		.933 (.57,1)	.026 (0,.19)	.698 (.49,1)	.731 (.47,1)	.010 (0,.21)
Med Horses	.792 (.55,1)	.830 (.59,1)	.098 (0,.33)	.950 (.66,1)		.026 (0,.17)	.821 (.59,1)	.872 (.56,1)	.009 (0,.23)
Med Deer	.109 (0,.35)	.274 (0,1)	.923 (.58,1)	.126 (0,.34)	.114 (0,.32)		.019 (0,.28)	.018 (0,.24)	.372 (0,1)
Low Cattle	.765 (.55,1)	.778 (.57,1)	.104 (0,.36)	.827 (.63,1)	.903 (.69,1)	.116 (0,.36)		.785 (.55,1)	.016 (0,.19)
Low Horses	.642 (.47,1)	.667 (.50,1)	.077 (0,.32)	.826 (.61,1)	.915 (.67,1)	.102 (0,.33)	.913 (.70,1)		.008 (0,.22)
Low Deer	.068 (0,.28)	.158 (0,.35)	.481 (0,1)	.079 (0,.31)	.067 (0,.30)	.644 (.41,1)	.085 (0,.29)	.062 (0,.29)	

uniformly higher than Morisita's $\hat{C}_\lambda$. In every case, the hypothesis of complete overlap of cattle and horse diets cannot be
rejected. Except in cases where the variability in the diets gave
rise to confidence intervals of (0, 1), deer diets are significantl‍y
different from cattle and horse diets. The overall significance
of these comparisons cannot be assessed because these are non-
simultaneous confidence intervals.

In Table 2 a similar analysis is given using all 40 diet
components. For Morisita's index the confidence intervals tend
to be wider, but for Horn's R_0 the results are mixed. The same
general conclusions can be drawn although the actual magnitudes
of the indices are changed.

2.3 Multivariate Analysis. If the diet components are uncorrelated or equicorrelated, a univariate analysis such as above
might be justified. The error mean squares and cross-products
matrix from the factorial in blocks MANOVA is an estimate, $\underset{\sim}{S}$,
of the underlying covariance matrix $\underset{\sim}{\Sigma}$. The matrix $\underset{\sim}{S}$ appears
in Table 3. A likelihood ratio test of sphericity $H_0: \underset{\sim}{\Sigma} = \sigma^2 \underset{\sim}{I}$
gave an asymptotic chi-square of $\chi^2 = 1122$ with 209 degrees of
freedom which is highly significant (see Anderson, 1958, p. 263),
leading to rejection of H_0 .

A test of equicorrelation, $H_0: \rho_{ij} = \rho$ (Morrison, 1967, p.
252) gave a chi-square value of $\chi^2 = 250.9$ with 189 degrees of
freedom which is significant but not to the same extent as the
sphericity test. Further insight into the lack of equicorrelation
is given by considering the eigenvalues of $\underset{\sim}{S}$. Under the equicorrelation hypothesis the roots of the correlation matrix asso-
ciated with $\underset{\sim}{\Sigma}$ are 19 $(1 - \rho)$'s and a root equal to $1 + 19\rho$.
There is not a single dominant root in the sample and, in fact,
there are eight roots greater than unity (Figure 3).

MANOVA tests for interaction, main effect of animals, and
main effect of altitude were each significant ($p < 0.01$, largest
root test). It is clear from the diet means (Table 4) that these
global hypotheses should be rejected. The deer mean diet vectors
of treatments 7, 8, and 9 are almost orthogonal to the cattle and
horse diets. Projections of the largest root test for the 9 mean
diets were used for multiple comparison. All 36 pairwise comparisons were made for (a) individual diet components, (b) total
grasses, (c) total nongrasses, and (d) grasses minus nongrasses.
No significant differences were found on a component-by-component
basis. However, for (b), (c), and (d) there were significant

TABLE 2: *Observed overlaps and 90% confidence intervals based on 40 diet species (S = 40). Horn's values below the diagonal and Morisita's above.*

	High Cattle	High Horses	High Deer	Medium Cattle	Medium Horses	Medium Deer	Low Cattle	Low Horses	Low Deer
Hi Cattle		.954 (.49,1)	.033 (0,1)	.610 (.31,1)	.592 (.31,1)	.020 (0,1)	.491 (.27,1)	.299 (0,1)	.007 (0,.24)
Hi Horses	.944 (.55,1)		.162 (0,1)	.712 (.38,1)	.709 (.39,1)	.154 (0,1)	.603 (.35,1)	.393 (.23,1)	.041 (0,1)
Hi Deer	.122 (0,.26)	.278 (0,1)		.025 (0,.23)	.022 (0,.27)	.988 (.50,1)	.018 (0,.15)	.009 (0,.28)	.275 (.15,1)
Med Cattle	.798 (.46,1)	.847 (.51,1)	.119 (0,.29)		.933 (.51,1)	.026 (0,1)	.698 (.39,1)	.731 (.40,1)	.010 (0,.15)
Med Horses	.798 (.47,1)	.841 (.51,1)	.102 (0,.29)	.958 (.56,1)		.026 (0,1)	.820 (.47,1)	.872 (.48,1)	.009 (0,.15)
Med Deer	.112 (0,.24)	.281 (0,1)	.932 (.54,1)	.133 (0,.28)	.119 (0,.29)		.019 (0,.17)	.018 (0,.29)	.372 (.20,1)
Low Cattle	.771 (.47,1)	.787 (.50,1)	.107 (0,.25)	.836 (.51,1)	.909 (.56,1)	.118 (0,.27)		.785 (.44,1)	.016 (0,.16)
Low Horses	.648 (.40,1)	.678 (.43,1)	.077 (0,.26)	.834 (.50,1)	.920 (.56,1)	.102 (0,.29)	.921 (.57,1)		.008 (0,.19)
Low Deer	.068 (0,.26)	.158 (0,1)	.481 (.29,1)	.079 (0,.24)	.067 (0,.24)	.644 (.37,1)	.085 (0,.24)	.062 (0,.22)	

TABLE 3: Error matrix S from MANOVA (rounded to nearest integer).

	1	2	3	4	5	6	7	8	9	10	11	12	13	14	15	16	17	18	19	20
1	3154																			
2	-576	1896																		
3	-291	-244	1650																	
4	-529	-101	-116	337																
5	-364	-271	-651	-12	1038															
6	38	295	-43	-56	-101	477														
7	47	-1079	-539	236	366	-421	1579													
8	-111	24	73	-21	-18	-4	-32	42												
9	-91	-32	198	251	-173	-54	148	26	1825											
10	248	-106	87	-93	-23	-72	-107	-18	-509	1000										
11	-129	-60	-65	52	74	12	110	-6	-594	-296	1035									
12	-22	81	-31	2	-3	26	-30	6	47	-71	-82	66								
13	69	-5	29	-28	7	-12	-50	3	-181	-19	16	15	82							
14	9	8	-30	1	26	-3	-36	-11	-123	36	74	-41	-2	106						
15	-1	-17	15	-8	14	-6	-10	4	-40	15	-13	0	14	12	13					
16	-2	10	-9	3	1	3	-3	-3	46	-48	-15	2	-1	-1	-4	19				
17	-25	-7	8	5	4	1	6	2	32	-21	-11	1	-3	-3	4	9	11			
18	-21	33	3	-14	-7	25	-18	6	4	-38	-2	12	0	3	2	-3	-5	22		
19	18	-4	6	-5	4	-3	-13	-0	-15	-40	-6	6	26	-4	5	1	1	-2	11	
20	-596	107	1	107	73	-97	-130	25	89	85	-73	-3	27	-5	4	-9	-5	-5	8	373

TABLE 4: *Average diets (%) for 5 ridges.*

	1	2	3	4	5	6	7	8	9
Variable	Cattle Hi	Cattle Med	Cattle Low	Horse Hi	Horse Med	Horse Low	Deer Hi	Deer Med	Deer Low
Grasses and grasslikes									
1. Sedge	59.69	21.40	13.29	45.59	19.58	6.01	1.39	0.26	0.09
2. Needleandthread	5.49	21.92	11.40	7.96	24.31	31.57	0.11	1.12	0.25
3. Wheatgrasses	9.18	30.66	9.33	10.24	17.50	10.34	0.29	0.25	0.18
4. Prairie Junegrass	8.54	8.53	18.30	11.62	9.77	6.98	0	0.04	0.05
5. Bromes	0.74	7.52	13.72	1.41	11.62	13.39	0	0.22	0.07
6. Indian ricegrass	2.09	3.32	7.50	4.42	8.89	8.31	0.08	0.12	0.03
7. Bluegrass	5.11	0.35	10.50	1.16	3.64	13.20	0	0	0.02
8. Fescue	2.51	0.88	0.84	2.52	0.83	0.46	0.03	0.56	0.19
Forbs and shrubs									
9. Utah serviceberry	0.86	0.65	0.43	7.83	0.58	0.19	73.51	67.88	13.44
10. Pinyon pine	0	0.03	0.11	0	0	0.03	3.31	11.03	39.53
11. Junipers	0	0	0.34	0	0	0.03	1.48	2.85	37.50
12. Bladderpod	0.11	0.49	0.10	0.09	0.20	0.10	2.99	3.12	2.93
13. Barberry	0	0.09	0.03	0	0	0	3.88	2.71	0.79
14. Big sagebrush	0	0	0	0.12	0	0	0.22	3.43	2.89
15. True mountainmahogany	0	0.04	0	0.03	0	0	2.57	1.71	0.34
16. Snowberry	0.03	0	0.04	0	0	0	3.12	0.83	0.11
17. Bluebells	0.10	0.03	0	0.06	0	0	2.49	0.64	0.07
18. Rabbitbrushes	0.05	0.28	0.33	0.85	0.34	0.62	0.31	1.54	1.18
19. Gambel oak	0	0	0	0.06	0	0	1.78	0.17	0.02
20. Common winterfat	4.82	2.54	12.56	3.91	1.97	7.31	0.48	0.17	0.30

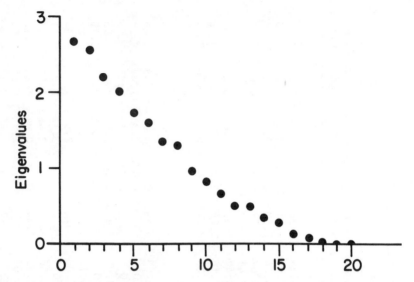

FIG. 3: Eigenvalues of the error correlation matrix.

differences in cattle and horse diets relative to deer diets
(Table 5). The large differences required for significance indicate that there is a relatively large variation within an animal-
altitude combination. The difference between the differences in
(a) and (b) is attributable to the 20 minor species. Since (c)
is a direct function of (a) and (b), the three columns give different views of the same phenomena.

Multivariate analysis provides simultaneous tests of the
same differences which the overlap indices addressed in a one-
pair-at-a-time fashion. We are able to say with 95% confidence
that 1) none of the diets are statistically different for any
single component, but 2) relative to the differential use of
grasses and nongrasses the diet of deer are different from horses
and cattle; finally 3) none of the horse and cattle diets differ
even at different altitudes.

3. SUMMARY AND DISCUSSION

Overlap measures are used in two contexts. If the overlap
is between organisms or organism-locations, then the vector
variates are resources such as nutrient needs, time use, food
size, distance above or below the ground, temperature, and so on.
In this context overlap is termed niche, habitat, or ecotope depending on the nature and range of the resource variables included.
Overlap of communities is a question of similarity of two sample
locations in terms of the relative frequency of organisms occurring

TABLE 5:   *Treatment contrasts.*

Treatment	vs	Treatment		(a) Grasses	(b) Nongrasses	(c) (a) - (b)
Cattle, Hi		Cattle,	Med	-1.23	1.84	-3.07
			Low	8.47	-7.96	16.43
		Horse	Hi	8.43	-6.98	15.41
			Med	-2.79	2.90	-5.69
			Low	3.08	-2.30	5.38
		Deer,	High	91.46	-90.15	181.61
			Med	90.78	-90.11	180.89
			Low	92.48	-93.12	185.59
Cattle, Med		Cattle,	Low	9.70	-9.80	19.50
		Horse,	Hi	9.66	-8.82	18.48
			Med	-1.56	1.06	-2.62
			Low	4.32	-4.13	8.45
		Deer,	Hi	92.69	-91.98	184.68
			Med	92.01	-91.94	183.96
			Low	93.71	-94.95	188.66
Cattle, Low		Horse,	Hi	-0.04	0.98	-1.02
			Med	-11.26	10.86	-22.12
			Low	-5.39	5.67	-11.05
		Deer,	Hi	82.99	-82.18	165.17
			Med	82.31	-82.14	164.45
			Low	84.01	-85.15	169.16
Horse, Hi		Horse,	Med	-11.22	9.88	-21.10
			Low	-5.35	4.68	-10.03
		Deer,	Hi	83.03	-83.17	166.20
			Med	82.35	-83.13	165.48
			Low	84.04	-86.14	170.18
Horse, Med		Horse,	Low	5.87	-5.19	11.07
		Deer,	Hi	94.25	-93.04	187.29
			Med	93.57	-93.00	186.57
			Low	95.27	-96.01	191.28
Horse, Low		Deer,	Hi	88.38	-87.85	176.23
			Med	87.70	-87.81	175.51
			Low	89.39	-90.82	180.21
Deer, Hi		Deer,	Med	-0.68	0.04	-0.72
			Low	1.02	-2.97	3.99
Deer, Med		Deer,	Low	1.70	-3.01	4.71
Difference Required				70.30	69.91	139.82

at each. The vector variates are numbers, biomass, or energy
contribution of each of S species. In the same sense that there
is a continuum of habitats in a given geographic region based on
the distribution of resources in the region, there is also a
conceptual continuum of communities based on the distribution of
organisms. At a particular point, the community consists of a
collection of organisms which may interact at several scales. In
animal communities, some animals may be far ranging such as birds
or fish; others may be largely fixed such as soil microfauna or
coral reef organisms. In plant communities, there may be large
scale patterns due to soil and geologic configurations as well
as smaller scale interplays including competition for sunlight in
the canopy or competition for soil nutrients. In either plant or
animal associations, communities are distinct if the interactive
collection of organisms at each point can be distinguished objec-
tively on the basis of empirical evidence.

Possible overlap measures include a) traditional indices,
b) components of variance analysis, and c) multivariate statistics
including MANOVA and volume of intersection of hyperellipsoids.

Traditional overlap measures ignore the information content
of the covariances between resource or organism variables. They
are quite arbitrarily defined and the choice of a particular overlap measure reflects largely one's personal preference. The pairwise comparisons are difficult to interpret on an experiment-wide
basis because tests for significant overlap are defined only for
individual comparisons. Components of variance analysis is most
useful when one is investigating overlap along a single dimension.
Consideration of the overlap problem as a multivariate one provides
several benefits:

1. It involves a number of techniques whose statistical
properties are well established (such as MANOVA, discriminant
analysis, cluster analysis, principle component analysis, and
factor analysis).

2. It allows a simultaneous assessment of intra- and intercommunity hypervolume overlap.

3. The relative importance of various resource or organism
variables in establishing overlap or nonoverlap is determined
by using multivariate multiple comparison techniques, principle component analysis, canonical analysis, and stepwise techniques such as stepwise discriminant analysis.

4. Niche, habitat, or community breadth can be simply defined
because breadth concerns the volume of the hyperellipsoid
which is proportional to the square root of the determinant
of the covariance matrix.

In the original sense, overlap ought to be defined in terms of the coincidence of the ellipsoids describing the activity space of the organisms. Or, in measuring the similarity of communities, it ought to be defined in terms of the coincidence of the ellipsoids describing the relative frequency of different organisms in the community. This is probably the intent of the various authors of indices, but their measures have involved only scalar comparisons of the ellipsoid centers. The only solution is to assess the overlap of niches and habitats and the similarity of communities in multivariate terms.

REFERENCES

Abderson, T. W. (1958). *An Introduction to Multivariate Statistical Analysis*. Wiley, New York.

Bray, J. R. and Curtis, J. T. (1957). An ordination of the upland forest communities of southern Wisconsin. *Ecological Monographs*, 27, 325-349.

Cormack, R. M. (1979). In *Spatial and Temporal Analysis in Ecology*, R. M. Cormack and J. K. Ord, eds. Satellite program in Statistical Ecology, International Co-operative Publishing House, Fairland, Maryland.

Diggle, P. J. (1976). A spatial stochastic model of inter-plant competition. *Journal of Applied Probability*, 13, 662-671.

Dudzinski, M. L. and Arnold G. W. (1973). Comparisons of diets of sheep and cattle grazing together on sown pastures on the southern tablelands of New South Wales by principal components analysis. *Australian Journal of Agricultural Research*, 24, 889-912.

Garratt, M. W. and Steinhorst, R. K. (1976). Testing for significance of Morisita's, Horn's, and related measures of overlap. *American Midland Naturalist*, 96, 245-251.

Green, R. H. (1974). Multivariate niche analysis with temporally varying environmental factors. *Ecology*, 55, 73-83.

Green, R. H. (1971). A multivariate statistical approach to the Hutchinsonian niche: bivalve molluscs of central Canada. *Ecology*, 52, 543-556.

Horn, H. S. (1966). Measurement of 'overlap' in comparative ecological studies. *The American Naturalist*, 100, 419-424.

Hubbard, R. E. and Hansen, R. M. (1976). Diets of wild horses, cattle, and mule deer in the Piceance Basin, Colorado. *Journal of Range Management*, 29, 389-392.

Hummon, W. D. (1974). A similarity index based on shared species diversity, used to assess temporal and spatial relations among intertidal marine gastrotricha. *Oecologia*, 17, 203-220.

Hutchinson, G. E. (1958). Concluding remarks. *Cold Spring Harbor Symposium on Quantitative Biology*, 22, 415-427.

Morisita, M. (1959). Measuring of interspecific association and similarity between communities. *Kyushu University Faculty of Science Memoirs, Series E, Biology*, 3, 65-80.

Morrison, D. F. (1967). *Multivariate Statistical Methods*. McGraw-Hill, New York.

Oosting, H. J. (1956). *The Study of Plant Communities*. Freeman, San Francisco.

Orians, G. H. and Horn, H. S. (1969). Overlap in foods and foraging of four species of blackbirds in the potholes of central Washington. *Ecology*, 50, 930-938.

Poole, R. W. (1974). Measuring the structural similarity of two communities composed of the same species. *The Society of Population Ecology*, 16, 138-151.

Shugart, H. H. and Patten, B. C. (1972). Niche quantification and the concept of niche pattern. B. C. Patten, ed. *Systems Analysis and Simulation in Ecology*, Academic Press, New York.

Werger, M. J. A. (1973). On the use of association-analysis and principal component analysis in interpreting a Braun-Blanquet phytosociological table of a Dutch grassland. *Vegetatio*, 28, 129-144.

Whittaker, R. H. (1972). Evolution and measurement of species diversity. *Taxon*, 21, 213-251.

Whittaker, R. H. and Levin, S. A., eds. (1975). *Niche, Theory and Application*. Dowden, Hutchinson and Ross, Inc., Stroudsburg, Pennsylvania.

Whittaker, R. H., Levin, S. A., and Root, R. B. (1973). Niche, habitat, and ecotope. *The American Naturalist*, 107, 321-338.

Wiens, J. A. (1974). Habitat heterogeneity and avian community structure in North American grasslands. *American Midland Naturalist*, 91, 195-213.

L. Orloci, C. R. Rao, and W. M. Stiteler, (eds.),
Multivariate Methods in Ecological Work, pp. 279-300. All rights reserved.
Copyright © 1979 by International Co-operative Publishing House, Fairland, Maryland

MULTIVARIATE STATISTICS WITH APPLICATIONS IN STATISTICAL ECOLOGY

WILLIAM M. STITELER

SUNY
College of Environmental Science and Forestry
Syracuse, New York 13210 USA

SUMMARY. A brief introduction to multivariate statistical
methods is given. Examples are given of the application of the
techniques to problems in the biological sciences.

KEY WORDS. multivariate, discriminant analysis, canonical
correlation, principal components analysis, factor analysis.

1. INTRODUCTION

The purpose of this paper is to provide a brief introduction
to basic multivariate statistics. It is not meant to serve as a
substitute for a textbook on the subject since far too many details have been omitted. It is hoped, however, that the material
will be adequate to serve as a general foundation on which the
other papers of this volume can build.

2. VECTOR RANDOM VARIABLES

In general, a multivariate technique involves a sample of
size n from a population for which p variables are of interest
and are measured. This means that for each individual in the
sample we obtain a vector of p values -- one for each of the p
variables. We can denote the general observation then, by the
$p \times 1$ vector

$$X = \begin{bmatrix} X_1 \\ X_2 \\ \cdot \\ \cdot \\ \cdot \\ X_p \end{bmatrix}$$

3. MULTIVARIATE NORMAL DISTRIBUTION

We will see throughout that the development of multivariate techniques closely parallels that of their univeriate counter-parts. It should be no surprise, therefore, that it is convenient in the beginning to assume that each of the variables is normally distributed. In fact we make the slightly stronger assumption that the random vector X has a multivariate normal distribution with probability density function

$$f(X) = (2\pi)^{-p/2} |\Sigma|^{-\frac{1}{2}} \exp[-\frac{1}{2}(X-\mu)'\Sigma^{-1}(X-\mu)]$$

for $-\infty < X_1, X_2, \cdots, X_p < +\infty$

where $\mu = \begin{bmatrix} \mu_1 \\ \mu_2 \\ \cdot \\ \cdot \\ \cdot \\ \mu_p \end{bmatrix}$ is called the *mean vector*

or *centroid* and the symmetric matrix

$$\begin{bmatrix} \sigma_{11} & \sigma_{12} & \cdots & \sigma_{1p} \\ \sigma_{21} & \sigma_{22} & \cdots & \sigma_{2p} \\ \cdot & \cdot & & \cdot \\ \cdot & \cdot & & \cdot \\ \sigma_{p1} & \sigma_{p2} & \cdots & \sigma_{pp} \end{bmatrix}$$

is called the *covariance matrix*. Another name sometimes given to this matrix is the *variance-covariance matrix* to emphasize the

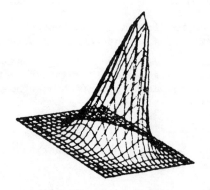

FIG. 1: Bivariate normal density function.

fact that the diagonal elements are the variances of the p
variables. To be consistent with ordinary correction σ_{ii} will
also be denoted by σ_i^2 .

There are, therefore, a total of $p + p(p+1)/2$ parameters
involved in the distribution of $\underset{\sim}{X}$.

It is convenient to picture the vector $\underset{\sim}{X}$ as defining a point
in p-dimensional space. Then, just as the univariate normal
density function describes the distribution of points along the
real line, the p-variate normal density function describes the
distribution of points in p-dimensional space. Figure 1 shows
the case where $p = 2$.

If we take any region R on the plane the probability of $\underset{\sim}{X}$
being contained in the region is given by the volume above the
region R and under the bell. The shape and location of this bell
and hence the volume above R is dependent on the $p + p(p+1)/2$
parameters.

4. PROBABILITY ELLIPSOIDS

The smallest region which contains $\underset{\sim}{X}$ with a specified
probability will have the shape of a p-dimensional ellipsoid.
In general, the mean vector alone specifies the location of the
center of the ellipsoid while all information about the shape and
orientation of the ellipse is contained in the parameters which
make up the co-variance matrix. The size of the ellipse is, of
course, determined by the particular value of the specified
probability.

5. CHARACTERISTIC ROOTS AND VECTORS

The extraction of the information on shape and orientation of the ellipsoid from Σ is an interesting application of characteristic roots and vectors. Because of its fundamental importance to the visualization and understanding of multivariate techniques, a brief review of this particular concept from matrix algebra is given here.

Given the $k \times k$ matrix $\underset{\sim}{A}$, it is clear from the rules of matrix multiplication that the result of post multiplying $\underset{\sim}{A}$ by a $k \times 1$ vector $\underset{\sim}{V}$ will be a product which is a $k \times 1$ vector. This gives rise to the question as to the possible existence of a vector $\underset{\sim}{V}$ with the property that $\underset{\sim\sim}{AV} = \underset{\sim}{V}$. If such a vector exists, it could be said to 'characterize' the matrix $\underset{\sim}{A}$. As it turns out, it is expecting a bit too much for such a vector to exist; but we can come close. There does exist a vector $\underset{\sim}{V}$ and a scalar c such that $\underset{\sim\sim}{AV} = c\underset{\sim}{V}$.

The vector $\underset{\sim}{V}$ is called a *characteristic vector* and the scalar c is called a *characteristic root*. $\underset{\sim}{V}$ and c are sometimes called the *latent vectors* and *latent roots* respectively and sometimes *eigenvectors* and *eigenvalues* respectively.

There are k solutions for a matrix of order k. If we take as a special case the $p \times p$ covariance matrix Σ, we will have p characteristic vectors, each of which has associated with it a characteristic root.

The p (orthogonal) characteristic vectors of Σ give the directions of the axes of the ellipsoid and the lengths of the axes are proportional to the square root of the associated characteristic root. In fact, the exact lengths of the p axes of the ellipsoid which contains $\underset{\sim}{X}$ with probability $1-\alpha$ are given by $2\sqrt{c_i}x_p^2(1-\alpha)$.

6. LINEAR COMBINATIONS

Linear combinations of random variables play an important role in multivariate statistics. It is useful to graphically interpret a linear combination of the p variables as a projection of a point from p-dimensional space to one-dimensional space (the real line). It is this projection which is often used to develop a multivariate technique by collapsing a p-variate problem to a univariate problem as we will see in the next section.

A particularly useful result is that if X has a p-variate normal distribution and a is a pxl vector of constants, then $a'X = a_1X_1 + a_2X_2 + \cdots + a_pX_p$ is a scalar and has a univariate normal distribution with mean $a'\mu$ and variance $a'\Sigma a$.

7. MULTIVARIATE T-TEST AND TWO GROUP DISCRIMINANT ANALYSIS

Perhaps the best known of the univariate statistical techniques is the t-test which evaluates the hypothesis that populations A and B have the same mean (i.e., H_o: $\mu_A = \mu_B$) . This hypothesis is evaluated by computing the statistic

$$t = \frac{\bar{X}_A - \bar{X}_B}{S_p \dfrac{1}{n_1} + \dfrac{1}{n_2}}$$

which in absolute value is interpreted as the number of standard deviations by which $\bar{X}_A$ differs from $\bar{X}_B$.

In the multivariate counterpart to this problem we have the hypothesis H_o: $\mu_A = \mu_B$ where both μ_A and μ_B are p × 1 vectors. The derivation of a test for this hypothesis is provided by the union-intersection principle (Roy, 1953). To start, we establish the fact that H_o: $\mu_A = \mu_B$ is true *if and only if* H_o: $a'\mu_A = a'\mu_B$ is true for every non-null vector a .

Note that the second hypothesis in this equivalence relation-ship involves a linear combination of the means which is a scalar. This means that this latter hypothesis can be tested for a par-ticular vector a by the ordinary t-statistic

$$t(a) = \frac{n_1 n_2}{n_1 + n_2} \frac{a'\bar{X}_a - a'\bar{X}_B}{\sqrt{a'S a}}$$

where S is the matrix of pooled estimates of covariances. This follows from the preceding results on linear combinations of random variables.

Using this test statistic we would (for this particular a) reject H_o in favor of the two-tailed alternative H_1: $\mu_A \neq \mu_B$

if $t(\underset{\sim}{a})$ exceeds the tabulated t value at the specified significance level with $n_1 + n_2 - 2$ degrees of freedom. It is well
known that the square of a student's t random variable with
$n-1$ degrees of freedom has an F distribution with 1 and $n-1$
degrees of freedom, so we could as well compute

$$t^2(\underset{\sim}{a}) = \frac{n_1 n_2}{n_1 + n_2} \frac{(\underset{\sim}{a}'\bar{X}_A - \underset{\sim}{a}'\bar{X}_B)^2}{\underset{\sim}{a}'S\underset{\sim}{a}}$$

and compare with the tabulated F with 1 and $n_1 + n_2 - 2$
degrees of freedom.

This tells us how to test our hypothesis for a particular
vector $\underset{\sim}{a}$ and if, by chance, we discover a vector $\underset{\sim}{a}$ for which
the second hypothesis would be rejected, then we would (by the
equivalence) reject also the multivariate hypothesis. It is clear,
however, that we could not entertain the thought of running through
all possible $\underset{\sim}{a}$ vectors.

What we need to know is the largest possible value that
$t^2(\underset{\sim}{a})$ can take over all possible $\underset{\sim}{a}$. This is available through
a relatively straightforward application of the calculus. This
largest value is known as Hotelling's T^2 after its inventor and
is given by:

$$T^2 = \frac{n_1 n_2}{n_1 + n_2} (\bar{X}_A - \bar{X}_B)'\underset{\sim}{S}^{-1}(\bar{X}_A - \bar{X}_B)$$

It is interesting to note the correspondence to the square of the
univariate t-statistic.

The quantity

$$\frac{n_1 + n_2 - p - 1}{(n_1 + n_2 - 2)p} T^2$$

has an F distribution with degrees of freedom p and $n_1 + n_2 -
p - 1$.

We can, therefore, test H_o: $\underset{\sim}{a}'\mu_A = \underset{\sim}{a}'\mu_B$ for the 'worst'
vector $\underset{\sim}{a}$ and whatever action we take relative to that hypothesis
we must also take for the multivariate hypothesis.

Example 1: A study was conducted (Eav, Lillesand, and Manion 1978) to determine the feasibility of using aerial color photography to detect diseased trees. A sample of eleven diseased trees and a sample of eleven healthy trees was selected. Each tree was measured for image densities on red (X_1) , green (X_2) , and blue (X_3) layers of the color film. The sample centroids were:

$$\bar{X}_{\sim D} = \begin{bmatrix} 1.42 \\ 1.16 \\ 1.79 \end{bmatrix} \qquad \bar{X}_{\sim H} = \begin{bmatrix} 1.06 \\ 0.88 \\ 1.69 \end{bmatrix}$$

and the pooled sample covariance matrix was:

$$S_{\sim} = \frac{1}{20} \begin{bmatrix} 0.1939 & 0.2745 & 0.2984 \\ 0.2745 & 0.5054 & 0.5639 \\ 0.2984 & 0.5639 & 0.7245 \end{bmatrix}$$

The value of $\{(N_1 + N_2 - p - 1)/[(N_1 + N_2)p]\}T^2$ was 46.81 which is much larger than the tabulated F value with 3 and 18 degrees of freedom at the .01 level of significance. The hypothesis of equality of population centroids is, therefore, rejected with the conclusion that diseased and healthy trees have different image densities in red, green, and blue layers of color film.

Example 2: A study was conducted (Lea, 1979) to determine if there is a difference in the concentrations of nutrients in the foliage of sugar maple trees grown on poor sites from those grown on good sites. A sample of foliar concentrations of N , P , and K was taken on 6 poor sites and on 5 good sites. The sample centroids were:

$$\bar{X}_{\sim p} = \begin{bmatrix} 1.99 \\ 0.10 \\ 0.56 \end{bmatrix} \qquad \bar{X}_{\sim G} = \begin{bmatrix} 2.74 \\ 0.20 \\ 0.72 \end{bmatrix}$$

and the pooled sample covariance matrix was:

$$S = \frac{1}{9} \begin{bmatrix} 0.4390 & 0.0250 & 0.1536 \\ 0.0250 & 0.0078 & 0.0071 \\ 0.1536 & 0.0071 & 0.0964 \end{bmatrix}$$

The value of $\{(N_1 + N_2 - p - 1(/[N_1 + N_2 - 2)p]\}T^2$ was 12.94 which is larger than 8.45, the tabulated value of F with 3 and 7 degrees of freedom at the .01 level of significance. The conclusion is, therefore, that the foliar concentrations of N, P, and K differ between poor and good sites.

Notice that rejection of the null hypothesis in the multivariate t-test leaves us wondering which of the p variables are responsible for the difference. In Example 1 we might wonder if the diseased and healthy trees differ in all three of the colors. Procedures for answering this question are available but in the interest of economy of space will not be discussed in this paper. Morrison (1976) provides a good discussion of the procedure.

8. DISCRIMINANT ANALYSIS

In the preceding discussion we were concerned only with finding the largest possible value of $t^2(a)$ without regard for what particular vector, a, gave rise to that maximum value. It turns out that this particular vector is useful in a technique called discriminant analysis. As we saw in Section 6 the linear compound $a'X$ is interpreted as a projection from p-dimensional space to the real line. The vector a which maximizes $t^2(a)$ is equivalent to the projection which gives the maximum separation between the two populations. Alternatively we could say that it maximizes the distance between the projected sample means. In Figure 2a we see that the vector a associated with the second projection would give the maximum separation between populations A and B. This would suggest that if one wanted to classify an individual as to membership A or B based on the two variables contained in X it would be best done on the basis of $a'X$.

The advantage of taking a 'multivariate' view of the classification problem is apparent in Figure 2b. It is clear that if we consider either of the two variables alone we will have a substantial overlap in the values taken by members of A with members of B. As a result we would not be very successful in classifying an individual. However, by taking the appropriate linear combination $a'X$ we could get a very good separation of the two groups.

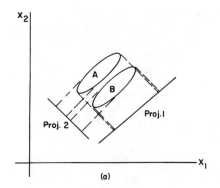
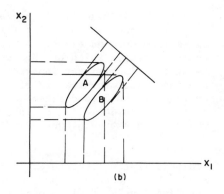

FIG. 2: (a) *Projections of two bivariate populations;* (b) *bivariate discrimination versus single-variable discriminations.*

The vector $\underset{\sim}{a}$ which provides this maximum separation is given by $\underset{\sim}{\Sigma}^{-1}(\underset{\sim}{\mu}_1 - \underset{\sim}{\mu}_2)$. Ordinarily the population parameters would be unknown so it would be necessary to use $\underset{\sim}{S}^{-1}(\underset{\sim}{\bar{X}}_1 - \underset{\sim}{\bar{X}}_2)$.

In order to classify a particular individual as to which of the two groups it belongs, the procedure would be, after measuring the appropriate variables on that individual, to compute $\underset{\sim}{a}'\underset{\sim}{X}$ and classify the individual as a member of A if $\underset{\sim}{a}'\underset{\sim}{X}$ is closer to $\underset{\sim}{a}'\underset{\sim}{\bar{X}}_1$ than it is to $\underset{\sim}{a}'\underset{\sim}{\bar{X}}_2$.

This procedure is overly simplified and assumes two things: (1) That the loss incurred by incorrectly classifying a member of A as being a member of B is the same as the loss incurred by incorrectly classifying a member of B as being a member of A . (2) There is an equal *a priori* chance of membership in the two groups. For a discussion of the general procedure, see Morrison (1976).

Example 3: In Example 1 we saw that there was a highly significant difference between the population centroids of diseased and healthy trees where the variables were image densities of red (X_1) , green (X_2) , and blue (X_3) layers of color film. The discriminant function based on these data would be $4.24X_1 +$ $0.33X_2 - 1.87X_3$. The mean value scored by the healthy trees was 1.65 and the mean value scored by diseased trees was 3.06. Assuming equal *a priori* probabilities of membership and equal losses the classification procedure would call any tree with a score $\leq$ 2.355 healthy and any tree with a score > 2.355 diseased.

Example 4: In Example 2 a study indicated that concentrations of N , P , and K in foliage of sugar maple were lower on poor sites than on good sites. Given a foliage sample, then, one could determine which of the two site categories it came from by using the discriminant function -1.99N - 8.56P + 2.13K. The average discriminant score of the six poor sites was -3.62 and the average of the 5 good sites was -5.66. A foliage sample with a score of < -4.64 would be classified as coming from a poor site while a score > -4.64 would indicate a good site.

There are a few examples of applications of discriminant analysis in the literature. One paper by Edwin White and Donald Mead (1971) describes the use of discriminant analysis of foliar nutrient levels of slash pine to discriminate between trees with normal 'green' foliage and those with abnormal winter 'yellow' foliage. This example illustrates the point that discriminant analysis is sometimes used as much to study the importance of the variables in the classification process as it is to actually classify individuals. We don't need measurements of foliar nutrients to see if a tree is 'green' or 'yellow.'

9. MULTIVARIATE ANALYSIS OF VARIANCE

In many problems more than two groups are involved which means that for testing equality of means (centroids) we need a multivariate counterpart to the single response analysis of variance. There are, surprisingly, three such tests in common use and as of now we have no good information on which is 'best.' As we might expect, all of the tests are based on the 'ratio' of the between groups covariance matrix to the within groups covariance matrix. If we denote these two matrices by $\underset{\sim}{B}$ and $\underset{\sim}{W}$ respectively and let $C_1, C_2, \cdots, C_p$ be the characteristic roots of $\underset{\sim}{W}^{-1}\underset{\sim}{B}$, then the three test statistics are:

$$(1) \quad |\underset{\sim}{W}|/|\underset{\sim}{B} + \underset{\sim}{W}| = 1/\sum_{i=1}^{p} C_i \quad ;$$

$$(2) \quad \sum_{i=1}^{p} C_i \quad ;$$

$$(3) \quad \text{Max}\{C_i\} \quad .$$

For an example of the application of multivariate analysis of variance see the paper by Burton Barnes (1975). In that paper leaf, bud, and twig characters were compared for several populations of trembling aspen from several locations in the United States and Canada.

10. CANONICAL CORRELATION

Canonical correlation is a technique for studying the relationship between two groups of variables. It is an extension of simple correlation for relating two variables and multiple correlation for relating a single variable to a group of other variables.

In forestry, for example, it is customary to describe the quality of a site for some purpose such as timber production by a single number called a site index. A study of the relationship between several environmental variables and site quality might involve the multiple correlation coefficient between this site index and the environmental variables. It might, however, be more reaonable to describe site quality by several variables which necessitates a technique for correlating two groups of variables.

This division of the variables into two groups results in a partitioning of the random vector $\underset{\sim}{X}$:

$$\underset{\sim}{X} = \begin{bmatrix} \underset{\sim}{X}^{(1)} \\ \underset{\sim}{X}^{(2)} \end{bmatrix}$$

where $\underset{\sim}{X}^{(1)}$ has r elements and $\underset{\sim}{X}^{(2)}$ has s elements and $r+s = p$.

This partitioning of $\underset{\sim}{X}$ results in the corresponding partitioning of $\underset{\sim}{\mu}$ and $\underset{\sim}{\Sigma}$ as follows:

$$\underset{\sim}{\mu} = \begin{bmatrix} \underset{\sim}{\mu}^{(1)} \\ \underset{\sim}{\mu}^{(2)} \end{bmatrix}$$

$$\underset{\sim}{\Sigma} = \begin{bmatrix} \underset{\sim}{\Sigma}_{11} & \underset{\sim}{\Sigma}_{12} \\ \underset{\sim}{\Sigma}'_{12} & \underset{\sim}{\Sigma}_{22} \end{bmatrix} = \begin{bmatrix} \sigma_{11} & \cdots & \sigma_{1r} & \sigma_{1r+1} & \cdots & \sigma_{1p} \\ & & & & & \\ & & & & & \\ \sigma_{r1} & \cdots & \sigma_{rr} & \sigma_{rr+1} & \cdots & \sigma_{rp} \\ \sigma_{r+1} & \cdots & \sigma_{r+1r} & \sigma_{r+1r+1} & \cdots & \sigma_{r+1p} \\ & & & & & \\ & & & & & \\ \sigma_{p1} & \cdots & \sigma_{pr} & \sigma_{pr+1} & \cdots & \sigma_{pp} \end{bmatrix}$$

and of their estimated values:

$$\underset{\sim}{\bar{X}} = \begin{bmatrix} \underset{\sim}{\bar{X}}^{(1)} \\ \underset{\sim}{\bar{X}}^{(2)} \end{bmatrix} \qquad \underset{\sim}{S} = \begin{bmatrix} \underset{\sim}{S}_{11} & \underset{\sim}{S}_{12} \\ \underset{\sim}{S}'_{12} & \underset{\sim}{S}_{22} \end{bmatrix} .$$

A related problem is to test H_o: $\underset{\sim}{\Sigma}_{12} = 0$, the hypothesis of independence between the variables in $\underset{\sim}{X}^{(1)}$ and the variables in $\underset{\sim}{X}^{(2)}$. The derivation of a test for this hypothesis using the union-intersection principle involves taking linear combinations of the two groups of variables. If $\underset{\sim}{a}$ is an $r \times 1$ non-null vector of constants and $\underset{\sim}{b}$ is an $s \times 1$ non-null vector of constants then, taking the linear combinations $\underset{\sim}{a}'\underset{\sim}{X}^{(1)}$ and $\underset{\sim}{b}'\underset{\sim}{X}^{(2)}$ we have two scalars and can compute their ordinary correlation coefficient. The largest possible correlation coefficient obtained over all vectors $\underset{\sim}{a}$ and $\underset{\sim}{b}$ would be used to test the hypothesis H_o: $\underset{\sim}{\Sigma}_{12} = 0$ since that hypothesis is equivalent to the hypothesis that $\underset{\sim}{a}'\underset{\sim}{X}^{(1)}$ and $\underset{\sim}{b}'\underset{\sim}{X}^{(2)}$ are uncorrelated for all $\underset{\sim}{a}$ and $\underset{\sim}{b}$.

This largest correlation coefficient is called a *canonical correlation coefficient* and the linear compounds $\underset{\sim}{a}'\underset{\sim}{X}^{(1)}$ and $\underset{\sim}{b}'\underset{\sim}{X}^{(2)}$ are called *canonical variates*.

As it turns out this largest correlation coefficient is given by the largest characteristic root of $\underset{\sim}{S}_{11}^{-1}\underset{\sim}{S}_{12}\underset{\sim}{S}_{22}^{-1}\underset{\sim}{S}'_{12}$. In fact the other characteristic roots of $\underset{\sim}{S}_{11}^{-1}\underset{\sim}{S}_{12}\underset{\sim}{S}_{22}^{-1}\underset{\sim}{S}'_{12}$ can also be interpreted as canonical correlation coefficients of $q-1$ additional pairs of independent linear compounds where $q = \min(r,s)$.

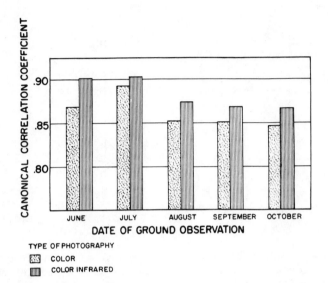

FIG. 3: Canonical correlation coefficients between ground and photo variables.

In other words, we can find q rx1 vectors $\underset{\sim}{a}_1, \underset{\sim}{a}_2, \cdots, \underset{\sim}{a}_q$ and q sx1 vectors $\underset{\sim}{b}_1, \underset{\sim}{b}_2, \cdots, \underset{\sim}{b}_q$ such that $\underset{\sim}{a}_1' \underset{\sim}{x}^{(1)}$ has the maximum correlation with $\underset{\sim}{b}_1' \underset{\sim}{x}^{(2)}$, $\underset{\sim}{a}_2' \underset{\sim}{x}^{(1)}$ has the maximum correlation with $\underset{\sim}{b}_2' \underset{\sim}{x}^{(2)}$ under the condition that $\underset{\sim}{a}_2' \underset{\sim}{x}^{(1)}$ is is independent of $\underset{\sim}{a}_1' \underset{\sim}{x}^{(1)}$ and $\underset{\sim}{b}_2' \underset{\sim}{x}^{(1)}$ is independent of $\underset{\sim}{b}_2' \underset{\sim}{x}^{(2)}$, etc.

Example 5: Eav, Lillesand, and Manion (1978) used canonical correlation as a basis for deciding whether color or color infra-red film would be best in detecting tree stress from disease. A group of characteristics measured on aerial photographs was correlated with a group of variables measured on the trees. The results are shown graphically in Figure 3. Color infrared film gave consistently higher canonical correlations between the ground variables and the photograph variables. The results also indicated that the best time to do the aerial photography is in July.

11. PRINCIPAL COMPONENTS ANALYSIS

Factor analysis and principal components analysis are two techniques for reducing the dimensionality of a problem. That is

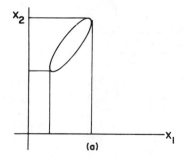
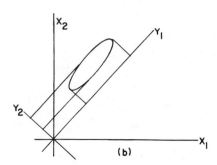

FIG. 4: (a) *Concentration ellipse for bivariate populations;* (b) *principal components analysis with two variables.*

to say they are generally used to condense all or most of the in-formation contained in the p variables into a smaller number of 'factors.' Because the two techniques are used for the same pur-pose and because of similarities in the terminology and computa-tional procedures there has tended to be a considerable amount of confusion involved in the use of the two techniques. The two techniques do differ considerably, however, in their basic model.

It is convenient to discuss principal components analysis first because it has a simpler model and because it, unlike factor analysis, can be represented graphically. The basic idea can in fact be illustrated nicely if we consider the case where $p = 2$ and a concentration ellipse as shown in Figure 4a.

Notice that there is a fair amount of dispersion in either X_1 or X_2 considered alone but that, due to the correlation be-tween the two variables much of the dispersion in their joint dis-tribution occurs along the major axis of the ellipse. In fact, if we rotate the X_1 , X_2 axes so that they are parallel to the axes of the ellipse and call the new axes Y_1 and Y_2 we can see from Figure 4b that most of the variation is accounted for by Y_1 and we might perhaps think of dropping Y_2 .

In other words, since there is very little variation in Y_2 , we might eliminate it, thereby reducing the number of dimensions from two to one without a substantial loss of information.

In the general p-variable case we may find that only a rela-tively few dimensions are necessary, after an appropriate rotation, to represent most of the information contained originally in p dimensions.

The formal model for principal components analysis can be written

$$Y_1 = a_{11}X_1 + a_{12}X_2 + \cdots + a_{1p}X_p$$

$$Y_2 = a_{21}X_1 + a_{22}X_2 + \cdots + a_{2p}X_p$$

$$\cdot \qquad \cdot \qquad \cdot \qquad \qquad \cdot$$
$$\cdot \qquad \cdot \qquad \cdot \qquad \qquad \cdot$$
$$\cdot \qquad \cdot \qquad \cdot \qquad \qquad \cdot$$

$$Y_p = a_{p1}X_1 + a_{p2}X_2 + \cdots + a_{pp}X_p$$

where the coefficients a_{ij} form an orthogonal matrix $\underset{\sim}{A}$. In matrix notation this model can be written $\underset{\sim}{Y} = \underset{\sim}{A}'\underset{\sim}{X}$.

As we know from elementary linear algebra, this formula with $\underset{\sim}{A}$ as an orthogonal matrix represents a rigid rotation (possibly followed by a reflection about the origin) of the X-coordinate axes so that a particular point represented by $\underset{\sim}{X}$ in the original space is now represented by $\underset{\sim}{Y}$ in the new (rotated) space. Our objective is to rotate the axes in such a way that they are parallel to the axes of the ellipsoid, The solution is to take $\underset{\sim}{A}$ to be the matrix made up of the normalized characteristic vectors as columns.

Now since $\underset{\sim y}{\Sigma} = \underset{\sim}{A}\underset{\sim x}{\Sigma}\underset{\sim}{A}'$ it follows from a property of characteristic roots and vectors that

$$\underset{\sim y}{\Sigma} = \begin{bmatrix} C_1 & 0 & \cdots & 0 \\ 0 & C_2 & \cdots & 0 \\ \cdot & \cdot & \cdot & \cdot \\ \cdot & \cdot & \cdot & \cdot \\ \cdot & \cdot & \cdot & \cdot \\ 0 & 0 & \cdots & C_p \end{bmatrix}$$

and since $\underset{\sim}{A}$ is orthogonal we have $\underset{\sim}{A}^{-1} = \underset{\sim}{A}'$ so that $\underset{\sim x}{\Sigma} = \underset{\sim}{A}'\underset{\sim y}{\Sigma}\underset{\sim}{A}$. Also, since $\underset{\sim y}{\Sigma}$ is a diagonal matrix, we can write

$$\Sigma_{\~y} = \begin{bmatrix} \sqrt{c_1} & 0 & \cdots & 0 \\ 0 & \sqrt{c_2} & \cdots & 0 \\ \cdot & \cdot & & \cdot \\ \cdot & \cdot & & \cdot \\ \cdot & \cdot & & \cdot \\ 0 & 0 & \cdots & \sqrt{c_p} \end{bmatrix} \cdot \begin{bmatrix} \sqrt{c_1} & 0 & \cdots & 0 \\ 0 & \sqrt{c_2} & \cdots & 0 \\ \cdot & \cdot & & \cdot \\ \cdot & \cdot & & \cdot \\ \cdot & \cdot & & \cdot \\ 0 & 0 & \cdots & \sqrt{c_p} \end{bmatrix}$$

so that $\Sigma_{\~x} = \underset{\~\~}{BB'}$ where

$$\underset{\~}{B} = \begin{bmatrix} C & C & & C \\ H & H & & H \\ A & A & & A \\ R & R & \cdots & R \\ V & V & & V \\ E & E & & E \\ C & C & & C \\ T & T & \cdots & T \\ 1 & 2 & \cdots & p \end{bmatrix} \begin{bmatrix} \sqrt{c_1} & 0 & \cdots & 0 \\ & & & \\ 0 & \sqrt{c_2} & \cdots & 0 \\ \cdot & \cdot & & \cdot \\ \cdot & \cdot & & \cdot \\ \cdot & \cdot & & \cdot \\ 0 & 0 & \cdots & \sqrt{c_p} \end{bmatrix}.$$

We have assumed that X has a multivariate normal distribution so the vector Y would also have a multivariate normal distribution and the fact that all correlations are zero would imply that the Y_i's are independent random variables.

Furthermore, it is a property of characteristic roots that their sum for a particular matrix is equal to the trace of that matrix. In other words, the sum of the variances of the principal components is equal to the sum of the variances of the original variables. It is customary, therefore, to report the characteristic roots in decreasing order of magnitude and to express them as a ratio of the total variance multiplied by 100 to obtain the percentage of total variance accounted for by each component. This gives a basis for making some judgement as to which components, if any, can be discarded.

The computational procedure involved in principal components analysis is then, simply to extract from the covariance matrix of X the characteristic roots which give the variances of the components and the associated characteristic vectors which give the linear combination making up the component.

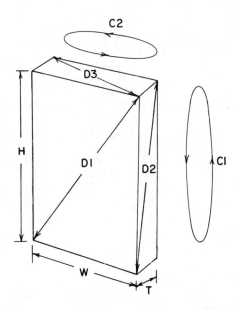

FIG. 5: Eight variables measured on sample of textbooks.

Because principal components analysis is nothing more than a description of the probability ellipsoid, the role of units of measurement must be considered. One can manipulate the ellipsoid at will by choosing an appropriate scale and, therefore, a principal components analysis can be 'adjusted' to obtain a desired result. For this reason, unless all variables are measured in the same units or in comparable units, it is customary to standardize the variables.

Example 6: We will use here an example which involves a sample of textbooks on each of which eight variables were measured. It is an example from ecology only in that the sample happened to come from the shelf of a graduate student in forest management. It serves as a nice example, however, because it is not complex and because of the unexpected outcome. The eight variables measured on each book are illustrated in Figure 5.

It would seem that because a textbook is a three-dimensional object which can be described completely by the three variables labeled T , W , and H , and because any of the remaining five variables can be constructed from them, that this technique would give those three variables as principal components. This is not the case, however, as we can see in Table 1. Only two components account for nearly all of the variation in the eight variables. In fact, only the first four components are given in Table 1 because the remaining four had such a small contribution,

TABLE 1: *Principal components analysis on eight dimensions of* 25
books. Correlation coefficients in parenthesis.

Variable	1	2	3	4
1-T	-.15(-.39)	.94(.91)	-.26(-.12)	.03(0)
2-W	-.38(-.99)	-.15(-.15)	-.09(-.04)	-.27(-.04)
3-H	-.38(-.99)	-.11(-.11)	-.15(-.07)	-.46(-.06)
4-D1	-.38(-.99)	-.13(-.13)	-.11(-.05)	.19(.02)
5-D2	-.38(-.99)	-.08(-.08)	-.16(-.07)	.49(.06)
6-D3	-.38(-.99)	-.10(-.10)	-.10(-.04)	-.48(-.06)
7-C1	-.34(-.89)	.20(.19)	.92(.41)	.06(.01)
8-C2	-.38(-.99)	.03(.03)	-.10(-.04)	-.46(-.06)
Char. Root	6.8399	.9410	.1977	.0172
Cum. %	85.50	97.26	99.73	99.05

accounting for only 0.05% of the total variance. Examination of
the correlations between the eight original variables and the first
two principal components indicates that the first component could
be called 'size of cover' and the second component could be called
'thickness' of the book. The reason that there are only two components is that the majority of textbooks have a fixed ratio between width and height of something like 0.7 to 1.0. While textbooks vary considerably in the size of the cover, the 'shape' is
fairly constant so that it is redundant to measure both width
and height.

Even this simple example illustrates an important fact about
principal components analysis and factor analysis. The person interpreting the results must have an intimate understanding of the
particular system being studied. A statistician with a thorough
understanding of the technique but no biological training would
have no more success in applying the technique to a biological
problem than would a biologist with no understanding of the
technique.

Another important lesson to learn from this example is that
an *a priori* hypothesis can sometimes stand in the say of a successful interpretation.

12. FACTOR ANALYSIS

Factor analysis is similar to principal components analysis
in purpose as well as in several aspects of the model. The model
is given by:

$$X_1 = a_{11}F_1 + a_{12}F_2 + \cdots + a_{1k}F_k + E_1$$

$$X_2 = a_{21}F_1 + a_{22}F_2 + \cdots + a_{2k}F_k + E_2$$

$$\vdots \qquad \vdots \qquad \vdots \qquad \qquad \vdots \qquad \vdots$$

$$X_p = a_{p1}F_1 + a_{p2}F_2 + \cdots + a_{pk}F_k + E_p \ .$$

In matrix notation this can be written as $\underset{\sim}{X} = \underset{\sim}{A}\underset{\sim}{F} + \underset{\sim}{E}$, where

$$\underset{\sim}{A} = \begin{bmatrix} a_{11} & a_{12} & \cdots & a_{1k} \\ a_{21} & a_{22} & \cdots & a_{2k} \\ \cdot & \cdot & & \cdot \\ \cdot & \cdot & & \cdot \\ \cdot & \cdot & & \cdot \\ a_{p1} & a_{p2} & \cdots & a_{pk} \end{bmatrix} \qquad \underset{\sim}{F} = \begin{bmatrix} F_1 \\ F_2 \\ \cdot \\ \cdot \\ \cdot \\ F_k \end{bmatrix} \qquad \underset{\sim}{E} = \begin{bmatrix} E_1 \\ E_2 \\ \cdot \\ \cdot \\ \cdot \\ E_p \end{bmatrix} .$$

Here the model specifies that each of the original variables X_i can be represented as a linear combination of a set of variables F_1, F_2, $\cdots$, F_k $(k \leq p)$, called *common factors*, plus a variable E_i called the *specific factor* which represents a component unique to that particular X_i and not shared by the other X's .

We notice one difference between factor analysis and principal components analysis immediately. In principal components analysis the number of components or 'factors' is, by virtue of the model, always p . Whether or not some are discarded later is, of course, another matter. In factor analysis, on the other hand, the number of factors k must be specified ahead of time to complete the model. In actual practice we are not always prepared to make a decision on the value of k ahead of time so most of the standard computer library packages incorporate some quantitative criterion in order to automate this decision-making process.

At this point it is useful to make some assumptions about the terms in the model. If we assume that the common factors are independent of each other as well as being independent of the specific factor, letting V stand for variance:

$$V(X_i) = V(a_{i1}F_1) + V(a_{i2}F_2) + \cdots + V(a_{ik}F_k) + V(E_i)$$

and since a_{ij} is constant

$$= a_{i1}^2 V(F_1) + a_{i2}^2 V(F_2) + \cdots + a_{ik}^2 V(F_k) + V(E_i) \quad .$$

Without loss of generality, we can assume in addition that each of the F_j has a unit variance since any scaling constant needed to accomplish this can be accounted for in the a_{ij} . We have then

$$V(X_i) = a_{i1}^2 + a_{i2}^2 + \cdots + a_{ik}^2 + V(E_i) \quad .$$

The total variance of an original variable X_i is partitioned into two parts. The term $\sum_j a_{ij}^2$ is the amount of variance accounted for by the k common factors and is called the *communality*. The term $V(E_i)$ is the variance which cannot be explained by the common factors and is called the *specificity*.

If we add the further assumption that the E_i's are independent of each other, we can express the covariance of $\underset{\sim}{X}$ as $\underset{\sim}{\Sigma}_X = \underset{\sim\sim}{AA}' + \underset{\sim}{\Sigma}_E$ where

$$\underset{\sim}{\Sigma}_E = \begin{bmatrix} V(E_1) & 0 & \cdots & 0 \\ 0 & V(E_2) & \cdots & 0 \\ \cdot & \cdot & & \cdot \\ \cdot & \cdot & & \cdot \\ \cdot & \cdot & & \cdot \\ 0 & 0 & \cdots & V(E_p) \end{bmatrix} \quad .$$

If we compare this with the factorization of the covariance matrix resulting from the model for principal components analysis, we see another fundamental difference between the two techniques. In principal components analysis we require all of the variance as well as all of the covariances to be accounted for by the p components. In factor analysis we require that all of the covariance (correlation) between the variables be accounted for by the (common) factors but allow a portion of the variance of each variable to go unexplained.

The factorization of $\underset{\sim}{\Sigma}_X$ is not unique because for any matrix $\underset{\sim}{A}$ of factor loadings the matrix $\underset{\sim\sim}{AP}$, where $\underset{\sim}{P}$ is orthogonal, also serves the purpose since

$$\underset{\sim\sim}{AP}(\underset{\sim\sim}{AP})' + \underset{\sim E}{\Sigma} = \underset{\sim\sim\sim}{APP}'\underset{\sim}{A}' + \underset{\sim E}{\Sigma} = \underset{\sim\sim}{AA}' + \underset{\sim E}{\Sigma} \ .$$

The matrix product $\underset{\sim\sim}{AP}$ is interpreted graphically as a rotation of the m axes represented by the matrix $\underset{\sim}{A}$. There would be p points in this m–dimensional coordinate system (one for each of the p variables) representing the rows of $\underset{\sim}{A}$.

Because any of the infinite number of solutions to a factor analysis are equally good in terms of satisfying the model, the one which is most easily interpreted would be preferred. It is of course no easy task to define what is meant by 'most easily interpreted' and as a result there are several commonly applied analytical procedures. One procedure, for example, maximizes the variance of the loadings within factors and is called varimax rotation.

There are many examples of the application of factor analysis in the literature. See, for example, the papers by Davidson (1975) and LaBastide and Van Goor (1970).

13. PRINCIPAL FACTOR ANALYSIS

A hybrid offspring of factor analysis and principal compon-ents analysis is a model sometimes referred to as principal factor analysis (Harris, 1975). In this technique the main diagonal entries of $\underset{\sim}{\Sigma}$ are replaced by estimated communalities and a

principal components analysis is performed on the resulting matrix. This technique actually fits the factor analysis model since it allows for a specificity term. It has the convenience of being relatively simple computationally but the estimated parameters may not necessarily have the best properties. For a brief discussion of the computation problems of factor analysis in general and of the maximum likelihood approach in particular, see Morrison (1976).

This may well be the most frequently employed procedure in factor analysis since both BMD and SPSS computer packages use it.

REFERENCES

Anderson, T. W. (1958). *Introduction to Multivariate Statistical Analysis*. Wiley, New York.

Barnes, B. V. (1975). Phenotypic variation in trembling aspen in Western North America. *Forest Science*, 21, 319–328.

Davidson, J. (1975). Use of principal components, factor analysis, and varimax rotation to describe variability in wood of *Eucalyptus deglupta*. *Wood Science and Technology*, 9, 275-291.

Eav, B. B., Lillesand, T. M., and Manion, P. D. (1978). Development and evaluation of a photographic remote sensing system for the detection and quantification of urban tree stress. *Proceedings of the Society of Photogrammetry, 44th Annual Meeting*, Washington, D. C.

Green, P. E. and Carroll, J. D. (1976). *Mathematical Tools for Applied Multivariate Analysis*. Academic Press, New York.

Harris, R. J. (1975). *A Primer of Multivariate Statistics*. Academic Press, New York.

La Bastide, J. G. A. and Van Goor, C. P. (1970). Growth-site relationships in plantations of *pinus elliottii* and *araucaria angustifolia* in Brazil. *Plant and Soil*, 32, 349-366.

Lea, R. (1979). Responses in a Northern New York Fagus-Betula-Acer Stand to Fertilization. Ph.D. thesis, SUNY College of Environmental Science and Forestry, Syracuse, New York.

Morrison, D. F. (1976). *Multivariate Statistical Methods*. McGraw Hill, New York.

Roy, S. N. (1953). On a heuristic method of test construction and its use in multivariate analysis. *Annals of Mathematical Statistics*, 24, 220-238.

White, E. H. and Mead, D. J. (1971). Discriminant analysis in tree nutrition research. *Forest Science*, 17, 425-427.

[Received August 1978. Revised February 1979]

L. Orloci, C. R. Rao, and W. M. Stiteler, (eds.),
Multivariate Methods in Ecological Work, pp. 301-308. All rights reserved.
Copyright © 1979 by International Co-operative Publishing House, Fairland, Maryland

NEW COMPUTER-ORIENTED METHODS FOR THE STUDY OF NATURAL AND
SIMULATED VEGETATION STRUCTURE

ZOLTÁN SZŐCS

Research Institute for Botany
Hungarian Academy of Sciences
H-2163 Vacratot, Hungary

SUMMARY. Two computer-based sampling and pattern simulation
methods are described. They are intended to stimulate development in the following fields: studies of elementary spatial
structure in the vegetation, comparative studies of sampling
procedures in vegetation surveys, optimization of vegetation
sampling, study of plant population processes both in real
space and time.

KEY WORDS. vegetation, structure, pattern, sampling, simulation,
photocomputations.

1. INTRODUCTION

It is a salient feature of the statistical ecology of past
decades that a great deal of effort has been spent to develop
and apply multivariate methods. Yet, the field has almost
completely lost sight of the unsolved problems of sampling. As
a consequence of this, there is a striking contrast between the
exactness of the methods which analyze the data and the roughness
of the procedures by which the vegetation is sampled. It is
quite obvious, however, that the quality of sampling nas influence
and limiting power on the information flow in the course of a
complex vegetation study. There is no reason to analyze data
sets, obtained in an uncontrolled, inaccurate and non-optimized
way, by means of a sophisticated, highly organized mathematical
apparatus, for the methods' precision in the analysis cannot
alleviate the problems flowing from weaknesses in the data.

The choice of sampling parameters, to be considered in the
sequal, is a matter of subjective judgement, traditionally
without optimizations of any kind. But, this is not surprising
at all, considering that the theoretical basis for such an
optimization has yet to be developed. Perhaps Juhasz-Nagy's
general theory of association processes in space and characteristic areas may serve as a starting point in the development
of a theory for vegetation sampling (Juhasz-Nagy, 1967, 1972,
1973, 1976; Devai, Horvath, and Juhasz-Nagy, 1971; Juhasz-Nagy,
Devai and Horvath, 1973).

The exact influence of sampling on the results of data
analysis remains unknown and therefore uncontrolled. Who can
tell, for example, without experiments from the results of
component analysis (PCA) of interspecific associations what
changes would occur if sampling units of another size, shape
and arrangement were used?

The purpose of this paper is to give an additional impetus
to the development of the neglected theory and practice of
vegetation sampling. For this reason, I describe a new approach
to the study of vegetation structure and new methods for
practical applications. The presentation is not intended to give
a full account of the results obtained by these methods. The
aim is only to show specific aspect selected from a broad
range of possibilities.

2. THE CONCEPT OF VEGETATION STRUCTURE

First of all, let us define the concept of structure [S].
We consider [S] of an entity (Juhasz-Nagy, 1972) as the given
arrangement of given elements of the entity in a given space.
Depending on the choice of elements and space, we may have, of
course, different particular cases of the general definition.
Every entity has indeed a series of concurrent structures
(principle of structural simultaneity; Juhasz-Nagy, 1972).

One may apply the general definition to the vegetation
in several ways. It is reasonable to use the simplest case.
Let us take a well-bounded part of a real plant community, to
be referred to as a stand, as an entity. Let the living plant
individuals existing in the stand be designated as its elements
and the real, two-dimensional surface soil as the topographic
space. This kind of [S] of the vegetation may be referred to
as the elementary vegetational structure [EVS]. In the following, we shall consider [EVS] only. For a certain given real
[EVS], we have a given arrangement of elements and every
structural question should refer to this.

To analyze such an arrangement, we may proceed in two ways:
(a) in an inductive way starting from a real situation
(arrangement) trying to uncover the underlying, basic properties
(Section 3), or (b) in a deductive way starting from a
theoretically defined artificial arrangement trying to approximate, more and more closely the real situation by means of
transformations and superpositions (Section 4). This paper
presents new methods for both approaches.

3. THE PHOTOCOMPUTATIONAL METHOD [PCM]

3.1 Advantages. The main idea of [PCM] is this: the properly
transformed and digitalized [EVS] can be stored in computer core.
The whole procedure of vegetation sampling can then be made
automatically by an appropriate program. This solution offers
many advantages in that it is highly flexible, avoids the trampling effect, repeatable, fast, exact, and free from subjective
distortions. [PCM] consists of fixation, transformation,
digitalization, entry and pseudo-simulation sampling.

Although [PCM] was developed by the author independently,
he discovered later that [PCM] is similar to a method of
analysis of plant dispersion based on low-level aerial photographs used by others (West and Goodall, 1971; Goodall and
West, 1972).

3.2 Fixation. The fixation of the real, primary picture [PP]
of the vegetation can be made in several ways, but the most
desirable is a distribution map showing the locations of the
individuals within the entity. For this purpose, we used
ground-level vertical photographs, color diapositives, with good
results. It is important that the individuals be clearly
distinguishable and identifiable to species on the diapositives.
Colored plastic markers proved to be quite helpful for this
purpose. Two man-days work by skilled persons were sufficient
for the fixation of an area of 100-120 m^2 in an open grassland
containing 15 species and about 4000 individuals. There are
many possibilities to increase the technical efficiency and
accuracy of this solution and for adaptation to other vegetation
types, e.g., using stereophotograms (Pierce and Eddleman, 1969),
artificial light, flexible permanent frame for cameras, etc.

3.3 Transformation. In order to digitize [PP] must be transformed into a geometric image [GI]. There are three main types
of practical solutions:

[GI] is a two-dimensional spatial point distribution [SPD].

The points are localized in the insertion point or in the
approximate center of the area covered by the individual
(Dacey, 1973).

[GI] is a composite of regular two-dimensional geometrical
figures (circles, ovals, quadrats, etc.) scattered in a plane
(Mohn and Stavem, 1974).

[GI] is a partition of the area of the entity, i.e., a
mosaic map of two or more phases (Pielou, 1974; Jumars, Thistle
and Jones, 1977).

I have chosen, for simplicity, the [SPD] solution. There
are many problems to solve even in this simple case. The other
types of [GI] are built, at least in part, on [SPD].

Transformation is a step of considerable importance in
the course of [PCM]. It is also inevitable in each case when
units of vegetation have to be sampled. It can be easily
proven that some kind of transformation is implied as a part in
even the most traditional sampling procedures (e.g., estimation
of species cover by the Braun-Blanquet method).

3.4 Digitization. [GI] must be numerically coded for computer
analysis and storage. In the case of [SPD], coding is very simple
to accomplish. We take the usual Cartesian coordinates of the
points and use these as the primary data set [PDS]. The
digitization of other types of [GI] may be much more difficult,
requiring the services of a skilled programmer.

3.5 Pseudosimulation Sampling. This written [PSS], is at the
very core of [PCM]. The practical solutions of the previous
steps may be modified, improved or substituted, but [PSS] is
essential. Every sampling procedure should consist of a
series of well-defined steps (an algorithm). Such a procedure
can be programmed, and processed by a computer, assuming that
the population studied is a real data set stored in the same
computer core. After the data set [PDS] is entered and stored
in computer core, only a suitable program is needed for the
implementation of sampling.

The program CROCUS was written in FORTRAN IV for this
purpose by the author. CROCUS takes [PDS] as input and offers
a broad variety of choice for a combination of sampling
parameters, such as shape (circle, oval, quadrat, line, point),
size, number, orientation, and arrangement (random, systematic)
of the sampling units. The result is a matrix of coincidences

[CM]. This is a primary matrix where rows and columns represent species and sampling units respectively. Besides [CM], CROCUS performs several optional initial matrix manipulations, including transformation of [CM] into a presence-absence matrix, calculation of different marginals, subtotals, means, variances, etc. The secondary data sets [SDS] are suitable as input for subsequent analysis in multivariate or other methods. Optional printout of the results or any part of the [SDS] may be requested.

CROCUS can perform any distance-based sampling method (nearest neighbor, random pairs, etc.) and also, the combinatorial, linear, and sequential methods by variants of CROCUS.

4. THE PATTERN SIMULATION METHOD [PTSM]

It is possible to generate every well-defined deterministic or stochastic [GI] as a representation of the corresponding hypothetical [EVS] by a computer. The simulated [EVS] can be studied, sampled, and analyzed in the same way as if it were a real [EVS], fixed, transformed and digitized in the course of a [PCM] procedure. The sampling is, therefore, by [PSS]. So [PTSM] consists of only two steps: pattern generation and [PSS]. Both are in one program, SCILLA written in FORTRAN IV.

One may produce and superimpose by SCILLA random, clustered, or regular [SPD]s of any type. It is also possible to choose the overall density of [SPD], the within-cluster and between-cluster densities, pattern of clusters, or points. These [SPD]s may be plotted, if necessary, but generally they are retained in memory core in the form of an [SDS], ready for the sampling by [PSS].

I have chosen first an [SPD], as the simplest type of [GI] for SCILLA, but the generation of other types of [GI] can also be programmed with some invention. The basic type of [SPD] is a random one [PSPD]. It can be regarded as a realization of a two-dimensional spatial Poisson-process. For this reason, [RSPD] is a theoretical zero-point of the different [SPD]s. It is very important to know and measure the deviation of a real [SPD] from the corresponding [RSPD] with the same density. This deviation can be measured several ways. The best of them is, probably, the chi-square test on two frequency distributions of the numbers of individuals per sampling unit. This test, however, is influenced by the sampling parameters. One of the benefits of [PTSM] is the possibility to see how exactly this deviation depends on the density and the sampling parameters. It seems to me that there is no other practical way to analyze the involved complex interrelationships.

The results of [PTSM] have a similar form as [PCM]including
tabular listings and permanent record files of [SDS].

5.   IMPORTANCE OF [PCM] AND [PTSM]

[PCM] can be regarded as a sampling procedure of real
practical value in field studies with exactness commensurable
with requirements of any multivariate and other high level
data analysis. [PCM] and [PTSM] offer pathways to find the
optimal sampling strategy for a given vegetation unit and a
given objective.

[PSS] in the frame of [PCM] or another method, is the
first to open the way to exact comparative studies of different
sampling methods. [PSS] is non-destructive, repeatable, and
makes possible to apply the different sampling methods to the
same vegetational object. These could be done by no other means
because of the inevitable trampling effect.

Almost every problem involved and to be solved by [PCM]
has its counterpart of [PTSM]. The inductive and deductive
approaches of Section 2 are most fruitful when they are used
in combination. For example, if we try to study the deviation
of a real [SPD], obtained by [PCM], from a random [SPD], we
may choose a direct way, comparing the empirical frequency
distribution to the corresponding Poisson distribution. But
the result will be more realistic when we simulate the proper
[RSPD] with the same overall density, in [PTSM], and then apply
the very same method of sampling to both. Applications of
[PCM] and [PTSM] are not restricted to the field of plant
ecology. They are expected to be useful in every case when
the task of analyzing the structure of an entity arises, similar
to [EVS] and transformable into a proper [GI].

There is a wide gap between plant demography (plant
population dynamics) and the structural investigations (pattern
analysis, ordination, etc.). These properties of the vegetation
are, in reality, time or space dependent. But population
processes are both temporal and spatial, such as succession,
seed dispersal, migration, vegetative reproduction, density-
and age-dependent birth and death processes, competitive
inhibition of germination, etc. This is also clear from Levin's
(1976) review of the problems of patchiness of the environment
and patch formation in communities.

REFERENCES

Dacey, M. F. (1973). Some questions about spatial distributions. In *Directions in Geography*, R. J. Chorley, ed. Methuen, London.

Devai, I., Horvath, K., and Juhasz-Nagy, P. (1971). Some problems of model-building in synbotany. Part I. Spatial diversity processes of the binary type in a simple situation. *Annales Universitatis Budapestiensis, Sectio Biologica*, 13, 19-32.

Goodall, D. W. and West, N. E. (1972). An integrated set of computer programs for studying plant dispersion patterns. (Mimeographed). A paper presented at the Annual Meeting of ESA, Minneapolis.

Juhasz-Nagy, P. (1967). On some characteristic areas of plant community studies. *Proceedings of the Colloquium on Information Theory*. Debrecen, Hungary, 269-282.

Juhasz-Nagy, P. (1972). Structure of vegetation. Part I. *Botanikai Kozlemenyek*, 59, 1-6.

Juhasz-Nagy, P. (1973). Structure of vegetation. Part II. *Botanikai Kozlemenyek*, 60, 35-41.

Juhasz-Nagy, P., Devai, I., and Horvath, K. (1973). Some problems of model-building in synbiology. Part 2. Association processes in a simple situation. *Annales Universitatis Budapestiensis, Sectio Biologica*, 15, 39-51.

Juhasz-Nagy, P. (1976). Spatial dependence of plant populations. I. Equivalence analysis (an outline of a new model). *Acta Botanica Academiae Scientiarum Hungariae*, 22, 61-78.

Jumars, P. A., Thistle, D., and Jones, M. L. (1977). Detecting two-dimensional spatial structure in biological data. *Oecologia*, 28, 109-123.

Levin, S. A. (1976). Population dynamic models in heterogeneous environments. *Annual Review of Ecology and Systematics*, 7, 287-310.

Mohn, E. and Stavem, P. (1974). On the distribution of randomly placed discs. *Biometrics*, 30, 137-156.

Pielou, E. C. (1974). *Population and Community Ecology*. Gordon and Breach, New York.

Pierce, W. R. and Eddleman, L. E. (1969). A field stereo-
photographic technique for range vegetation analysis.
Journal of Range Mangement, 22, 218-220.

West, N. E. and Goodall, D. W. (1971). Analysis of plant
dispersion patterns from low level aerial photographs.
(Mimeographed). A paper presented at the Annual Meeting
of ESA, Fort Collins.

[*Received July* 1978. *Revised January* 1979]

L. Orloci, C. R. Rao, and W. M. Stiteler, (eds.),
Multivariate Methods in Ecological Work, pp. 309-535. All rights reserved.
Copyright © 1979 by International Co-operative Publishing House, Fairland, Maryland

ECOLOGICAL APPLICATIONS OF CANONICAL ANALYSIS

R. GITTINS

Department of Plant Sciences
University of Western Ontario
London, Ontario Canada N5A 6B7

SUMMARY. The problem of investigating relationships between two
(or more) sets of variables arises in many branches of Science.
Yet, despite its importance and generality, the problem has
received less attention from statisticians and others than it
would seem to deserve. The method which above all seems
to address this question most directly is canonical correlation
analysis - here referred to for convenience as canonical analysis.
Nevertheless, canonical analysis is often cooly received, despite
a lack of suitable alternatives.

   After first reviewing the nature and properties of canonical
analysis an assessment of the method as a tool for use in connection with exploratory studies in ecology is made. The results of
applications of canonical analysis to several sets of ecological
data are described and discussed with this objective in mind. The
paper also aims to serve a didactic purpose by drawing attention
to the value of canonical analysis as a means of unifying several
otherwise seemingly unrelated methods of data analysis widely used
in ecological work.

KEY WORDS. association in r × c tables, canonical correlation,
canonical variates analysis, multiple discriminant analysis, multivariate analysis of variance, dummy variables, soil-vegetation
relationships, spatial variation, vegetation-herbivore relationships.

TABLE OF CONTENTS

CHAPTER I: INTRODUCTION

CHAPTER 1

INTRODUCTION

Many questions of interest to ecologists call for the investigation of relationships between variables of two distinct but associated kinds. This may involve, as, for example, in plant ecology, connections between the occurrence of plant communities and their component species, on the one hand, and soil or other environmental variables, on the other; or perhaps, as in palaeoecology, investigation of relationships between fossil pollen samples and their modern counterparts from known vegetation types. In more general terms, the question which arises calls for the exploration of relationships between any two (or more) sets of variables of ecological interest. Despite the importance and generality of the question, relatively few formal models for the investigation of relationships of this kind have been developed. The available approaches are those of multivariate cross-spectral analysis and canonical correlation analysis, or related methods such as multiple pattern analysis (Noy-Meir and Anderson, 1971) and multiple predictive analysis (Macnaughton-Smith, 1965; Orlóci, 1978). Substantive applications of these methods in ecology are few. In practice all too often recourse is made to multiple regression (e.g. Vuilleumier, 1970; Noy-Meir, et al. 1973; Schall and Pianka, 1978; Teeri, Stowe, and Murawski, 1978). In such cases, the set of criterion or response variables is first dismembered into p univariate responses (p being the total number of response variables in the set), and a series of p multiple regression analyses then performed. This procedure however results in the discarding of information on inter-relationships within the response set, the consequences of which cannot be dismissed lightly (see for example, Bock, 1975, p. 20). One sometimes also finds principal components analysis pressed into service to investigate relationships between two sets of variables, either jointly or separately (Austin, 1968; Gittins, 1969; Barkham and Norris, 1970; Goldsmith, 1973). Again, except perhaps in special circumstances, such as where there is a need to condense the number of variables, the practice is difficult to justify.

The method which above all others seems to address the question of the interrelatedness of two measurement domains most directly is canonical correlation analysis. For convenience, we shall refer to this technique in what follows simply as canonical analysis. Part of the appeal of canonical analysis is that it does operate on both sets of variables *simultaneously*, focusing, in particular, on the covariance or correlation structure between them. Yet the method is not well known among ecologists, and, moreover, when it has been used has not infrequently given disappointing results (Austin, 1968; Cassie and Michael, 1968; Barkham and Norris, 1970; Gauch and Wentworth, 1976; cf. also W. T. Williams, 1976, p. 66). Consequently, canonical analysis is often viewed with indifference or even suspicion by ecologists. The linear, orthogonal nature of the solution, in particular, is often held to be unrealistic in ecological contexts and largely responsible for many of these failures (Austin, 1972; Dale, 1975; Gauch and Wentworth, 1976). W. T. Williams (1976), however, considers the high unique variances characteristic of many biological data sets to be the likely cause. Certainly, both explanations seem plausible.

Having regard for the intrinsic appeal of certain features of canonical analysis on the one hand, and of the widespread disquiet expressed by ecologists concerning the method together with the paucity of suitable alternatives on the other, it is clear that attempts should be made to better define what can be accomplished by canonical analysis. The present paper is a contribution towards this end. The principal aims of the paper are to review the nature and properties of canonical analysis and to provide a basis for an assessment of the method as an analytical tool for use in connection with exploratory studies in ecology. A subsidiary objective is to show that canonical analysis is also valuable as a means of unifying several seemingly unrelated methods of data analysis widely used in ecological investigations. With these objectives in mind, seven examples of canonical analysis are reported. The first two examples are in the nature of experiments designed to throw light on the ability of the method to recover *known* relationships from among sets of ecological variables. These are followed by five further analyses which illustrate something of the flexibility and power of the method which arises in part from various specializations on the nature of the variables employed.

As mentioned above, relationships among sets of variables may also be pursued by means of cross-spectral analysis (Jenkins and Watts, 1968; Box and Jenkins, 1976). This procedure has considerable ecological relevance and promise in connection with the analysis of spatial pattern. Attention here however is confined to canonical analysis. Canonical analysis places less restriction on the structure of the data required for analysis and accordingly is the more broadly applicable of the two procedures. Such a limitation

on coverage serves to keep the presentation within reasonable
bounds. Before turning to a formal treatment of canonical analysis,
we first sketch in general terms something of the essential nature
and features of the method.

1. CANONICAL ANALYSIS: OVERVIEW

The notion that field observations in ecology are multivariate
in character and are therefore conveniently expressed and treated
algebraically as *vectors* is gaining general acceptance (Noy-Meir,
1971; Orlóci, 1975; Pielou, 1977). In studies involving variables
of two kinds, the sample observations give rise to *partitioned*
vectors. Such vectors consist of two subvectors which may for
example be made up of, say, p ecological variables and q environmental or other variables, respectively. Interest in
vector variables of this kind centers on the nature of relationships among the variables comprising each set, although within-set
relationships also merit attention. All the available information
about linear relationships both within and between sets is
summarized in the covariance matrix, or in some similar scalar-
product matrix of the vector variables. If the elements of a
partitioned (p + q)-vector are expressed in standard form
with zero mean and unit variance, then the covariance matrix of
the vector is identically the correlation matrix of the original
variables. Denoting the partitioned vector of standardized
variables for the j*th* sample as $\underset{\sim}{z}_j' = [\underset{\sim}{z}_j^{(x)}, \underset{\sim}{z}_j^{(y)}]$, where $\underset{\sim}{z}_j^{(x)}$
and $\underset{\sim}{z}_j^{(y)}$ represent subvectors of ecological and environmental
variables respectively, and denoting the covariance matrix of
$\underset{\sim}{z}_j$ as $\underset{\sim}{R}$, we may represent the composition of $\underset{\sim}{R}$ as follows:

$$\underset{\sim}{R} = \frac{1}{N} \sum_{j=1}^{N} \underset{\sim}{z}_j \, \underset{\sim}{z}_j'$$

$$= \left[\begin{array}{c|c} \underset{\sim}{R}_{11} & \underset{\sim}{R}_{12} \\ \hline \underset{\sim}{R}_{21} & \underset{\sim}{R}_{22} \end{array} \right]$$

where, $\underset{\sim}{R}_{11}$: p × p matrix correlations between the ecological
              variables

$\underset{\sim}{R}_{22}$: q × q matrix correlations between the environmental
          variables

$\underset{\sim}{R}_{12} = \underset{\sim}{R}_{21}'$: p × q matrix correlations between the variables
          of each set

The internal structure of a matrix of this kind is rarely evident on inspection, even where there are strong grounds for supposing the existence of an intrinsic structure of some kind corresponding to suspected interrelationships among the variables. The principal task of analysis is therefore to operate on the matrix R in some way such that the structure of the matrix becomes manifest. It will perhaps be appreciated that the task of comprehending *between-set* relationships will be made easier if we can at the same time also disentangle and simplify the *within-set* correlations between variables. Thus, the starting-point for a canonical analysis may be regarded at the correlation matrix $\underset{\sim}{R}$, or some similar matrix. The aim of analysis is to summarize and clarify the internal structure of $\underset{\sim}{R}$ and in this way throw light on the nature of relationships between the two sets of variables of interest.

The strategy of canonical analysis is analgous to that of principal components analysis (PCA). It entails rotation of the coordinate frame of the sample space determined by the data to a new position in which some feature or property of the data is emphasized. In components analysis, the feature of interest is the total *variance* associated with a single set of variables; in canonical analysis it is the *covariance* between two sets of variables. Notice, however, that in canonical analysis the data can be regarded as generating two sample spaces, of p and q dimensions respectively, which correspond to the two sets of variables. Nevertheless, it is possible to simultaneously rotate the coordinate frames of each space to new positions in which the covariance or correlation structure of the data is clearly revealed. Moreover, the new coordinate systems are such that relationships within each set of variables are disentangled (much as in components analysis), while at the same time facilitating a reduction in dimensionality (again, much as in PCA). Algebraically, the rotations are equivalent to finding linear transformations of each set of variables, $U_k = \underset{\sim}{a}_k' \ \underset{\sim}{z}^{(x)}$ and $V_k = \underset{\sim}{b}_k' \ \underset{\sim}{z}^{(y)}$ such that the simple correlation between the transformed variables U_k and V_k is maximized. The new variables U_k and V_k which correspond to the new axes of the coordinate frame, are called *canonical variates*, while the familiar product-moment correlation coefficient between the canonical variates U_k and V_k is referred to as a *canonical correlation coefficient*. The explanation for this terminology will shortly become apparent. In all there will be s pairs of such linear transformations, $k=1,\cdots,s$ where s is equal to the smaller of p and q, successive pairs of canonical variates after the first being uncorrelated with the preceding variates. The procedure is illustrated diagrammatically in Figure 1.1 for the first pair of canonical variates. Hotelling (1935, p. 142) has provided an alternative diagrammatic representation of the nature of the method.

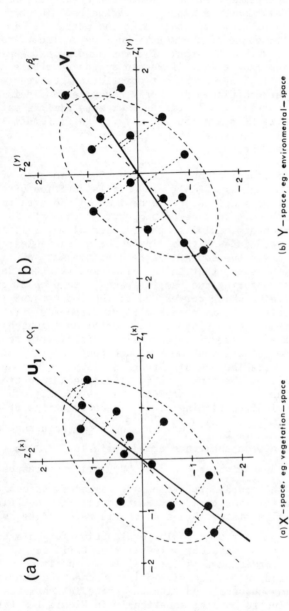

(a) X − space. eg. vegetation−space

(b) Y − space. eg. environmental−space

FIG. 1.1: Diagrammatic illustration of the principle of canonical analysis. A sample (N = 15) and concentration ellipse are shown (a) in vegetation-space (p = 2) and (b) in an environmental-space (q = 2). The canonical variates U_1 and V_1 are by definition those linear combinations of the variables of each space for which the simple correlation coefficient, r_1, between the projected points is a maximum over all possible choices for U and V; r_1 is the first or largest canonical correlation. For comparison, the first principal components, α_1 and β_1, of each measurement domain are also indicated. In general, there will be s = min(p,q) pairs of such canonical variates, U_k and V_k, and canonical correlation coefficients, r_k.

In passing, mention should also be made of the fact that some aspects of canonical analysis are best appreciated by thinking conceptually of the sample as a galaxy of 2N points in a single space of (p + q)-dimensions.

Ordinarily, most of the covariation between the measurement domains will be concentrated in the first two or three pairs of canonical variates, U_k, V_k, k = 1, $\cdots$, 3 (say). The remaining s - 3 pairs of canonical variates can then often be neglected with little or no loss of useful information. In this way a reduction in the dimensionality of the data may be achieved. Tests of significance are available which in appropriate circumstances may be helpful in determining dimensionality. Where a reduction in dimensionality is possible, mapping of the sample into the 2- or 3- dimensional space of the retained canonical variates facilitates visual representation and examination of the data. Graphic displays of this kind are generally immensely rewarding in the insight into the covariance structure of the observations which they convey.

It is instructive to consider the correlation matrix $R_{\sim uv}$ of the transformed variables. This matrix has a particularly simple and appealing form. Confining our attention for the moment to the non-null canonical correlations and variates, of which in general there will be s, $R_{\sim uv}$ may be represented as follows:

$$R_{\sim uv} = \begin{array}{cc} & \begin{array}{cc} u & v \end{array} \\ \begin{array}{c} u \\ v \end{array} & \begin{bmatrix} I_{\sim} & \Lambda_{\sim} \\ \Lambda_{\sim}' & I_{\sim} \end{bmatrix} \end{array} \qquad (1.1)$$

where

$$I_{\sim} = \begin{bmatrix} 1 & & 0 \\ & \ddots & \\ 0 & & 1 \end{bmatrix}, \qquad \Lambda_{\sim} = \begin{bmatrix} \lambda_1 & & 0 \\ & \ddots & \\ 0 & & \lambda_s \end{bmatrix}.$$

Here $I_{\sim}$ is the identity matrix of order s while $\Lambda_{\sim}$ is an s × s diagonal matrix whose non-zero elements, λ_k, are the canonical correlations between the kth pair of canonical variates U_k and V_k, k = 1, $\cdots$, s. It is convenient to arrange the correlations λ_k so that $\lambda_1 \geq \cdots \geq \lambda_s$. The correlation structure within and between the new variables may therefore be characterized briefly as follows: (a) all *within-set* correlation is reduced to zero; (b) all linear correlation *between* the two sets is channeled through the canonical correlation coefficients λ_k. We therefore see from (1.1) that the transformation to canonical variates achieves the joint objectives of (a) disentangling the correlations within each set of variables while (b) emphasizing and clarifying linear relationships between the two measurement domains. The question of the substantive meaning and interpretation of the canonical variates, however, remains. We shall return to this question in a later section. It is the correlation matrix $\underset{\sim}{R}_{uv}$ which gives its name to canonical analysis. $\underset{\sim}{R}_{uv}$ is a *canonical form* of the correlation matrix $\underset{\sim}{R}$ between the original variables. In other words, $\underset{\sim}{R}_{uv}$ expresses important or interesting information about the data represented by the original correlation matrix $\underset{\sim}{R}$. The same information is present in each case, it is simply more conveniently conveyed by the matrix $\underset{\sim}{R}_{uv}$ of correlations between the canonical variates. Thus analysis does not alter the initial data in any way, apart perhaps from a possible adjustment for example to standard score form. Canonical analysis merely provides a convenient coordinate frame within which to examine the data for a particular purpose.

We have sketched the derivation of canonical analysis above in an informal way by pointing to analogies between it and the better-known procedure of principal components analysis. But canonical analysis is probably even more closely related to regression than to components analysis. In fact, we may regard canonical analysis as a natural generalization of multiple regression. The distinction between the latter procedures is that canonical analysis is concerned with predicting a set of response variables rather than a single response, as in multiple regression. In other words canonical analysis involves the regression of a *vector* of response variables on a vector of predictors. Moreover, as the designation of variables in canonical analysis as, for example, dependent and independent is arbitrary, it follows that the analysis is symmetric with respect to both sets of variables. This means that we may think of either set of variables being used to predict the other, or indeed, of both sets of variables simultaneously predicting each other. Thus canonical analysis can be thought of as an analysis in which two multiple regressions are embedded. This point is best appreciated by noting that if the first canonical variate of, say, the X-domain, U_1, is regressed onto the q variables of the Y-domain, then the resulting regression coefficients are identically the coefficients comprising the canonical variate V_1. Similarly, multiple regression of the first canonical variate V_1 of the Y-domain onto the set of p X-variables leads to regression weights which are identically the elements of U_1. Unlike, multiple regression analysis, however, where there will be at most one regression relationship between the variables, in canonical analysis there will in general be s pairs of such relationships, although these are unlikely to be of equal value or interest. Thus we see that canonical analysis shares features in common with both components and regression analysis. Corresponding members of a pair of canonical variates may be thought of as components of their respective measurement domains somewhat akin to principal components in nature which are related in a regression sense across domains. It should be understood however, that, while on occasion canonical variates may indeed be collinear with principal components, in general it is not to be expected. Corresponding to these shared affinities are differences in emphasis which may be stressed in different applications. In certain cases the method is used purely descriptively to condense and expose mutual relationships among variables of different kinds, while in other applications it is the predictive aspect which is of most value.

The variables entering into a canonical analysis are not required to be continuous and moreover need not be directly measured or observed in the usual sense at all. Various kinds of specialization of the variables of either or both sets are admissible. We shall see that this property enhances the flexibility of canonical analysis very considerably. One of the commonest specializations involves the use of binary-valued dummy variables. Variables of this kind are easily constructed and are helpful where a classificatory structure exists or can be superimposed on a p-variate sample in order to utilize the information inherent in the classification. Where such information is relevant to the problem in hand, it is readily retrieved by the construction of dummy variables which are then taken to comprise one set of variables in a canonical analysis. A large class of problems in ecology lend themselves to formulation and analysis in this way. Sometimes it is possible to cross-classify the sample with respect to two criteria. In such cases the procedure described can be extended, leading to a canonical analysis in which both sets of variables are binary-valued dummy variables.

Other forms of specialization involve the use of more elaborately coded dummy variables, principal components of either or both sets of observed variables and the spatial coordinates of observed vector-valued variables having a spatial distribution.

As with other multivariate techniques, canonical analysis may be employed as a statistical method with the objectives of estimation and testing, or less formally, as a means of exploratory data analysis. In the latter case, distributional assumptions of the data are not required. Nevertheless, canonical analysis is specified by a linear, additive model and, moreover, requires a continuous data structure for its proper use. The realities of ecological data, which frequently embody nonlinearities and discontinuities, should warn us against applying the method where these conditions are not at least approximately met. Fortunately, as with other forms of regression, a certain degree of non-linearity can be accommodated by the model, while steps can be taken to unfold more complicated nonlinear structures prior to analysis. It is also possible that the model itself might be modified to render it more robust in the presence of heterogeneous data. Thus, while the conditions necessary for the proper use of canonical analysis even as a means of exploratory data analysis are fairly stringent, steps can be taken to enhance compatability between the realities of the data and the requirements of the model. The method is best-suited to the exploratory phase of research where the need is simply to detect the presence of certain relationships or effects and to determine their direction. To go beyond this point stronger models based on some concept of the actual processes underlying the data are required. The wider aim of canonical analysis is to provide a starting point for the development of models of this kind by screening out all but a small number of the most promising variables or processes for consideration.

The terminology associated with canonical analysis is not firmly established. We should therefore take steps to make our own usage plain and to relate it to the terminology of other workers. As employed here, canonical analysis is synonymous with canonical correlation analysis. By canonical correlation analysis we mean a technique of multivariate analysis which provides linear functions of two sets of variables which have special properties in terms of correlations *irrespective* of the nature of the variables comprising either set. With this understanding in mind, canonical analysis includes as special cases both *canonical variates analysis*, in which one set of variables consists of binary-valued dummy variables designating class membership, and the (canonical) analysis of *association in* r × c *contingency tables*, in which both sets of variables are binary in character. Canonical variates analysis in the above sense is formally equivalent to *multiple discriminant analysis*, the distinction between the two resting principally on the use to which the analysis is put: where interest centers on the structure or mutual relationships among k multivariate universes we speak of canonical variates analysis; where the aim is to allocate one or more p-variate samples among k mutually exclusive categories or universes we speak of multiple discriminant analysis. We also note the formal equivalence of the canonical analysis of association in r × c tables to *correspondence analysis*. The term canonical analysis is used by many workers (e.g. Bartlett, 1947, 1965; Seal, 1964; Pearce, 1969; Goldstein and Grigal, 1972; Kowal, Lechowicz and Adams, 1976) in a narrower sense than understood here which corresponds to canonical variates analysis as

defined above.   On the other hand, Pielou (1977) uses the term
canonical variate analysis in a wider sense than adopted here to
refer to what we have called canonical correlation analysis.

The theory of canonical correlation was developed by Hotelling
(1935, 1936) as a means of identifying the most predicable
p-variate criterion, given the availability of several predictor
and criterion variables.   We shall see that the theory developed
has since proved to have other applications.

CHAPTER II

THEORY

2. CANONICAL CORRELATIONS AND CANONICAL VARIATES

2.1 The Data Matrix. Let $x_{\sim j}$, $j = 1, \cdots, N$ be a sample of size
N from a $(p + q)$-variate distribution. Furtherfore, let $x_{\sim j}$
be partitioned into two subvectors $x_{\sim j}^{(x)}$ of p ecological
variables and $x_{\sim j}^{(y)}$ of q environmental or other variables, or
vice versa $(p + q = n)$. Then

$$x_{\sim j} = \begin{bmatrix} x_{\sim j}^{(x)} \\ \\ x_{\sim j}^{(y)} \end{bmatrix} , \quad j = 1, \cdots, N ,$$

where $[x_{\sim j}^{(x)}]' = [x_1, \cdots, x_p]$, $[x_{\sim j}^{(y)}]' = [x_{p+1}, \cdots, x_n]$. For
convenience we assume $p \geq q$.

The observed or assigned sample values on the variates may
be written collectively in matrix form as

$$\underset{\sim}{X} = [x_{ij}] ,$$

where x_{ij} stands for the value of the *ith* variable in the *jth*
sample $(i = 1, \cdots, n)$. The vector of sample means is denoted
$\bar{x}' = (\bar{x}_1, \cdots, \bar{x}_n)$ and the diagonal matrix of sample variances
$\underset{\sim}{S} = \text{diag} (s_1^2, \cdots, s_n^2)$. Without loss of generality we shall assume
throughout that the variables are centered and standardized to
zero mean and unit variance. We shall denote the *jth* vector of
standardized variables by $z_{\sim j}$ where

$$\underset{\sim}{z}_j = \underset{\sim}{S}^{-\frac{1}{2}}(\underset{\sim}{x}_j - \bar{\underset{\sim}{x}}) \quad .$$

The variance-covariance matrix of $\underset{\sim}{z}_j$ is the $n \times n$ matrix $\underset{\sim}{R}$ where

$$\underset{\sim}{R} = \frac{1}{N} \sum_{j=1}^{N} \begin{bmatrix} z_{\sim j}^{(x)} \\ z_{\sim j}^{(y)} \end{bmatrix} \begin{bmatrix} z_{\sim j}^{(x)} & z_{\sim j}^{(y)} \end{bmatrix} = \begin{bmatrix} \underset{\sim}{R}_{11} & \underset{\sim}{R}_{12} \\ \underset{\sim}{R}_{21} & \underset{\sim}{R}_{22} \end{bmatrix}$$

Here the $p \times p$ matrix $\underset{\sim}{R}_{11}$ is the covariance matrix of $z_{\sim j}^{(x)}$, the $q \times q$ matrix $\underset{\sim}{R}_{22}$ is the covariance matrix of $z_{\sim j}^{(y)}$ and $\underset{\sim}{R}_{12} = \underset{\sim}{R}_{21}'$ of order $p \times q$ is the covariance matrix between $z_{\sim j}^{(x)}$ and $z_{\sim j}^{(y)}$.

Inter-relationships within and between the variables comprising $z_{\sim j}^{(x)}$ and $z_{\sim j}^{(y)}$ are specified by the variance-covariance or correlation matrix $\underset{\sim}{R}$ of $\underset{\sim}{z}_j$. The internal structure of $\underset{\sim}{R}$, and the connections between the variables, however, are rarely evident on inspection. We therefore seek linear transformations of each set of variables to new variates U_k and V_k (the canonical variates) for which the correlation matrix has a particularly simple and appealing form. We require that:

(a) all the U_k be uncorrelated with one another;

(b) all the V_k be uncorrelated with one another; and

(c) that the pairs of canonical variates U_k, V_m for
   $k, m = 1, \cdots, s$ where $s = \min(p,q)$, be maximally
   correlated for $k = m$ and zero otherwise.

The dependencies between the original variables are thus disentangled and reduced to their simplest possible form. The *within-set* correlations are all zero, while the *between-set* correlations (the canonical correlations) are maximized between s pairs of variates and reduced to zero between all other pairs.

2.2 Derivation of Canonical Correlations and Variates. We seek linear combinations of $z_{\sim j}^{(x)}$ and $z_{\sim j}^{(y)}$

$$U = a_1 z_{1j} + \cdots + a_p z_{pj} \qquad = \underset{\sim}{a}' \underset{\sim}{z}_j^{(x)}$$

$$V = b_{p+1} z_{p+1j} + \cdots + p_n z_{nj} = \underset{\sim}{b}' \underset{\sim}{z}_j^{(y)} \quad,$$

where the coefficients $\underset{\sim}{a}' = (a_1, \cdots, a_p)$ and $\underset{\sim}{b}' = (b_{p+1}, \cdots, b_n)$ are chosen to maximize the simple correlation, r, between U and V. The correlation r between $\underset{\sim}{a}$ and $\underset{\sim}{b}$ expressed as a function of $\underset{\sim}{a}$ and $\underset{\sim}{b}$ is

$$r = cov(\underset{\sim}{a}'\underset{\sim}{z}^{(x)}, \underset{\sim}{b}'\underset{\sim}{z}^{(y)}) \Big/ \Big\{ var(\underset{\sim}{a}'\underset{\sim}{z}^{(x)}) \quad var(\underset{\sim}{b}'\underset{\sim}{z}^{(y)}) \Big\}^{\frac{1}{2}}$$

$$= \underset{\sim}{a}'\underset{\sim}{R}_{12}\underset{\sim}{b} \Big/ \Big\{ (\underset{\sim}{a}'\underset{\sim}{R}_{11}\underset{\sim}{a}) (\underset{\sim}{b}'\underset{\sim}{R}_{22}\underset{\sim}{b}) \Big\}^{\frac{1}{2}} \quad. \tag{2.1}$$

The weights $\underset{\sim}{a}$ and $\underset{\sim}{b}$ are determined only up to a constant of proportionality, and, for simplicity, it is customary to choose $\underset{\sim}{a}$ and $\underset{\sim}{b}$ so that U and V have unit variance. Then

$$var(U) = \underset{\sim}{a}'\underset{\sim}{R}_{11}\underset{\sim}{a} = 1 \quad, \quad var(V) = \underset{\sim}{b}'\underset{\sim}{R}_{22}\underset{\sim}{b} = 1 \quad. \tag{2.2}$$

Expression (2.1) now reduces to

$$r = \underset{\sim}{a}'\underset{\sim}{R}_{12}\underset{\sim}{b} \quad. \tag{2.3}$$

Thus the algebraic problem is to find $\underset{\sim}{a}$ and $\underset{\sim}{b}$ to maximize (2.3) subject to (2.2). Writing $f(\underset{\sim}{a},\underset{\sim}{b})$ for the function to be maximized under constraints, the problem may be represented

$$f(\underset{\sim}{a},\underset{\sim}{b}) = \underset{\sim}{a}'\underset{\sim}{R}_{12}\underset{\sim}{b} - \tfrac{1}{2}\lambda(\underset{\sim}{a}'\underset{\sim}{R}_{11}\underset{\sim}{a} - 1) - \tfrac{1}{2}\mu(\underset{\sim}{b}'\underset{\sim}{R}_{22}\underset{\sim}{b} - 1) \quad,$$

where λ and μ are Lagrange multipliers.

2.3 Solving for the Canonical Correlations and Canonical Variates.
To maximize $f(\underset{\sim}{a},\underset{\sim}{b})$ we differentiate $f(\underset{\sim}{a},\underset{\sim}{b})$ first with respect to the elements of $\underset{\sim}{a}$ and then of $\underset{\sim}{b}$. On setting the results to zero we have

$$\frac{\partial f(a,b)}{\partial a} = R_{12}b - \lambda R_{11}a = 0 \qquad (2.4)$$

$$\frac{\partial f(a,b)}{\partial b} = R'_{12}a - \mu R_{22}b = 0 \quad . \qquad (2.5)$$

Premultiplying (2.4) by a' and (2.5) by b' gives

$$a'R_{12}b - \lambda a'R_{11}a = 0$$

$$b'R'_{12}a - \mu b'R_{22}b = 0 \quad . \qquad (2.6)$$

As $a'R_{11}a = b'R_{22}b = 1$, (2.6) reduces to

$$a'R_{12}b = \lambda = \mu \quad . \qquad (2.7)$$

From (2.5), and noting that $\lambda = \mu$ we have

$$R'_{12}a = \lambda R_{22}b \quad . \qquad (2.8)$$

Multiplying (2.4) on the left by $R_{21}R_{11}^{-1}$ gives

$$R_{21}R_{11}^{-1}R_{12}b - \lambda R'_{12}a = 0 \qquad (2.9)$$

and substituting (2.8) into (2.9) yields the equation in b

$$(R_{21}R_{11}^{-1}R_{12} - \lambda^2 R_{22})b = 0 \quad . \qquad (2.10)$$

Nontrivial solutions for λ^2 and b exist provided that

$$|R_{21}R_{11}^{-1}R_{12} - \lambda^2 R_{22}| = 0 \quad , \qquad (2.11)$$

where the determinant on the left is a polynomial in λ^2 of degree s. Thus (2.11) may be solved for s roots, which we may arrange such that $\lambda_1^2 \geq \lambda_2^2 \geq \cdots \geq \lambda_s^2$. As it is the maximum correlation between U and V that is required we take $r = \sqrt{\lambda_1^2}$. Substitution of the numerical value of λ_1^2 into (2.10) enables the equations to be solved for the coefficient vector b_1 corresponding to λ_1^2. The coefficients of a_1 are then readily found using (2.4), as

$$\underset{\sim}{a}_1 = (R_{\sim 11}^{-1}R_{\sim 12}\underset{\sim}{b}_1) \, / \, \lambda_1 \; . \tag{2.12}$$

We therefore see that the positive square root of the largest eigenvalue of (2.10), $r = \sqrt{\lambda_1^2}$, is the maximum correlation between $U = \underset{\sim}{a}_1' \underset{\sim}{z}_j^{(x)}$ and $V = \underset{\sim}{b}_1' \underset{\sim}{z}_j^{(y)}$. Furthermore, the required vectors of weights $\underset{\sim}{a}_1$ and $\underset{\sim}{b}_1$ are given by the pair of eigenvectors associated with λ_1^2.

There are $s - 1$ further correlations, r_k, given by the square roots of the remaining eigenvalues, λ_k^2, $k = 2, \cdots, s$. Substituting the values of these λ_k^2 into (2.10) and of λ_k and $\underset{\sim}{b}_k$ into (2.12) yields the coefficient vectors $\underset{\sim}{a}_k$ and $\underset{\sim}{b}_k$ of the $s - 1$ further linear relationships between $\underset{\sim}{z}_j^{(x)}$ and $\underset{\sim}{z}_j^{(y)}$ corresponding to the correlations. The s non-zero, positive square roots $r_k = \lambda_k^2$ of (2.10) are called the *canonical correlations* between the *canonical variates* $U_k = \underset{\sim}{a}_k' \underset{\sim}{z}_j^{(x)}$ and $V_k = \underset{\sim}{b}_k' \underset{\sim}{z}_j^{(y)}$ for $k = 1, \cdots, q \le p$.

The correlation matrix of the canonical variates is given by

$$
\begin{bmatrix} \underset{\sim}{U} & \underset{\sim}{0} \\ \underset{\sim}{0} & \underset{\sim}{V} \end{bmatrix}
\cdot
\begin{bmatrix} R_{\sim 11} & R_{\sim 12} \\ R_{\sim 21} & R_{\sim 22} \end{bmatrix}
\cdot
\begin{bmatrix} \underset{\sim}{U}' & \underset{\sim}{0} \\ \underset{\sim}{0} & \underset{\sim}{V}' \end{bmatrix}
=
\begin{bmatrix} I_{\sim p} & \Gamma_{\sim} \\ \Gamma_{\sim}' & I_{\sim q} \end{bmatrix}
$$

where $\Gamma_{\sim} = [\Lambda_{\sim}|0]$, $\Lambda_{\sim} = \mathrm{diag}(\lambda_1, \lambda_2, \cdots, \lambda_s)$ and $\lambda_1 \ge \cdots \ge \lambda_s$ are the canonical correlations. The within-set correlations of the transformed variables are all zero while all correlation between the original variables is channeled into s canonical correlations between the new variables. The structure of this correlation matrix is manifestly simpler than that of the correlation matrix between the original variables. In practice with sample data the canonical correlation coefficients are all distinct and the corresponding canonical variates therefore uniquely determined up to scale factors. In terms of regression, transformation to canonical variates results in the predictable criterion variance being reallocated among the transformed variables. The transformation leaves the total predictable variance unchanged but enables prediction to be achieved in terms of the smallest possible number of variables.

The canonical variates provide a convenient coordinate frame within which the covariance or correlation structure of the data is particularly clearly displayed. The canonical weights are sometimes helpful in the interpretation of this coordinate system. We turn now to review the nature and properties of canonical correlation coefficients and of canonical weights and variates, and in the succeeding section (Section 4) consider the use of significance tests in canonical analysis.

3. PROPERTIES OF CANONICAL CORRELATION COEFFICIENTS, WEIGHTS AND VARIATES

3.1 Properties of Canonical Correlation Coefficients. We have seen that canonical correlations, r_k, are product-moment correlation coefficients between the kth pair of canonical variates, U_k and V_k. It is therefore not surprising that they should share many properties in common with the familiar correlation coefficient. The principal distinction between the two is that canonical correlations are maximized rather than estimated quantities, such as the product moment correlation coefficient as it is generally encountered. A number of properties of canonical correlation coefficients are summarized below.

(i) Canonical correlations assume values in the range -1 to $+1$. It is, however, customary to work with and report only the absolute value $|r_k|$. The magnitude of r_k expresses the degree of linear correlation between U_k and V_k.

(ii) Canonical correlation coefficients are also interpretable as multiple correlation coefficients between a particular canonical variate of one domain and the complete set of variables of the other.

(iii) Canonical correlations are dimensionless quantities and hence are invariant under nonsingular linear transformations of the variables of either or both sets.

(iv) Like the square of a simple correlation coefficient, which is a ratio of explained to total variance, a squared canonical correlation coefficient, r_k^2, represents the ratio of two determinants or generalized variances – namely the ratio of the generalized explained or regression variance to the generalized total variance (e.g. see Bartlett, 1947, 1965). The square of a canonical correlation is also interpretable as the proportion of variance in one variate, U_k, say, that is predictable from, or common to, the corresponding variate V_k.

(v) In statistical applications, the canonical correlation coefficients r_k are regarded as sample estimates of corresponding population quantities, ρ_k (say). In such cases the k*th* sample canonical correlation coefficient r_k is a biased estimate of the k*th* population canonical correlation ρ_k.

The invariance of canonical correlations under changes of origin and scale has some important consequences. In particular, canonical correlation coefficients calculated from the sums of squares and products matrix, the variance-covariance matrix and the correlation matrix for a particular set of data are identical. Furthermore, prior transformation of either or both sets of variables to their principal components leaves the canonical correlations unchanged.

The biased nature of the sample correlations arises from the fact that even after ordering the sample and population correlations r_k and ρ_k respectively in descending order of magnitude, there is no certainty that the k*th* sample canonical correlation has been generated by the k*th* population canonical correlation ρ_k. In this connection we also observe that the magnitude of the sample correlations r_k does depend to an appreciable extent on the relative numbers of variables and samples involved. As the number of variables $(p + q)$ approaches the sample size N, the value of r_1 tends rapidly to unity. In the special case of the multiple correlation coefficient, R, it can be shown that even when the criterion and predictor variables are uncorrelated in the population $(\rho = 0)$ the expected value of R^2 approaches 1 as $(p + q)$ and N converge (e.g. see Morrison, 1976, p. 108). Moreover, when $(p + q) > N$ one or more canonical correlations of unity will inevitably arise. This is so because N samples or points can generate at most a space of $N - 1$ dimensions. Consequently, in such cases p-space and q-space have $[(p + q) - (N - 1)]$ dimensions in common and these common dimensions will be represented by canonical correlation coefficients of unity. Thus the magnitude of canonical correlation coefficients unsupported by probability statements as to their significance can be misleading as indices of the extent of linear relationship between the measurement domains in question. These connections between the number of variables, sample size and the magnitude of the canonical correlations need to be borne in mind in assessing the latter.

Notice that r_k is a measure of relationship between *linear composites* of each measurement domain rather than of the measurement domains themselves. If U_k and V_k were invariably collinear with the principal components of their respective domains, then U_k and V_k would optimally account for the variance of the variables of which they are linear functions. But in many cases canonical variates are not collinear with principal components and do not account for a high proportion of the total variance of their particular domain - canonical variates are required only to be maximally correlated in pairs across domains. Consider, as an example, an analysis in which just one variable of a domain is highly correlated with a single variable of a second domain. A high canonical correlation will inevitably result, since it can be shown that the magnitude of at least one canonical correlation must exceed the absolute magnitude of the largest observed simple correlation between variables of different domains. Yet such an occurrence could hardly be taken to indicate a high degree of relationship between the measurement domains themselves. For this reason a canonical correlation coefficient is a highly unreliable index of relationship between the original variables, tending to overstate the true extent of the relationship. We shall see, however, that interpretive devices which better indicate the extent to which two measurement domains are related have been developed.

If on statistical or substantive grounds r canonical correlations are judged worthy of further consideration, then r may be called the *rank* of the model for the data. In other words, the dimensionality of the linear relationship between the measurement domains is r.

Apart from measuring the relationship between canonical variates, canonical correlations have two uses. First, providing an indication of the dimensionality of linear relationship between the measurement domains $z^{(x)}$ and $z^{(y)}$, and secondly, in the construction of further interpretive indices.

3.2 Properties of Canonical Weights. The elements of the coefficient vectors a_k and b_k are analogous to the standard partial regression weights of multiple regression. They therefore share many properties in common with regression weights of this kind, some of which we summarize below.

(i) The magnitude and sign of canonical weights can be used as an indication of the presence of certain variables or effects and of their direction.

(ii) The numerical values of canonical weights depends
on the selection of variables as well as on their
scales. Addition or deletion of variables in either
set is likely to produce major alterations in the
remaining coefficients. Standardization of the observed variables to zero mean and unit variance will
remove the scaling effects but the interdependencies
still remain.

(iii) The weights tend to be highly unstable in replicate
samples drawn from the same population. Several
factors contribute to the instability, notably
measurement errors in the observations, multicollinearity
within the variables of each set and, in some cases,
inadequate sample size.

The scaling of the coefficient vectors $\underset{\sim}{a}_k$ and $\underset{\sim}{b}_k$ is purely a
matter convenience. It is common practice to scale $\underset{\sim}{a}_k$ and $\underset{\sim}{b}_k$
such that the canonical variates $U_k = \underset{\sim}{a}'_k \underset{\sim}{z}_j^{(x)}$ and $V_k = \underset{\sim}{b}'_k \underset{\sim}{z}^{(y)}$
have zero mean and unit variance. Once this has been accomplished,
however, it is sometimes useful to re-scale $\underset{\sim}{a}_k$ and $\underset{\sim}{b}_k$ to unit
length or so that their largest element is unity. Alternatively,
the coefficient vectors can be standardized such that the sum of the
squares of their elements is equal to r_k^2. This procedure has
the advantage that the coefficient vectors reflect the magnitude
of the latent root to which they refer.

We have mentioned that the magnitude and sign of the canonical
weights may indicate the relative importance of the variables
comprising $\underset{\sim}{z}_j^{(x)}$ and $\underset{\sim}{z}_j^{(y)}$ to the canonical correlation r_k. More
often, however, substantive interpretation of the overall pattern
of the weights is difficult. There are several reasons why this
should be so. First, as Bock (1975, p. 393) has pointed out, the
weights represent a sort of compromise, under the constraint of
orthogonality, between maximizing between-set covariation while
minimizing within-set variation. Secondly, as mentioned above,
canonical weights like regression coefficients, depend on the
selection of variables for investigation and on their scales.
Finally, the weights also depend on sample-specific variation and
on a variety of other factors such as those mentioned under (iii)
above. For these reasons the substantive value of the canonical
weights is often small. Fortunately, the correlations between the
observed variables and the canonical variates provide a useful
alternative to the canonical weights in the interpretation of
canonical variates. Furthermore, steps can sometimes be taken to
try to improve the interpretability of the weights themselves.

These steps include transformation of the observed variables
to their principal components prior to analysis in order to remove
dependencies, estimation of the weights by ridge regression rather
than least squares procedures (Carney, 1975), correcting for
measurement errors in the observations (Meredith, 1964) and the
use of larger samples. In studies where the canonical weights
do prove to have substantive value, graphic display of the weights
is usually worthwhile. Two or three dimensional plots of the
weights against the canonical variates are effective and easily
made.

We noted above that canonical correlation coefficients
are invariant under nonsingular linear transformation of the
original variables. Consequently, canonical correlations calculated for example from such different matrices as the sums of
squares and products matrix, $\underset{\sim}{S}$, the variance-covariance matrix,
$\underset{\sim}{V}$, and the correlation matrix, $\underset{\sim}{R}$, are indentical. The corresponding vectors of weights, however, differ. $\underset{\sim}{S}$ and $\underset{\sim}{V}$ both
give rise to weights which are expressed in the same metric as
the original variables, while $\underset{\sim}{R}$ leads to canonical weights which
are dimensionless. Notice, however, that it is possible to pass
simply and directly between the weights calculated from any of
these different matrices (e.g. see Finn, 1974, p. 190).

In view of the inherent difficulty of their interpretation,
the principal use of canonical weights is in the evaluation of the
canonical variates.

3.3 Properties of Canonical Variates. Canonical variates are
linear combinations of observations constructed so as to maximize
the simple correlation coefficient between the composite variables.

In several respects, canonical variates resemble principal
components. Both are linear combinations of variables chosen to
maximize a particular quantity. In the case of principal components it is the variance of the derived variables which is maximized;
in canonical analysis it is the covariance or correlation between
corresponding members of a pair of derived variables which is
maximized. The solution in each case is provided by the eigenstructure of a symmetric and a nonsymmetric matrix, respectively.
In geometric terms, principal components represent a rotation of
the coordinate frame of sample space to a new position in which
the axes coincide with the directions of maximum scatter in the
sample. Similarly, canonical variates represent separate rotations
of the coordinate systems of the two (superimposed) sample spaces
determined by the data. Each system of axes is rotated to a new
position in which the angle between corresponding members of a
pair of axes across domains is as small as possible. The cosine

of this angle is the canonical correlation between the variates
(axes) in question.

While a canonical variate may be collinear with a principal
component of the same domain, there is no requirement that it
must be so. To see this recall that, for a chosen reference inner
product, principal components are uniquely defined; the canonical
variates of a specified measurement domain on the other hand,
depend critically on the composition of the 'second' domain. The
class of variables which might be used for the second domain is
almost infinitely large. The canonical variates of the specified
domain are correspondingly varied. This fact draws attention to
the opportunities provided by canonical analysis for partitioning
the total variance of a particular domain in a great variety of
ways. Notice, however, that in the case of *two* specified sets
of variables, the canonical variates are uniquely defined up to
a scale factor.

Canonical variates calculated from the correlation matrix,
R, are dimensionless. They are assessed accordingly in terms
of the standardized variables comprising $z^{(x)}$ and $z^{(y)}$. On
the other hand, unless the variables of each set are all expressed
in the same metric, canonical variates calculated both from the
sums of squares and products matrix, S, and the variance-covariance
matrix, V, have meaningless dimensions. This follows from the
nature of their construction as weighted sums of variables expressed in different units. Thus, where either S or V is the
starting point for analysis, it is usually necessary to scale the
elements of the weight vectors a_k and b_k by the standard
deviations of their respective variables before calculating the
variates. The canonical variates are then assessed in terms of
the standardized observed variables. The resulting canonical
variates turn out to be those which would have been obtained had
analysis been based on the correlation matrix. These considerations
lead in practice to canonical analysis being performed most frequently on the correlation matrix R. The general reliance on R,
however, should not allow one to overlook the possibility that
analyses based on such alternative scalar product matrices as S
and V, or, indeed, on quite different matrices, might not also
be profitable under certain circumstances.

The scaling of canonical variates is arbitrary. It is
customary to scale the variates to zero mean and unit variance.
However there is much to be said for scaling U_k and V_k such
that their respective variances are equal to the square of the

corresponding canonical correlation, r_k^2. The canonical variates then reflect the magnitude of the canonical root to which they refer. In certain applications, however, in which the sample is made up of a number of distinct groups, as in canonical variates analysis, it is advantageous to scale the variates so that the average within-groups variance on each canonical variate is unity. This procedure equalizes or 'spherizes' the within-groups variance on the canonical variates and maximizes the squared distances between group means.

Canonical variates are interpreted in much the same way as principal components. Interpretation therefore involves consider-ation of both the contribution of the original variables to a canonical variate and the pattern of variation of a variate over the sample. The contribution of the observed variables to a canonical variate has in the past generally relied on examination of the canonical weights, the magnitude and sign of which were taken to indicate the presence and direction of the effect of particular variables. However, particularly where the variables comprising each domain are even moderately intercorrelated, as they generally are in ecological contexts, the possibility of interpreting the canonical variates by inspection of the canonical weights is practically nil (Meredith, 1964). Realization of this difficulty is increasingly leading to the weights being supplanted by the correlations between the original variables and the canonical variates as aids in the interpretation of the latter. The variable/canonical variate correlations are more stable than the canonical weights in replicate samples; for this and for other reasons they are more reliable indicators of the contribution of particular variables to the canonical variates.

Inspection of the sample values (or scores) on a canonical variate may immediately suggest a substantive interpretation for the variate. Graphics form a valuable adjunct to the scores in interpretation. Examination of the sample in one or more sub-spaces associated with the analysis frequently proves rewarding. In particular, mapping and display of the sample in the following two-dimensional spaces associated with the canonical variates may be illuminating:

(i) $\underset{\sim}{U}_k$ against $\underset{\sim}{U}_m$; $\underset{\sim}{V}_k$ against $\underset{\sim}{V}_m$, for $k \neq m$ and $k, m = 1, \cdots, r$;

(ii) $\underset{\sim}{U}_k$ against $\underset{\sim}{V}_k$, $k = 1, \cdots, r$;

(iii) difference $(\underset{\sim}{U}_k - \underset{\sim}{V}_k)$ against difference $(\underset{\sim}{U}_m - \underset{\sim}{V}_m)$, for $k \neq m$ and $k, m = 1, \cdots, r$.

Such displays are effective means of summarizing different aspects
of the covariance structure of the data. Where both members of a
pair of canonical variates are at the same time also principal
components of their respective domains, plots of $U_{\sim k}$ against $U_{\sim m}$
and $V_{\sim k}$ against $V_{\sim m}$ may be expected to resemble those of separate
components analyses of each domain. In other cases, graphical
analysis can be expected to draw attention to quite different
features of ecological interest. The precise nature of these
features will depend both on the measurement domains involved and
the relationship between them. More penetrating graphical methods
are available. Stereograms (Rohlf, 1968; Orlóci, 1978, p. 171)
enable the sample to be viewed after mapping into the space associated with three canonical variates, while Andrews' (1972) procedure
is available for use in connection with mappings involving still
higher-dimensional spaces.

Where an analysis involves spatially distributed samples, the
simple device of plotting the spatial distribution of a canonical
variate can be useful. Such distribution maps are often effective
aids in the interpretation of canonical variates.

The function of the canonical variates is to provide a
parsimonious summarization of a body of data which simultaneously
draws attention to the presence and nature of any linear relationship which may exist between the measurement domains investigated.
It is hoped and expected that by mapping the sample into the space
of the canonical variates, greater insight into the covariance
structure of the data will be obtained than by examining either
the initial observations themselves or the matrix of interset
correlations between the measurement domains. From the viewpoint
of regression, canonical variates are useful in specifying those
weighted linear combinations of the variables of one set which are
most *predictable* from the variables of the other set.

4. TESTS OF SIGNIFICANCE

Up to this point we have assumed a sample of (p + q)-variate
observations but one whose characteristics are otherwise unspecified. If the variables comprising at least one of the p- or
q-element *subvectors* which make up the observed vector-valued
variables can reasonably be regarded as having a joint multinormal
distribution, and, provided the samples are independently drawn,
then significance tests based on the canonical correlations may
be made.

4.1 Joint Nullity of All s Canonical Correlations. It is usually
of interest to test the simultaneous departure of the canonical
correlation coefficients from zero. Joint nullity of all s
canonical correlations would indicate the absence of any linear
relationship between the X-variables and Y-variables. The
appropriate null and alternative hypotheses are

$$H_0 : \rho_1 = \rho_2 = \cdots = \rho_s = 0$$
$$(4.1)$$
$$H_1 : \rho_k \neq 0 \text{ for at least one k, } k = 1, \cdots, s,$$

where the ρ_k stand for the population quantities corresponding
to the sample canonical correlation coefficients r_k. Hypothesis
(4.1) may be expressed equivalently in terms of the independence of
the two sets of variables. Writing $\underset{\sim}{\Sigma}_{12}$ for the matrix of intercorrelations in the population:

$$H_0 : \underset{\sim}{\Sigma}_{12} = \underset{\sim}{0}$$

$$H_1 : \underset{\sim}{\Sigma}_{12} \neq \underset{\sim}{0}$$

A test of both forms of the hypothesis based on the canonical
correlations is provided by Wilks' likelihood-ratio criterion Λ,
where

$$\Lambda = \prod_{k=1}^{s} (1 - r_k^2) \qquad (4.2)$$

The range of Λ is 0 to 1. From (4.2) it is apparent that Λ
may be regarded as the product of the proportion of variance left
unexplained by the s canonical correlations. If there is little
correlation between the two sets of variables Λ will therefore
be close to unity, while if they are closely correlated Λ will
approach zero. Although the distribution of Λ has been tabulated
(Wall, 1968), tables of the distribution are not yet widely
available (see however Timm, 1975, p. 624). A satisfactory and often
more convenient approximation to the distribution of Λ, at least
for large samples, is provided by Bartlett's χ^2 transformation

$$\chi^2 = - [(N - 1) - \tfrac{1}{2}(p + q + 1)] \log_e \Lambda, \qquad (4.3)$$

where χ^2 under the null hypothesis is distributed approximately
as a chi-squared variate with pq degrees of freedom. The null
hypothesis of (4.1) is rejected if $\chi^2 > \chi_\alpha^2 (pq)$. It can be shown

(e.g., see Bartlett, 1947; Finn, 1974, p. 192) that the test
(4.2) of the joint nullity of the canonical correlations is equivalent to a test of the contribution of all q predictor variables
to regression. This equivalence is a further reflection of the
close relationship of canonical analysis to both correlation and
regression.

If the overall hypothesis (4.1) of no relationship or of independence can be rejected, it is generally of interest to remove
the contribution of the largest root, the first two roots, and
so on, to Λ and then to assess the significance of the remaining
canonical correlations.

4.2 Joint Nullity of the Smallest s-k canonical correlations. A
general criterion for testing the joint nullity of correlations k
to s is given by

$$\Lambda_k = \prod_{i=k}^{s} (1 - r_i^2) \ . \tag{4.4}$$

The significance of Λ_k may be assessed by the chi-square transformation

$$\chi^2 = -[(N - 1) - \tfrac{1}{2}(p + q + 1)] \log_e \Lambda_k, \tag{4.5}$$

where χ^2 is distributed approximately as a chi-squared variate
with $(p - k + 1)(q - k + 1)$ degrees of freedom. The partition of
(4.2) which leads to (4.4) is based on the assumption that the
$k - 1$ canonical correlations removed do in fact correspond to non-
zero population values. Thus Λ_k provides a test of the *residuals*
after the effects of the preceding correlations have been removed.
Although hypothesis (4.1) is seldom of direct interest, it is the
appropriate null hypothesis when the effects of canonical variates
assumed to represent the association between the two sets of
variables have been eliminated. In this sense the hypothesis of
no relationship is of central importance in significance testing
(Williams, 1967).

By letting $k = 2, \cdots, s$ in (4.4), sequential tests of all
but the first canonical correlation, all but the first two
correlations, and so on are possible. Acceptance of H_0 is
commensurate with concluding that any association between the two
sets is concentrated in the first or preceding $k - 1$ canonical
variates. In this way, it may be possible to isolate one or a
small number of linear composites of the measures that describe
all significant relationship between the two sets. In such cases

an indication of the rank or dimensionality of the relationship
is obtained. The tests for significant canonical correlations
other than the first, however, are very conservative unless the
correlations removed are close to 1.

Despite the widespread use of partitioned-Λ tests in canonical
analysis and more generally, these tests have been strongly criticised
(e.g. see Harris, 1975, p. 111; 1976). The fundamental problem is
that there is no way of knowing which of the s population
canonical correlation coefficients, ρ_k, has generated the k*th*
largest sample correlation r_k. Harris has pointed out that two
difficulties arise from the indeterminancy. First, as it is
generally claimed that if the overall test is significant then
r_1 at least must be significant, the significance of r_1 itself
is never directly tested. But such a claim is not well-founded, a
significant overall Λ indicating only that *some* pair of canonical
variates in the population have a non-zero correlation. Secondly,
the partitions of Λ do not in fact have the chi-squared distributions generally attributed to them. This follows from
Lancaster's (1963) proof that the partition corresponding to the
first canonical root, r_1^2, in particular, does not possess the
χ^2 distribution frequently ascribed to it. For these reasons
partitioned-Λ tests should be approached with some degree of
caution. Fortunately, as we shall shortly see, an alternative
procedure based on the null distribution of the canonical roots
themselves, which is free from the difficulties mentioned, is
available.

4.3 The Significance of Individual Canonical Correlation Coefficients.
An alternative to the likelihood-ratio tests of association considered above is arrived at through the application of Roy's union-
intersection principle of test construction. In this context it
is convenient to write the multivariate hypotheses (4.1) as an
intersection of composite univariate hypotheses and as a union of
corresponding alternative hypotheses respectively:

$$H_0 : \bigcap_{\underset{\sim}{a},\underset{\sim}{b}} (\rho(\underset{\sim}{a},\underset{\sim}{b}) = 0)$$

$$H_1 : \bigcup_{\underset{\sim}{a},\underset{\sim}{b}} (\rho(\underset{\sim}{a},\underset{\sim}{b}) = 0)$$

(4.6)

for all nonnull p- and q-element vectors $\underset{\sim}{a}$ and $\underset{\sim}{b}$. The
test criterion for hypothesis (4.6) is the square of the maximum
attainable correlation between the linear composites $\underset{\sim}{a}'\underset{\sim}{z}_j^{(x)}$ and
$\underset{\sim}{b}'\underset{\sim}{z}_j^{(y)}$ for all choices of $\underset{\sim}{a}$ and $\underset{\sim}{b}$ subject to the normalizing

conditions $a'R_{11}a = b'R_{22}b = 1$. We have seen that this quantity is the largest root, r_1^2, of (2.10). The significance of r_1^2 may be tested by referring it to critical points of the greatest characteristic root (gcr) distribution (e.g. see Harris 1975, p. 300; Timm, 1975, p. 607; Morrison, 1976, p. 379). The null hypothesis is accepted at the level α if

$$r_1^2 \leq \theta_\alpha(s, m, n)$$

and rejected otherwise. Here $\theta_\alpha(s, m, n)$ is the upper 100α percentage point of the gcr distribution with parameters $s = \min(p,q)$, $m = (|p - q| - 1)/2$ and $n = (N - p - q - 2)/2$.

If the overall hypothesis (4.6) of independence is rejected, the significance of the k*th* root may be tested ($k = 2, \cdots, s$). The procedure is identical to the overall test using r_1^2, but an adjustment to the first degree-of-freedom parameter however is required. The parameter s_k is given by $s_k = \min(p - k + 1, q - k + 1)$. Like the corresponding likelihood-ratio tests, these tests of the r_k, excepting r_1^2, are very conservative.

4.4 The Contribution of Particular Variables. The significance of the contribution of particular *variables* to the canonical relationship can be tested by a modification of the likelihood ratio procedure described above. The test is accomplished by comparing the χ^2 values which result when the variables of interest are first included and then omitted.

Λ is first calculated by (4.2) with all the variables included. Now suppose k_1 X's and k_2 Y's are omitted. Λ is then recalculated for the remaining $(p - k_1)$ plus $(q - k_2)$ variables. Denoting this second value by Λ^*, the quantity

$$\chi^2 = -[(N - 1) - \tfrac{1}{2}(p + q + 1)] \log_e \Lambda^*$$

is distributed approximately as a chi-squared variate with $(p - k_1)$ $(q - k_2)$ degrees of freedom. The test statistic is then

$$\chi^2 = -[(N - 1) - \tfrac{1}{2}(p + q + 1)] \log_e (\Lambda/\Lambda^*),$$

which may be referred to the chi-square distribution with $pk_2 + qk_1 - k_1k_2$ degrees of freedom. Marriott (1974) has pointed out that by putting $k_1 = 1$ and $k_2 = 0$, a significance test of one of the coefficients in the canonical variates results. It is then possible by equating significance levels to obtain rough estimates of the standard errors of the canonical coefficients.

In view of the doubtful validity of partitioned-Λ tests it is worth noticing that the effects of variables could similarly be tested by means of Roy's largest-root criterion. The squared canonical correlations obtained in separate analyses including and then omitting the variables of interest would in such cases be referred to the gcr distribution.

5. FURTHER INTERPRETIVE DEVICES

Canonical analysis is concerned with the number of linear relationships between two measurement domains with the nature of these relationships. We have seen that canonical correlation coefficients go some way towards establishing the number and strength of the relationships present, while canonical weights and variates may be helpful in clarifying their nature. However, canonical correlations and weights each leave something to be desired as interpretive devices; canonical correlation coefficients express relationships between linear composites which, though perhaps highly correlated, may be relatively unimportant components of their respective domains, while the interpretation of canonical weights has all the problems attendant on the interpretation of beta weights in multiple regression. For these reasons a number of other interpretive indices has been developed which supplement those originally put forward by Hotelling (1936). We turn now to consider some of these newer indices, in particular: (i) the correlations between canonical variates and the original variables; (ii) the proportion of the total variance of a measurement domain associated with a canonical variate; (iii) the redundancy of a measurement domain; and (iv) the communality of a variable.

5.1 Correlations Between Canonical Variates and the Original Variables. The sign and magnitude of the variable/canonical variate correlations are useful in showing (a) which variables contribute most heavily to a canonical variate and the direction of their effects; and (b) the existence of affinities or contrasts among variables in their relationship with a canonical variate. The correlations thus contribute towards establishing the nature of relationships which may be present between domains. The correlations are more stable than either the raw or standardized canonical weights under the addition or deletion of variables and in replicate samples drawn from the same population. Moreover, correlations are more readily translated into meaningful terms than are weights such an canonical or regression weights (Dempster, 1969, p. 160). Thus the square of a variable/canonical variate correlation expresses the proportion of the variance of a variable which is directly associated with a particular canonical variate. For convenience, we shall on occasion refer to variable/canonical variate correlations as *structure correlations*. Variable/canonical variate correlations are of two kinds: (i) *intraset correlations*: correlations between canonical variates and observed variables of the *same* domain; and (ii) *interset correlations*: correlations between canonical variates of one domain and the observed variables of the *other*.

5.1.1 Intraset correlation coefficients. Such correlations express the contribution of the variables of a domain to the canonical variates of the same domain. In so doing they help to establish substantive interpretations for these canonical variates. There are two sets of intraset correlations corresponding to the two measurement domains. For the variables and canonical variates of $\underset{\sim}{z}^{(x)}$ the intraset correlations are given by

$$\text{cor}(\underset{\sim}{z}_j^{(x)}, U_k) = \text{cor}(\underset{\sim}{z}_j^{(x)}, \underset{\sim}{a}_k'\underset{\sim}{z}_j^{(x)})$$

$$= \frac{1}{N} \sum_{j=1}^{N} \underset{\sim}{z}_j^{(x)} [\underset{\sim}{z}_j^{(x)}]' \underset{\sim}{a}_k$$

$$= \underset{\sim}{R}_{11} \underset{\sim}{a}_k = \underset{\sim}{s}_k^{(x)} \text{ (say)} , \qquad (5.1)$$

where $\underset{\sim}{s}_k^{(x)}$ os the $p \times 1$ vector of correlations between the k*th* canonical variate of $\underset{\sim}{z}^{(x)}$ and the observed variables of $\underset{\sim}{z}^{(x)}$.

In a similar way for the Y-domain

$$\text{cor}(\underset{\sim}{z}_j^{(y)}, V_k) = \underset{\sim}{R}_{22} \underset{\sim}{b}_k = \underset{\sim}{s}_k^{(y)} \text{ (say)}.$$

5.1.2 Interset correlation coefficient. The interset correlations characterize interrelationships between the canonical variates of one measurement domain and the observed variables of the other. There are two sets of interset correlations. For the correlations of the variables of $\underset{\sim}{z}^{(x)}$ with the canonical variates of $\underset{\sim}{z}^{(y)}$ we have

$$\text{cor}(z_{\sim j}^{(x)}, V_k) = \text{cor}(z_{\sim j}^{(x)}, b_{\sim k}' z_{\sim j}^{(y)})$$

$$= \frac{1}{N} \sum_{j=1}^{N} z_{\sim j}^{(x)} [z_{\sim j}^{(y)}]' b_{\sim k}$$

$$= R_{\sim 12} b_{\sim k} = s_{\sim k}^{(xv)} \quad (\text{say}) \quad,$$

where $s_{\sim k}^{(xv)}$ is the $p \times 1$ vector of interset correlations between the $k\mathit{th}$ canonical variate, V_k, of $z_{\sim}^{(y)}$ and the observed variables of $z_{\sim}^{(x)}$.

From (2.4) it is easy to see that

$$R_{\sim 12} b_{\sim k} = r_{\sim k} R_{\sim 11} a_{\sim k} \quad,$$

which shows how the interset correlations between $z_{\sim}^{(x)}$ and V_k can be obtained directly from (5.1).

Similarly

$$\text{cor}(z_{\sim j}^{(y)}, U_k) = R_{\sim 21} a_{\sim k}$$

$$= r_k R_{\sim 22} b_{\sim k} = s_{\sim k}^{(yu)} \quad (\text{say}) \quad,$$

where $s_{\sim k}^{(yu)}$ is the $q \times 1$ vector of correlations between the $k\mathit{th}$ canonical variate, U_k, of $z_{\sim}^{(x)}$ and the observed variables in $z_{\sim}^{(y)}$.

Interset correlation coefficients closely resemble multiple correlation coefficients. The distinction between the two is that the composite variable entering into a multiple correlation coefficient is determined under the restriction that it be maximally correlated with the response variable; in the case of an interset correlation the maximization is with respect to a linear *composite* of response variables. This results in the interset correlation of a variable being smaller than its multiple correlation with the same set of predictor variables, except in the unlikely event that the variables' intraset correlation is unity.

The square of an interset correlation coefficient specifies the proportion of the variance of a variable which is predictable by a canonical variate of the 'other' domain.

While variable/canonical variate correlations are open to interpretation by inspection, graphical analysis is a valuable aid in comprehending relationships among them. Two or three dimensional plots of the correlations are a most effective means of obtaining insight into the covariance structure of the data. *Simultaneous* plots of the correlations of both sets of variables against selected canonical variates often prove to be particularly rewarding.

Apart from the uses of variable/canonical variate correlations noted above, we shall see that they are also used in the construction of other interpretive devices, notably in the variance extracted by a canonical variate and in the variable communalities.

5.2 Variance Extracted by a Canonical Variate. The proportion of the total variance of a measurement domain which is associated with a canonical variate is referred to as the (measurement) variance extracted or accounted for by the variate. This quantity represents the amount of 'overlap' or shared variance common to a measurement domain and a particular canonical variate. It is therefore an index of the extent to which a canonical variate explains the total variance of the domain of which it is a linear composite. The variance extracted by a canonical variate is calculated as the mean of the squared variable/canonical variate correlations between the variate and the variables on which it is defined.

For the k*th* canonical variate U_k of $z^{(x)}$, the variance extracted, U_k^2, is given by

$$U_k^2 = \sum_{i=1}^{p} s_{ik}^2 \, / \, p \; , \tag{5.2}$$

where U_k^2 is the mean of the p squared intraset correlations, s_{ik}^2, with U_k. Similarly, for the variance extracted by the k*th* canonical variate of $z^{(y)}$, V_k^2, we have

$$V_k^2 = \sum_{h=1}^{q} s_{hk}^2 \, / \, q \; . \tag{5.3}$$

We note that in general $U_k^2 \neq V_k^2$. This is so because, while U_k may be a major dimension of its measurement domain, V_k may correspond to a much smaller dimension of $z^{(y)}$, or *vice versa*.

5.3 Redundancy. Redundancy is the proportion of the total variance of a measurement domain predictable from a linear composite of the *other* domain, given the availability of the second domain. The term redundancy is therefore synonymous with *explained variance*.

The redundancy, $V^2_{x|v_k}$, of $z^{(x)}$ with respect to the k*th* canonical variate $V_k = b'_k z^{(y)}$ of $z^{(y)}$, given the availability of $z^{(y)}$, is given by the mean of the squared interset correlations between the elements of $z^{(x)}$ and V_k. That is

$$V^2_{x|v_k} = \sum_{i=1}^{p} s^2_{ik} / p , \qquad (5.4)$$

where s_{ik} is the correlation between the i*th* variable of $z^{(x)}$ and the k*th* canonical variate of $z^{(y)}$. The similarity of (5.4) to the expression for the variance extracted by a canonical variate given by (5.2) will be plain. Evidently, redundancy is the between-set analog of the variance extracted by a canonical variate. In a similar way, the redundancy, $U^2_{y|u_k}$, of $z^{(y)}$, given the availability of $z^{(x)}$, is given by

$$U^2_{y|u_k} = \sum_{h=1}^{q} s^2_{hk} / q ,$$

where s_{hk} is the correlation between the h*th* variable of $z^{(y)}$ and the k*th* canonical variate of $z^{(x)}$.

Redundancy can be approached from a different direction. Redundancy may be regarded as the product of the within-set variance times the between-set variance accounted for by a canonical variate. In other words, redundancy is the product of the variance extracted by a canonical variate from its own domain times the variance which the canonical variate shares with its counterpart of the other domain. We may therefore write

$$V^2_{x|v_k} = U^2_k r^2_k \qquad (5.5)$$

for the redundancy of $z^{(x)}$, given the availability of the set $z^{(y)}$, and

$$U^2_{y|u_k} = V^2_k r^2_k \qquad (5.6)$$

for the redundancy of $z^{(y)}$ given the availability of $z^{(x)}$.

In general $v^2_{x|v_k} \neq u^2_{y|u_k}$. This follows on relating expressions (5.5) and (5.6) and recalling the inequality $u^2_k \neq v^2_k$. The redundancy associated with a canonical variate is an important index of the predictive or explanatory power of the canonical variate in relation to the 'other' domain. An alternative measure of this property has been proposed by Finn (1974, p. 191).

5.4 Total Redundancy. The sum of the redundancy indices for each domain over the r retained canonical variates of the rank r model, yields an overall or total redundancy index for each domain. For the total redundancy in the set $z^{(x)}$, given the canonical variates $V_1, \cdots, V_r$ of $\underset{\sim}{z}^{(y)}$, we have

$$v^2_{x|v_1}, \cdots, v_r = \sum_{k=1}^{r} \left(\sum_{i=1}^{p} s^2_{ik} / p \right)$$

$$= \sum_{k=1}^{r} v^2_{x|v_k} .$$

Similarly, the total redundancy in $z^{(y)}$, given the canonical variates $U_1, \cdots, U_r$ of $\underset{\sim}{z}^{(x)}$, may be written

$$u^2_{x|v_1}, \cdots, u_r = \sum_{k=1}^{r} \left(\sum_{h=1}^{q} s^2_{hk} / q \right)$$

$$= \sum_{k=1}^{r} u^2_{y|u_k} .$$

Total redundancy provides an overall measure of the variance explained in one variable set by the variables of the other. Unlike a canonical correlation coefficient, which expresses the relationship between linear composites of each domain, total redundancy is a direct expression of the interrelatedness of the measurement domains themselves. Total redundancy is asymmetric between domains, so that in general $v^2_{x|v_1, \cdots, v_k} \neq u^2_{y|u_1, \cdots, u_k}$.

Total redundancy has application in establishing the rank, r, of the canonical model judged to give the most acceptable fit to the data. The explanatory power of the model for various trial values of r is assessed informally in conjunction with the substantive value of the corresponding canonical variates and the overall reduction in dimensionality which will result. Total redundancy thus helps to establish the number of noteworthy linear relationships between domains.

5.5 Variable Communalities. The communality of a variable expresses the proportion of the variance of the variable which is accounted for by the retained canonical variates of a fitted model. Two communalities are associated with each variable, corresponding to the two sets of canonical variates, U_k and V_k:

(i) *intraset communality:* the proportion of variance accounted for by the retained canonical variates of the variables' *own* set; and (ii) *interset communality:* the proportion of variance accounted for by the retained canonical variates of the *'other'* set.

5.5.1 Intraset communalities. These are conveniently obtained as the sum of squared intraset correlations between a variable and the retained canonical variates. For the ith variable of $\underset{\sim}{z}^{(x)}$, the intraset or within-set communality, h_{wi}^2, is therefore

$$h_{wi}^2 = \sum_{k=1}^{r} s_{ik}^2,$$

where s_{ik} is the intraset correlation of variable i with the kth canonical variate of $\underset{\sim}{z}^{(x)}$. Likewise, for the within-set communality of the hth variable of $\underset{\sim}{z}^{(y)}$, we have

$$h_{wh}^2 = \sum_{k=1}^{r} s_{hk}^2.$$

5.5.2 Interset communalities. These are readily calculated as the sum of squared interset correlations between a variable and the retained canonical variates. The interset or between-set communality, h_{bi}^2, of the ith variable of $\underset{\sim}{z}^{(x)}$ with the retained canonical variates of $\underset{\sim}{z}^{(y)}$ is therefore given by

$$h_{bi}^2 = \sum_{k=1}^{r} s_{ik}^2 ,$$

where s_{ik} is the interset correlation of variable i of $z^{(x)}$ with the kth canonical variate of $z^{(y)}$. In a similar way, the interset communality of the hth variable of $z^{(y)}$ with the retained canonical variates of $z^{(x)}$, is

$$h_{bh}^2 = \sum_{k=1}^{r} s_{hk}^2 ,$$

where s_{hk} denotes an interset correlation coefficient.

Interset communalities may be interpreted as the proportions of variance which the variables of one set have in common with the space spanned by the canonical variates of the other set. The information provided is useful in deciding which variables of either set to retain or delete in efforts to enhance the overall explanatory power or fit of a canonical model.

5.6 *Concluding Remakrs.* We have seen that part of the interpretive function previously associated with canonical correlation coefficients and canonical weights is now effectively performed by other quantities. Redundancy provides a measure which is in some ways a more satisfactory expression of the interrelatedness of two sets of variables than a canonical correlation coefficient; similarly, the correlations between a canonical variate and the original variables often prove to be more trustworthy indicators of those variables which contribute most heavily to the variate, than are canonical weights. We have also seen that variable communalities are useful indices of the extent to which the variance of a variable is accounted for by an analysis. Furthermore, total redundancy and variable communalities are helpful in establishing the rank r of the model judged to best satisfy the joint requirements of explanatory power and parsimony. In certain applications, however, the number of statistically significant canonical correlation coefficients might be used for this purpose.

Much effort continues to be devoted to the improvement and extension of canonical analysis. In Section 7 recent developments in this area are reviewed.

6.   PSEUDO-VARIABLES IN CANONICAL ANALYSIS

Considerable freedom exists as to the nature of the variables which may be employed in canonical analysis. While the variables can be continuous there is no requirement that they must be so.

Moreover, it is not necessary for the variables used to be observed or measured in the usual sense at all, or indeed, that they be real-valued. Thus the class of admissible variables in canonical analysis is very wide. We shall see that this property not only greatly enhances the flexibility of the method in practice but also provides a means of relating many familiar statistical procedures whose mutual affinities are not always appreciated.

6.1 The Practical Value of Pseudo-variables. Ecological surveys or experiments frequently give rise to p-variate samples which possess an inherent classificatory structure. Thus a sample may consist for example of units belonging to distinct plant communities, plant or animal species or to different experimental treatments. Provided the categories recognized are mutually exclusive and collectively exhaustive, the classification can be represented by a set of artificial variables generally known as pseudo-variables or dummy variables. Pseudo-variables are vector-valued quantities whose elements consist of arbitrary numbers to which meanings can be assigned. Various coding schemes are in use. The most widespread practice involves giving a 'score' to each item in a sample to signify the membership or non-membership of items in a particular class. Thus, for example, in a vegetation survey a *binary-valued* dummy variable could be created to distinguish between samples of two plant communities by assigning a score of 1 to all samples of one community and a score of 0 to all samples of the other. More generally, where a sample consists of k classes k - 1 binary valued dummy variables prove sufficient to represent the classification; a k*th* variable is not required because a sample-unit not belonging to any of k-1 classes must belong to the k class. Sample-units belonging to one of the first k=1 classes, say, could then be assigned a score of 1 on the dummy variable corresponding to the appropriate class and 0 on all remaining dummy variables; this would lead to sample-units belonging to the k*th* class being characterized by a score of 0 on all k-1 dummy variables. The procedure is illustrated in Table 6.1. The example refers to N samples classified into k = 5 classes designated a,b,$\cdots$,e. The classification is expressed in coded form by the four dummy variables Y_i. The set of four scores associated with each sample-unit uniquely specifies the position of the units in the classification. In this way dummy variables can be used to operationalize a classification enabling it to be entered into and operated on during the course of an analysis. The class specified by a score of 0 on all dummy variables may be thought of as a reference group in terms of which the other groups are compared or related. The choice of group which is signified in this way is unrestricted; it is not in fact necessary that it be the last group. Apart from efficiency, the use of k - 1 rather than k variables has the advantage of allowing the matrix of binary-valued variables to be inverted by standard methods during analysis.

TABLE 6.1:   Representation of a classification with k = 5 *groups by* k - 1 *binary-valued dummy variables.   (See text for explanation).*

Dummy-variable	Group				
	a	b	c	d	e
Y_1	1	0	0	0	0
Y_2	0	1	0	0	0
Y_3	0	0	1	0	0
Y_4	0	0	0	1	0

The relevance of binary coded variables in the present context is that the dummy variables together with the p original observed or measured variables can jointly provide the input necessary for canonical analysis. This version of canonical analysis has in fact proved to be the most widely used form of the method in ecology and related areas of biology (see Seal, 1964, Ch. 7; Pearce, 1969; Grigal and Goldstein, 1971; Blackith and Reyment, 1971, Ch. 8; Goldstein and Grigal, 1972; Kowal, Lechowicz and Adams, 1976; Orlóci, 1978, p. 241; cf. however Kessel and Whittaker, 1976). Canonical variates analysis, multiple discriminant analysis and the multivariate analysis of variance may all be regarded fundamentally as canonical analysis in which one set of variables consists of binary-valued dummy variables. Notice however that the use of binary coded variables in such analyses need not necessarily be explicit. Where dummy variables are not explicitly used, the equivalence of an analysis to canonical analysis may not be self-evident. The principal objectives which can be addressed by canonical analysis of this kind may be summarized as follows: (a) obtaining the 'best' reduced-dimensional representation of k p-variate samples; (b) deriving optimal allocation rules for assigning one or more p-variate sample-units among k pre-existing categories; and (c) assessing differences which may exist between k p-variate group means.

Sometimes it is possible to cross-classify a sample with respect to two criteria. The procedure described can then be generalized, leading to a canonical analysis in which both sets of variables are pseudo-variables. Despite the ease and speed with which suitable data can be obtained, applications of this form of canonical analysis, though increasing, are less frequent than in the case of the previously mentioned form. Ecological applications have been made however by Hatheway (1971) and Hill (1973).

Correspondence analysis and a particular version of contingency
table analysis are each essentially canonical analyses in which
both sets of variables are pseudo-variables. Once again however
such analyses can be performed without the explicit presence of
pseudo-variables. The substantive issues which can be approached
by analysis of this sort are: (a) the simultaneous scaling of
individuals and attributes; (b) the analysis of association in
r × c tables.

Generalization of a quite different kind arises from the use
of coding schemes other than binary coding. Kerlinger and
Pedhazur (1973, Ch. 7) give a clear account of some of the possibilities and of the properties of the solutions based on them.
The opportunities provided by such alternative coding schemes are
not explored here; numerical examples have, however been given by
Williams (1967), Hope (1968, p. 129) and Dempster (1969, p. 223).
These examples together with the applications of binary coded
variables reported in Sections 15-17 below, illustrate something
of the richness imparted to canonical analysis by the use of
dummy variables of different kinds.

In view of their binary character, binary coded variables
can hardly be expected to have a joint multivariate normal distribution. The question of the effect of any departure from multivariate normality on the tests of significance described earlier
(Section 4) therefore arises. Fortunately, the consequences of the
use of dummy variables on the significance tests are not serious.
Precise tests of significance in canonical analysis call for the
joint normality of only one of the two sets of variables. Thus,
provided the set of observed or measured variables can reasonably
be assumed to be jointly multinormal, the tests remain valid for
analyses involving a single set of pseudo-variables. Furthermore,
even where the variables of both sets are dummy variables the
tests are asymptotically correct. Therefore, at least for large
samples, the tests will still be informative. Bartlett (1965) has
expressed the view that in such cases the test procedures are
believed to be "broadly correct provided the tests of significance
are not taken too precisely."

6.2 Pseudo-variables and Affinities Between Statistical Methods.
The role of dummy variables in extending the applicability of canonical analysis can be regarded from a different viewpoint. By
the use of dummy variables it can be shown (e.g. see Green, 1976,
pp. 270 and 285) that almost all of the methods of classical
multivariate analysis are merely special cases of canonical analysis.
Furthermore, where a restriction is placed on the number of variables
entering one or both sets, then the same can be demonstrated for
many familiar univariate statistical procedures also. Table 6.2
summarizes some special cases of canonical analysis which arise under

TABLE 6.2: *Special cases of canonical analysis which arise under restrictions on the kind and/or number of variables.*

Nature of variables		Number of criterion variables	
Predictor	Criterion	(a) One	(b) More than one
Binary	Continuous	Analysis of variance	Multivariate analysis of variance; T^2
Continuous	Binary	Discriminant analysis; T^2	Multiple discriminant analysis; canonical variates analysis
Mixed binary & continuous	Continuous	Analysis of covariance	Multivariate analysis of covariance
Binary	Binary	Binary regression	Analysis of association in r × c tables
Continuous	Continuous	Multiple regression	Canonical correlation analysis

restrictions on the kind and/or number of variables entering an analysis. The methods of the right-hand side of the Table are more general than those of the left-hand side in that no restriction is placed on the number of variables. Similarly, the methods of the last row are more general than those above them in the sense that they embody all of the preceding methods by the use of dummy variables. Table 6.2 therefore shows something of the generality of canonical analysis. By restricting both the criterion and predictor variables to a single member (p = q = 1), the Table can be extended to incorporate several additional univariate statistics (cf. Knapp, 1978). Though the various methods are frequently interchangeable, even in computation, it is not the intention to suggest that in practice the methods of Tabel 6.2 necessarily be treated as canonical analysis *per se*. There may be circumstances however when it is advantageous to do so (cf. the general-purpose program MULTIVARIANCE of Finn, 1974, pp. 10 and 397; also Bock, 1975, p. xii).

Statistical methods, both univariate and multivariate, are all too often presented as if they existed in separate categories. A more balanced view of the subject however would result if this custom were to be complimented by a demonstration that a majority of parametric methods are in fact expressions of a single underlying model. While the multivariate general linear model would provide a yet more comprehensive framework for this purpose than canonical analysis, the latter is nevertheless a convenient means of accomplishing much the same task. A recent introductory text which does however go a considerable way towards achieving the goal of a unified account of multivariate statistical methods through the use of dummy variables in canonical analysis is that of Green (1978).

Thus, we see that by the use of dummy variables: (a) the range of applicability of canonical analysis is extended; and (b) a significant unification of the body of statistical methods is achieved. We may therefore conclude that the use of dummy variables in canonical analysis has a valuable contribution to make to both the teaching and the practice of quantitative ecology.

Several of the special cases of canonical analysis appearing in Table 6.2 will be familiar to ecologists. Little would therefore be gained by illustrating their use here. Others among them are less well-known. Accordingly, it may be worthwhile to provide examples of at least some of these methods. Ecological applications of canonical correlation analysis, canonical variates analysis, the multivariate analysis of variance and of the analysis of association in r × c tables form the substance of Chapter III. Binary regression is also deserving of wider recognition among ecologists. Noy-Meir *et al.* (1973), however have given a comprehensive account of this procedure in an ecological context, which may be consulted for details of the method and its applications.

7. EXTENSIONS AND GENERALIZATIONS

The value of canonical analysis has sometimes been questioned because of the problem of interpretability (e.g. see Kendall and Stuart, 1968, p. 395; Dempster, 1969, p. 179; Mulaik, 1972, p. 423; Marriott, 1974, p. 31; Kendall, 1975, p. 69). Williams (1976, p. 66) has gone further in expressing the view that canonical correlation analysis in the form described by Hotelling (1935, 1936) cannot be said to have "been useful in agriculture, or indeed in biology in general." Canonical analysis, however, is the focus of much current research; the subject is expanding and worthwhile proposals for strengthening the method continue to be put forward. Much activity has been stimulated in particular by difficulties inherent in the interpretation of canonical correlation coefficients and canonical weights. In this Section we review attempts to overcome or circumvent these and other difficulties which are sometimes encountered and describe also the generalization of canonical analysis to more than two sets of variables. We first consider, however, an alternative to canonical analysis put forward by Wollenberg in 1977.

7.1 Redundancy Analysis - an Alternative to Canonical Analysis.
A shortcoming of a canonical correlation coefficient is its failure
to take account of the within-set variance associated with a canonical variate. This limitation led to the development of the concept
of redundancy (Stewart and Love, 1968). Unlike a canonical correlation coefficient, redundancy takes account of both the within-set
and the between-set variance associated with a given canonical
variate. Hence redundancy may be regarded as a two-part explained-
variance index. Recently, Wollenberg (1977) has suggested maximizing
redundancy itself, rather than the canonical correlation coefficient
or simply the between-set component of explained-variance. The procedure, which is referred to as *redundancy analysis,* is certainly an
appealing idea. Redundancy, unlike a canonical correlation, is a
measure of directed relationship, being asymmetric between domains.
A new index (C^2) of *directed relationship,* related to Mahalanobis'
D^2, has been proposed by Coxhead (1974). A useful feature of C^2
is that it is readily transformed to a quantity having an F-distribution, so that unlike redundancy, its significance can be assessed.
For further discussion of the concepts of redundancy, C^2 and some
alternative measures of multivariate association, see Cramer and
Nicewander (1979).

7.2 Improving the Interpretability of Canonical Weights. Among
the chief sources of difficulty in the interpretation of canonical
weights are collinearity or near-collinearity of the within-set matrice
R_{11} or R_{22},   errors of measurement in the observations and sample-
specific variation. Strategies for dealing directly or indirectly
with one or other of these issues have been put forward by Bargman
(1962), Meredith (1964), Carney (1975), Williams (1976), and Maxwell
(1977b).

7.2.1 The problem of multicollinearity. Where one or both sets of
measurements exhibit near-collinearity, the canonical weights will
be ill-determined. The condition becomes progressively more acute
as $|R_{11}|$ or $|R_{22}| \to 0$;   ultimately, one (or perhaps both) deter-
minants may vanish, so that one or other of the inverse matrices
R_{11}^{-1} or R_{22}^{-1} does not exist and the canonical roots and vectors
cannot ordinarily be obtained (cf. equations (2.11) and (2.10)). In
regression analysis it is known that the stability and interpretability of regression weights in the presence of collinearity is improved by the use of ridge estimates rather than their more familiar
least squares counterparts (e.g. see Hoerl and Kennard, 1970). The
ridge estimates, though slightly biased, have greatly reduced error
variances in comparison with their ordinary least squares counterparts. Carney (1975) has extended the principle of *ridge regression*
to canonical analysis. The resulting weights are referred to as
canonical ridge weights and Carney was able to show that at least in
certain circumstances these are an improvement over the usual canonical weights.

Multicollinearity can be entirely eliminated by an *orthogonal transformation* of the observed variables of either or both sets prior to analysis. The observed variables may then be replaced by, for example, their principal components or some subset of them before embarking on canonical analysis. Williams and Lance (1968) have put forward a proposal of this kind. Although the procedure is straightforward, it sometimes has the disadvantage that the substantive meaning of the components themselves is not clear. This inevitably leads to difficulties in the interpretation of the canonical analysis (cf. Kendall, 1975, p. 97).

In the presence of multicollinearity, canonical weights are often highly dependent on the particular sample in hand; addition or deletion of one or more observations then produces marked differences in the values of the canonical weights, including changes in algebraic sign. One solution which is sometimes practicable in such circumstances is to *increase sample size*. Another rather easier alternative is to *discard* one or more of the variables which appear to be responsible for the collinearity. This can usually be achieved without markedly affecting the overall fit of the model. More is said about sample-specific variation in Section 7.4 below while Willan and Watts (1978) have provided a valuable review of meaningful measures of multicollinearity.

7.2.2 The problem of measurement error. The effect of errors of measurement on the dependence of two sets of variables has been investigated by Meredith (1964). Provided the error variances are known or can be estimated reliably, then the canonical correlations and variates can be *corrected for attenuation* due to measurement error. The procedure calls for replacing the variance–covariance or correlation matrix $\underset{\sim}{R}$ by the reliability adjusted matrix

$\underset{\sim}{R}^* = \underset{\sim}{R} - \underset{\sim}{\Delta}$, where $\underset{\sim}{\Delta}$ is an $n \times n$ diagonal matrix of error variances of the variables in $\underset{\sim}{x}^{(x)}$ and $\underset{\sim}{x}^{(y)}$. Correction for measurement error can dramatically improve the outcome of an analysis. Even where the error variances are unknown and the sample size too small for their satisfactory estimation, steps can still be taken to correct for the effects of errors of measurement. In situations of this kind, Meredith (1964) has proposed performing a *factor analysis* on each set of variables separately before going on to conduct a canonical analysis across the common-factor spaces of each domain. In so doing, the unique (specific plus error) variance of each system is effectively separated from the non-error variance and removed prior to canonical analysis.

A similar suggestion has been made by Williams (1976, p. 67). Williams has suggested that a principal coordinates analysis be first carried out on each set of variables. The 'coordinates' corresponding to the smaller latent roots are discarded and the 'meaningful' principal coordinates then used as the input for canonical analysis. The procedure is referred to as *canonical coordinates analysis*. Williams (1976) has observed that in the context of agricultural experiments at least "this version [of canonical analysis] has invariably proved informative" and in practice has been found more useful than a direct canonical analysis of the original variables. Except perhaps in the rather special case of designed experiments, however, canonical analysis across the common-factor spaces of each domain would be more appropriate; in general, it would be incorrect to assume, as often seems to be the case, that error variance is reflected only in the smaller roots of a components or similar dimension-reducing analysis, as Maxwell (1977a, p. 59) has pointed out.

The effect of measurement error on canonical ridge estimates also deserves consideration. Maxwell (1977b), in the context of multiple regression, has argued that where measurement error is likely to be appreciable, as in ecology, poorly conditioned matrices of an extreme kind occur only infrequently. In such circumstances multicollinearity will escape detection. The practical value of ridge regression where measurement error is considerable is therefore nil. Accordingly, Maxwell has suggested a prior *factor analysis* of the data before embarking on regression analysis. This approach was shown to be more realistic in its underlying assumptions than ridge regression and to be unaffected by 'near-zero' roots in the matrix $\underset{\sim}{R}$ of intercorrelations among the predictor variables (Maxwell, 1977b; see also Lawley and Maxwell, 1973). The latter point is valuable where ill-conditioning is manifest. Thus, where high unique variances are likely, the use of a factor model to estimate regression coefficients may provide a more effective means of dealing with multicollinearity than ridge regression. These observations have a direct bearing on ecological applications of canonical analysis.

7.2.3 An alternative to canonical weights. Bargmann (1962) and Meredith (1964) have suggested using the correlations between the original variables and the canonical variates as a means of interpreting the latter. Such *structure correlations* have the advantages over canonical weights for this purpose of smaller standard errors and greater stability in replicate samples. Furthermore, correlations are generally easier to interpret than regression or canonical weights.

7.3 Rotation of Canonical Variates. Canonical analysis is one of a class of multivariate methods which involves solving a characteristic equation for its latent roots. For some members of this class it is well-known that rotation of the composite-variables specified by the latent vectors can improve interpretability while leaving certain optimized properties of the solution unchanged.

It has been shown that this is true of canonical analysis also. Cliff and Krus (1976) have demonstrated in particular that *rotation* by the normalized varimax criterion preserves (a) the sum of the canonical correlations; and (b) the total predictable variance of each measurement domain. Furthermore, rotation was found to enhance the interpretability of both the canonical weights and the structure correlations. For these reasons Cliff and Krus have recommended rotation as a routine procedure in canonical analysis. It is of interest to note that, while the sum of the canonical correlations is preserved under rotation, this quantity is more evenly distributed among corresponding pairs of canonical variates than in the unrotated solution. Scott and Koopman (1977) have also investigated rotation in canonical analysis, with encouraging results.

7.4 Cross-validation in Canonical Analysis. Canonical analysis involves maximization of the canonical correlation coefficients with respect to the variables of a particular sample. The rationale of maximization is that it facilitates prediction. However, as the maximization is performed on a particular sample, it follows that canonical analysis capitalizes on chance variation and covariation specific to the sample in hand. Consequently, if canonical weights are applied to a sample other than that on which they were determined, the correlations between corresponding pairs of linear composites ('canonical variates') will be smaller than the original canonical correlations, often appreciably so. In other words, weights which are optimal with respect to one sample may not have good general validity. In order to determine the extent of sample-specific covariation and so have some check on the general validity of an analysis, *cross-validation* is necessary (Thorndike and Weiss, 1973; Thorndike, 1976, 1977). Cross-validation involves first dividing the sample randomly into two sub-samples, not necessarily of equal size. The canonical weights developed in the larger sample are then applied to the standardized variables of the hold-out sample. The correlations between corresponding pairs of linear composites are finally compared with the canonical correlations of the larger sample. Unless fairly close agreement is found, no general validity can be claimed for the results obtained. Whenever the sample size is large enough to permit it and particularly where the number of variables in relation to sample size is large, cross-validation is recommended by Thorndike (1977) as a routine procedure.

7.5 Predicting a Criterion of Maximum Utility. One of the principal tasks of canonical analysis is to identify the most predictable criterion. Cronbach (1971) has pointed out, however, that the value of a criterion is not necessarily related to ones ability to predict it; the goal should rather be the most *useful* criterion. Thus a potential conflict may exist between the objective of greatest utility and the mathematics of least-squares prediction.

Cronbach's view has resulted in efforts to maximize the criterion
of greatest practical value rather than the most predictable
criterion in some applications of canonical analysis. For this
purpose it is necessary to have, on the basis of either theoretical
insight or practical experience, a reasonably clear idea of the
composition of the criterion composite required or expected.
Canonical analysis may then be used to find that subset of
predictor variables which best approximates the desired criterion.
Thorndike (1977) has described a *canonical prediction strategy* for
this purpose. The technique involves a stepwise approach to the
selection of predictor variables such that those predictors which
best predict the desired criterion are retained. More generally,
even where detailed insight into the nature of a criterion of
greatest utility is lacking, as in exploratory studies, the same
stepwise approach to the most predictable criterion may still be
informative.

A more rigorous approach to model construction is available
where the criterion variables can be logically ordered. This is
the *step-down analysis* of J. Roy (1958). Step-down analysis
enables noninformative and redundant dependent variables to be
discarded. At each stage the unique contribution of one
variable to the analysis is estimated and tested, providing a
basis for either retaining or partialing out the variable.

7.6 Generalizations of Canonical Analysis. Various generalizations
of canonical analysis to more than two data matrices have been
proposed. Broadly, such generalizations are of two kinds, namely
those in which (a) one or more of the m (m > 2) data matrices
consists of variables whose effects are to be *removed* before considering relationships between the two measurement domains of primary
interest; and those in which (b) the status of all m matrices is
alike. Generalizations of quite different kinds have been described
by Van de Geer (1971) and Miyata (1970) (see Sections 7.6.3 and
7.6.4 below).

7.6.1 Partial, part, and bipartial canonical analysis. Roy (1957,
p. 26), Cooley and Lohnes (1971), Timm (1975), and Timm and Carlson
(1976) have shown how relationships between two sets of variables
$\underset{\sim}{X}$ and $\underset{\sim}{Y}$ can be investigated after first partialing out the effects
of a third set of variables $\underset{\sim}{Z}$ from both sets. The procedure,
which can be regarded as a straightforward extension of simple
partial correlation and of canonical analysis, is known as *partial
canonical correlation analysis*. The objective is to find linear
combinations $U = \underset{\sim}{a}'\underset{\sim}{e}_X$ and $V = \underset{\sim}{b}'\underset{\sim}{e}_Y$ of unit variance of the
variables $\underset{\sim}{e}_X = \underset{\sim}{X} - \hat{\underset{\sim}{X}}$ and $\underset{\sim}{e}_Y = \underset{\sim}{Y} - \hat{\underset{\sim}{Y}}$, where $\underset{\sim}{e}_X$ and $\underset{\sim}{e}_Y$ are
residual vectors obtained from the regression of $\underset{\sim}{X}$ on $\underset{\sim}{Z}$ and

of Y on Z, such that the simple correlation between U and V is maximized. The resulting partial canonical correlations and variates thus refer to the residuals after the effects of a third set of variables is removed. Two modifications of the technique give rise respectively to *part* and *bipartial canonical analysis*. Part canonical analysis deals with relationships between X and Y after the linear effects of variables in Z have been removed from X but not Y. Bipartial canonical analysis on the other hand provides a means of assessing the linear relationship between X and Y after first removing the effects of variables in Z from X and of variables in W from Y. Tests of the significance of the partial, part of bipartial independence of X and Y can be made provided the usual requirements of independence and multivariate normality are met. Lee (1978) has further generalized the concepts of partial, part and bipartial canonical analysis.

7.6.2 Relationships among m sets of variables (m > 2). This problem has been considered by Horst (1961), McKeon (1965), Carroll (1968) and Kettenring (1971). Horst has suggested selecting a linear composite of the variables of each set such that the *sum* of the correlations between all the linear composites is maximized. More recently, Carroll has proposed finding sample canonical variates and an auxillary sample variate such that the *sum* of the squared correlations between the canonical variates and the auxillary is a maximum. Both methods reduce to Hotelling's classical procedure when m = 2. Kettenring (1971) has discussed these and some related proposals for *generalized canonical analysis*.

7.6.3 Maximizing the covariance between two measurement domains. The possibility of maximizing some function other than the correlation between linear transformations of the variables of two measurement domains was encountered above in connection with redundancy analysis (Section 7.1). A further suggestion of the same kind has been made by Van de Geer (1971, p. 169). Van de Geer has proposed maximizing the *covariance* between linear composites of two sets of variables, rather than the correlation between them. The solution provided has some interesting properties which are described by Van de Geer (*loc. cit.*) and Green (1978, p. 280). The method is useful in the context of pattern matching, as for example in the rotation of one matrix towards a fixed 'target' matrix.

7.6.4 Canonical analysis of complex variables. Miyata (1970) has generalized canonical analysis to *complex-valued variables* in a study of linear relationships between the time histories of two collections of points at sea-level (marigrams). The theory developed is applied in the frequency domain to investigate the power spectra of time series. We note that an account of the application of canonical analysis to time series has also been given by Brillinger (1975).

The efforts to improve and extend canonical analysis reviewed above should have the effect of increasing the appeal and useful-ness of canonical analysis to ecologists and others.

8. COMPUTATION

8.1 General Principles. Canonical analysis, like many forms of multivariate analysis, is fundamentally a singular value decomposition problem (see Stewart 1973, p. 317; Green 1976, p. 230). In canonical analysis it is the decomposition of the quadruple matrix product $R_{22}^{-1}R_{21}R_{11}^{-1}R_{12}$ which is required. This matrix derives from the generalized eigenequation (2.10).

The matrix R_{22} in (2.10) is in general nonsingular and on premultiplying (2.10) by R_{22}^{-1} we arrive at

$$(R_{22}^{-1}R_{21}R_{11}^{-1}R_{12} - \lambda^2 I)b = 0 \quad , \tag{8.1}$$

where the quadruple matrix product on the left is a nonsymmetric matrix of order q. In this way the generalized eigenequation (2.10) is reduced to the standard form (8.1). Provided the condition

$$|R_{22}^{-1}R_{21}R_{11}^{-1}R_{12} - \lambda^2 I| = 0 \tag{8.2}$$

is satisfied, nontrivial solutions of (8.1) exist.

Two procedures for calculating the eigenstructure of a non-symmetric matrix are in common use – the latent roots and vectors may be obtained either by direct polynomial expansion or after first transforming the matrix to a symmetric matrix (Ashton, Healy and Lipton, 1957; cf. also McDonald, 1968). Computer subroutines for finding the eigenstructure of a nonsymmetric matrix are widely avail-able and the direct solution of (8.2) and (8.1) is therefore straight-forward. However, subroutines for extracting the latent roots and vectors of symmetric matrices are in a more advanced state of development; there is therefore much to be said for transforming $R_{22}^{-1}R_{21}R_{11}^{-1}R_{12}$ into a symmetric matrix before extracting its roots.

Provided a nonsingular transformation is used for this purpose,
the latent roots will be preserved under the transformation (cf.
Section 3.1). In outlining the procedure, for covenience it
will be helpful to re-express (2.10) as

$$(\underset{\sim}{H} - \lambda^2\underset{\sim}{T})\underset{\sim}{b} = \underset{\sim}{0} \quad , \tag{8.3}$$

where $\underset{\sim}{H} = \underset{\sim}{R}_{21}\underset{\sim}{R}_{11}^{-1}\underset{\sim}{R}_{12}$ is a symmetric matrix of order q and
$\underset{\sim}{T} = \underset{\sim}{R}_{22}$ is a symmetric positive definite matrix of order q. Now,
as $\underset{\sim}{T}$ is symmetric and positive definite, it may be factored into
the product of a rank-q matrix and its transpose

$$\underset{\sim}{T} = \underset{\sim\sim}{CC'} \quad ,$$

where $\underset{\sim}{C}$ is the Cholesky factor of $\underset{\sim}{T}$ (Stewart, 1973, p. 389;
Maindonald, 1977). Substituting for $\underset{\sim}{T}$ in (8.3) we obtain

$$(\underset{\sim}{H} - \lambda^2\underset{\sim\sim}{CC'})\underset{\sim}{b} = \underset{\sim}{0} \quad . \tag{8.4}$$

Pre- and postmultiplying (8.4) within the brackets by $\underset{\sim}{C}^{-1}$ and
$[\underset{\sim}{C}^{-1}]'$, respectively, we have

$$(\underset{\sim}{C}^{-1}\underset{\sim}{H}[\underset{\sim}{C}^{-1}]' - \lambda^2\underset{\sim}{I})\underset{\sim}{b}* = \underset{\sim}{0} \quad .$$

Here the triple matrix product on the left is a symmetric matrix
of order q whose roots are identically those of $\underset{\sim}{R}_{22}^{-1}\underset{\sim}{R}_{21}\underset{\sim}{R}_{11}^{-1}\underset{\sim}{R}_{12}$.
The latent vectors b of (8.3) are however altered by the trans-
formation; but they are readily recovered by the relation

$$\underset{\sim}{b} = [\underset{\sim}{C}^{-1}]'\underset{\sim}{b}* \quad .$$

Efficient subroutines for the Cholesky decomposition of a
symmetric positive definite matrix (Healy, 1968; Barrett and Healy,
1978; Brent, 1970) and for solving the eigenstructure of a real
symmetric matrix (Ortega, 1967; Stewart, 1970; Sparks and Todd, 1973)
are widely available. Together, they provide the most stable method
numerically of the currently available procedures for the computation
of canonical analysis. Once the canonical roots λ^2 and weights
vectors b are obtained, the companion weight vectors $\underset{\sim}{a}$ (cf. 2.12),
canonical variates $U = \underset{\sim}{a}'\underset{\sim}{z}^{(x)}$ and $V = \underset{\sim}{b}'\underset{\sim}{z}^{(y)}$, and this interpretive indices of Section 5 are readily computed by simple matrix
operations.

In conclusion, we mention two alternative proposals for evaluating canonical variates. Horst (1961) has put forward a direct orthogonalization procedure, while Gower (1966) has described a Q-technique which sometimes has advantages over the methods outlined above.

8.2 Some Practical Considerations. Although the solution of canonical analysis is nowadays in principle straightforward, surprizingly few comprehensive computer programs for its implementation are available. Many of the most widely distributed and accessible programs leave much to be desired in terms of the interpretive quantities computed and reported. Of the readily available programs, I have found that by Cooley and Lohnes (1971, p. 194) to be satisfactory for a wide range of general ecological purposes. Cooley and Lohnes' program CANON is well-documented, easily implemented on most machines and the output provided is well-labelled and reasonably complete. (Problems may however be encountered on IBM Series 360 machines and on 36-bit machines in connection with subroutine HOW.) Nevertheless, the program is now rather out of date in terms of completeness and in the significance test procedure used. However, the FORTRAN coding of the main program is clear and easily followed, so that the program can readily be extended to provide additional indices or adapted for specific purposes. We note in particular that without modification to rescale the canonical variates, the program as it stands is not well-suited to canonical variates analysis (cf. Kowal, Lechowicz, and Adams, 1976). In view of the shortcomings of many current programs, it is well worthwhile ensuring before embarking on canonical analysis that a program does in fact report the quantities necessary for proper interpretation. Alternatively, consideration should be given to writing ones own program. Given access to the subroutines mentioned, writing a program ought not to be a difficult task. It is unfortunately difficult to avoid the conclusion that many ecologically unrewarding applications of canonical analysis have been due, at least in part, to insufficient care in the selection of a program or to unwillingness to write one.

Computational problems are sometimes encountered in canonical analysis because of the singularity of one or perhaps both correlation matrices R_{11} and R_{22}. In such cases, the inverses R_{11}^{-1} and R_{22}^{-1} do not exist and equation (8.2) cannot therefore ordinarily be solved. Problems of this kind are most likely to be encountered when the sample size is less than the number of variables in the larger set (e.g. see Gauch and Wentworth, 1976). Khatri (1976) has shown how by using Rao's g-inverse, R_{11}^{-} or R_{22}^{-}, in place of the regular inverse the computational problem can be overcome. Thus singularity in itself need not be a barrier in

canonical analysis.   Just what value is to be placed on the results
of such an analysis, however, remains unanswered (Dempster, 1969,
p. 241; Finn, 1974, p. 97).   An alternative strategy which may
be helpful in these circumstances is to condense the variables of
either or both sets prior to canonical analysis by a principal
coordinates or factor analysis, for example, as Williams and
Lance (1968) have suggested.   The original variables may then be
replaced by some subset of the composite variables, while ensuring
in so doing that   N > max (p,q).

CHAPTER III

APPLICATIONS

9. INTRODUCTION

Seven applications of canonical analysis are described. The
first two applications were in the nature of experiments. Their
purpose was twofold: (a) to enable an assessment of the ability
of canonical analysis to recover known relationships between sets
of variables of ecological interest to be made; and (b) to show
how the results of canonical analysis may be interpreted and the
kind of information provided. The first experiment dealt with
relationships between the abundances of three plant species and
the geographical position of the samples in which they occurred.
The second study concerned relationships between certain soil
variables and the representation of several plant species in a
limestone grassland community. Both investigations were purposely
characterized by the extreme simplicity of the ecological relationships involved. The intent here was to make an appraisal of
canonical analysis possible in fairly clear-cut terms. If analysis
failed to recover the simple relationships expected, then any
future role of the method in ecology would appear to be small. On
the other hand if successful, analysis would demonstrate that
canonical analysis could contribute usefully where the ecological
relationships involved were of a similar order of complexity to
those encountered here; moreover, the studies would then provide
a firm foundation from which further exploration of the opportunities
offered by canonical analysis could confidently be based.

The remaining applications were more realistic in character.
The purpose of these analyses was threefold: (a) to illustrate
the kind of opportunities offered by canonical analysis in the
analysis of ecological data and so contribute towards an improved
definition of the role of the method in ecology; (b) to show
something of the range and flexibility imparted to canonical
analysis by various specializations on the nature of the variables
employed; and (c) to draw attention to connections between canonical
analysis and other methods of data analysis which are frequently

encountered in ecological work. The first two of this group of
analyses concerned rain forest vegetation. The first analysis dealt
with relationships between soil properties and the occurrence of
different forest communities, and the second with relationships
between the seedling and mature tree composition of several different
forest communities. In both analyses all the variables used were
continuous.

These studies of rain forest vegetation were followed by two
analyses which illustrate the use of canonical analysis to explore
structural relationships among a number of multivariate universes.
The first analysis dealt with affinities between three plant
communities on the basis of their overall species composition; the
second with mutual relationships among eight grass species in terms
of their responses to added treatment levels of nitrogen in a
mineral nutrition experiment. In each of these analyses, one
measurement domain consisted of binary-coded dummy variables.
Certain connections between canonical analysis of this kind,
multivariate analysis of variance and multiple discriminant analysis
are explored in the context of the second study.

The final analysis concerned relationships between a number of
study areas and the presence of seven species of large herbivore
in East African rangeland. Both sets of variables in this study
consisted of binary-valued dummy variables. The formal equivalence
of this type of canonical analysis to the analysis of association
in r × c contingency tables and to correspondence analysis is
demonstrated in connection with this example.

9.1 Experimental Studies. In view of doubts which have been expressed concerning the suitability and usefulness of canonical
analysis in ecology it seemed desirable to first attempt to
establish the soundness of the method in a variety of ecological
contexts. With this purpose in mind a number of experimental
analyses were made. These analyses were based on ecological data
for which the major relationships between the variables involved
had already been established with some degree of confidence. It
was therefore possible to assess the ability of canonical analysis
to recover the *known* interrelationships among the variables. Two
investigations of this kind are reported here. Both studies involve relatively few variables (n = 6 and n = 11, respectively).
The results obtained therefore also provide a convenient means of
showing how to go about interpreting a canonical analysis. Accordingly, the results are worked through in some detail.

10.   EXPERIMENT 1:   AN INVESTIGATION OF SPATIAL VARIATION

10.1 Introduction. The experiment concerned the joint pattern of spatial variation common to three species belonging to a limestone grassland community in Anglesey, North Wales. The species were *Phleum bertolonii* DC., *Dactylis glomerata* L., and *Galium verum* L. From previous work, the species were known to share the same relatively simple generalized distribution pattern, which was centered on the south-western extremity of the study area. Furthermore, the direction and extent of the departure of the distribution of each species from the overall trend was also known. *P. bertolonii* was the most circumscribed of the species, being strictly confined to the south-western corner. *G. verum*, on the other hand, was the most widespread species, occurring throughout the study area although its maximum representation occurred in the south-west. *D. glomerata* was less extreme with a distribution intermediate between those of the other species, but resembling *P. bertolonii* more closely than *G. verum*.

Estimates of the abundance of the species in a random sample of N = 45   10 × 10 m stands were available. These data provided one set of variables for canonical analysis while the geographical coordinates of stands comprised a second set. Stand position was determined in relation to an arbitrary (X,Y)-coordinate system whose origin was located close to the center of the area surveyed. The X-coordinate specified stand position in the east-west direction and the Y-coordinate in the north-south direction. In addition, the cross-product term, XY, between X and Y was also calculated. In view of the known south-westerly character of the species' distributions it was anticipated that the inclusion of the cross-product term would improve the overall fit of the model. Thus, in all three spatial measures were used. The purpose of the experiment was to determine to what extent canonical analysis could recover known facts about the spatial variability of the species examined. Briefly, these facts were: (a) that the species share a common distribution pattern centered on the southwestern extremity of the area studied; and (b) that there exist departures from the joint trend in the detailed distribution of the species, *P. bertolonii* being the most circumscribed species and *G. verum* the most widespread.

For the analysis, N = 45 random samples, p = 3 ecological variables, q = 3 spatial variables. The results are summarized in Tables 10.1 and 10.2. We first consider the question of the *dimensionality* of the relationship between the two sets of variables.

10.2 Results.

10.2.1 Dimensionality. Table 10.1 reports the canonical correlations r_k together with two alternative assessments of significance
based on them. The first canonical correlation coefficient (.79)
is appreciably larger than the second (.46) while the third is very
small (.09). Roy's largest-root criterion provides a test of the
significance of individual correlations; Bartlett's χ^2 statistic
provides tests of the joint nullity of the residuals after the
larger roots are successively removed. Both procedures lead to
rejection (p << .01) of the overall null hypothesis of independence.
It appears that species abundance does vary with geographical
position. In order to pursue the nature of the connection we need
to establish the number of significant relationships between the
variables and the interpretation of these relationships. From Table
10.1 (a) and (b) it is apparent that no significant relationship
remains after the first two canonical correlations are eliminated.
Thus it appears that two dimensions or, equivalently, the canonical
variates associated with r_1^2 and r_1^2, are required in order to
fully account for the relationships present. To interpret the
canonical variates ecologically, we require the correlations between the variates and the original variables. These are shown in
Table 10.2, together with other interpretive indices derived from
them.

*TABLE 10.1: Limestone grassland, Anglesey. Canonical analysis of
spatial variation in the representation of three selected species.
Canonical correlation coefficients and related tests of significance.*

(a) *The canonical correlation coefficients r_k and their significance.
Approximate critical values of Roy's largest-root criterion $\theta_\alpha(s,m,n)$
are shown for $\alpha = .05$ and $\alpha = .01$ (m = ½; n = 18½).*

k	s	r_k	r_k^2	$\theta_{.05}(s,m,n)$	$\theta_{.01}(s,m,n)$	p
1	3	.795	.631	.631	.350	<<.01
2	2	.462	.214	.198	.268	<.05
3	1	.088	.008	.175*	.243*	>.05

*From the U-distribution: $\theta_\alpha(1,m,n) = 1 - U_\alpha(p,1,n)$.

(b) *Bartlett's approximate test of the joint nullity of the
smallest s - k canonical correlations.*

k	Roots	Chi-square	df	p
0	1, 2, 3	50.47	9	<.0001
1	2, 3	10.04	4	.04
2	3	.31	1	.58

10.2.2 Intraset correlations. The intraset correlations appear
in the left-hand side of Table 10.2 and are helpful in establishing
the nature of the canonical variates defined on each set of variables. The square of each intraset correlation coefficient represents the proportion of the variance of a variable which is
directly associated with a particular canonical variate.

The correlations of the spatial variables with $V_1 = \underset{\sim}{b}_1' \underset{\sim}{z}^{(y)}$
differ in sign. This suggests that V_1 represents a *contrast* of
east (X) and north (Y) against southwest/northwest (XY). XY shows
the strongest correlation (-0.70) with V_1 closely followed by
Y (.62). The corresponding correlations of the species with $U_1 =$
$\underset{\sim}{a}_1' \underset{\sim}{z}^{(x)}$ are all alike in sign, indicating that U_1 expresses some
feature or trait common to all three species with respect to the
linear composite of spatial variables, V_1. We notice that the
strength of the species correlations varies appreciably, from that
of *G. verum* (-.43) to that of *P. bertolonii* (-.98). V_1 accounts
for 37 percent of the variance common to the spatial domain
($V_1^2 = .370$), while U_1 accounts for 64 percent of the variance
of the species measures ($U_1^2 = .640$).

Turning to the second pair of canonical variates we see that
the correlations tend to be smaller than those of the variables
with U_1 and V_1. The intraset correlations of the spatial
variables with V_2 are alike in sign; V_2 evidently expresses
some characteristic *common* to the three spatial variables in
relation to the linear composite of the species represented by U_2.
The strongest correlation (.53) belongs to XY. V_2 therefore
appears to represent in part some aspect of 'southwest/northeasterly-
ness' which is, presumably, uncorrelated with that previously
specified by V_1. From the intraset species correlations with
U_2 we see that the strongest relationship is with *D. glomerata*
(-.48), *P. bertolonii* and *G. verum* being characterized by quite
small positive correlations. U_2 evidently represents a subsidiary
distinction among the species previously unified by U_1. In terms
of U_2 at least, *D. glomerata* is seen to be equally distinct from
the remaining species. Although it is not a well-defined component,
U_2 is characterized above all by *D. glomerata*, expressing what is
unique to the distribution of this species, namely its approximately
equal relationship to the distribution of *both* the other species.

The second pair of canonical variates account for only small amounts of the variability of their respective domains, the variances of V_2 and U_2 being $V_2^2 = .152$ and $U_2^2 = .092$ respectively. V_2 and U_2 are evidently appreciably weaker components of their measurement domains than the first pair of canonical variates.

The third canonical variate V_3 of the spatial variables is characterized by sizeable correlations which contrast the variable X (.86) against Y (−.68) and XY (−.48). The corresponding variate U_3 of the species domain very clearly distinguishes G. verum from the remaining species. U_3 in fact is essentially a pure, though inverse, measure of G. verum. It is of some interest to find that V_3 is the strongest canonical variate of the spatial measures, accounting for 48 percent of the variance of this domain $(V_3^2 = .479)$; U_3 although somewhat weaker $(U_3^2 = .268)$, is nevertheless a sharply defined and relatively strong component of the species domain.

Before proceeding it may be worthwhile to summarize the salient characteristics of the canonical variates here. The first canonical variate of the species domain, U_1, clearly expresses some aspect of distribution *common* to all three species which is exemplified above all by *P. bertolonii*. U_2 and U_3, by way of contrast, are relatively pure measures of *D. glomerata* and *G. verum* respectively, and therefore specify features *unique* to the distribution of each. The nature of the V_k of the spatial domain are less easily characterized in this way. However, their dominant features can very roughly be subsumed under the headings of 'southwest/northeast', 'southwest/northeast' and 'east plus south', respectively. The ecological significance of relationships between the two sets of canonical variates is taken up in Section 10.2.4 below.

10.2.3 Intraset communalities. The intraset communalities (Table 10.2, h_w^2) of the variables of both sets are all essentially unity. Thus, the variances of the variables of both sets are fully accounted for by the canonical variates. This is a consequence of the fact that in this analysis there are an equal number of variables in each set (p = q = 3) and that Table 10.2 reports the full rank solution (r = 3). The intraset communalities are therefore of little substantive interest here; however, the *profile* of each variable over the canonical variates is useful in showing how the total variance of the variable is distributed among the variates.

TABLE 10.2: *Limestone grassland, Anglesey. Canonical analysis of spatial variation in the representation of three selected species: correlations between the original variables and the canonical variates.*

Canonical variate	U_1	U_2	U_3	h_w^2	V_1	V_2	V_3	h_b^2
Species								
G. verum	-.433	.126	-.893	1.000	-.344	-.058	-.078	.128
D. glomerata	-.877	-.479	-.027	.999	-.697	-.221	-.002	.535
P. bertolonii	-.982	.175	.077	1.000	-.780	.081	.007	.615
Variance extracted	.640	.092	.268	1.000	.404	.020	.002	.426
Redundancy	.404	.020	.002	.426	.404	.020	.002	.426

Canonical variate	V_1	V_2	V_3	h_w^2	U_1	U_2	U_3	h_b^2
Spatial coordinate								
X (East)	.483	.164	.860	1.000	.384	.076	.075	.159
Y (North)	.625	.378	-.684	1.000	.497	.175	-.060	.270
XY	-.697	.534	-.479	1.000	-.554	.247	-.042	.370
Variance extracted	.370	.152	.479	1.000	.234	.032	.004	.270
Redundancy	.233	.032	.004	.269	.234	.032	.004	.270

10.2.4 Interset correlations. The interset variable/canonical variate correlations appear in the right-hand side of Table 10.2 and are helpful in clarifying the nature of interrelationships among the two sets of variables. The square of each interset correlation coefficient is analogous to the coefficient of determination, R^2, of multiple regression. Thus each squared correlation expresses the proportion of the variance of a variable which is predictable from a particular canonical variate of the 'other' measurement domain. In interpreting these correlations it is necessary to bear in mind the interpretation of the canonical variates arrived at from consideration of the intraset correlations.

For completeness two sets of interset correlations appear in the right-hand side of Table 10.2, although in this analysis only one set could have any possible ecological significance - namely those between the linear composites V_k of the spatial measures and species' abundance; relationships in the other direction, that is of the 'effect' of species' abundance on spatial position, are clearly devoid of ecological meaning. In many applications, however, the *directed* nature of the present analysis is lacking, being supplanted by an interest which is symmetric with respect to variables of both sets.

The species' correlations with V_1 show that the species all respond in the same sense to the spatial contrast represented by V_1. From the negative signs of the correlations, and recalling the nature of V_1, it is apparent that species vary directly with XY, the spatial variable most strongly correlated with V_1. In other words, all three species tend to increase in abundance south-westwards and north-eastwards with increase in XY. Similarly, it is clear that from the signs of the correlations that the three species vary inversely with X (east) and Y (north). This is equivalent to saying that the species *increase* in abundance west-wards and southwards from the center of the study area. The absolute values of the correlations with V_1 show that *P. bertolonii* is the species most strongly (.78) associated with the linear composite represented by V_1, closely followed by *D. glomerata* (.70).

Before proceeding, it will be helpful to clarify the meaning of the cross-product term XY. The area of vegetation examined has an irregular shape and the disposition of the community in relation to the coordinate system used is such that the northeast quadrant is virtually without samples. This simplifies interpretation of the XY term. In view of the virtual absence of samples NE from the origin, it is clear that the XY term can account for variation only in a south-westerly direction from the origin. We

can now see that jointly the spatial and species variables give
rise to a consistent and intuitively reasonable picture: the
abundances of all three species increase westwards, southwards and
above all south-westwards from the center of the area surveyed.
From Table 10.2 (upper right-hand section) we see that the proportion of variance common to the species measures accounted for by
the first canonical variate $V_1 = \underset{\sim}{b'_1} \underset{\sim}{z}^{(y)}$ of the spatial domain
is $V^2_{x|v_1}$ = .404. This is an appreciable amount. Similarly, we
note that the first canonical variate $U_1 = \underset{\sim}{a'_1} \underset{\sim}{x}^{(x)}$ of the
species domain accounts for 23 percent of the variance of the
spatial set $(U^2_{y|u_1}$ = .234).

We may now summarize the relationships between the ecological
and spatial variables as defined by the first pair of canonical
variates. *P. bertolonii, D. glomerata* and *G. verum* all increase
in a generally south-westerly direction from the origin. But
there are differences among the species in the degree of their
response, *P. bertolonii* having the strongest relationship and
G. verum the weakest. In all, 40 percent of the total variation
of the species domain is accounted for in this way.

The interset correlations between the species and the second
canonical variate of the spatial measures, V_2, are all small in
absolute terms (< 0.23). The strongest relationship is the weak
(-.22) inverse correlation of *D. glomerata*, the correlations of
the remaining species with V_2 being close to zero. Recalling
that V_2 perhaps represents some aspect of 'southwesterly-ness',
the inverse relationship with *D. glomerata* implies that the
species increases in abundance, at least to some extent, in the
opposite direction. This result indicates that, while the species may be predominantly southwestern in distribution, as shown
by its relationship to V_1, it is by no means strictly so,
being 'centered' somewhere between the southwestern extremity and
the center of the area studied. Further examination of the interrelationships involved is hardly called for in view of the negligible
explanatory power of V_2 in the species domain $(V^2_{x|v_2}$ = .02).

In summary, the second canonical relationship is interpretable
as a weak though largely uncontaminated expression of *D. glomerata*.
The relationship appears to identify what is unique about the distribution of this species, notably that the distribution is not
strictly southwestern in character, and, because of this, bears
equal resemblance to the relatively more extreme distributions
of both *P. bertolonii* and *G. verum*. Moreover, the more extreme

species are alike in being virtually uncorrelated with V_2; hence they are totally lacking in the 'intermediate' type of distribution characterizing $D.$ $glomerata.$ The relationships involved are evidentally rather fine points of detail, V_2 accounting for only 2 percent of the variance common to the species measures.

The low explanatory power of the third canonical variate of the spatial measures, V_3, in the species domain $(v^2_{x|v_3} = .002)$ indicates that there is little to be gained from the third canonical relationship between the spatial and species measures. We may therefore safely disregard the interset correlations of the species with V_3.

10.2.5 Redundancy. Redundancy expresses the explanatory power of a canonical variate of one domain with respect to the observed variables of the other.

The percentage of variance common to the first pair of canonical variates U_1 and V_1 is about 63 percent, since $r^2_1 = .631$ (Table 10.1 (a)). From Table 10.2, however, it can be seen that while 64 percent of the total variance of the species comprising $z^{(x)}$, on which U_1 is defined, is accounted for by this canonical variate, only 40 percent of the variance in $z^{(x)}$ is accounted for by the corresponding canonical variate V_1 of the *spatial* domain. Similarly 37 percent of the total variance of the spatial variables comprising $z^{(y)}$ is accounted for by the first canonical variate V_1 of this domain, while the corresponding canonical variate U_1 of the *species* domain accounts for only 23 percent of the variance in $z^{(y)}$. Evidently U_1 and V_1 are weaker explanatory constructs across domains than within domains. It is also apparent that the explained variance across domains of each canonical variate is appreciably less than one might be led to expect from the square of the first canonical correlation coefficient (cf. Section 3.1).

It is readily verified from Table 10.2 that redundancy is also given by the within-set variance (U^2_k or V^2_k) of a canonical variate attenuated by the between-set variance (r^2_k) of the variate. Both derivations help to explain why redundancy should provide a more acceptable measure of the interrelatedness of domains than the square of a canonical correlation coefficient. For U_1 and V_1 we see from Table 10.2 that

$$U^2_{y|u_1} = V^2_{1(y)} \; r^2_1 = (.370) \; (.631) = .233$$

$$V^2_{x|v_1} = U^2_{1(x)} \; r^2_1 = (.640) \; (.631) = .404.$$

These quantities are the redundancies of the spatial (Y)- and species (X)- domains, respectively; the redundancy in $z^{(y)}$, given the availability of U_1 of the species domain, is .233 while the redundancy in $z^{(x)}$, given the availability of V_1 of the spatial domain, is .404. These results indicate that a higher proportion of the variance of the species set is predictable from the first canonical variate V_1 of the spatial domain than *vice versa*.

Turning to the second pair of canonical variates we find (Table 10.1(a)) that the percentage of variance common to U_2 and V_2 is about 21 percent ($r^2_2 = .214$). From Table 10.2 it can be seen that U_2 accounts for 9 percent of the variance of the species domain on which it is defined, and for 3 percent of the variance of the spatial domain. Similarly, we see that V_2 accounts for 15 percent of the variance of the spatial domain of which it is a linear composite, and for only 2 percent of the variance of the species domain. Although moderately correlated ($r_2 = .46$), U_2 and V_2 are nevertheless rather weak components of their respective domains. Both features are reflected in the redundancies associated with these canonical variates:

$$U^2_{y|u_2} = V^2_{2(y)} \; r^2_2 = (.152) \; (.214) = .032$$

$$V^2_{x|v_2} = U^2_{2(x)} \; r^2_2 = (.092) \; (.214) = .020 \;\; .$$

The explanatory power of both canonical variates across domains is small. A similar result arises in connection with the third pair of canonical variates, although for rather different reasons. U_3 and V_3, in contrast to the second pair of variates, are relatively strong components of their respective domains but are themselves only weakly correlated ($r_3 = .09$); hence the shared variance common to U_3 and V_3 is very small ($r^2_3 = .008$). The redundancies associated with U_3 and V_3 are, respectively:

$$U^2_{y|u_3} = V^2_{3(y)} \quad r^2_3 = (.268)\ (.008) = .004$$

$$V^2_{x|v_3} = U^2_{3(x)} \quad r^2_3 = (.479)\ (.008) = .002$$

The effect of the virtual absence of covariation between U_3 and V_3 ($r^2_3 = .008$) on the explanatory power of these canonical variates across domains is obvious.

The redundancy indices suggest that the linear relationship between domains is, for all practical purposes, effectively accounted for by the first pair of canonical variates. This is rather a different conclusion from that arrived at through the statistical tests of significance based on the canonical correlation coefficients (Section 10.2.1). But it is important to realize that the two indices, redundancy and the canonical correlation coefficient, refer to different aspects of the analysis. In general, redundancy is likely to be of greater substantive value than a canonical correlation coefficient simply because it takes account of the strength of a canonical variate within the domain on which it is defined. Therefore, where redundancy and a canonical correlation coefficient lead to conflicting interpretations, that based on redundancy will tend to override that of the canonical correlation, irrespective of any statistical significance which may attach to the latter. Accordingly, in the present analysis, we conclude that there are grounds for neglecting the second and third pairs of canonical variates. This is equivalent to declaring that a model of rank $r = 1$ is considered to provide the most acceptable fit to the data.

Total redundancy expresses the proportion of the total variance of one measurement domain which is predictable from all the canonical variates of the other. We see from Table 10.2 that 42.6 percent of the variance of the species domain is predictable from the spatial measures, while 27 percent of the variance of the latter is predictable by the species. However, recalling the acceptance of a model of rank 1, we ought properly to adjust the total redundancy estimates to take account of this. Acceptance of a rank 1 model leads to revised total redundancies which in this case are the redundancies generated by V_1 and U_1 alone. Thus the revised quantities are 40 and 23 percent for the X- and Y-domains respectively. We conclude that 40 percent of the observed species variability is attributable to the spatial measures used.

10.2.6 Interset communalities. The interset species communalities (Table 10.2, h^2_b) show that sizeable proportions of the variance of

P. bertolonii (.61) and *D. glomerata* (.53) are accounted for by
the spatial measures. On the other hand, the variance of *G. verum*
is less well explained (.13). This is not altogether surprising
as *G. verum* is much more widely distributed than either of the
other species. It is to be expected that the *joint* distribution
pattern common to all three species will be strongly influenced by
the close similarity of the distributions of *P. bertolonii* and *D.
glomerata.* Consequently, we may expect the somewhat disparate
distribution of *G. verum* to be less adequately accounted for than
the distributions of the other species.

Notice that the interset communalities of Table 10.2 refer
to the full rank 3 model. In the present case, however, the
communalities of the rank 1 model do not differ to any appreciable
extent from those reported here.

10.2.7 The canonical variates. The interpretation of the canonical
variates has to some extent already been considered (Sections 10.2.2,
10.2.4, and 10.2.5). As a final step in their interpretation
graphical representation proved informative. In order to conserve
space, however, the scores of samples on the canonical variates
are not themselves reported.

Figure 10.1 shows the sample of N = 45 stands mapped into
the space of the first pair of canonical variates, U_1 and V_1.
The correlation between U_1 and V_1 is r_1 = 0.79 and this is
expressed in the Figure by the tendency of points to fall on the
NE–SW diagonal. There are two distinctive features to the plot:
(a) the very noticeable scatter of points extending outwards along
the S.W. diagonal; and (b) the bunching of points and the truncation
of the scatter on the positive side of both axes close to the origin.
The first feature reflects the correlation between the abundance of
the species and stands with a large XY coordinate, i.e. stands
located at the south-western extremity of the area. The second
characteristic arises both from the scarcity of stands in the north-
eastern quarter of the survey area and from the absence of two of
the species and the comparative rarity of the third from virtually
all stands except those in the southwest. Consequently, potential
sample points in the First (NE) quadrant of the sample space of
Figure 10.1 are unrealized in the sample.

Figure 10.1 very effectively summarizes the known facts about
the joint distribution of the species provided the meaning of the
canonical variates is kept in mind (Section 10.2.2).

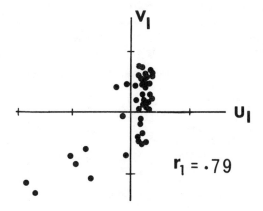

FIG. 10.1:   Limestone grassland, Anglesey.   Canonical analysis of spatial variation in three selected species.   Sample of N = 45 stands mapped into the subspace of the first pair of canonical variables, U_1 and V_1.

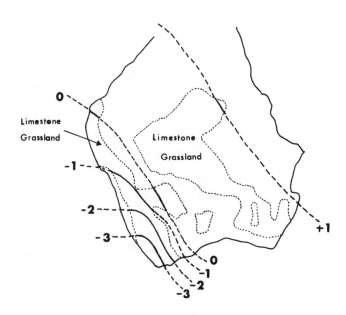

FIG. 10.2:   Limestone grassland, Anglesey.   Canonical trend-surface analysis.   Spatial distribution of the first canonical variate, U_1, of the species domain.

An alternative representation of the same phenomenon is provided by mapping the field distribution of canonical variate U_1
of the species' domain. In Figure 10.2 the canonical variate scores
for samples on U_1, the first variate of the species set, are
plotted on a map of the survey area at the position of the samples.
Contours have been inserted to show the form of the corresponding
canonical trend-surface. The principal feature of interest is the
clear 'centering' of the canonical trend-surface on the south-
western extremity of the map. The level nature of the surface over
much of the remainder of the area reflects the absence of two of the
species and the comparative rarity of the third over a large part
of the region. The canonical trend-surface summarizes that part of
the spatial distribution of the species which is common to all
three species. In other words the surface represents the generalized
joint distribution pattern of the species. From the direction of
the species' correlations with U_1 and from the shape of the canonical-
trend it is apparent that species' abundance increases towards the
southwest. Notice that the surface does not show the absolute
abundance of the species, the trend being expressed in terms of
arbitrary units. The question of *why* the canonical trend should
have the form that it does is a question of considerable ecological
interest. Pursuit of questions of this kind would take us too far,
however, and fall outside the scope of the present review.

10.3 Conclusions. The first canonical relationship recovers both:
(a) the joint southwesterly trend in spatial variation common to
all three species; and (b) departures in the distribution of particular species from the joint trend, *G. verum* being the species
least closely associated with the overall trend and *P. bertolonii*
the species most closely associated with it. It will be recalled
that these are precisely the characteristics which at the outset
were considered to comprise the salient ecological features of
the data. The relationship accounts for 40 percent of the
species' variation. This can be considered to be a satisfactory
result, especially if the somewhat disparate nature of the distribution of *G. verum* is borne in mind.

The second and third canonical relationships also appear to
distinguish meaningful ecological aspects of the data. However,
in comparison with the first pair of canonical variates the contributions of the subsequent variates are insignificant. The second
and third canonical relationships apparently focus on rather fine
points of detail, singling out features unique to the distributions
of *D. glomerata* and *G. verum* respectively.

In some circumstances it might be considered desirable to try
to improve the fit of the model. Two steps could be taken in
efforts to achieve this: (a) addition of further polynomial terms

of the spatial variables, X and Y; and (b) deletion of variables from the 'response' set - *G. verum* would be an obvious candidate for removal; deletion of predictor variables (polynomial terms in X and Y) found not to contribute usefully to the overall fit.

11.   EXPERIMENT 2:   SOIL-SPECIES RELATIONSHIPS IN A LIMESTONE GRASSLAND COMMUNITY

11.1 Introduction. The second experiment was concerned with connections between species' abundance and associated soil character- istics in a limestone grassland community in Anglesey, North Wales. Previous work (Gittins, 1969) suggested that the factor-complexes of soil moisture and soil fertility were influential in determining the representation of most species present. The precise nature of the soil-species relationships involved provided a basis against which the results of a series of experiments designed to examine the ability of canonical analysis to recover previously established relationships of this kind could be assessed. One such experiment is described here.

Estimates of the composition of the community and of the intensity of three soil properties based on a random sample of forty-five 10 × 10 m stands were available. Thirty-three species were encountered in all. The soil variables examined were depth (d), a surrogate for the more direct probable ecological control of soil moisture, extractable phosphate (P) and exchangeable po- tassium (K). For the purpose of the experiment eight species were selected from the thirty-three for which data were available. Selection was guided primarily by the desire to obtain species with the widest possible range of response to variation in soil depth. The species selected, in approximate order of their response, were: *Helictotrichon pubescens* (Huds.) Pilger, *Trifolium pratense* L., *Poterium sanguisorba* L., *Phleum bertolonii* DC., *Rhytidiadelphus squarrosus* (Hedw.) Warnst., *Hieracium pilosella* L., *Briza media* L., and *Thymus drucei* Ronn. Species at the *H. pubescens* end of the sequence tended to be most abundant on the deeper soils, while those at the *T. drucei* end tended to be most abundant on the shallower soils; species towards the middle were relatively independent of soil depth.

This choice of species was expected to generate an essentially one-dimensional 'data structure'. The inclusion of *P. bertolonii* and *P. sanguisorba,* however, was purposely designed to enrich this fundamentally simple structure, the representation of both species being believed to be influenced by soil P in addition to depth. Furthermore, there was evidence to suggest that these species responded in opposite senses to variation in P, *P. bertolonii* increasing sharply in abundance with increase in P over the range encountered at the study area, while *P. sanguisorba* tended to be most abundant

on soils deficient in P. The remaining species were all considered
to be largely independent of variation in soil P over the range
encountered in the field. Soil potassium, on the other hand, unlike
either depth or phosphate, was believed to have a negligible effect
on the behavior of *all* the species examined. The resulting data
structure was accordingly expected to be characterized by two
features: (a) a major component corresponding to differential
species' response to variation in soil depth; and (b) a subsidiary
component corresponding to the contrasting behavior of *P. bertolonii*
and *P. sanguisorba* in relation to soil P. In reality, the relationships involved were known to be somewhat less clear-cut than
for reasons of simplicity they are stated here. Complications arise
because in the field soil depth and phosphate tend to be correlated
to some extent, the deeper soils in general also being those of
highest phosphate status. Thus species may respond more to the
interaction of depth x phosphate than to either variable acting
alone. Such considerations indicated that it might perhaps be
worthwhile to include interaction terms between all three soil
variables in the analysis. Consequently, it was decided to make
use of the interaction terms $d \times P$, $d \times K$ and $P \times K$ as well as
the three original soil variables, giving a total of six soil
measures in all.

For the canonical analysis we therefore have a random sample
of $N = 45$ stands, $p = 8$ estimates of species' abundance and
$q = 6$ soil variables. Interest centers on the nature of the relationship between the two sets of variables. For the analysis
to be judged successful, it would need to identify the connections
between the soil properties and species described above, and, in
so doing, draw attention to the difference in magnitude between
the overall effects of soil depth and phosphate on the species as
a whole. Moreover, the totally subordinate role of soil potassium
would also need to be exposed.

The results are summarized in Tables 11.1 and 11.2. We
consider first the dimensionality of the relationship between the
variables.

11.2 *Results*.

11.2.1 *Dimensionality*. The canonical correlation coefficients
and associated tests of significance are reported in Table 11.1.
Three of the correlations exceed 0.5 in magnitude, the values of the
first two being $r_1 = .92$ and $r_2 = .70$. Only the first of these
however proves to be statistically significant ($p \ll .01$).

TABLE 11.1: Limestone grassland, Anglesey. Canonical analysis of relationships between six soil properties and eight selected species. Canonical correlation coefficients and related tests of significance.

(a) The canonical correlation coefficients r_k and their significance. Approximate critical values of Roy's largest-root criterion $\theta_\alpha(s,m,n)$ are shown for $\alpha = .05$ and $\alpha = .01$ (m = ½; n = 14½).

k	s	r_k	r_k^2	$\theta_{.05}(s,m,n)$	$\theta_{.01}(s,m,n)$	p
1	6	.923	.851	.557	.616	<< .01
2	5	.696	.484	.511	.575	> .05
3	4	.521	.271	.457	.525	> .05
4	3	.374	.140	.392	.466	> .05
5	2	.257	.066	.313	.391	> .05
6	1	.218	.048	.280*	.350*	> .05

*From the U-distribution: $\theta_\alpha(1,m,n) = 1 - U_\alpha(p,1,n)$.

(b) Bartlett's approximate test of the joint nullity of the smallest s - k canonical correlations.

k	Roots	χ^2	df	p
0	1,2,3,4,5,6	115.06	48	< .0001
1	2,3,4,5,6	45.35	35	.11
2	3,4,5,6	21.33	24	.62
3	4,5,6	9.77	15	.83
4	5,6	4.28	8	.83
5	6	1.78	3	.62

Roy's largest-root criterion, r_1^2, and Bartlett's χ^2 approximation of Λ both lead to rejection of the overall hypothesis of no association, $H_0: \Sigma_{12} = 0$ (Sections 4.1 and 4.3). From Table 11.1 (a) we find that $\theta_s = .851 > \theta_\alpha$ (6, ½, 14½) = .616 for $\alpha = .01$; similarly, from Table 11.1 (b) we see that $\chi_s^2 = 115.06 > \chi_\alpha^2(48) = 93.22$ for $\alpha = .0001$. We conclude that the species and soil variables are linearly related. Proceeding to the tests of the subsidiary hypotheses (Sections 4.2 and 4.3) we see from Table 11.1 (a) and (b) that none of the remaining canonical correlation coefficients can be inferred on the basis of either test criterion to be statistically significant. It therefore appears that a single dimension is sufficient to fully account for the relationship between the variables.

11.2.2 Intraset correlations. Table 11.2 contains the correlations between the original variables and the canonical variates corresponding to the first three canonical correlation coefficients. We consider first the correlations between the soil variables and the canonical variates $V_k = b'_k z^{(y)}$ defined on them.

The soil variables all contribute in the same direction to V_1 and, with one exception (K), all have sizeable correlations with this variate. V_1 is however characterized particularly by $d \times P$ (.91) and d (.83). Above all, therefore, V_1 seems to be an expression of soil depth together with its interaction with phosphate. The correlations of the soil variables with V_2 are noticable weaker than their correlations with V_1; the strongest correlations are those of P (−.47), $P \times K$ (−.35) and $d \times P$ (−.27). Accordingly, V_2 may be regarded essentially as an expression of P. In a similar way we see that V_3 is predominantly an expression of K, the strongest relationships being those of K (.95), $P \times K$ (.82) and $d \times K$ (.55). The first three canonical variates of the soil domain can therefore very roughly be taken to be expressions of d together with $d \times P$, P and K respectively.

From the variances of the V_k we see that V_1 and V_3 each account for some 40 percent of the total variance of soil variables, while V_2 accounts for only 9 percent ($V_2^2 = .088$). Together, the three canonical variates 'extract' 90.4 percent of the variance of the soil domain; the V_k (k = 1,···,3) therefore efficiently summarize the variance of the soil domain on which they are defined.

We next consider the corresponding canonical variates $U_k = a'_k z^{(x)}$ of the species domain. The opposing signs of the species correlations with U_1 indicate that U_1 expresses some characteristic of the species which differentiates among them with respect to the linear composite of soil variables represented by V_1. A majority of the species have sizeable correlations with U_1, the strongest being those of *H. pubescens* (.97) and *T. drucei* (−.89), though in opposite senses. Only *R. squarrosus* (.02) is uncorrelated with U_1. On arranging the species in terms of the magnitude of their correlations with U_1 we see that the arrangement closely approximates the order given by the response of the species to soil depth described above (Section 11.1). The second component U_2 is characterized largely by the contrast between *P. bertolonii*

(-.67) and *P. sanguisorba* (.45). Three species (*H. pubescens*, *T. pratense* and *T. drucei*) make virtually no contribution to this canonical variate. The last component, U_3, is basically a weakly-defined contrast of *T. pratense* (-.49) and *R. squarrosus* (-.39) against *P. bertolonii* (.35).

Of the three canonical variates U_1 is much the strongest, accounting for 38 percent of the total variance of the species domain; U_2 and U_3 account for some 12 percent and 8 percent of the species variance, respectively. Collectively, the three canonical variates 'extract' 58.3 percent of the variance of the species domain. They are therefore somewhat less representative of their domain than are the corresponding variates V_k of the soil domain.

11.2.3 Intraset communalities. The intraset communalities (h_w^2) of the soil variables for the rank 3 model also appear in Table 11.2. With the exception of that for $d \times K$ (.74), the communalities all exceed .90. Thus the V_k $(k = 1, \cdots, 3)$ uniformly account for substantial proportions of the variances of each of the soil variables. The equivalent species communalities are more variable. Nevertheless, they show that in general the U_k do account for sizeable proportions of the species' variances.

11.2.4 Interset correlations. Ecological interest in the present analysis is confined strictly to the possible effects of the soil variables on species' abundance. Thus the analysis is *directed* in nature. For this reason we may confine our attention to the correlations between the species and the V_k of the soil domain. These correlations appear in the upper right-hand section of Table 11.2.

From the species correlations with V_1 it is clear that considerable differences exist between species with respect to V_1. Several species, notably *H. pubescens* (.89) and *P. bertolonii* (.46), vary directly with V_1 while other species, notably *T. drucei* (-.83) and *B. media* (-.70) are inversely related to V_1. There are also considerable differences in the strength of the relationships, ranging from those of *H. pubescens* and *T. drucei* to that of *R. squarrosus* (.01). Moreover, the order of species in terms of the size of their correlations with V_1, is substantially that of the approximate order of the species' responses to

TABLE 11.2: Limestone grassland, Anglesey. Canonical analysis. Relationships between six soil characteristics and eight selected species: correlations between the original variates and canonical variates.

Canonical variate	U_1	U_2	U_3	h_w^2	V_1	V_2	V_3	h_b^2
Species
H. pubescens	.968	-.018	-.008	.937	.893	-.010	-.004	.798
P. bertolonii	.499	-.666	.346	.812	.460	-.464	.180	.459
T. pratense	.492	-.050	-.490	.485	.454	-.035	-.255	.272
P. sanguisorba	.381	.449	-.277	.423	.352	.313	-.144	.243
R. squarrosus	.015	-.316	-.386	.249	.014	-.220	-.201	.089
H. pilosella	-.296	.433	-.037	.276	-.273	.301	-.019	.165
B. media	-.759	.200	.150	.639	-.700	.139	.078	.515
T. drucei	-.895	-.013	-.197	.840	-.826	-.009	-.103	.693
Variance extracted	.380	.122	.081	.583	.323	.059	.022	.404
Redundancy	.323	.059	.022	.404	.323	.059	.022	.404

Canonical variate	V_1	V_2	V_3	h_w^2	U_1	U_2	U_3	h_w^2
Soil factor
d	.834	.291	-.348	.901	.769	.203	-.181	.665
P	.639	-.466	.553	.931	.590	-.324	.288	.536
K	.160	.031	.952	.933	.148	.022	.496	.268
d × P	.910	-.275	.234	.958	.840	-.191	.122	.757
d × K	.636	.186	.552	.737	.587	.117	.288	.441
P × K	.414	-.346	.821	.965	.382	.246	.428	.390
Variance extracted	.422	.088	.394	.904	.360	.042	.107	.509
Redundancy	.360	.042	.107	.509	.360	.042	.107	.509

variation in soil depth mentioned previously (Section 11.1). Re-calling that V_1 may broadly be interpreted as an expression of soil depth together with d × P, it is clear that the first canonical relationship largely recovers the relationship between the species and soil depth (and d × P) described at the outset.

   P. bertolonii (-.46), *P. sanguisorba* (.31) and *H. pilosella* (.30) are the species most closely related to V_2. Recalling that V_2 primarily represents soil P, and taking note of the directions of the correlations, we see that *P. bertolonii* varies directly with soil P while *P. sanguisorba* and *H. pilosella* vary inversely with P. Thus the second canonical relationship retrieves the subsidiary relationship between *P. bertolonii* and *P. sanguisorba* described above. From the magnitude of the correlations it is clear that the relationships involved are considerably weaker than those of several species with V_1.

   The similarity of *H. pilosella* and *P. sanguisorba* was not anticipated. Reference back to the original data confirmed that *H. pilosella* does indeed exhibit a tendency towards greater abun-dance on P deficient soils. Thus the analysis properly directs attention to this previously overlooked, though minor, relationship.

   The majority of the species correlations with V_3 do not depart appreciably from zero. The strongest correlations are those of *T. pratense* (-.25) and *R. squarrosus* (-.20). Evidently none of the species has much in common with the linear composite of soil variables specified by V_3, which represents essentially K.

11.2.5 Redundancy. The redundancy in the species domain generated by each of the canonical variates V_k of the soil domain is $V^2_{x|v_1}$ = .323, $V^2_{x|v_2}$ = .059 and $V^2_{x|v_3}$ = .022, respectively (Table 11.2). It is clear that the explanatory power of the V_k falls regularly across the variates. This is rather a different situation from that obtaining within the soil domain. There we saw that V_2 accounts for less than a quarter of the variance accounted for by either V_1 or V_3.

   From the redundancies it is clear that V_1 (which represents d and d × P), accounts for much the greater part of the explained variance of the species examined. It is also of interest to note that V_2 (roughly soil P) accounts for considerably more (6%) of the species' variance than is explained by V_3 (2%), despite the strength of V_3 (roughly soil K) in the soil domain. These points may conveniently be summarized as follows: (a) that variation in d and d × P appears to have a profound effect on species abundance, explaining 32 percent of the variation in the representation of the species studied; (b) that quite small differences in P appear to have relatively large effects on the representation of certain species and account for 6 percent of the total species' variance; and (c) that relatively large differences associated with K appear to have a negligible effect on the species studied, explaining some 2 percent of the total variation in species' abundance.

The redundancies associated with the V_k are broadly consistent with the earlier finding based on the significance of the canonical correlation coefficients that all linear relationship between domains is concentrated in a single dimension. However, we should not overlook that the canonical variates corresponding to r_2 also proved to have considerable ecological interest. Bearing this in mind it is clear that there are grounds for overriding the earlier judgement concerning the number of trustworthy relationships between the variables. The explanatory power of V_2 across domains $(V^2_{x|v_2} = .06)$ together with the ecological interest attaching to the relationship suggest that a model of rank 2 would provide a more acceptable combination of fit, insight and parsimony than the original model (r = 1). For this reason we opt for a model of r = 2.

The total redundancy of the species' domain, given the first three canonical variates of the soil domain, is 40.4 percent (Table 11.2). For the rank 2 model the revised total redundancy is 38.2 percent. In order to see how the explained variance represented by this figure is distributed among the species we turn to the interset species communalities.

11.2.6 Interset communalities. The interset species communalities (Table 11.2, h^2_b) vary considerably, from those of *H. pubescens* (.80), *T. drucei* (.69) and *P. bertolonii* (.46) to that of *R. squarrosus* (.09). Thus the explanatory power of the model is unevenly distributed among the species. Species at either end of the sequence specified at the outset (Section 11.1) are those which on *a priori* grounds can be considered to be most susceptable to variation in the soil properties examined. From the interset communalities it is clear that variation in the abundance of such

species is at least moderately-well accounted for. Species towards
the center of the sequence, on the other hand, are those whose
representation can, from general biological considerations, be shown
to be least related to the explanatory variables used. This point
is borne out by briefly considering $R.$ $squarrosus,$ whose variance
is scarcely accounted for at all by the model (h_b^2 = .09). $R.$

$squarrosus$ is unique among the species studied in that it is a
bryophyte. As such, it is physiologically and morphologically quite
distinct from the remaining species and in its ecology may be expected
to be no less different from them also. On the basis of these considerations it seems likely that the representation of $R.$ $squarrosus$
will be influenced more by variables such as humidity and competition,
which were not employed, than by any of the soil factors actually
used. It therefore comes as no surprise to find that the variation
of this species remains virtually unaccounted for. The same general
argument can be extended to other species whose variation is
similarly poorly explained, notably $H.$ $pilosella.$

While on the whole the model may be held to possess a reasonable
degree of explanatory power, its lack of fit may itself be turned
to good advantage. The lack of fit, as expressed by the interest
communalities, directs attention to the most promising additional
variables which might be used in efforts to improve explanatory
power.

11.2.7 The canonical variates. In order to conserve space the
scores of samples on the canonical variates are not reported here.
However, Figure 11.1 displays the sample mapped into one of the
subspaces associated with the analysis, namely that of the first two
canonical variates U_1 and U_2 of the species' domain. Notice
that strictly speaking U_1 and U_2 are not orthogonal, although
for convenience in the Figure the axes representing them are drawn
at right angles, as is customary. The normalized vectors of weights,
$\underset{\sim}{a}_1$ and $\underset{\sim}{a}_2$, show that U_1 and U_2 intersect at an angle of 72°
in species-space; nevertheless the projections of the samples onto
U_1 and U_2 are correlated. Reference to the original data
showed that U_1 distinguishes between sample-sites in which $H.$
pubescens and $T.$ *pratense* attain their maximum representation (e.g.
sites 5, 13 and 35) and sites in which the same is true of $T.$ *drucei*
and $B.$ *media* (sites 1, 3, and 43). Having regard for the differences
in response of these species in relation to soil depth, it is clear
that U_1 represents a contrast between sites with respect to the
depth of their soils. Similarly, U_2 contrasts sites in which $P.$
sanguisorba and $H.$ *pilosella* are especially abundant (sites 4, 8
and 30) from sites in which the maximum representation of $P.$ *bertolonii*

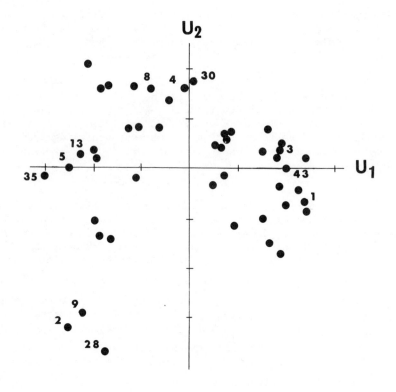

FIG. 11.1: Limestone grassland, Anglesey. Canonical analysis of
relationships between six soil properties and eight selected species.
N = 45 stands mapped into the subspace of the first two canonical
variates U₁ and U₂ of the species domain. The canonical variates
have beeb normalized to r²ₖ. (Direction of   reflected).

occurs (sites 2, 9 and 28). U_2 evidently corresponds therefore to differences between sites with respect to soil phosphorus status. Figure 11.1 clearly shows that U_2 affects essentially only the deeper-soil sites (sites with negative scores on U_1); sites with shallow soils by comparison are largely invariant with respect to soil P.

In conclusion we mention that the corresponding soil canonical variates V_1 and V_2 are open to equivalent interpretations in terms of soil properties; in addition V_3 very sharply distinguishes between sites of conspicuously high or low potassium status. Examination of the sample after mapping into other spaces associated with the analysis, together with maps of the spatial distribution of selected canonical variates also proved informative in emphasizing particular aspects of the relationships between the two sets of variables.

11.3 Conclusions. The canonical variates corresponding to the first two canonical correlations substantially retrieve the structure given to the data by the selection of species for analysis. Specifically: (a) the species correlations with V_1 identify the anticipated differences between species with respect to variation in soil depth and depth × phosphate; (b) the species correlations with V_2 identify the anticipated responses of species to variation in soil P; and (c) the explanatory power of V_1 and V_2 in the species' domain is consistent with expectations based on prior knowledge of the relative importance of soil depth (and d × P), and soil P, respectively, on the behavior of the species examined.

The variance of those species considered to be most influenced by the soil factors studied is adequately accounted for by the results obtained; conversely, the variance of species considered to be largely independent of the soil variables used remains unaccounted for. The analysis also draws attention to a point of some ecological interest concerning *H. pilosella*, previously overlooked, and confirms the anticipated negligible ecological influence of soil K. Thus there are grounds for concluding that a canonical model (r = 2) effectively identifies and communicates the salient ecological features of the data analysed. Furthermore, it seems likely that the explanatory power of the model could fairly easily be increased by the addition or deletion of appropriate variables.

12. EXPERIMENTAL ANALYSES: APPRAISAL AND CONCLUSIONS

Canonical analysis identified the expected relationships in both experiments. In each case analysis resulted in some 40 percent of the variance of the variable-set of interest being accounted for. Although the percentage of explained variance is not high, it is well to remember that in comparable procedures such as multiple regression, where explained variances (R^2) of perhaps 70 percent are not uncommon, the ratio of predictor variables to the criterion variable is often of the order of 5:1. In both experiments the ratio of predictor to criterion variables was nearer 1:1. In such circumstances it is perhaps only to be expected that in general the percentage of predictable variance will fall. The results obtained nevertheless demonstrate that canonical analysis is well able to efficiently recover relationships of ecological interest between two sets of variables. In one case analysis also led to the recognition of a previously overlooked ecological feature of some interest. Moreover, each analysis completely extracted the known ecological content of the data analyzed, notwithstanding the rather modest level of explained variance achieved. In ecological terms, therefore, I think that there can be little doubt as to the success of the analyses and hence of the effectiveness of the method in these applications. The apparent discrepancy between the assessment of the analyses ecologically and in terms of explained variance suggests that the variance *not* accounted for may be wholly attributable to the high unique variances which characterize ecological systems generally (Goodall, 1970, p. 102; Williams, 1976, p. 67).

A feature of both analyses which seems likely to have made a substantial contribution to their success is the comparative lack of vegetational and environmental diversity of the study area common to both investigations. The lack of diversity reflects the small (5 ha) geographical extent of the study area and its overall physical homogeneity. Consequently, the data structures analyzed are thought likely to have been substantially linear and continuous. While the structure of a body of field data in practice can seldom be known *a priori* with any degree of certainty, it is often possible to arrive at certain general statements concerning it. In both investigations, complications arising from major discontinuities can be discounted on the basis of field evidence; however the presence of a partial environmental and vegetational discontinuity in the field affecting two sites suggests that the data structures are unlikely to be uniformly continuous (Gittins, 1965). There is also evidence which suggests that some degree of nonlinearity may be present. The evidence is provided by the proportion of zero entries in the data analyzed (see Swan, 1970; Austin and Noy-Meir, 1971; Austin, 1972). Two data structures arise in canonical analysis generated by correlations within each of the two variable-sets involved; in both experiments the proportion of zero values in one of the two

sub-matrices corresponding to the variable-sets is rather high.
These proportions for the 'species' sub-matrix of Experiments 1
and 2 are (54/135) = .40 and (164/360) = .46, respectively and are
certainly suggestive of some degree of nonlinearity in the data.
Thus, while the data structures involved are thought to have been
broadly linear and continuous, there is evidence to suggest that
in each case they were neither strictly linear nor continuous.
The analysis however, appears to have been robust in the presence
of presumed departures of the magnitude indicated. In connection
with nonlinearity, it may be that the use of cross-product terms
contributed to the efficiency of the analyses to some extent.

On the strength of the results obtained it seems reasonable
to conclude that, at least in studies where the diversity of the
material of interest is small, the use of canonical analysis in
ecology may be rewarding. At any rate, the results would seem
to justify efforts directed towards a more penetrating ecological
assessment of the method. The experimental analyses described
above, based as they are on selected data sets of limited diversity,
can hardly be considered to be representative of real ecological
investigations. Accordingly, it seemed appropriate to proceed to
examine the performance of canonical analysis in the context of a
varied range of ecological problems of the kind encountered in
practice. Five such applications are considered below (Sections
13-17). In performing these analyses, where the purpose was
exploratory data analysis, it was necessary to assume only that
the data structures involved did not depart too drastically from
linear point-clusters of uniform density; where statistical tests
were envisaged, the assumptions required were more stringent –
namely, that the samples used were independently drawn from some
multivariate normal universe.

13. SOIL-VEGETATION RELATIONSHIPS IN A LOWLAND TROPICAL RAIN FOREST

13.1 Introduction. The aim of the following application of canonical
analysis was to clarify relationships between soils and vegetation
in a 1 km^2 area of rain forest in the Bartica Triangle region of
Guyana, South America. The analysis was based on field observations
kindly made available by Dr. J. Ogden of the Australian National
University, Canberra. The field data consisted of estimates of
the composition of 25   100 × 100 m stands of forest vegetation
together with determinations of selected soil characteristics of
the sites. Stands were arranged on a square lattice at intervals
of 200 m and, in describing their vegetation, attention was confined
to the predominant, woody component of the vegetation. One hundred
and seventy species of woody plant were encountered in all, the
representation of these species being estimated and expressed in

terms of basal area. Ten soil samples were collected in each
site and pooled to form a single, composite sample on which determinations of twenty physical and chemical properties were later made
in the laboratory. A comprehensive account of the vegetation and
soil sampling procedures used has been given by Ogden (1966).

Before embarking on the canonical analysis it was necessary
to re-express the data in a form better suited to analysis of this
kind. The most pressing need in view of the small size of the
sample (N = 25) and large number of variables (n = 190) was to
reduce the number of variables in some way consistent with the
overall objective of the study. It was also necessary to arrive
at some more comprehensive description of site *vegetation* than
that provided by a statement simply of species representation,
given the stated aim of the analysis.

A reduction in species number was achieved by ranking species
in terms of their independent contributions to the total species
sum of squares, using a modified form of the program RANK developed
by Orlóci (1978, p. 26) for this purpose. After examining the
ordering of species obtained in the light of field knowledge of
the species (Ogden, 1977) and in relation to a stress index calculated from the data (Orlóci, 1978, p. 34), all but the top-ranking
thirty-three species were eliminated. In a separate analysis the
soil variables were similarly reduced from twenty to nine.

The overall vegetational composition of the stands was next
determined in relation to the thirty-three retained species. Non-
centered principal components analysis of site-normalized data
followed by varimax rotation of the components was used for this
purpose. The choice of method here was guided by evidence that
the data structure determined by the field observations was not
uniformly continuous (Gittins and Ogden, 1977) and by the declared
interest in soil-vegetation relationships (see Noy-Meir, 1973b;
Noy-Meir *et al.*, 1975). Six non-centered varimax components were
found to account for 97 percent of the total stand sum of squares,
each component corresponding to a recognizable and reasonably
distinct forest community. These components were considered to
adequatley characterize the vegetation and accordingly were used as
one set of variables in the canonical analysis. The forest
community defined by each component was named after a representative species; the communities identified in this way, together with
an indication of the 'importance' of each, as expressed by the
percentage of the total sum of squares associated with the
defining varimax component, were: 'Greenheart' (*Ocotea rodiaei*
(Schomb.) Mez.)(44%); 'Wallabba' (*Eperua falcata* Aubl.) (22%);
'Morabukea' (*Mora gonggrijpii* (Kleinh.) Sandw.) (15%); *Mora (Mora
excelsa* Bth.) (6%); *Pentaclethra (Pentaclethra macroloba* (Willd.)
Kze.) (8%); and *Eschweilera (Eschweilera sagotiana* Miers.) (3%).

The nine retained soil variables comprised the second set of variables in the canonical analysis. These variables were: pH, base saturation (%), phosphorus (ppm P), sand (% particles 0.05 - 0.20 mm), silt (% particles 0.002 - 0.05 mm), active acidity (KCl acid), potassium (m-equiv./K 100 g air dry soil), potential cation exchange capacity (CeCg) and moisture retaining capacity (% H_2O in air dry soil). Details of the soil analytic procedures used in the determinations have been described by Ogden (1966).

The canonical analysis thus came to be based on a systematic sample of N = 25 stands of rain forest vegetation, p = 6 generalized vegetation variables and ·q = 9 soil properties. The ecological objective was to achieve a preliminary understanding of how forest differentiation in the area surveyed might be related to the soil properties examined. The results of the analysis are summarized in Tables 13.1 and 13.2.

The field observations on which the analysis was based were collected during the course of an expedition which was in the field for a period of six weeks. The expedition's vegetation research program had several objectives and the design of the vegetation survey was influenced by a need to satisfy the sometimes conflicting requirements of different objectives. For this reason the data obtained are not ideally suited to canonical analysis, even following the modifications described above. In this connection the small size of the sample is viewed with particular concern, while the lack of independence among samples might also be queried. Provided these points are borne in mind in interpreting the results, however, the analysis may nevertheless prove useful in illustrating the kind of insight which canonical analysis can yield in conjunction with exploratory ecological surveys of the kind involved.

13.2 Results.

13.2.1 Dimensionality. The canonical correlation coefficients and some related indices appear in Table 13.1. The quantities $1 - r_k^2$ express the proportion of variance which remains unaccounted for by a particular root; these entries enable the effectiveness of the analysis to be assessed from an alternative viewpoint to that provided by the canonical roots r_k^2.

TABLE 13.1: Lowland tropical rain forest, Guyana. Canonical analysis. Relationships between nine soil properties and six forest communities. Canonical correlation coefficients r_k and related quantities.

k	r_k	r_k^2	$1 - r_k^2$
1	.926	.857	.143
2	.904	.817	.183
3	.745	.556	.444
4	.653	.427	.573
5	.468	.220	.780
6	.265	.070	.930

The magnitude of the canonical correlation coefficients suggest the existence of several strong linear relationships between the two measurement domains. In the absence of independent samples and hence of significance tests, it seems reasonable from the size of the first canonical correlation, $r_1 = .93$, in particular, to tacitly accept that the measurement domains are indeed related, notwithstanding the inflationary effect of the comparatively large number of variables in relation to sample size (see Section 3.1). In order to estimate the dimensionality of the relationship, reliance was placed on the total redundancy of the vegetation domain and on the contributions of the individual canonical variates V_k of the soil domain to this (Table 13.2, upper-right). Notice that Table 13.2 reports the structure correlations and related indices for the canonical variates corresponding to the roots r_1^2, r_2^2, r_3^2 and r_5^2; the redundancies associated with the remaining soil canonical variates are $V_{x|v_4}^2 = .030$ and $V_{x|v_6}^2 = .011$, respectively. The contributions of V_4 and V_6 to the explained variance of the vegetation domain are evidently very small. In interpreting the results, it is clear that we may safely neglect these components. This finding, together with a preliminary ecological assessment of the interset structure correlations corresponding to all six canonical variates V_k, $k = 1, \cdots, 6$, were instrumental in the selection of the rank 4 model reported in Table 13.2 for the analysis.

To interpret the retained canonical variates we turn to the structure correlations (Table 13.2). The table reports for completeness the structure correlations of both sets of canonical variates, U_k and V_k, although only those of the V_k are of immediate interest.

13.2.2 Intraset correlations. We see from the correlations of the soil variables with the V_k (Table 13.2, lower-left) that V_1 is characterized principally by phosphorus (.89), although silt (.43) and potential cation exchange capacity (.38) are to some extent also related in a direct sense to this canonical variate; we note also the inverse correlation between sand (-.31) and V_1. The second canonical variate, V_2, is essentially a contrast of percentage sand (.77) against cation exchange capacity (-.77) and moisture retaining capacity (-.51). In a similar way we see that for simplicity V_3 may roughly be characterized by pH (-.45), while V_5 can be equated approximately with base saturation (-.73) and active acidity (.68).

Of the four canonical variates V_2 is the strongest, absorbing 22 percent of the total variance of the soil domain ($V_2^2 = .223$) followed by V_1 and V_5 which each account for some 17 percent; V_3 is a much weaker composite accounting for only 5 percent of the variance of the soil variables. Collectively, the four canonical variates account for 61 percent of the total variance of the soil domain. The intraset variable communalities range between those of potassium (.35) and phosphorus (.95); on the whole they indicate that the variance of a majority of the soil variables is at least moderately-well explained by the canonical variates retained.

13.2.3 Interset correlations. From the correlations of the forest communities with V_1 it can be seen (Table 13.2, upper-right) that for the most part of the communities are all rather closely related to V_1; the signs of the correlations, however, indicate that the communities respond to variation in V_1 in different ways. *Mora* (.74), *Pentaclethra* (.44), and *Eschweilera* (.39) forests all respond in a direct sense to variation in V_1, while Wallaba (-.48) and Greenheart (-.20) are inversely related to V_1. Only Morabukea (.16) appears to be substantially independent of variation in this canonical variate. Recalling that V_1 may roughly be equated with soil P and silt, it is clear that *Mora* and to some extent also *Pentaclethra* and *Eschweilera* forests tend to be associated with silty soils of relatively high P status, while in contrast Wallaba and Greenheart tend to be found on comparatively sandy soil deficient in P.

The forest communities are also sharply differentiated by their correlations with V_2. Wallaba (.72) and Greenheart (-.63) clearly respond in opposite senses to variation in V_2, while, with the exception of Morabukea (-.37) forest, the remaining communities are largely invariant with respect to V_2. Thus V_2 represents essentially a contrast between the forest communities of P deficient sites. Recalling the earlier interpretation of V_2 (Section 13.2.2) and observing the signs of the structure correlations, we see that the occurrence of Wallaba is associated with sandy, infertile soils of low water retaining capacity, while Greenheart is inversely related to soils of this kind, i.e. Greenheart tends to be found on comparatively base rich, clayey soils of relatively high water retaining capacity. Furthermore, we see that Morabukea forest resembles Greenheart in its association with such soils, at least to some extent.

The third canonical variate, V_3, is not well-defined in the vegetation domain, its strongest correlation being +0.5 with Morabukea. As V_3 can crudely be equated with pH and potential cation exchange capacity, it appears that Morabukea forest shows some tendency to be found on acid soils of low potential cation exchange capacity. In addition, Greenheart (-.30) and *Mora* (-.27) forests are weakly associated with soils of the same kind, though in an inverse sense.

It is clear from the magnitude of the interset structure correlations of V_5 that this canonical variate is an appreciably weaker explanatory construct in the vegetation domain ($V^2_{x|v_5}$ = .039) than in the soil domain on which it is defined ($V^2_{5(y)}$ = .171). The strongest interset correlations of V_5 are with *Pentaclethra* (.31) and *Eschweilera* (.22) forests. Bearing in mind the earlier interpretation of V_5, these correlations suggest that the differentiation of these forest communities may be influenced, at least in some degree, by the occurrence of soils of low base saturation and high active acidity.

In conclusion, it is worth noticing that in seeking to comprehend the relationships between soils and vegetation it is instructive to examine the structure correlations of Table 13.2 from two different but related points of view - namely, by columns and by rows.

(a)   Columnwise the structure correlations express the apparent 'effect' of the soil factor-complex corresponding to a particular canonical variate on the vegetation as a whole; while

(b) row-wise the structure correlations express the apparent
     'response' of a particular community to the soil factor-
     complexes corresponding to all the canonical variates.
     Thus the *profile* of a community over the canonical
     variates is helpful in identifying the edaphic controls
     of the community.

We have for example seen that V_1 distinguishes between communities
of P rich and P depleted soils, respectively, while V_2 operates
within the vegetation of P deficient soils to differentiate
communities in relation to the particle size composition *etc.* of
their associated soils. This process of specifying the apparent
effect of the different soil factor-complexes is readily extended
to the remaining soil canonical variates, V_k. In a similar way,
from the profile of Wallaba forest, for example, it is clear that
the edaphic correlates of this community are P deficient, sandy
soils of low potential cation exchange and water retaining
capacities.

    Thus the structure correlations can lead to recognition both
of the apparent effects of different soil factor-complexes on
the vegetation and of the edaphic correlates or perhaps controls of
particular communities.

13.2.4 Redundancy. From the redundancy in the vegetation domain
attributable to the canonical variates V_k of the soil domain
we see (Table 13.2, upper-right) that the explanatory power of the
V_k is concentrated largely in V_1 and V_2. Together these
canonical variates account for 38 percent of the total variation
in the forest communities. Thus, variation in the soil factor-
complexes corresponding to V_1 and V_2, which for simplicity we
have characterized respectively by P and by particle size,
appears to have pronounced effects on forest composition. On the
other hand, V_3 and V_5 are weaker explanatory constructs; they
account respectively for 8 percent and 4 percent of the total
variance of the vegetation domain. It appears, therefore, that
variation in the factor-complexes represented by V_3 (roughly pH)
and V_5 (base saturation) is associated with comparatively small
changes in forest composition. It is interesting to observe in
this connection, however, that small differences in soil pH appear
to have a proportionately greater influence on the development and
composition of the vegetation than appreciably larger differences
in the factor-complex characterized by base status; this is apparent
on comparing the variances associated with V_3 and V_5 in the soil
and vegetation domains, respectively.

TABLE 13.2: *Lowland tropical rain forest, Guyana. Canonical analysis. Relationships between nine soil properties and six forest communities: correlations between the original variables and canonical variates.*

Canonical variate	U_1	U_2	U_3	U_5	h_w^2	V_1	V_2	V_3	V_5	h_b^2
Community										
Greenheart	-.218	-.697	-.398	.441	.886	-.203	-.630	-.297	.207	.569
Wallaba	-.516	.798	.023	-.169	.932	-.481	.721	.017	-.079	.758
Morabukea	.168	-.412	.671	-.335	.760	.157	-.372	.500	-.157	.438
Mora	.798	.172	-.368	-.225	.852	.744	.152	-.274	-.105	.664
Pentaclethra	.470	-.065	-.090	.662	.671	.438	-.059	-.067	.310	.296
Eschweilera	.419	-.006	.379	.478	.548	.391	-.005	.283	.224	.283
Variance extracted	.229	.221	.149	.175	.775	.198	.181	.083	.039	.501
Redundancy	.197	.181	.083	.039	.500	.198	.181	.083	.039	.501

Canonical variate	V_1	V_2	V_3	V_5	h_w^2	U_1	U_2	U_3	U_5	h_b^2
Soil factor										
pH	.347	-.076	-.451	.420	.506	.323	-.069	-.336	.197	.261
Base saturation	-.092	.370	-.004	-.732	.681	-.086	.335	-.003	-.343	.237
Phosphorus	.888	.238	-.157	.285	.951	.828	.215	-.117	.134	.763
Sand	-.311	.769	-.202	-.001	.729	-.290	.695	-.151	-.000	.590
Silt	.427	-.249	.048	-.351	.370	.398	-.225	.036	-.164	.237
Active acidity	.321	-.212	-.175	.676	.636	.299	-.192	-.130	.317	.244
Potassium	.204	-.514	.001	-.208	.349	.190	-.465	.001	-.097	.262
Cation exchange	.378	-.766	-.351	.178	.885	.352	-.693	-.262	.083	.680
Moisture factor	.119	-.507	-.110	.304	.376	.111	-.458	-.082	.142	.249
Variance extracted	.167	.223	.049	.171	.609	.144	.182	.027	.038	.391
Redundancy	.143	.182	.027	.038	.390	.144	.182	.027	.038	.391

The V_k account collectively for 50 percent of the total variance of the vegetation domain. Thus the fit of the model may be considered to be reasonably satisfactory. Finally, from the interset communalities of the communities we note that the explanatory power of the model is rather evenly distributed among all six communities.

13.2.5 The canonical variates. Figure 13.1 displays the sample after its projection into three subspaces associated with the analysis.

Figures 13.1 (a) and (b) show the sample in the two-dimensional subspaces of the canonical variates corresponding to r_1^2 and r_2^2, respectively. The plots illustrate the linear correlation between the forest communities and the soil factor-complexes characterized respectively by (a) phosphorus (V_1) and (b) particle size (V_2). The differentiation of the vegetation apparently in response to variation in these factor-complexes is clearly evident; *Mora* forest in particular stands out as a very distinctive community while the trends in forest composition associated with changes in V_1 and V_2 are another notable feature of the plots.

Figure 13.1 (c) portrays the sample mapped into the space of the first three canonical variates, U_k (k = 1, $\cdots$,3), of the vegetation domain. The canonical variates in this instance have been normalized to be proportional to the redundancy (explained variance) associated with each V_k of the soil measures in the *vegetation* domain. The stereogram enables the differentiation of the forest communities in apparent response to variation in all three major soil factor-complexes to be viewed simultaneously. Specifically, the Figure shows the differentiation of:

(a) *Mora, Pentaclethra* and *Eschweilera* forests in response to variation in the factor-complex corresponding to V_1;

(b) Wallaba, Greenheart, and Morabukea forests within the vegetation of P impoverished sites, in response to variation in V_2; and of

(c) Morabukea forest, and to some extent also, Greenheart, *Eschweilera* and *Mora*, in response to variation in V_3.

Moreover, the sterogram enables the relative importance of the separate components of the overall covariation between soils and vegetation to be appreciated, as well as the mutual affinities of the forest communities themselves with respect to the soil properties examined.

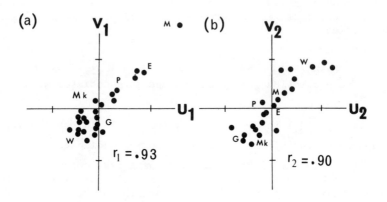

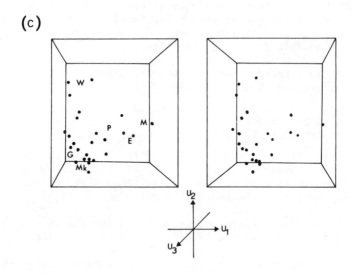

FIG. 13.1: Lowland tropical rain forest, Guyana. Canonical analysis of relationships between nine soil properties and six forest communities. N = 25 stands mapped into subspaces defined by canonical variates (a) U_1 and V_1; (b) U_2 and V_2; (c) U_k, k = 1, $\cdots$, 3 of the vegetation domain. In (a) and (b) the canonical variates normalized to unity. Key as in Fig. 14.1.

In the following Subsection we briefly integrate and assess the ecological significance of the results described.

13.2.6 Ecological significance of the results. The soil-vegetation relationships established can themselves be related to more fundamental processes. The area surveyed consisted physiographically of a dissected plateau with a relief some 25 m in amplitude. The topographic gradient from plateau top to creek bottom has been recognized by Ogden (1966) as a major determinant of the soil-vegetation relationships of the area. Ogden showed that with the topographic gradient was associated a catenary soil series, which in turn appeared to influence the composition of the vegetation at different positions on the gradient. The canonical variates of the rank r = 4 model can all be related to features of the topographic and catenary soil/vegetation system. The interset structure correlations of V_1 reflect the soil/vegetation changes associated with the global topographic gradient from plateau surface to valley floor, while the structure correlations of the remaining V_k (k > 1) correspond to local contrasts within the overall system at different locations on the gradient in topography.

It is not claimed on the basis of the results obtained that analysis has succeeded in identifying soil properties which are themselves necessarily among the immediate ecological or physiological controls of particular communities. We need only consider that, in the initial selection of soil variables or in reducing their number prior to analysis, we may well have overlooked or eliminated variables of greater physiological significance than any of those actually used to appreciate this point. What the analysis has achieved is a reasonably concise specification of the covariance structure of the soil properties and plant communities examined. Moreover, the interpretation of this structure proved useful in directing attention to those soil-vegetation relationships which seem most likely to repay attention in subsequent efforts which may be made to explain the occurrence and distribution of the forest communities of the area studied.

13.3 Conclusions. The salient features of ecological interest may be summarized as follows:

(a) the soil factor-complex represented by phosphorus is associated with major changes in forest composition;

(b) the factor-complex represented by particle size is associated with major changes in the composition of the vegetation of phosphorus depleted sites;

(c) soil pH, base saturation, and active acidity appear
to be associated with the occurrence of the smaller
forest communities.

The analysis also enabled some of the edaphic influences or controls
of particular communities to be recognized, at least in a provisional
way.

Before attempting an appraisal of the analysis in ecological
terms it is necessary to consider possible effects of the size and
design of the sample on the results obtained (cf. Section 13.1).

Sample size. The small size of the sample is a disquieting feature
of the analysis; its consequences may have been to have:

(a) inflated the numerical values of the canonical and
structure correlation coefficients and the interpretive indices derived from them, bearing in mind the
relatively large number of variables used;

(b) produced results which are highly dependent on the sample
in hand.

The first point should warn us against unguarded acceptance of the
strength of the relationships established, while the second has a
bearing on their *validity* beyond the sample examined.

The ratio of variables to samples (15/25 = .60) in the present
analysis strongly suggests that the numberical values of the interpretive indices of interest will have been boosted to an undetermined
extent by a factor from this source. The theoretical understanding
necessary to place the interpretation of analyses of this kind on
a firm footing is at present lacking. Until the theory can be
developed, in cases where an unavoidable restriction is placed on
sample size, interpretation needs to be guided by an awareness that
the use of a larger sample than that actually employed would almost
certainly result in smaller values of the indices of interest than
those in fact obtained.

The question of the sample-dependency of the results, on the
other hand, causes less concern. The 25 stands studied comprise 25
percent of the area of interest and, for the purposes of vegetation
survey, would seem to constitute an adequate sample; any replicate
sample of the same size might reasonably be expected to substantially
reproduce the results described above. Sample-dependency is therefore
considered unlikely to have influenced the results obtained to any
appreciable extent.

Thus, the principal consequence of sample size which needs to
be taken into account here is its effect in exaggerating the strength
of the soil-vegetation relationships established; for this reason we
regard this aspect of the results with some skepticism.

Sampling design. The lack of independence among samples is not regarded as inappropriate to the aims of the investigation. Indeed, Williams (1971; 1976, p. 131) has argued convincingly that systematic sampling designs are the preferred designs in most ecological survey work. As a consequence of the systematic sampling, the results obtained are expected to reflect chiefly the *overall* trends and relationships which were the focus of interest in the present survey.

We can now take up the question of the success of the analysis in ecological terms. The results obtained are broadly consistent with those of analyses based on the same data by Ogden (1966) and with the results of independent investigations by Davis and Richards (1934) and Schultz (1960) in nearby areas. This similarity gives confidence in the ability of canonical analysis to identify relationships of ecological significance. The results obtained, however, do differ from those of previous studies in specifying the covariance structure of interest in greater clarity and detail than has hitherto been achieved. Further work in the field would be necessary to substantiate the additional information provided. Nevertheless, on the basis of the evidence available, Ogden (1977) has expressed the view that the analysis represents a useful contribution to the description and comprehension of soil-vegetation relationships in the area surveyed.

14. DYNAMIC STATUS OF A LOWLAND TROPICAL RAIN FOREST

14.1 Introduction. The stability of vegetation over time is frequently of interest to plant ecologists. Stability, however, is a property of vegetation which is not readily studied, because, as Austin (1977) has pointed out, its definitive investigation calls for sequential observations over time. Observations of this kind are not easily acquired, especially in the case of forest vegetation where the time-scales involved are normally prohibitively long. Field data collected by Dr. J. Ogden during the course of the expedition to the Bartica Triangle region of Guyana referred to above, and kindly placed at my disposal, however, suggested an alternative means of obtaining insight of a preliminary kind into the dynamic status of vegetation. Observations on forest composition in terms of trees and seedlings had been made and it seemed that an examination of relationships between these separate components might be used to provide at least an initial assessment of the stability or otherwise of the vegetation. Canonical analysis would be an appropriate means of analyzing the data for this purpose. Any sizeable discrepancy which might be revealed between the representation of trees and seedlings would suggest that successional changes were indeed taking place.

Estimates of forest composition in terms of the density of
woody species (a) greater than 5 cm in diameter at breast height
('trees') and (b) less than 30 cm in height ('seedlings') had been
made for 25 100 × 100 m systematically arranged samples of forest
vegetation. The size and number of individual trees and seedlings
encountered differed appreciably, and, for this reason, different
procedures had been used in their estimation. Estimates of tree
density were based on observations obtained by the point-centered
quarter method; twenty-five points were located on a lattice within
each stand and the distances from the points to the three nearest
trees in the four 'spatial quadrants' about each point measured and
the trees identified. From the 300 such records in each stand an
estimate of the relative density per 100 m^2 of each species was made.
Seedling density per stand, on the other hand, was estimated by
the complete enumeration and identification of all seedlings in
ten 5 m^2 quadrats systematically arranged in each stand. Though
not strictly comparable, the tree and seedling estimates are
believed to be satisfactory for the purpose in mind. One hundred
and seventy species of woody plant were encountered in all. A
detailed account of the sampling procedures used has been given
by Ogden (1966).

In order to express the data in a form suitable for canonical
analysis, the field observations were modified by first reducing the
number of species to thirty-three and then specifying site vegetation by reference to a small number of composite variables
derived from the thirty-three retained species. The methods
described in Section 13.1 above were used for this purpose. In
separate analyses, six non-centered varimax components were found
to account for 93 percent of the sum of squares of the site-normalized
tree data and for 89 percent of the corresponding quantity for
seedlings. Thus a total of twelve components were found to adequately summarize the separate contributions of trees and seedlings
to the total vegetation of the sites. The two sets of components
were accordingly used as the variables in the canonical analysis.
The communities recognized, together with an indication of the
importance of each in terms of the percentage of the total sum of
squares association with their defining varimax components, were:

Community	% of total SS associated with defining component	
	Tree density	Seedling density
Greenheart	23	23
Wallaba	20	17
Pentaclethra	13	9
Morabukea	13	14
Mora	} 17	6
Eschweilera		20
Jessenia	7	-
Total	93	89

It is clear that by and large the same tree and seedling communities were recognized in each analysis and that the communities were of comparable size in terms of plants of both kinds.

The canonical analysis therefore came to be based on a systematic sample of $N = 25$ stands of rain forest vegetation, each characterized by $p = 6$ and $q = 6$ generalized variables which expressed the contribution of seedling and tree communities, respectively, to the total vegetation. The ecological aim of the analysis was to provide a preliminary assessment of the dynamic status of the forest. More specifically, the objective was to discover whether the field observations were consistent with the view that the forest was likely to be replaced by vegetation of substantially the same composition, as the individual trees composing it aged and died and were progressively replaced by the present generation of seedlings.

As it is reasonable in ecological terms to regard the existing tree vegetation as at least to some extent determining the composition of the seedling vegetation associated with it, rather than the converse, we may properly regard the analysis as *directed* in character.

14.2 Results.

14.2.1 Dimensionality. The canonical correlation coefficients (Table 14.1) are decidedly large, four of them exceeding 0.85 in magnitude. Thus the correlations suggest that the domains of interest may well be linked by several linear relationships, even when allowance is made for the effect of the large number of variables (n = 12) relative to sample size (N = 25) on their magnitude. From the size of the first canonical correlation,

Table 14.1: Lowland tropical rain forest, Guyana. Canonical analysis of tree-seedling relationships. Canonical correlation coefficients r_k and related indices.

k	r_k	r_k^2	$1 - r_k^2$
1	.988	.977	.023
2	.956	.914	.086
3	.946	.894	.106
4	.871	.760	.240
5	.651	.424	.576
6	.056	.003	.997

$r_1 = .99$, in particular, it seems reasonable to accept tacitly that the measurement domains are in fact related. In proceeding to consider the dimensionality of this relationship, the total redundancy of the seedling domain and the contributions of the individual V_k of the tree domain to this index proved helpful. Table 14.2 (upper-right) contains the relevant quantities for the $r = 4$ rank model; the redundancies associated with those canonical variates of the tree domain not reported in Table 14.2 are $V_{x|v_5}^2 = .013$ and $V_{x|v_6}^2 = .000$, respectively. Clearly, V_5 and V_6 are likely to contribute little or nothing to our understanding of tree/seedling relationships. In view of this, and noting the sizeable explanatory power of the preceding V_k ($k = 1, \cdots, 4$), the dimensionality of the association may be taken to be four. Accordingly, in interpreting the results we shall confine our attention to the rank 4 solution reported in Table 14.2.

14.2.2 Intraset correlations.

The first canonical variate, V_1, of the tree domain (Table 14.2, lower-left) is dominated by a sharp contrast between Wallaba (-.94) and Morabukea (.75) forests. The remaining communities resemble Morabukea in having sizeable positive correlations with V_1 and hence also participate in this relationship to a greater or lesser extent. Thus V_1 expresses

(a) a sharp compositional distinction between Wallaba and the other communities, particularly Morabukea;

(b) some degree of compositional similarity among the non-Wallaba communities, at least in relation to that of Wallaba itself.

The second canonical variate, V_2, operates entirely within the non-Wallaba vegetation. There, V_2 differentiates between *Pentaclethra* (.74) and Greenheart (-.56) forest, and at the same time suggests some degree of compositional affinity between Greenheart *Mora/Eschweilera* (-.30) and *Jessenia* (-.24). V_3 and V_4 are open to interpretation in broadly the same terms. Thus compositional differences are suggested between Greenhart and Morabukea and between *Mora/Eschweilera* and Morabukea, while other communities appear to be related at least to the extent of the direction of their involvement with a particular variate.

The overwhelming impression is of a sharp contrast in composition between Wallaba and Morabukea forests in relation to the vegetation as a whole, while within the non-Wallaba vegetation the distinctiveness of Morabukea, *Pentaclethra* and Greenheart forests is striking. Thirty-six percent of the total variance associated with the tree communities is absorbed by V_1. This canonical variate is twice as strong as each of the remaining variates, which themselves are of roughly equal strength; evidently the distinction between Wallaba and the non-Wallaba communities constitutes a major compositional feature of the forest vegetation examined. Collectively, the V_k (k = 1,$\cdots$,4) account for 84 percent of the total variance of the tree domain and hence can be considered to effectively summarize this variation.

14.2.3 Interset correlations. We find from the interset correlations (Table 14.2, upper-right) that the first canonical variate, V_1, of the tree domain is characterized in the seedling domain above all by the contrast in sign between Wallaba (-.94), Morabukea (.63), and, to a lesser extent, *Eschweilera* (.41) forests; the correlations of the three remaining seedling communities with V_1 do not depart appreciably from zero. Thus, to a considerable degree, V_1 bears the same relationship to the seedling communities as to the tree communities on which it is defined. Glancing at the structure correlations of the seedling communities with the remaining V_k (k = 2,3,4) and comparing these with the intraset correlations of corresponding tree communities, reveals that covariation between respective seedling and tree communities and the V_k is generally of the same sign and magnitude. Indeed, the fidelity of the intraset and interset correlations of the V_k across domains in both direction and strength is a remarkable feature of the results. Interpretation is facilitated by noting that by column the interset correlations may be thought of as expressing the 'effect' of a particular V_k of the tree domain on

the seedling vegetation, while by row the correlations express the
'response' of a particular seedling community to the different
components of the tree vegetation represented by the V_k collectively
It will be recalled (Section 14.2.2) that columnwise the intraset
correlations of the V_k express compositional contrasts or affinities
among the tree communities. Bearing these points in mind, it is clear
that:

(a)   each V_k has substantially the same 'effect' on the
seedling vegetation as it does in relation to the tree
vegetation on which it is defined;

(b)   the 'responses' or profiles of corresponding tree and
seedling communities in the space of the V_k are substantially identical.

Eventually, the seedling communities are related among themselves in
much the same way as are the tree communities, while tree and seedling
counterparts of particular communities vary directly with one another.
It is also apparent, however, that the relationship between equivalent tree and seedling communities falls short of an exact one-to-
one correspondence.

14.2.4 Redundancy.   The redundancy in the seedling domain attributable to the canonical variates V_{k_2} of the tree domain falls sys-
tematically across the V_k from $V_{x|v_1}^2 = .248$ to $V_{x|v_4}^2 = .136$.
Nevertheless, each canonical variate contributes usefully to our
understanding of tree/seedling relationships. Collectively, the
V_k (k = 1,$\cdots$,4) account for 76 percent of the total variance of the
seedling communities, so that the fit of the rank 4 model is reasonably
good. The interset communalities of the seedling communities show
that a high proportion (>.80) of the variance of a majority of the
seedling communities is accounted for by the model; only the variance
of *Mora* (.33) and *Eschweilera* (.67) are, by comparison, poorly explained. It will be recalled that these communities have no exact
counterparts among the tree communities, where they were merged in
a single component.

14.2.5 The canonical variates.   Figure 14.1 displays the sample
mapped into subspaces associated with particular canonical variates
and summarizes selected aspects of the results.

Figure 14.1 (a) shows the sample after projection onto the first
pair of canonical variates, U_1 and V_1. The Figure provides a
view of the sample chosen to reveal the extent of the relationship

TABLE 14.2: Lowland tropical rain forest, Guyana. Canonical analysis of tree-seedling relationships. Correlations between the original variables and the canonical variates.

Canonical variate	U_1	U_2	U_3	U_4	h_w^2	V_1	V_2	V_3	V_4	h_b^2
Seedling community										
Greenheart	.038	-.542	.763	.286	.959	.038	-.519	.721	.249	.853
Wallaba	-.952	.047	-.256	.153	.997	-.941	.045	-.242	.133	.964
Pentaclethra	.198	.847	.326	-.205	.905	.196	.810	.308	-.179	.821
Morabukea	.639	-.179	-.624	.405	.994	.631	-.171	-.590	.353	.900
Mora	.026	.040	-.010	-.662	.441	.026	.038	-.009	-.577	.335
Eschweilera	.411	-.469	.250	-.567	.773	.406	-.449	.236	-.494	.666
Variance extracted	.254	.211	.201	.179	.845	.248	.193	.180	.136	.757
Redundancy	.248	.193	.180	.136	.757	.248	.193	.180	.136	.757

Canonical variate	V_1	V_2	V_3	V_4	h_w^2	U_1	U_2	U_3	U_4	h_b^2
Tree community										
Greenheart	.292	-.560	.730	.211	.976	.289	-.536	.690	.184	.881
Wallaba	-.936	.000	-.264	.206	.988	-.925	.000	-.250	.180	.951
Pentaclethra	.438	.742	.285	-.300	.914	.433	.710	.269	-.261	.832
Morabukea	.754	-.173	-.504	.372	.991	.745	-.165	-.477	.324	.915
Mora/Eschweilera	.548	-.297	.078	-.661	.832	.542	-.284	.074	-.576	.712
Jessenia	.410	-.241	.000	-.355	.352	.405	-.230	.000	-.309	.312
Variance extracted	.365	.173	.157	.146	.842	.357	.159	.141	.111	.767
Redundancy	.356	.159	.141	.111	.767	.357	.159	.141	.111	.767

between tree and seedling communities to maximum possible effect.
It is plain that stands of Wallaba and Morabukea contribute most
to this relationship, though in opposite senses. Provided that the
ecological significance of the canonical variates is borne in mind
(e.g. see the structure correlations of U_1 and V_1 in Table 14.2),
it is apparent that stands with high positive scores on V_1 are
characterized above all by Morabukea tree vegetation *and* that it is
precisely these stands whose seedling vegetation is also characterized chiefly by Morabukea; similarly, it can be seen that stands with
large negative scores on V_1 consist predominantly of Wallaba forest
and that such stands are precisely those whose seedling vegetation
is also composed largely of Wallaba. As a corollary, it is abundantly
clear that Morabukea and Wallaba forests, in particular, are most
unlikely to be in any way dynamically related.

The relationships between the remaining communities appear
from Figure 14.1 (a) to be closer. Before such an assessment can
be accepted as established, however, it is necessary to examine
the relationships involved in a wider perspective. Figure 14.1 (b)
shows the sample after projection onto the (V_1, V_2)-plane of the tree
space of the analysis and thus provides a somewhat more comprehensive
view. The canonical variates here although drawn at right angles
can in fact be shown to intersect at an angle of 70° in variable
space and have been scaled to reflect the explanatory power of each
in the *seedling* domain. The Figure shows the expected polarity
of stands of Wallaba and Morabukea forest in relation to V_1. In
addition attention is drawn to the relative polarity of *Pentaclethra*
and Greenheart stands with respect to V_2. The tendency of stands
to form clusters which correspond to particular communities is in
fact a striking feature of the plot. This tendency of stands to
from clusters which correspond to particular communities is in fact
a striking feature of the plot and extends also to V_3
and V_4 (not illustrated) in relation to which Greenheart and *Mora*
stands, respectively, are each sharply differentiated from Morabukea.
Thus, in general, stands corresponding to particular communities
exhibit some tendency towards being represented by rather widely
separated and internally coherent groups of points. The diffuseness
and implied heterogeneity of the small *Pentaclethra* community,
however, is notable exception to this remark.

In view of the strength of the correlations between U_1 and V_1
$(r_1 = .99)$ and between U_2 and V_2 $(r_2 = .96)$, it will be
appreciated that the relative positions of samples in the (U_1, U_2)-
plane of seedling space will be substantially the same as in the

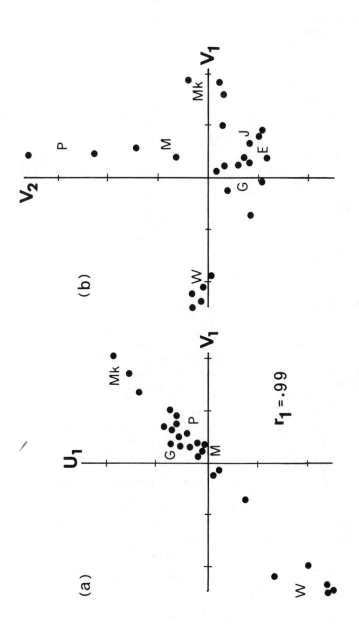

FIG. 14.1: *Lowland tropical rain forest, Guyana. Canonical analysis of tree-seedling relationships. N = 45 stands mapped into subspaces defined by canonical variates (a) U_1 and V_1, normalized to unity; (b) V_1 and V_2 of the tree domain normalized to redundancy generated by each in seedling domain. E: Eschweilera, G: Greenheart, J: Jessenia, M: Mora, Mk: Morabukea, P: Pentaclethra, W: Wallaba.*

(V_1, V_2)-plane of Figure 14.1 (b). A *simultaneous* plot of the sample
in relation to both pairs of canonical variates would therefore
emphasize the strong association between tree seedling counterparts
of particular communities.

14.2.6 Ecological significance of the results. It appears from
the results obtained that the vegetation examined consists of
communities which tend to be compositionally distinct but between
the tree and seedling counterparts of which there exists a high
degree of floristic and vegetational similarity. It was also found
that the correspondence between tree and seedling counterparts of
particular communities falls short of an exact one-to-one correspondence. The pronounced inverse compositional relationships between
Wallaba and Morabukea, *Pentaclethra* and Greenheart, Greenheart and
Morabukea, and between *Mora/Eschweilera* and Morabukea forests
suggest that successional changes between any of these communities
is highly improbable. On the other hand, communities which were
not polarized in this way, notably *Pentaclethra, Mora, Eschweilera,*
and *Jessenia* seem to be those among which successional relationships
might most profitably be sought. In general, however, it seems
probable that each forest community will be replaced by a community
of like composition as the individuals composing it die and are
progressively replaced by developing seedlings. In a word, the
dynamic status of the forest appears to be stable.

It is important to note that the results throw little if any
light on any medium- or long-term changes which may be taking place.
The limitation arises because the time interval represented by the
difference in size between the trees and seedlings examined is short
relative to any possible longer-term changes in the vegetation. As
such relatively long-term changes are generally those of greatest
interest to ecologists, the analysis hardly represents a significant
contribution to the study of forest dynamics. What the analysis
has done is provide a description of the covariance structure of
tree and seedling counterparts of the vegetation. The ecological
interpretation of this structure proved rewarding in pointing to
the likelihood of the stability of the vegetation in the short-
term while at the same time indicating communities between which
dynamic shifts in composition seem most plausible in the longer-term.

14.3 Conclusions. The principal finding of ecological interest is
of likely short-term stability of the vegetation.

Before turning to assess the success of the analysis in ecological
terms, it is necessary to refer again to the effect of the relationship
between the number of variables and samples on the results obtained.
The ratio of variables to samples is decidedly high (12/25 = .48),

and almost surely will have inflated the canonical correlations
and interpretative indices derived from them. Accordingly, it
would be prudent to regard the relationships described as being in
fact weaker than the quantities reported in Tables 14.1 and
14.2 might at first sight suggest.

In appraising the analysis ecologically it is helpful to
briefly compare the results provided by canonical analysis with
those obtained by other workers. In a study of tree/seedling
relationships in the same area using a different approach, Ogden
(1966) was led to an identical conclusion concerning the similarity
of the tree and seedling components of the vegetation and the short-
term stability which this implies. Moreover, the distinctive
vegetational composition of Wallaba and Morabukea forests, which
was one of the most striking features suggested by the results of
the canonical analysis, has been commented on by several previous
workers (Davis and Richards, 1933; Fanshawa, 1952; Ogden, 1966).
These similarities sustain the view expressed above (Section 13.3)
that canonical analysis applied to 'real' data is well able to
identify relationships of ecological significance. Furthermore,
the results of the canonical analysis go well beyond those of previous studies in the detailed picture of tree/seedling relationships
which they provide. Further field work would be necessary to
establish the worth of these additional insights. Ogden (1977),
however, considered that in the light of the available information th
canonical analysis has merit in its considerable clarification of
tree/seedling relationships in the vegetation examined.

In a wider context the analysis is instructive in drawing
attention to the fundamental problem of obtaining satisfactory
field observations for the direct study of forest succession. The
problem arises because of the very lengthy time-scales necessarily
involved. The appropriateness of simulation studies in this field
is therefore plain and deserves greater emphasis. Yet, from the
results described above, it seems possible that canonical analysis
may prove to be a useful addition to the *dynamic* procedures
reviewed by Austin (1977) for the study of time-dependent changes
in vegetation, in the case of short-lived communities, and an
alternative to Austin's *static* methods in the case of forest
vegetation.

15.   THE STRUCTURE OF GRASSLAND VEGETATION
IN ANGLESEY, NORTH WALES

15.1 Introduction. Many ecological endeavors lead naturally to
comparisons between plant communities with respect to their overall
composition or to comparisons between plant or animal species in
terms of their responses to particular experimental treatments. The
objective of *comparative studies* of this kind is to obtain, through

a description of community or species' differences, insight into
the processes responsible for the differences and an indication of
the avenues along which more intensive study might be rewarding.
In this Section we illustrate the use of canonical analysis in
connection with a comparative study of this sort. Previous work
on the composition and structure of grassland vegetation overlying
a small (5 ha) area of Carboniferous Limestone in Anglesey, North
Wales, showed the vegetation to be composed of three comparatively
distinct communities (Gittins, 1965). The communities recognized
were described as limestone, neutral, and eutrophic grassland.
With the aim of obtaining further insight into the structure of
the vegetation a clearer understanding of interrelationships among
the communities was sought. It seemed to me that examination of
the sample data in a geometric space chosen to emphasize both the
distinctiveness of the communities and the mutual relationships
between them would be appropriate for this purpose. Such a mapping
could be achieved by a canonical analysis in which one set of
variables consisted of estimates of species representation in the
stands surveyed, while the variables of the second set consisted
of dummy variables which specified the community affiliation of
stands.

The original vegetation survey was based on a random sample
of $N = 45$ 10 × 10 m stands. Thirty-three species were encountered
during the survey and stand composition was expressed in terms of
estimates of the abundance of these species. Preliminary analysis
which included a transformation to principal components showed that
a majority of the stands could be referred unequivocally to one or
other of the three communities mentioned above; six stands, however,
which were transitional between limestone and neutral grassland
could not readily be assigned to either community. Before the
canonical analysis could be undertaken, it was necessary to allocate
the transitional stands in some acceptable way among the communities.
For this purpose a decision rule based on a linear discriminant
function calculated between limestone and neutral grassland was
established. Application of the rule led to four of the transitional
stands being assigned to limestone grassland and two of them to
neutral grassland. In this way the total sample came to be
partitioned into $k = 3$ mutually exclusive and exhaustive groups
which corresponded to limestone, neutral, and eutrophic grassland.
Something of the composition of the communities can be obtained
from Table 15.1. The table reports the mean representation of
eight selected species chosen to emphasize differences in
composition between the communities. Comparison of the rows of
the table, that is of the community mean vectors or *centroids*,
reveals substantial differences in composition between the three
grassland types.

TABLE 15.1: *Structure of grassland vegetation, Anglesey. Canonical variates analysis. Mean values of three grassland communities on eight selected variables (species).*

g	Grassland community	Sample size,N_g	Species*							
			X_1	X_2	X_3	X_4	X_5	X_6	X_7	X_8
1	Limestone	27	70.2	44.4	61.1	16.3	20.6	30.1	2.4	0.1
2	Neutral	16	14.7	4.7	71.5	75.0	72.7	9C.4	35.5	11.0
3	Eutrophic	2	0.0	2.0	6.0	4.0	56.0	86.0	83.5	80.0
	Average of means		28.3	17.0	14.7	48.5	49.8	68.8	40.5	30.4

*Species $X_1, \cdots, X_8$ are respectively: *Pseudoscleropodium purum* (Hedw.) Fleisch., *Thymus drucei* Ronn., *Carex flacca* Schreb., *Agrostis tenuis* Sibth., *Trifolium repens* L., *Helictotrichon pubescen* (Huds.) Pilger, *Dactylis glomerata* L., *Phleum bertolonii* DC.

Because the number of species (3) encountered was large relative to the size of the sample (N = 45), the two rarest species were omitted from the analysis in order to reduce the ratio of variables to samples, if only slightly. The thirty-one remaining species were then taken to comprise one set of variables in the canonical analysis. Pseudo-variables representing the vegetation classification comprised the variables of the second set and were constructed in the following way. A binary-valued variable, Y_1, designated to correspond to limestone grassland, was established by assigning to each limestone grassland stand a value 1 on this variable while each non-limestone grassland stand received a value of 0. A second dummy variable, Y_2, corresponding to neutral grassland, was created in a similar way by assigning a score of 1 to each neutral grassland on this variable and a score of 0 to all other stands. The variables Y_1 and Y_2 together completely accounted for the three-way stand classification, stands of eutrophic grassland being uniquely specified by a score of 0 on both Y_1 and Y_2. We may conveniently summarize the creation of the dummy variables Y_i by defining the value of the ith dummy variable (i = 1,$\cdots$, k - 1) in the jth stand (j = 1,$\cdots$,N) to be

$$Y_{ij} = \begin{cases} 1 & \text{if the } j\text{th stand belonged to community i} \\ 0 & \text{otherwise.} \end{cases}$$

Thus the canonical analysis came to be based on a random sample
of N = 45 10 × 10 m stands of grassland vegetation, p = 31
estimates of species' abundance, and q = 2 binary-valued dummy
variables corresponding to k = 3 grassland communities. The
ecological objectives were to describe the mutual relationships of
the communities and to elucidate the environmental factors or
other generating processes responsible for the compositional
differences between them. In a word, that is, to describe and
account for the structure of the vegetation. The term structure
is used here in the spirit of Orlóci's (1975, p. 3) definition to
refer to an abstract property of vegetation dependent on stand
resemblance as measured by some objective function of stand
composition. If the communities were found to differ in composition, then considerable interest would attach to the dimensionality
of the solution. A single non-zero root, in particular, would
indicate not only that there were differences in composition between the communities but also that the relationship between them
was one of *collinearity*. Ecologically, collinearity would imply
that the communities could be regarded as displacements along a
single underlying environmental gradient or other generating
process. The results of the analysis are summarized in Tables
15.2 and 15.3.

Before turning to consider the results, we pause to make a
number of general observations concerning a canonical analysis
of this kind in which one set of variables consists of binary-
valued dummy variables. We first note that the null hypothesis
(4.1) of no association can in this context be regarded as an
hypothesis of the equality of the communities in average composition.
Writing $\underset{\sim}{\mu}_g$ (g = 1,···,k) for the p-vector of species' means in
the population corresponding to the gth sample community, hypothesis
(4.1) may be restated as

$$H_0: \quad \underset{\sim}{\mu}_1 = \cdots = \underset{\sim}{\mu}_k$$

$$H_1: \quad \text{Not all } \underset{\sim}{\mu}_g \text{ are equal.}$$

(15.1)

As with hypothesis (4.1), the largest squared canonical correlation,
r_1^2, or Wilks' Λ provide convenient test criteria for hypothesis
(15.1). Rejection of H_0 in a test of size α would be equivalent to declaring that the communities differed in average composition with confidence $1 - \alpha$. In such cases the smaller roots,
r_k^2, (k > 1) would be helpful in locating where, or between which
communities, the differences lay. The point of interest here is
that in a comparative study the squared canonical correlations are
interpretable as indicating whether the communities (or other groups)
are separable on the p observed variables - the larger the root, the
greater the separation. Examination of the sample in the space of

the canonical variates corresponding to the roots judged to be statistically significant or otherwise worthy of attention, would then summarize and provide insight into the structure of the vegetation or other material examined. The rôle of the dummy variables in the analysis is to account for the variation among the communities, and, in particular, to isolate the $q = k - 1$ degrees of freedom associated with them.

Thus, we see that a canonical analysis in which one set of variables consists of dummy variables is open to interpretation in rather different terms from such an analysis in which both sets of variables are continuous. The immediate objective of canonical analysis in a comparative study can in fact often be regarded quite simply as to obtain the best low-dimensional representation of the structure of the p-variate groups studied. In view of these distinguishing features, canonical analysis of this sort is often referred to as *canonical variates analysis*, following Rao (1952, p. 364).

15.2 Results.

15.2.1 Dimensionality.

The canonical correlation coefficients and related significance tests appear in Table 15.2. There are two canonical correlations, both of which exceed 0.95 in size.

Roy's largest-root criterion and Bartlett's χ^2-approximation of Λ both lead to rejection of the null hypothesis (15.1) of no differences in average community composition. From Table 15.2(a) we find that $\theta_s = .963 > \theta_\alpha$ (2,14,5) $= .926$ for $\alpha = .01$; similarly, from Table 15.2(b) we see that $\chi_s^2 = 158.95 > \chi_\alpha^2$ (62) $= 112.2$ for $\alpha = .0001$. Accordingly, we are led to reject H_0 and to conclude that the average composition of the communities differs at the $\alpha = .01$ level of significance at least. To determine whether both canonical roots are necessary to account for the differences, we require a test of the residual association which remains after r_1^2 is eliminated. Bartlett's χ^2-approximation of Λ_k (k = 2) shows that $\chi_s^2 = 69.94 > \chi_\alpha^2$ (30) $= 67.6$ for $\alpha = .0001$ (Table 15.2(b)). Thus it appears that both r_1^2 and r_2^2 will be required in order to locate the differences between communities. This is equivalent to asserting that the community mean vectors are hardly likely to be collinear.

In turning to examine the relationships among the communities we will first consider the scatter of the communities along the canonical variates corresponding to r_1^2 and r_2^2 and then the correlations of the species with these canonical variates.

TABLE 15.2: Structure of grassland vegetation, Anglesey. Canonical variates analysis. Canonical correlation coefficients and related tests of significance.

(a) The canonical correlation coefficients r_k and their significance. Approximate critical values of Roy's largest-root criterion $\theta_\alpha(s,m,n)$ are shown for $\alpha = 0.05$ and $\alpha = 0.01$ (m = 14; n = 5).

k	s	r_k	r_k^2	$\theta_{.05}(s,m,n)$	$\theta_{.01}(s,m,n)$	p
1	2	.981	.963	.894	.926	<.01
2	1	.962	.925	<.01*	<.01*	<.01*

*
From the F-distribution; $F_s(30,12)$ = 4.933 > $F_\alpha(30,12)$ for α = .01.

(b) Bartlett's approximate χ^2-test of collinearity of community centroids; m = $\{(N-1) - \frac{1}{2}(p + q + 1)\}$.

Root	df	χ^2		p
r_1^2	p - q - 1 = 32	- m $\log_e(1-r_1^2)$	= 89.01	<.0001
r_2^2	(p-1)(q-1) = 30	- m $\log_e\{\Lambda/(1-r_1^2)\}$	= 69.94	<.0001
Total	pq = 62	- m $\log_e\Lambda$	= 158.95	<.0001

15.2.2 The canonical variates. Figure 15.1 displays the sample in the space of canonical variates U_1 and U_2 of the species' domain.

   In the Figure, the origin has been placed at the contrast of eutrophic grassland with itself, that is at the null contrast, while the canonical variates themselves have been scaled so that the average within-groups variance weighted by sample size on U_1 and U_2 is unity. This particular standardization has the property of maximizing the distances between communities. Moreover, the canonical variates have been drawn at right angles, as is customary, although they can in fact be shown to intersect at an angle of 119° in variable space. The circles, of radius $(\chi^2(2)/N_g)^{\frac{1}{2}}$, are the 99% confidence regions about the sample mean vectors and provide a rough indication of the separation and uncertainty of each of these vectors.

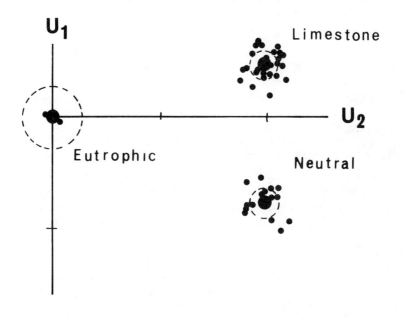

FIG. 15.1: *Structure of grassland vegetation, Anglesey. Canonical variates analysis. Three communities displayed in the subspace of the first two canonical variates U_1 and U_2 of the response variables. The origin is located at the null contrast while U_1 and U_2 have been scaled such that the average within-community variance weighted by sample size on each variate is unity. Circles depict the 99% confidence regions about the community centroids.*

The Figure shows three spherical stand clusters which correspond to limestone, neutral, and eutrophic grassland. Differences in the size of the clusters are apparent; while limestone and neutral grasslands are of roughly comparable size, each is appreciably larger than eutrophic grassland. Apart from the differences in size, the coherency and clear separation of the clusters and the departures of their centroids from collinearity are perhaps the most striking features of the plot. U_1 separates limestone from neutral grassland while U_2 distinguishes both these communities from eutrophic grassland. The latter is a highly distinctive community, characterized by both the direction and the extent of its separation. The confidence circles tightly enclose the community centroids, enhancing the distinctiveness and separation of the communities.

15.2.3 Intraset structure correlations. In order to conserve space, the structure correlations are not reported in full here; the correlations between U_1 and U_2 and selected species, however, appear in Table 15.3. It can be seen that species differ with respect to both the direction and the strength of their relationships with the canonical variates. *Trifolium repens* L. (-.88) and *Pseudoscleropodium purum* (Hedw.) Fleisch. (.87) are the species most strongly correlated with U_1, and hence are important species in the discrimination of limestone and neutral grassland which U_1 achieves. Similarly, *Phleum bertolonii* DC, (-.83) and *Carex flacca* Schreb. (.60) are the species most closely related to U_2 and are therefore likely to be important in the discrimination of eutrophic grassland from the remaining communities. The intraset communalities, h_w^2, of Table 15.3 express the extent of species' involvement in the overall discrimination of the communities; consequently h_w^2 indicates the potential value of particular species for this purpose.

15.2.4 Ecological significance of the results. The size, coherency, and spatial relationships of the stand groups are all directly interpretable in ecological terms.

Differences in the size of the groups reflects differences in the size of the corresponding communities (cf. Table 15.1). The coherency and lack of overlap among the stand clusters suggests that the communities represented are discrete entities. However, the effect of assigning transitional stands to either limestone or neutral grassland prior to the canonical analysis must be taken into account here. This procedure will have resulted in the

TABLE 15.3: Structure of grassland vegetation, Anglesey. Canonical variates analysis. Intraset structure correlations and communalities (h_w^2) *of selected species.*

Species	U_1	U_2	h_w^2
Pseudoscleropodium purum	.870	.341	.873
Thymus drucei	.750	.229	.615
Carex flacca	-.056	.597	.360
Agrostis tenuis	-.847	.233	.767
Trifolium repens	-.877	-.116	.783
Helictotrichon pubescens	-.850	-.205	.765
Dactylis glomerata	-.621	-.568	.708
Phleum bertolonii	-.270	-.827	.757

discreteness of these communities as well as the separation between them being exaggerated and needs to be borne in mind in interpreting the spatial relationships of the communities. The proximity of limestone and neutral grassland together with the isolation of eutrophic grassland indicates that limestone and neutral grassland resemble each other in composition to an appreciably greater extent than either community resembles eutrophic grassland; at the same time, it is abundantly clear that compositionally eutrophic grassland is a very distinct community. Furthermore, the pronounced departure of the group centroids from collinearity shows that structurally the vegetation can hardly be considered to consist of communities which represent displacements along a single underlying environmental gradient or equivalent factor.

In attempting to explain the structure of the vegetation ecologically, we rely on field knowledge of the behavior of species which contribute importantly to the discrimination of the communities. Thus, for example, of the species related to U_1, *Thymus drucei* Ronn. and *Agrostis tenuis* Sibth. are known to be strongly influenced by variation in soil depth within the range encountered in the area surveyed (Gittins, 1965). Moreover, it is known that these species differ in their response to soil depth, *T. drucei* tending to occur in greatest abundance on shallow, well-drained soils, while *A. tenuis* tends to be most abundant on deeper soils with more balanced water relationships. Now, provided the direction of the correlations of the species with U_1 (Table 15.3) are heeded, it is apparent that stands in which *T. drucei* is abundant will in general have positive scores on U_1, while the converse will hold for stands in which *A. tenuis* is abundant. It therefore appears that the discrimination between limestone and neutral grassland effected by U_1 may well reflect the influence of soil depth and the conditions which vary with it on these and other species which to a greater or lesser extent are also correlated with U_1. It is equally clear from the correlations of T. $drucei$ and A. $tenuis$ with U_2 that these species contribute little to the differentiation of communities associated with U_2. It can however be shown from field knowledge of the ecology of species which are strongly correlated with U_2, e.g. P. $bertolonii$, that the separation of eutrophic grassland from limestone and neutral grassland may be related to variation in soil phosphorous (see Gittins, 1965).

Thus, variation in the soil factor-complex exemplified by soil depth seems to explain at least in a provisional way the differentiation of limestone and neutral grassland, while soil phosphorous or possibly a related factor-complex similarly seems to explain the differentiation of eutrophic grassland from the remaining communities. Hence the canonical variates U_1 and U_2 appear to correspond to ecological factors in terms of which the structure of the vegetation can be comprehended.

In addition to clarifying certain aspects of the ecology of the vegetation, the results obtained also direct attention to lines along which further work might usefully proceed. The soil components of the factor-complexes of which soil depth and phosphorus seem likely to be part, and the effects of these components on individual species, would seem to be particularly deserving of attention in this regard.

Interpretation would not be complete without reference to the assumptions on which the tests of significance of Table 15.2 are based. In addition to independence and multinormality (Section 4), it is necessary for the variance-covariance, or dispersion, matrices of the grassland communities to be equal. Only if the data satisfy the assumptions, of course, will the tests be valid. The homogeneity of the dispersion matrices was not in fact examined; consequently, those aspects of the results which depend on the significance tests are open to question. However, a rough check of the homogeneity of the sample dispersions based on the within-group variance of the communities on each standardized canonical variate was performed following the canonical analysis. Univariate Bartlett's tests of the equality of the community variances on U_1 and U_2 showed them to be homogeneous. (For U_1, $\chi_s^2 = 2.442 < \chi_\alpha^2(2) = 2.773$ for $\alpha = .250$; for U_2, $\chi_s^2 = 0.489 < \chi_\alpha^2(2) = 0.713$ for $\alpha = .700$). Thus, as far as they go, these tests lend support to the likelihood of the relevant dispersion matrices being at least approximately equal. This in turn provides support for the validity of the tests of the canonical roots. Notice that the homogeneity of the population variance-covariance matrices has less bearing on the purely algebraic aspects of the analysis. Thus the extraction and descriptive use of the canonical roots and variates is largely unimpaired with respect to the sample in hand irrespective of the homogeneity or otherwise of the dispersions.

15.3 Conclusions. Before embarking on an ecological appraisal of the analysis, it will be helpful to summarize the salient points of ecological interest which have been established. We also have to consider the effect of the variable/sample ratio on the results obtained.

The chief findings of the ecological interest are that:

(a) the vegetation consists of three comparatively discrete communities, two of comparable size and closely related in composition, while the third is a smaller community sharply differentiated from the others; and that

(b) the structure of the vegetation can provisionally be interpreted in terms of environmental gradients in soil depth and soil phosphorus, or covariates of these properties.

The ratio of variables to samples (v/s) in the analysis is distinctly high (33/45 = .73). Once again, therefore, it is necessary to take into account an inflationary factor from this source in assessing the analysis. The magnitude of the canonical correlation coefficients, structure correlations, and communalities must therefore be accepted with some reserve. We observe, however, that in the present case steps could rather easily have been taken to surmount this somewhat disquieting feature of the analysis. By repeating the analysis after first having eliminated species found on the basis of the present analysis to contribute least to the discrimination of the communities, the vagueness introduced by the high v/s ratio could have been reduced substantially. Species with communalities $h_w^2 < 0.70$, say, might well have been regarded as candidates for removal.

We are now in a position to evaluate the analysis. The results obtained are corroborated by field knowledge of the vegetation and also by the results of previous work based on different methods (Gittins 1965; 1969). A principal components analysis, for example, of essentially the same 31 × 45 matrix of species' abundances as used in the canonical analysis, led to results which for all practical purposes are identical to those expressed at the beginning of this Section. The results of the present

analysis therefore demonstrate the ability of canonical analysis
to identify features of ecological significance in a comparative
study of plant communities. The results provided in fact surpass
those of the other methods used in the ease and efficiency with
which they enable the structure of the vegetation to be described
and communicated. For these reasons it seems justifiable to
conclude that canonical variates analysis is likely to prove
rewarding in the investigation of structural relationships
between plant communities. Although the vegetation involved was
structurally simple, nothing about the analysis suggested that
its usefulness would be confined to vegetation of this kind.
Indeed, on theoretical grounds and from applications in related
fields (Rao, 1952; Seal, 1964) it seems likely that the potential
value of canonical variates analysis will be fully realized only
in relation to vegetation of greater structural complexity
than encountered here. Thus, the opportunities afforded by
canonical variates analysis would seem to be particularly promising
in connection with large-scale vegetation survey, where the
heterogeneity of the vegetation examined is most likely to be
considerable.

It is instructive to compare briefly the results of the
canonical analysis with those of the principal components analysis
referred to in the preceding paragraph. The comparison is facilitated by reference to Figure 15.1 and to Figure 9 of Gittins (1969),
which summarize major features of the respective analyses. The
Figures show that while components analysis conveys considerable
detail about the internal structure of the three communities,
canonical analysis reveals little or nothing about this aspect of
the vegetation. On the other hand, canonical analysis draws
attention to the *overall* structure of the vegetation with much
greater economy and effect than is achieved by components analysis.
The comparison is illuminating because it draws attention to
distinctive features of the two analyses which themselves in turn
suggest that the *sequential* application of canonical variates and
principal components analyses in connection with vegetation survey
might sometimes be rewarding. Canonical variates analysis would
provide a convenient means of exploring the overall structure of
the vegetation while components analysis might then be employed to
examine the internal structure of particular communities. Notice
that the application of both procedures to the same matrix of
species' abundances, as in the present example, would in general
be inappropriate because of the fundamentally different conceptual-
izations of the sample which the use of each method implies.

16. THE NITROGEN NUTRITION OF EIGHT GRASS SPECIES

16.1 Introduction. Identification of the factors which influence
or control the occurrence and representation of plant species
in the field is a prime objective of much ecological work. Preliminary indications as to the identity of the operative factors
may be obtained either by direct observation of species in the
field or by use of exploratory multivariate techniques in the
analysis of survey data. The next step towards identification
usually involves attempts to substantiate provisional insights
of the kind provided. The most direct and convincing means of
substantiation results from examination of the responses of the
species of interest to treatments of one or more of the indicated
factors in a designed experiment. We describe in this Section
how canonical analysis may be used for the analysis of experiments
of this kind. The study concerns the comparative responses of
eight grass species to five treatment levels of nitrate nitrogen.
General field observation suggested that the species in question
differed in their responses to variation in soil fertility. The
experiment was designed to enable an assessment of the evidence
for the differential response of the species to one component
of the factor-complex of soil fertility, namely nitrate nitrogen,
to be made.

The species studied were *Lolium perenne* L., *Dactylis glomerata*
L., *Phleum bertolonii* DC., *Briza media* L., *Koeleria cristata* (L.)
Pers., *Festuca ovina* L., *Festuca rubra* L. and *Helictotrichon
pubescens* (Huds.) Pilger. Nitrogen treatments of 1, 9, 27, 81
and 243 ppm N were obtained by varying the amount of $NaNO_3$ in
culture solution, the level of other nutrients being held at or
near optimum concentration. The nigrogen treatments were chosen
to vary between what for the species was expected to be
from critically low to almost toxic. Species' response to the
treatments was expressed as dry weight yield in grams. The
ecological objective of the experiment was to assess the evidence
for differential response among the species and to characterize
the differences, if present.

Individuals of each species were grown separately in pots
under sand culture in an unheated greenhouse using a split-plot
experimental design. Main-plots correspond to species (8) and
sub-plots to treatments (5). There were five complete replications
(blocks) of the experiment. Species were randomly assigned to
the main-plots and treatments similarly assigned to the sub-
plots. The experimental unit was the individual pot and
corresponded to a particular species and a particular treatment
level of nitrogen. Plants were harvested and the dry weight yield
of the experimental units determined after a growth period of two
months. Numerical analysis was based on a logarithmic transformation (base 10) of dry weight yield.

An analysis of variance appropriate to a split-plot experimental design would have provided the simplest and most direct form of analysis. For our present purposes, however, it will be convenient to regard the experiment rather differently in order that we may take advantage of other forms of analysis more in keeping with our wider objectives. We shall see that by regarding the experiment as generating vector-valued rather than scalar responses, the results can be analyzed by both canonical analysis and the multivariate analysis of variance. This changed perspective of the experiment will therefore enable us to show how canonical analysis can be applied to the analysis of designed experiments and, in addition, to examine the nature of the relationship between canonical analysis and the multivariate analysis of variance. The necessary multivariate conceptualization of the experiment can be achieved quite simply by regarding the *species* rather than the individual pot as the experimental unit. The q-vector of responses associated with each species consists of the dry weight yield at each of the five nitrogen treatments separately administered to the representatives (pots) of the species. As before, there are b = 5 blocks each comprised of k = 8 species. Thus, the experiment can be regarded as an orthogonal k × b multivariate randomized block design with k = 8 conditions (species), b = 5 blocks and q = 5 responses. The ecological purpose of the experiment remains unchanged. As before, we wish to assess the comparative responses of the species to the treatments administered.

We begin by describing the application of the multivariate analysis of variance (Section 16.2). We will next consider the canonical analysis (Section 16.3) and then make a comparison of the methods based on the results obtained (Section 16.4). In Section 16.5 the ecological implications of the results will be taken up before we turn finally to an assessment of the analyses both ecologically and in more general terms (Section 16.6). As likelihood-ratio and union-intersection tests are widely used in connection with both forms of analysis, the application of tests of both kinds will be illustrated.

16.2 Multivariate Analysis of Variance. The species' mean vectors over the replicates for the five treatment responses are reported in Table 16.1. Differences between species in response are apparent. To test these differences formally a multivariate analysis of variance may be performed. It would take us too far to give a comprehensive account of this procedure here; while an outline of the method is given, reference should be made to one of the standard texts for the details (e.g. Timm, 1975, pp. 389-92; Morrison, 1976, pp. 182-85). As our ecological interest centers on differences between species in their mean response, an appropriate null hypothesis is that the species' mean vectors over the five variables, μ_g (g = 1,$\cdots$,k), are jointly equal:

$$H_o: \underset{\sim}{\mu}_1 = \cdots = \underset{\sim}{\mu}_k \quad . \qquad (16.1)$$

The alternate is that $\underset{\sim}{\mu}_i \neq \underset{\sim}{\mu}_j$ for at least one pair of species i and j.

TABLE 16.1: Nitrogen nutrition of eight grass species. Multi-variate analysis of variance and canonical variates analysis. Mean yield ($\log_{10}$ dry-weight (g) × 10) over five replicates of eight grass species grown in sand culture at five levels of nitrogen.

g	Species	Sample size, N_g	Variable (ppm N) 1	9	27	81	243
1	L. perenne	5	0.9631	1.5150	1.5656	1.9683	2.0663
2	D. glomerata	5	0.8808	1.4528	1.6527	1.8229	1.8651
3	P. bertolonii	5	0.8082	1.4114	1.5033	1.6404	1.6850
4	H. pubescens	5	0.7513	1.2754	1.1940	1.4057	1.3737
5	K. cristata	5	0.6938	1.1811	0.9786	1.2695	1.1784
6	B. media	5	0.7905	1.0702	1.1059	1.2870	1.1218
7	F. ovina	5	0.7479	1.0589	1.1433	1.2326	1.2161
8	F. rubra	5	0.3393	1.0426	0.8275	1.0418	1.3643
Average of means			0.7469	1.2509	1.2464	1.4535	1.4838

The likelihood-ratio criterion Λ provides a multivariate statistic for testing (16.1). Λ may be calculated as the determinantal ratio

$$\Lambda = \frac{|\underset{\sim}{Q}_e|}{|\underset{\sim}{Q}_e + \underset{\sim}{Q}_h|} = \frac{1}{|\underset{\sim}{I} + \underset{\sim}{Q}_e^{-1}\underset{\sim}{Q}_h|} \quad ,$$

where $\underset{\sim}{Q}_e$ and $\underset{\sim}{Q}_h$ are residual and hypothesis mean-corrected sums of squares and cross products (SSP) matrices of order q based on ν_e and ν_h degrees of freedom, respectively ($\nu_e = N - k - b$, $\nu_h = k - 1$). The significance of Λ may be tested by reference to tables of percentage points of its null distribution (Wall, 1968; Timm, 1975 p. 624) or, following transformation, to tables of the χ^2 or F distributions. Bartlett's χ^2 approximation of Λ, for example, is given by

$$\chi^2 = -\{\nu_e + \nu_h - \tfrac{1}{2}(q + \nu_h + 1)\} \log_e \Lambda \quad ,$$

and is distributed approximately as a chi-squared variate on
$q \cdot \nu_h$ degrees of freedom. Hypothesis (16.1) is rejected with
confidence $1 - \alpha$ if $\chi_s^2(q \cdot \nu_h) > \chi_\alpha^2(q \cdot \nu_h)$, where $\chi_\alpha^2(q \cdot \nu_h)$
is the 100α upper percentage point of the χ^2 distribution
with $q \cdot \nu_h$ degrees of freedom.

We note that Λ may also be calculated from the s roots
of $|Q_h - \lambda Q_e| = 0$ by the relation

$$\Lambda = \prod_{k=1}^{s} (1 + \lambda_k)^{-1} ,$$

where $s = \min(\nu_h, q)$.

An alternative to Λ as a test criterion is given by the
union-intersection principle. This involves an attempt to find
that test statistic which will lead to the best chance of rejecting
H_o. Application of the principle (Morrison, 1976, pp. 176-8)
leads to the largest root, λ_1, of the determinantal equation

$$|Q_h - \lambda Q_e| = 0$$

as the test statistic. The significance of λ_1 may be assessed
by referring

$$\Theta_s = \frac{\lambda_1}{1 + \lambda_1}$$

to a table of the percentage points of the gcr distribution
(Timm, 1975, p. 607; Morrison, 1976, p. 379). Hypothesis (16.1)
is rejected at the significance level α if $\Theta_s(s,m,n) > \Theta_\alpha(s,m,n)$,
where $\Theta_\alpha(s,m,n)$ is the 100α upper percentage point of the
gcr distribution when H_o is true. The distribution parameters
are:

$$s = \min(k-1, q) \quad m = \tfrac{1}{2}(|k-1-q|-1) \quad n = \tfrac{1}{2}(k(b-1)-b-q).$$

The likelihood-ratio and union-intersection tests of the
equality of the species' centroids both require the homogeneity
of the species' variance-covariance matrices. A test (Morrison,
1976, p. 252) of the equality of the dispersions showed them to
be nonhomogeneous (p < .0001), notwithstanding the initial
$\log_{10}$ transformation of the data. This result unfortunately
does not provide practical guidance in the present case, however,

because the homogeneity test is known to be sensitive to departures
from multivariate normality, an aspect of the data which regrettably
was not examined. Moreover, although knowledge of the robustness
of multivariate test procedures in general is sketchy, it is
believed that the multivariate analysis of variance may be reasonably robust in the presence of moderate nonhomogeneity of dispersion
where the sample sizes are equal (Bock, 1975, p. 236). Thus,
despite the apparent lack of homogeneity among the variance-covariance matrices, it is by no means clear that the tests of (16.1)
will be invalidated. In order to proceed, we shall therefore
accept the pooled within-species dispersion matrix as the best
available estimate of the corresponding population dispersion.
This will enable us to illustrate the test procedures described
above and in addition to go on to sketch the relationship between
the multivariate analysis of variance and canonical analysis. On
the other hand, it must also be accepted that the statistical
interpretation of the results obtained will need to be treated
with some circumspection.

16.2.1 Results. The results of the multivariate analysis of
variance are summarized in Tables 16.2 and 16.3.

Of primary interest in the present analysis is the test of
significance of differences between species' vector-means. From
Tables 16.2 and 16.3 (a) it can be seen that $\Lambda = 0.015691$.
Bartlett's transformation yields

$$\chi_s^2 = -\{28 + 7 - \tfrac{1}{2}(5 + 7 + 1)\}\log_e(0.015691)$$

$$= 118.4083, \quad \text{with } (5)(7) = 35 \text{ degrees of freedom.}$$

As $\chi_s^2(35) = 118.41 > \chi_\alpha^2(35) = 74.92$ for $\alpha = .0001$ we reject
the null hypothesis (16.1). It therefore seems that the species
differ in response and that the differences warrant interpretation.
We shall return to this aspect of the analysis later in this
section (p. 121).

It is readily verified from Table 16.3(b) that Λ may also
be calculated from the roots λ_k of the determinantal equation
$|Q_h - \lambda Q_e| = 0$. The Table reports the roots of this equation and
shows that the continued product of a simple function of the roots,
$\Pi(1 + \lambda_k)^{-1}$, leads within rounding error to the same value of Λ
as calculated in Table 16.3(a) as a determinantal ratio.

TABLE 16.2: *Nitrogen nutrition of eight grass species. Multivariate analysis of variance of species' response to five treatment levels of nitrogen.* Y_1: 1 ppm N, Y_2: 9 ppm N, $\cdots$, Y_5: 243 ppm N.

Source of dispersion	df	SSP (symmetric)					Univariate		Multivariate		
		Y_1	Y_2	Y_3	Y_4	Y_5	F	p	Λ	χ^2	p
Blocks	4	0.1195					1.91	.14	.337506	29.3267	.0815
		0.0844	0.1397				1.54	.22			
		0.0517	0.0876	0.3119			2.38	.08			
		-0.0416	0.0511	-0.0026	0.1868		1.19	.34			
		0.0366	0.0722	0.1752	0.0084	0.1042	.88	.49			
Species	7	1.1966					10.92	<.0001	.015691	118.4083	<.0001
		0.8732	1.2735				8.03	<.0001			
		1.5884	1.7873	3.0667			13.37	<.0001			
		1.7106	2.0525	3.1644	3.5920		13.10	<.0001			
		1.1885	2.1172	3.0431	3.5419	4.2375	20.31	<.0001			
Error	28	0.4383									
		0.0384	0.6341								
		0.0037	0.4344	0.9173							
		0.0932	0.1731	0.1046	1.0965						
		0.1853	-0.0382	-0.1439	0.3094	0.8347					
Total	39	1.7544									
		0.9960	2.0473								
		1.6438	2.3093	4.2959							
		1.7622	2.2767	3.2664	4.8753						
		1.4104	2.1512	3.0744	3.8597	5.1764					

$= Q_b$ (Blocks); $= Q_h$ (Species); $= Q_e$ (Error); $= Q_t$ (Total)

$|Q_e| = 1.158460 \times 10^{-1}$; $\quad |Q_e + Q_h| = 7.38306 \times 10^0$; $\quad |Q_e + Q_b| = 3.432418 \times 10^{-1}$

Although the block effects have little intrinsic interest, the effectiveness of the blocking is nevertheless worth examining (see Table 16.2). For this purpose we require

$$\Lambda_b = \left| \underset{\sim}{Q}_e (\underset{\sim}{Q}_e + \underset{\sim}{Q}_b)^{-1} \right| .$$

$$= \frac{0.1158460}{0.3432418} = 0.337506 .$$

Applying Bartlett's transformation we find

$$\chi_s^2 = -\{28 + 4 - \tfrac{1}{2}(5 + 4 + 1)\}\log_e(0.337506)$$

$$= 29.3267, \quad \text{with} \quad (5)(4) = 20 \quad \text{degrees of freedom.}$$

As $\chi_s^2(20) = 29.33 < \chi_\alpha^2(20) = 31.41$ for $\alpha = .05$ we are led to conclude that variation among the blocks is insignificant. In other words, the blocking procedure used does not seem to have been effective.

Table 16.3(c) reports the results of the test of (16.1) using Roy's largest-root criterion, λ_1. As $\Theta_s(5,\tfrac{1}{2},11) = 0.910 > \Theta_\alpha(5,\tfrac{1}{2},11) = 0.671$ for $\alpha = .01$ we are led to reject H_o, as before. A similar union-intersection test of the blocking effects employing the largest root, λ_1^*, of $\left| \underset{\sim}{Q}_b - \lambda^* \underset{\sim}{Q}_e \right| = 0$ could also be made.

Having shown H_o to be unsupported by the data we now wish to determine which among the species differ and to identify the variables responsible for the differences. A variety of methods exists for these purposes. The methods include pairwise comparisons between species, partition of the hypothesis SSP matrix $\underset{\sim}{Q}_h$, construction of simultaneous confidence bounds and assessment of the contribution of particular variables or combinations of variables to the rejection of H_o. It would take us too far to consider these ramifications of the multivariate analysis of variance here. For our purpose it is sufficient to note first that the choice of method is to some extent governed by whether the test of (16.1) was conducted as a likelihood-ratio or as a union-intersection test; and secondly that the union-intersection procedure leads to the closest affinities with canonical analysis. The union-intersection test of (16.1) extends naturally and directly to *discriminant analysis* as a means of pursuing the analysis beyond the acceptance decision. In order to facilitate further comparison, we give a brief description of discriminant analysis here. The elements of the characteristic vector $\underset{\sim}{\alpha}_1$ associated

TABLE 16.3: *Nitrogen nutrition of eight grass species. Multi-variate analysis of variance. Likelihood-ratio and union-intersection statistics for test of the overall hypothesis of the joint equality of the species' mean vectors.*

(a) Likelihood-ratio criterion, Λ, calculated as a determinantal ratio

$$\Lambda = \frac{|Q_e|}{|Q_e + Q_h|} = \frac{.115846}{7.383006} = 0.015691 \ .$$

(b) Likelihood-ratio criterion, Λ, calculated from the roots λ_k of $|Q_h - \lambda Q_e| = 0$.

k	λ_k	$(1 + \lambda_k)^{-1}$
1	10.058	.090432
2	2.623	.276014
3	0.424	.702247
4	0.103	.906618
5	0.013	.987167

$$\Lambda = \prod_{k=1}^{s} (1 + \lambda_k)^{-1}$$

$$= (.090432)(.276014) \cdots (.987167)$$

$$= .015688.$$

(c) Union-intersection criterion

$$\Theta_s(s,m,n) = \frac{\lambda_1}{1 + \lambda_1} = \frac{10.058}{1 + 10.058} = .910 \ ,$$

with distribution parameters

$$s = \min(k-1,q) \quad m = \tfrac{1}{2}(|k-1-q|-1) \quad n = \tfrac{1}{2}(k(b-1)-b-q)$$

$$= 5 \qquad\qquad = \tfrac{1}{2} \qquad\qquad = 11 \ .$$

with the largest root λ_1 of $|Q_h - \lambda Q_e| = 0$ are open to interpretation as weights which, following standardization, may reflect the contribution of particular variables to the rejection of H_o.

Moreover, the variable L_1 which results from the application of
the weights in the linear transformation $L_1 = \alpha_1' \, \bar{y}_h$ (h=1,$\cdots$,8),
where $\bar{y}_h$ is the 5×1 vector of treatment means for the hth
species, is helpful in locating where the differences between
species occur. L_1 is variously known as the $discriminant$
$function$ or $canonical\ variate$ associated with λ_1. The
characteristic vectors α_k which correspond to the s-1 remaining
roots λ_k (k > 1) of $Q_e^{-1}Q_h$ may similarly be employed to clarify
any further relationships which may exist beyond those accounted
for by the first discriminant function.

The results of a discriminant analysis applied to the nutrition
experiment are reported in Tables 16.3(b) and 16.4. The roots λ_k
of the characteristic equation $|Q_h - \lambda Q_e| = 0$ are given in Table
16.3(b), while Table 16.4 (a) and (b) contains the corresponding
characteristic vectors α_k and associated discriminant functions
L_k. The product–moment correlation coefficients between the
species' $scores$ on the discriminant functions and the response
variables are reported in Table 16.4(c). It will be instructive
to compare these quantities with the results of a canonical analysis
of the same experimental data in a later Section (Section 16.4).

Before turning to the canonical analysis, it is of interest to
note that univariate F tests of the contribution of the response
variables to the rejection of H_o are provided as a by-product of
the multivariate likelihood–ratio test of that hypothesis (Table
16.2). Judging by the univariate F-ratios, which are based
on 7 and 28 degrees of freedom, all five variables contribute
to the significant q–variate effect. Nevertheless, it is apparent
that Y_5 is the variable which best differentiates the species
while Y_2 is the variable in terms of which the species differ
least. These univariate tests, however, are not independent and
in rejecting a univariate hypothesis at the α nominal probability
level, we must bear in mind that the probability of committing a
Type I error may actually be greater than α.

16.3 Canonical Analysis. In Section 15.1 we saw how a canonical
analysis in which the variables of one set were binary variables
could be used to test the equality of sample mean vectors drawn
from several distinct, p–variate universes. Hypothesis (15.1)
in fact is precisely the hypothesis (16.1) of the multivariate
analysis of variance. Thus we have already used canonical analysis

TABLE 16.4: Nitrogen nutrition of eight grass species. Discriminant analysis. Discriminant weights, species' scores and structure correlations.

(a) Discriminant weights. The weights have been scaled so that the largest element in each vector α_k is unity.

Response variable	α_1	α_2	α_3	α_4	α_5
Y_1	.317	1.000	.267	-.593	.729
Y_2	.095	-.328	1.000	1.000	.400
Y_3	.889	.206	-.801	.557	-.024
Y_4	.349	.214	.184	-.267	-.999
Y_5	1.000	-.621	-.146	-.529	.450

(b) Co-ordinates of species' centroids on the discriminant functions, L_k.

Species	L_1	L_2	L_3	L_4	L_5
L. perenne	1.435	-0.232	-.468	-.511	.040
D. glomerata	1.175	-0.026	-.484	.197	-.133
P. bertolonii	0.666	-0.098	-.158	.442	.061
B. media	-0.694	1.078	-.220	-.187	-.156
K. cristata	-0.799	0.261	.885	.096	-.046
F. ovina	-0.574	0.660	-.701	-.190	.172
F. rubra	-0.992	-1.819	-.257	-.103	-.034
H. pubescens	-0.217	0.175	.466	.256	.095

(c) Correlation coefficients between the response variables and the discriminant functions, L_k.

Response variable	L_1	L_2	L_3	L_4	L_5
Y_1	.652	.636	.164	-.190	.229
Y_2	.795	-.032	.356	.427	.053
Y_3	.862	.201	-.192	.354	-.007
Y_4	.879	.115	.184	-.092	-.343
Y_5	.886	-.309	.040	-.238	.076

in a test of the hypothesis for which the multivariate analysis of variance is commonly used. Here, we employ canonical variates analysis to test hypothesis (16.1) of the joint equality of the species' centroids.

In canonical analysis, the likelihood-ratio statistic $\Lambda = |I - Q_t^{-1}Q_h|$ is normally calculated from the roots r_k^2 of the determinantal equation $|R_{21}R_{11}^{-1}R_{12} - r^2 R_{22}| = 0$ by the relation

$$\Lambda = \prod_{k=1}^{s} (1 - r_k^2) \ ,$$

where $s = \min(p,q)$. The test of (16.1) in terms of Λ may then be performed as described previously (Section 4.1). If the overall hypothesis can be rejected, then Λ may be partitioned and further tests based on the components of Λ conducted (Section 4.2). The union-intersection test of (16.1), on the other hand, uses the largest root, r_1^2, of $|R_{21}R_{11}^{-1}R_{12} - r^2 R_{22}| = 0$ as the test criterion (Section 4.3). If on the basis of this test H_o is rejected, further tests employing the smaller roots r_k^2 $(k > 1)$ will be informative.

The variables used in the canonical analysis consisted of binary-valued dummy variables to account for the species affiliation of samples, on the one hand, and the species' responses to the five nitrogen treatments on the other. The dummy variables X_i $(i=1,\cdots,k-1)$ corresponding to the $k = 8$ species were defined for the jth sample $(j=1,\cdots,N)$ by setting

$$X_{ij} = \begin{cases} 1 \text{ if the } j\text{th sample belonged to the } i\text{th species} \\ 0 \text{ otherwise} \ . \end{cases}$$

Helictotrichon pubescens was arbitrarily chosen as the species not explicitly represented by a particular pseudo-variable and which therefore functioned as the reference species in the analysis.

For the canonical analysis we therefore had $N = 40$ samples representing $k = 8$ species which were randomly assigned to $b = 5$ blocks, $p = 7$ binary-valued dummy variables corresponding to the species and $q = 5$ response variables corresponding to the experimental treatments. The nitrogen treatments were randomly assigned to five separate representatives (pots) of each species. The analysis had two goals. The first objective was ecological in nature and consisted of showing how canonical analysis could be used to clarify the comparative relationships among species with respect to their responses to the five experimental treatments considered simultaneously. The dimensionality of these relationships was likely to be of special interest. Collinearity of the species' mean vectors, in particular, would imply not only that there were differences in response between species, but, furthermore, that the species could be considered to form an 'ecological series' in terms of their response to increasing concentrations of soil nitrogen. The second objective of the analysis was to draw attention to relationships which exist between canonical analysis and one-way multivariate analysis of variance. To facilitate comparison, the canonical analysis was performed using sums of squares and cross products matrices rather than correlation matrices in order to put the two analyses on an equal footing. For the same reason it was also necessary to remove the block effects, whose only function was to eliminate extraneous variation, before embarking on the canonical analysis. Removal was accomplished by subtracting the block effects SSP matrix $Q_{\sim b}$ from the total

SSP matrix $Q_{\sim t}$ prior to analysis (see Table 16.2). Thus the

total SSP matrix used in the canonical analysis had been *adjusted* for blocks (cf. Bartlett, 1951; 1965).

The results of the analysis are summarized in Tables 16.5, 16.6 and 16.7.

16.3.1 Results. Dimensionality. The union-intersection and likelihood-ratio tests (Table 16.5(a) and (b)) lead decisively in both cases to rejection of the hypothesis of no treatment differences between species. From the Table we see that $\Theta_s(5,\frac{1}{2},11)$ = .910 > $\Theta_\alpha(5,\frac{1}{2},11)$ = .671 for α = .01 while $\chi_s^2(35)$ = 118.42 > $\chi_\alpha^2(35)$ = 74.926 for α = .0001. Accordingly, it seems reasonable to conclude that the species and their responses are associated, or, in other words, that the species differ in their responses to the experimental treatments. Having found H_o to be untenable, we pass to more specific tests based on the smaller canonical roots r_k^2 (k > 1). From Table 16.5(a) it is apparent that only r_1^2 and r_2^2 achieve significance. Bartlett's approximate test of collinearity (Table 16.5(c)), suggests that the three smallest roots r_k^2 (k > 2) reflect only random variation, and so provides a comparable result. Thus we may anticipate that the canonical variates associated with the first two canonical roots will be informative in attempting to establish how the species differ among themselves and in interpreting the differences. The analysis may be pursued, therefore, by examining the sample after it has been mapped into one or more of the subspaces associated with the first two pairs of canonical variates.

TABLE 16.5: Nitrogen nutrition of eight grass species. Canonical variates analysis. Canonical correlation coefficients and related test of significance.

(a) The canonical correlation coefficients r_k and their significance. Approximate critical values of Roy's largest-root criterion $\Theta_\alpha(s,m,n)$ are shown for $\alpha=.05$ and $\alpha=.01$ ($m=\frac{1}{2}$; $n=11$).

k	s	r_k	r_k^2	$\Theta_{.05}(s,m,n)$	$\Theta_{.01}(s,m,n)$	p
1	5	.954	.910	.605	.671	<.01
2	4	.851	.724	.554	.623	<.01
3	3	.546	.298	.482	.564	>.05
4	2	.305	.093	.395	.484	>.05
5	1	.113	.013	–	–	–

(b) Bartlett's approximate χ^2 test of the joint nullity of all s canonical correlation coefficients.

$$\Lambda = \prod_{k=1}^{s} (1 - r_k^2)$$

$$= (1 - .910)(1 - .724) \cdots (1 - .013)$$

$$= 0.015610.$$

$$\chi^2 = -\{(N - b) - \tfrac{1}{2}(p + q + 1)\} \log_e \Lambda$$

$$= -\{35 - \tfrac{1}{2}(7 + 5 + 1)\}(-4.159844)$$

$$= 118.5556, \text{ with } p.q = 35 \text{ df.}$$

(c) Bartlett's approximate χ^2-test of the collinearity of species' centroids; $m = \{(N - b) - \tfrac{1}{2}(p + q + 1)\} = 28.5$.

Root	df	χ^2	p
r_1^2	$p + q - 1 = 11$	$-m\ln(1 - r_1^2) = 68.63$	–
r_2^2	$p + q - 3 = 9$	$-m\ln(1 - r_2^2) = 36.69$	–
Remaining roots	$(p-2)(q-2) = 15$	$-m\ln\{\Lambda/(1-r_1^2)(1-r_2^2)\} = 13.24$	.5838
Total	$p.q = 35$	$-m\ln \Lambda = 118.56$	<.0001

The canonical variates.   Figure 16.1 displays the sample projected into the subspace of canonical variates V_1 and V_2 of the response variables.   The origin in the Figure has been placed at the null contrast, that is at the contrast of *H. pubescens* with itself, and the axes drawn at right angles even though they are not orthogonal in response space.   We note also that V_1 and V_2 have been standardized so that the average within-groups variance on each variate is unity, while the circles, of radius $(\chi^2(2)/N_g)^{\frac{1}{2}}$, depict the 95% confidence regions about the species' centroids.   The configuration is characterized principally by two features:

(a) a well-marked distinction between species in relation to V_1; three species have comparatively large positive scores, while the remaining species have either zero or negative values; and

(b) a contrast within the subset of species with zero or negative scores on V_1 in terms of their scores on V_2.

The quantities reported in Table 16.6 enable the contributions of the treatments to the differentiation of species effected by the canonical variates to be assessed.   The Table contains the structure correlations between the canonical variates and the original variables as well as several indices derived from them. We defer the ecological interpretation of these results, however, until the connections between the multivariate analysis of variance, discriminant analysis and canonical analysis have been examined.   Some further results of the canonical analysis which will be useful in comparing methods are summarized in Table 16.7.

16.4 Relationships Between Multivariate Analysis of Variance and Canonical Analysis.   The simplest means of establishing the connection between the methods is to compare the statistics on which the test of (16.1) is based in each case.   Accordingly, we will first consider (Section 16.4.1) the relationship between the methods in terms of the likelihood-ratio and union-intersection tests of (16.1).   We will then go on in Section 16.4.2 to sketch the connection between multivariate analysis of variance and canonical analysis beyond the acceptance decision arising from (16.1).

Before embarking on the comparisons it will be helpful to state two identities which exist between SSP matrices which are fundamental to each analysis.   For definiteness and without loss of generality we shall suppose that, as in the present analysis, the first set of variables of the canonical analysis consists of

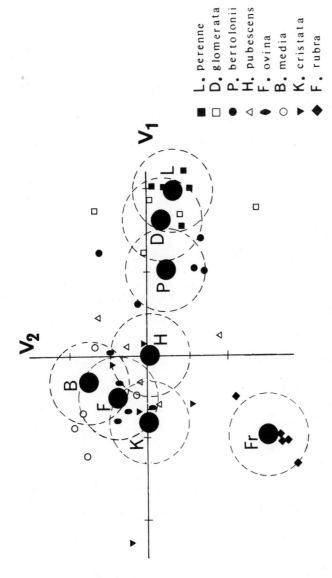

L. perenne
D. glomerata
P. bertolonii
H. pubescens
F. ovina
B. media
K. cristata
F. rubra

FIG. 16.1: Nitrogen nutrition of eight grass species. Canonical variates analysis. Species groups displayed in the subspace of the first two canonical variates V_1 and V_2 of the response variables. The origin is located at the null contrast while V_1 and V_2 have been scaled such that the average within-species' variance on each variate is unity. Circles depict the 95% confidence regions about the species' mean vectors.

dummy variables while the second set is made up of the response variables. With this designation of the variables, we state the following identities between the SSP matrices Q_h and $Q_{t*} = Q_e + Q_h$ of the multivariate analysis of variance (see Table 16.2) and the SSP matrices S_{11}, S_{12} and S_{22} used in canonical analysis:

$$Q_h = S_{21}S_{11}^{-1}S_{12}$$

$$Q_{t*} = S_{22} \quad .$$

For a proof of these relationships, see Krzyśko (1979, p. 233). Notice that Q_{t*} is used to denote the total SSP matrix after removal of the block effects, if any. With these simplifications, we can now write the determinantal equation (2.11) of canonical analysis as

$$|Q_{t*}^{-1}Q_h - r^2 I| = 0 \quad , \tag{16.2}$$

where for convenience we have replaced λ^2 by r^2.

It will also be advantageous at this point to have at our disposal three results from matrix algebra. The first two of these concern the characteristic roots of a matrix, which we now state. Let A be a square matrix of order p with characteristic roots λ_k ($k = 1, \cdots, p$). Then it can be shown (e.g. see Tatsuoka, 1971, pp. 126 and 140) that:

(a)   the product of the roots λ_k is equal to the determinant of A, $\Pi\lambda_k = |A|$; and that

(b)   the matrix $B = A + c I_p$, where c is an arbitrary scalar and I_p the identity matrix of order p, has the characteristic roots $\lambda_k + c$.

The third result involves determinants and states that the determinant of the product of two matrices A and B, is equal to the product of their determinants, that is $|AB| = |A||B|$.

We are now in a position to proceed to the comparisons.

16.4.1 Relationships between criteria for testing H_o. The likelihood-ratio criterion Λ in the multivariate analysis of variance is most frequently encountered as the determinantal ratio

$$\Lambda = \frac{|\underset{\sim}{Q}_e|}{|\underset{\sim}{Q}_e + \underset{\sim}{Q}_h|} \quad .$$

The corresponding quantity in canonical analysis is calculated from the canonical roots r_k^2 as

$$\Lambda = \prod_{k=1}^{s} (1 - r_k^2) \quad , \tag{16.3}$$

where the r_k^2 are the roots of (16.2).

The identity of the two expressions for Λ may be demonstrated using the properties of matrices stated above as follows. First, notice that from (16.2) we may write $\Pi r_k^2 = |\underset{\sim}{Q}_{t*}^{-1}\underset{\sim}{Q}_h|$. Similarly,

$$\prod_{k=1}^{s} (1 - r_k^2) = |\underset{\sim}{I} - \underset{\sim}{Q}_{t*}^{-1}\underset{\sim}{Q}_h|$$

$$= \frac{|\underset{\sim}{Q}_{t*} - \underset{\sim}{Q}_h|}{|\underset{\sim}{Q}_{t*}|} \tag{16.4}$$

$$= \frac{|\underset{\sim}{Q}_e|}{|\underset{\sim}{Q}_e + \underset{\sim}{Q}_h|} = \Lambda \quad ,$$

which establishes the connection. The equivalence of the test-criteria may readily be verified by reference to the numerical values of Tables 16.3(a) and 16.5(b).

Turning to the union-intersection tests, the test criterion in the case of multivariate analysis of variance is λ_1, where λ_1 is the largest root of $|\underset{\sim}{Q}_h - \lambda\underset{\sim}{Q}_e| = 0$. The corresponding quantity in canonical analysis is r_1^2, where r_1^2 is the largest root of $|\underset{\sim}{Q}_h - r^2\underset{\sim}{Q}_{t*}| = 0$. Although λ_1 and r_1^2 are not identical, each is readily expressed in terms of the other:

$$\lambda_1 = \frac{r_1^2}{1 - r_1^2} \quad , \quad \text{and} \quad r_1^2 \equiv \theta_s = \frac{\lambda_1}{1 + \lambda_1} \quad . \tag{16.5}$$

We have seen (Sections 16.2 and 16.3) that the significance of both λ_1 (following the transformation $\lambda_1/[1 + \lambda_1]$) and r_1^2 can be assessed by reference to tables of the percentage points of Roy's largest-root distribution. Thus for all practical purposes the two test criteria may be considered equivalents. The relationships described are easily verified by reference to the numerical values of Table 16.3(b) and (c) and 16.5(a).

The importance of the identities or relationships described is that they illustrate the *formal equivalence* between one-way multivariate analysis of variance and canonical variates analysis in relation to a test of hypothesis (16.1).

We complete this sketch of connections between the various test criteria by showing the identity of the likelihood-ratio criterion Λ as it arises in the multivariate analysis of variance as a determinantal ratio and as a product of characteristic roots. Again we shall need to make use of the properties of matrices stated above. Expressed as a ratio of determinants, the criterion is

$$\Lambda = \frac{|Q_e|}{|Q_e + Q_h|}$$

$$= \frac{1}{|I + Q_e^{-1}Q_h|}$$

$$= \prod_{k=1}^{s} (1 + \lambda_k)^{-1} ,$$

where λ_k is the kth characteristic root of $Q_e^{-1}Q_h$. This establishes the identity. Verification of the relationship is provided by the quantities reported in Tables 16.3(a) and (b).

16.4.2 Further relationships: discriminant analysis. Having outlined the formal equivalence of one-way multivariate analysis of variance to canonical variates analysis with respect to the test of (16.1), we briefly consider some relationships between the methods beyond the acceptance decision.

The roots λ_k of $|Q_h - \lambda Q_e| = 0$ together with the vectors α_k which are solutions of the homogeneous equations $(Q_h - \lambda_k Q_e)\alpha_k = 0$, provide the essential quantities for discriminant analysis (e.g. see Timm, 1975, p. 374). The corresponding quantities in canonical analysis are the roots r_k^2 of $|Q_h - r^2 Q_{t*}| = 0$ and vectors b_k which are solutions of the homogeneous equations $(Q_h - r_k^2 Q_{t*}) b_k = 0$. It can be shown that the roots λ_k and r_k^2 are connected by the simple relationships (16.5) - for a proof, see Timm (1975, p. 86) - while the corresponding vectors are identical. These various quantities are reported for the discriminant and canonical analyses in Tables 16.3(b), 16.4(a), 16.5(a) and 16.7(a).

Tables 16.3(b) and 16.5(a) contain the roots λ_k and r_k^2 $(k = 1, \cdots, s)$, respectively. It is readily verified that *all* s roots are connected by the relations (16.5). Consequently, both analyses yield identical values of the test criterion, $\theta_s(s,m,n)$. While the largest roots, λ_1 and r_1^2, may be used to test the initial hypothesis (16.1), the subsequent roots $(k > 1)$ enable the dimensionality of the respective solutions to be established, i.e. they provide guidance as to the number of discriminant functions or canonical variates necessary to adequately account for the differences between vector-means.

Tables 16.4(a) and 16.7(a) report the characteristic vectors α_k and b_k of the two analyses. The vectors have been scaled so that the largest element of each unity. Comparison of the vectors shows corresponding elements, within rounding error and a possible reflection in sign, to be identical. The vectors may be used to obtain further interpretive indices, a number of which are reported in Tables 16.4(b) and (c), 16.6 and 16.7(b). The scores of species' centroids on the discriminant functions $L_k = \alpha_k' \bar{y}_i$ and canonical variates $V_k = b_k' \bar{y}_i$, where $\bar{y}_i$ is the 5×1 vector of response means for the *ith* species (see Table 16.1) are shown in Tables 16.4(b) and 16.7(b). The corresponding scores differ only by a scale factor together with in some cases a reflection in sign; as the composite variables are defined only up to a scale factor however (cf. Section 3.3), we see that the two sets of scores are essentially the same. It may also be verified that the structure correlations of the two analyses are also identical up to a constant of proportionality (Tables 16.4(c) and 16.6, lower-right.

TABLE 16.6: Nitrogen nutrition of eight grass species. Canonical variates analysis. Correlations between the original variables and the canonical variates.

Canonical variate	U_1	U_2	U_3	h_w^2	V_1	V_2	V_3	h_b^2
Species								
L. perenne	.608	-.110	.347	.502	.580	-.094	.189	.381
D. glomerata	.498	-.013	-.358	.376	.475	-.011	-.195	.264
P. bertolonii	.282	-.047	-.117	.095	.269	-.040	-.064	.078
H. pubescens	-.092	.084	.344	.134	-.086	.071	.188	.048
B. media	-.294	.512	-.163	.375	-.280	.436	-.089	.276
K. cristata	-.339	.124	.656	.561	-.323	.106	.358	.244
F. ovina	-.244	.313	-.519	.427	-.233	.266	-.283	.205
F. rubra	-.420	-.863	-.190	.957	-.401	-.734	-.104	.710
Variance extracted	.143	.143	.143	.429	.130	.103	.043	.276
Redundancy	.130	.104	.043	.277	.130	.103	.043	.276

TABLE 16.6 (Continued).

Canonical variate	V_1	V_2	V_3	h_w^2	U_1	U_2	U_3	h_b^2
Treatment (ppm N)								
1 (Y_1)	.671	.654	.169	.907	.640	.556	.091	.727
9 (Y_2)	.818	-.033	.366	.804	.780	-.028	.200	.649
27 (Y_3)	.887	.207	-.197	.868	.846	.176	-.108	.758
81 (Y_4)	.904	.118	.190	.867	.862	.100	.104	.764
243 (Y_5)	.912	-.318	.041	.935	.870	-.271	.022	.831
Variance extracted	.711	.117	.048	.876	.647	.085	.014	.746
Redundancy	.646	.085	.014	.746	.647	.085	.014	.746

TABLE 16.7: Nitrogen nutrition of eight grass species. Canonical variates analysis. Weights for the responses variables and scores of species' centroids on the canonical variates of the response set.

(a) Canonical weights. The weights have been scaled so that the largest element in each vector $\underset{\sim k}{b}$ is unity.

Response variable	b_1	b_2	b_3	b_4	b_5
Y_1	.316	1.000	.267	.592	-.726
Y_2	.091	-.327	1.000	-1.000	-.401
Y_3	.890	.207	-.802	-.567	.025
Y_4	.347	.215	.185	.267	1.000
Y_5	1.000	-.621	-.147	.529	-.450

(b) Co-ordinates of species' centroids on the canonical variates V_k, of the response variables.

Species	V_1	V_2	V_3	V_4	V_5
L. perenne	1.503	-.271	-.482	.570	-.034
D. glomerata	1.227	-.048	-.497	-.174	.145
P. bertolonii	0.693	-.113	-.162	-.440	-.060
B. media	-.729	1.134	-.223	.169	.169
K. cristata	-.836	.288	.914	-.121	.050
F. ovina	-.604	.696	-.722	.177	-.180
F. rubra	-1.029	-1.874	-.274	.090	.006
H. pubescens	-.227	.188	.482	-.271	-.096

From these results it is clear that the formal equivalence noted earlier between the multivariate analysis of variance and canonical variates analysis extends beyond the acceptance decision to discriminant analysis and canonical variates analysis. A more detached viewpoint enables us to recognize that one-way multivariate analysis of variance and discriminant analysis are in fact simply particular cases of canonical analysis which arise under specialization of one of the two sets of variables in canonical analysis.

16.5 Ecological Interpretation of the Results. Having shown the results of the separate analyses to be for all practical purposes identical, we shall confine our attention in what follows to the results of the canonical analysis.

Figure 16.1 displays the sample in the (V_1, V_2)-plane of the response space. V_1 differentiates the species into two classes – those whose centroids have positive scores on V_1 (*L. perenne, D. glomerata,* and *P. bertolonii*), and the remainder with zero or negative scores. V_1, however, does not fully account for the relationships among species, the species' centroids clearly departing from collinearity. Evidently, the species cannot be considered to represent a simple 'ecological series' with respect to their responses to increased concentrations of N. V_2 operates almost entirely within the species having negative scores on V_1, defining in particular a sharp contrast between *B. media* and *F. rubra.* The diversity of species' response shown in the Figure is decidedly small, being, for example, appreciably less than in a comparable experiment which dealt with the phosphorus nutrition of substantially the same group of species (Gittins, 1975). For convenience, we may characterize the configuration by means of three species-classes, the identity of which is enhanced by the confidence regions about the vector-means. The classes consist of those species with positive scores on V_1, those with zero or negative scores on V_1 *and* non-negative scores on V_2, and a third class comprised of *F. rubra* alone which has negative scores on both V_1 and V_2. *F. rubra* emerges as a particularly distinctive species. The confidence regions also draw attention to differences in variation *within* species; *L. perenne* is the most homogeneous species, while *K. cristata* and *D. glomerata* are the most variable. The direction of maximum scatter of replicate samples of particular species is a feature of some interest concerning the more variable species; compare, for example, the dispersions of *D. glomerata* and *K. cristata.* Differences in homogeneity between species may well reflect inherent differences in the plant material used in the experiment. *L. perenne*

alone was grown from the seed of a selected agricultural strain
(S. 23). The remaining species, on the other hand, were all grown
vegetatively from tillers collected in the field, a conscious effort
being made to acquire as much genetic diversity as possible of each
within the confines of the community of interest.

To interpret the canonical variates we turn to the structure
correlations and other indices of Table 16.6. First, notice that
the correlations between the U_k and the dummy variables corresponding to the species on which the U_k are defined (Table 16.6,
upper-left), are proportional to the scores of species on the
canonical variates, V_k (Table 16.7(b)). These correlations
therefore add nothing to what has already been established concerning
the comparative relations among species. For this reason, these
intraset correlations need detain us only briefly. We notice
that U_1 discriminates reasonably efficiently among the species
as a whole, while U_2 discriminates chiefly between B. mdeia
and F. rubra. The U_k each uniformly account for 14.3 percent
of the variance of the species' domain, while in contrast, the extent
of the explained species' variance achieved by the U_k, k = 1,···,3
varies considerably between species.

The intraset correlations of the first canonical variate V_1
of the response domain (Table 16.6, lower-left) are noteworthy
on three accounts; the correlations indicate that the response
variables all contribute to V_1 in the *same direction* with
respect to the linear combination of species represented by U_1,
secondly that each variable makes a *sizable contribution* ($\geq$.67)
to the relationship, and, moreover, one which *increases systematically in strength* from that associated with Y_1 (.67) to that
associated with Y_5 (.91). V_1 therefore appears to represent
both species' response to N in any concentration together with
a progressively increasing response to increasing dosage. In a
similar way, it can be seen that V_2 apparently corresponds to
a contrast between the highest and lowest N treatments. Together,
V_1 and V_2 account for 82.8 percent of the total variance of
the response domain and therefore efficiently summarize the variance
of this domain. The intraset communalities h_w^2 of the V_k
(k = 1,2) show that together V_1 and V_2 account for a substantial part ($\geq$.67) of the variance of each response variable.
In other words it appears that *all* the response variables contribute
substantially to the differentiation of species shown in Figure 16.1.
It is interesting to observe that these communalities convey

essentially the same information concerning the relative importance of the variables in this regard as do the univariate F tests associated with the multivariate analysis of variance (cf. Table 16.2).

The interpretation of relationships among species based on the within-set structure correlations of the V_k is borne out by the interset correlations between U_1 and U_2 of the species domain and the variables Y_h, $h = 1, \cdots, q$ of the response domain (Table 16.6, lower-right). Furthermore, the interset correlations show U_1 and U_2 to account for sizeable proportions ($\geq .61$) of the variance of each of the Y_h. Bearing in mind the sign and magnitude of the species' correlations with U_1 and U_2, however, we see that the direction and strength of the contributions of individual species to the explained variances differ. Thus, in relation to U_1, *L. perenne, D. glomerata* and *P. bertolonii* are all related in a direct sense with the response variables while the remaining species for the most part tend to vary inversely with the Y_h (cf. Table 16.1). In a similar way, recalling that U_2 represents essentially a contrast between *B. media* (.51) and *F. rubra* (-.86), we see from the correlations of the response variables with U_2, that *B. media* is directly related to Y_1 (1 ppm N) and inversely to Y_5 (243 ppm N), while the converse is true of *F. rubra*. Reference to the experimental results of Table 16.1 shows this indeed to be the case; the response of *F. rubra* is conspicuously low at the 1 ppm N dosage and notably high, at least among the less-responsive species, at the 243 ppm N treatment, while the opposite is true of *B. media*. These relationships clarify the nature of the contrast between these two species specified by U_2. The redundancies generated by U_1 and U_2, are, respectively, $U^2_{y|u_1} = .640$ and $U^2_{y|u_2} = .085$. We see that the species' relationships specified by U_1 enable considerably more (7.6x) of the total variance of the response domain to be accounted for, than that specified by U_2. The contribution of U_3 in this regard is negligible $(U^2_{y|u_3} = .014)$. Together U_1 and U_2 account for 75 percent of the total response-domain variance. The corresponding interset communalities, h^2_w, show the explained variance to be equitably distributed among the response variables, though it increases systematically across them.

In short, it appears that:

(a)   the comparative relations among species displayed in Figure 16.1 can be comprehended in terms of their differential response to N (i) at all treatment levels but especially the higher dosages, and (ii) at the extreme treatment levels; and that

(b)   the discrimination among species effected by (i) and (ii) accounts for 75 percent of the total variation of the response variables.

Any attempt to generalize the sample results to some wider population is hazardous in view of the apparent lack of homogeneity among the within-groups dispersions in the population (Section 16.2). As a consequence of the nonhomogeneity, the decision to reject hypothesis (16.1) of the equality of the population vector-means can be questioned. It must therefore be accepted that inference based on the results obtained must either be curtailed or, alternatively, be guided by extra-statistical considerations. In an effort to throw further light on the extent of the nonhomogeneity of the dispersions, univariate tests of the homogeneity of the species' variances on V_1 and V_2, respectively, were performed following the canonical analysis. In relation to V_1 Bartlett's test gave χ_s^2 (7) = 12.47 < χ_α^2 (7) = 14.07 for α = .05, while in relation to V_2 the corresponding result was χ_s^2 (7) = 11.69 < χ_α^2 (7) = 12.02 for α = .10. Accordingly, there can be little doubt that the species' variances on each canonical variate, at least, are homogeneous. The tests therefore suggest that the species' dispersion matrices may not grossly violate the homogeneity condition and so lend support to the wider applicability of the results. Fortunately, the insight into the structural relationships among species conveyed by Figure 16.1, as well as other purely descriptive aspects of the analysis, hold, irrespective of the homogeneity or otherwise of the dispersion.

The relative number of variables to samples is a further aspect of the analysis which has a bearing on how the results are assessed or utilized. The v/s ratio in the present case is 12/40 = .30. This is a comparatively large value and can be expected to have had some inflationary effect on the canonical correlation coefficients and other interpretive devices. It would be wise, therefore, to accept the apparent strength of certain relationships with some reserve.

16.6 Conclusions. Before attempting to assess the worth of the analysis, it will be helpful to recall its purpose. The objectives were two-fold, namely (a) to clarify one aspect of the comparative ecology of the species; and (b) to show the existence and nature of connections between canonical analysis, on the one hand, and multivariate analysis of variance and discriminant analysis, on the other.

The salient ecological results may be summarized as follows.

(a) A fundamental distinction appears to exist between the species with respect to their overall response to the nitrogen treatments administered. *L. perenne, D. glomerata* and *P. bertolonii* are characterized by their responsiveness at *each* treatment level and by their systematic increase in performance with increasing dosage. *B. media, F. ovina, F. rubra, H. pubescens* and *K. cristata,* on the other hand, are comparatively unresponsive at *all* treatment levels.

(b) A subsidiary distinction occurs among the less-responsive species in relation to their behavior at the extreme treatment levels. Species exemplified by *B. media* perform comparatively poorly at the highest dosage administered and comparatively well at the lowest, while the converse is true of *F. rubra.*

(c) A substantial part of the total variance of the species' responses can be accounted for by the distinctions between species effected by the canonical variate U_1 and, to a lesser extent, by U_2.

The existence of two fundamentally different classes of species is consistent with existing views on the comparative ecology of the species. *L. perenne, D. glomerata* and *P. bertolonii* are all well-known constituents of fertile grassland, while the remaining species are known to thrive in grassland developed on improverished soils (Tansley, 1939; Hubbard, 1954). Furthermore, the ranking of the responsive species effected by U_1, namely *P. bertolonii* < *D. glomerata* < *L. perenne* (see also Figure 16.1) is also supported by field knowledge of the ecology of these species. Such external substantiation of the results shows canonical analysis to be capable at the very least of identifying and drawing attention to those relationships of ecological interest about whose existence we may be reasonably sure, even though their precise specification may previously have been lacking. The subsidiary distinction among the less-responsive species, in contrast, has not so far as I am aware, been commented on previously. This aspect of the results may therefore perhaps be an expression of the penetrative power of the method.

The result is useful in directing attention to an aspect of the comparative ecology of the species which could conceivably repay further attention.

The results obtained are useful in a number of other ways. They indicate, for example, that in any attempt to develop a deductive model of the nitrogen nutrition of this group of species, two parameters would be required to take account of the three different response-patterns among species. Moreover, the results illustrate the opportunities afforded by canonical analysis in the analysis of designed experiments. A distinctive and valuable property of the method in this context is the graphic description of the outcome of an experiment which the canonical variates provide (see Figure 16.1).

The principal methodological finding is that one-way multivariate analysis of variance and discriminant analysis are jointly equivalent to canonical analysis. The practical significance of this relationship for us is that it will enable us to better define the role or domain of canonical analysis in ecology. We can now appreciate, for example, that investigations which might call for analysis by multivariate analysis of variance or discriminant analysis, can be performed equally well by canonical analysis. Moreover, we have seen that an important class of univariate analyses of variance (split-plot analyses) can also be undertaken profitably by canonical analysis. In Section 18 we shall see that in exploratory studies the use of canonical analysis may in fact be more appropriate and more rewarding than the alternatives considered here.

In view of the points made above, there are grounds for concluding that the analysis has contributed to our understanding of the comparative ecology of the species examined and to our perception of what can be accomplished by canonical analysis.

17. HERBIVORE-ENVIRONMENT RELATIONSHIPS IN THE RUWENZORI NATIONAL PARK, UGANDA

17.1 Introduction. The Ruwenzori National Park lies astride the equator in the Western Rift Valley of Uganda. The Park contains a variety of plant communities, including forests, swamps and several kinds of grassland. The grasslands are noteworthy in that they support the highest recorded large mammal biomass of any natural area of the world (Bourlière, 1965; Coe, Cumming and Phillipson, 1976). An estimate of 294.9 kg ha^{-1} for the average year-round standing-crop herbivore biomass is given by Field and Laws (1970). Interrelationships between the herbivore species in particular and their environment - vegetation, climate, soils and other animal species - have been the subject of investigations by R. M. Laws and C. R. Field (e.g. see Field and Laws, 1970). These workers have established among other things that the overall distribution of herbivores in the Park shows considerable heterogeneity with respect to different vegetation types and in relation to the presence of standing water and the occurrence of fire. Table 2 of Field and Laws' (1970) paper contains estimates of the mean density km^{-2} of nine herbivore species in ten study areas. The data of this Table provided the starting point for the present study. Proceeding along rather different lines from Field and Laws, the data were examined for evidence of association between herbivores and study areas. The question of the existence of association was approached in two ways. The data were first analysed by contingency table analysis and then by canonical analysis. The purpose of the study was twofold, namely (a) to show the utility of canonical analysis in the context of an ecological problem involving the independence of two categorical variables; and (b) to illustrate the close connection between canonical analysis and the analysis of 2-way frequency tables.

The herbivore species of interest were elephant (*Loxodonta africana* Blumenbach), hippopotamus (*Hippopotamus amphibius* L.) warthog (*Phacochoerus aethiopicus* Pallas), buffalo (*Syncerus caffer* Sparrman), Uganda kob (*Adenota kob* Neumann), waterbuck (*Kobus defassa* Rüppell), reedbuck (*Redunca redunca* Pallas), bushbuck (*Tragelaphus scriptus* Pallas) and topi (*Damaliscus korrigum* Ogilby). Of these, hippopotamus alone was the subject of a management cropping program designed to reduce the numbers of this species in certain areas. The sites or study areas examined were selected principally for their accessibility and representativeness of the varied grassland communities and animal associations of the Park. A summary of certain physical characteristics of the sites is given in Table 17.1. On the basis of these and other characteristics, Field and Laws (1970) grouped the sites into four broad categories, namely: area 2; areas 7, 9 and 10; areas 1, 3, 5 and 6; and areas 4 and 8. Reference should be made to the original paper by Field and Laws (1970) for additional information concerning the sites and the distribution of particular herbivore species in relation to them. Notice that in view of the high degree of selectivity exercised in choosing sites, the sites examined did not comprise an unbiased sample of the biotic communities of the Park.

The density estimates of Field and Laws were modified prior to analysis in order to enhance compatability between the data and the requirements of the analytic techniques to be used with them. Modification involved (a) rounding estimates to their nearest integer values; and (b) eliminating the entries for two species (bushbuck, reedbuck) and one site (site 2). The items

eliminated were unusual in being either present in very small
numbers or in supporting a very low overall herbivore density and
it was considered that they might therefore exert a disproportionate
effect on the course of analysis (Anderberg, 1973, p. 221; Teil,
1975). It was however possible to re-introduce the items in
question into the analysis at a later stage, so that information
concerning them was not lost. The resulting table of counts or
frequencies is shown in Table 17.2 and forms the basis of
the analyses which follow. Ecological interest centers on the
existence of association between the herbivore species and the
study areas.

Table 17.2 clearly represents a 2-way contingency table.
The familiar χ^2-test for independence in r × c tables is
therefore readily identified as one analytic approach to the
declared ecological objective. If on the basis of the test the
existence of association can plausibly be demonstrated, then the
nature of the relationship can be pursued by partitioning the
overall chi-squared statistic for the table, χ_o^2. Various schemes
for partitioning χ_o^2 exist. The most widely used of these
involve either the exact partition of χ_o^2 into additive components,
each of which corresponds to a particular 2 × 2 table derived
from the original table; or harnessing subject-matter insight
to guide the partitioning. Here, however, in order to advance
our objective of illustrating the relationship of canonical
analysis to contingency table analysis, we shall adopt a less
well-known alternative based on the latent roots of a matrix
derived from the 2-way table of frequencies (Section 17.2). We
shall then go on to show how the same ecological objective can
be approached by canonical analysis (Section 17.3). In Section
17.4 a comparison of the two methods will be made based on the
results obtained, followed in Section 17.5 by the ecological
interpretation of the results. Finally, an assessment of the
analyses ecologically and in more general terms will be made
(Section 17.6).

Recalling the lack of independence among the study areas,
inference in a probabalistic sense is hardly possible. Yet the
data can nevertheless be profitably examined in a less formal,
exploratory spirit. Such a viewpoint provides the justification
for the substantive evaluation of results of Section 17.5. At the
same time, in order to allow a fuller account of the procedures
to be given and to facilitate comparison between them, we shall
also proceed *as if* the data did constitute a random sample.
This will enable various test criteria and their *nominal*
significance to be calculated.

TABLE 17.1: Ruwenzori National Park, Uganda. Physical characteristics of study areas.

Study Area	Vegetation	Proximity to standing water	Management (cropping)
1	Short grass/thicket	Close	Moderate
2	Tall grass with thicket	Distant	None
3	Short grass/thicket	Close	None
4	Short grass/thicket	Close	Substantial
5	Short grass/thicket	Moderate	None
6	Short grass/thicket	Close	None
7	Short grass	Distant	None
8	Short grass/thicket	Moderate	Substantial
9	Short grass	Distant	None
10	Short grass	Distant	None

17.2 Contingency Table Analysis. The general 2-way contingency
table arises where we have a sample of n observations classified
by two attributes A and B divided into $\{A_i\}$ and $\{B_j\}$
categories respectively. We shall denote a contingency table of
this kind by the matrix $A = [a_{ij}]$, where $i = 1, \cdots, r$ and
$j = 1, \cdots, c$ and represent the *ith* row and *jth* column totals of
A by $a_{i.}$ and $a_{.j}$, respectively. For convenience and without
loss of generality we assume that $r \leq c$. The usual null hypothesis
of independence states that in the absence of association the
expected frequencies in any row (column) are proportional to the
marginal frequencies. In terms of probabilities the null hypothesis
may be expressed

$$H_o: \quad \pi_{ij} = \pi_{i.} \; \pi_{.j} \; , \quad i = 1, \cdots, r \quad \quad (17.1)$$
$$j = 1, \cdots, c$$

where π_{ij} represents the probability in the population of an
observation belonging to the *ith* row category *and* the *jth* column
category. Similarly, $\pi_{i.}$ represents the corresponding probability
of an observation belonging to the *ith* row category of A and
$\pi_{.j}$ to the *jth* column category of A. The alternative hypothesis
is that the row and column attributes are not independent or, in other

words, are associated (H_1: $\pi_{ij} \neq \pi_{i.} \pi_{.j}$). The quantities in
(17.1) are readily estimated from sample data, and, as is well-
known, hypothesis (17.1) can be tested by determining the extent
of departures from expectation under H_o in the observed frequencies. The familiar test statistic is

$$\chi_o^2 = \sum_i \sum_j \frac{(a_{ij} - a_{i.} a_{.j}/n)^2}{a_{i.} a_{.j}/n} \tag{17.2}$$

where n is the total number of observations and a_{ij} and
$a_{i.} a_{.j}/n$ are the observed and expected frequencies respectively.
χ_o^2 may then be referred to the chi-squared distribution with
$(r-1)(c-1)$ degrees of freedom. Hypothesis (17.1) is rejected
with confidence $1-\alpha$ if $\chi_o^2 > \chi_\alpha^2(r-1)(c-1)$. Rejection of H_o
is normally followed by further analysis designed to throw light
on the nature of the relationship or relationships present.
Continued analysis of this kind involves the partition of χ_o^2.

The procedure used below for partitioning χ_o^2 derives
from Williams (1952). The method has been described and used
in ecological contexts by Hatheway (1971) and Orlóci (1978, p.
152). Nevertheless it remains relatively little-known. For
this reason we outline the method from a slightly different viewpoint before going on to apply it to the rangeland data. The
account in places relies heavily on Maxwell's (1973) particularly
clear exposition.

The procedure proposed by Williams (1952) for the analysis
of 2-way contingency tables is based on properties of the matrices
of observed and expected frequencies and of the matrix of departures
from expectation. In particular, use is made of an identity
between the trace of a matrix and the sum of its latent roots to
achieve an additive decomposition of χ_o^2. The rationale is as
follows. In the absence of association Williams (1952, p. 275)
has pointed out that the expected frequencies will form a matrix
of rank 1. The observed frequencies in practice will differ from
their expectations, if only because of sampling error. Under the
null hypothesis they will form a matrix of at most rank r, while
the departures from expectation will form a matrix of at most rank
$r-1$. The object of the usual significance tests is to decide
whether the departures from expectation are consistent with the
assumption under H_o that the expectations are of unit rank. If
the matrix of departures is not consistent with the expectations
being of unit rank, then the assumption of no association is untenable.

TABLE 17.2: Ruwenzori National Park, Uganda. Density km^{-2} *of nine large herbivore species in ten study areas.*

Herbivore	1	2*	3	4	5	6	7	8	9	10	Total
Elephant	1	2	3	3	1	4	2	2	0	0	16
Warthog	1	0	3	8	1	3	1	1	4	2	24
Hippopotamus	15	0	28	1	13	21	1	4	2	6	91
Reedbuck*	0	0	0	0	0	0	0	0	1	0	-
Waterbuck	7	0	3	10	1	4	2	1	0	0	28
Kob	9	1	2	0	6	0	44	10	78	71	220
Topi	0	0	0	0	0	0	0	0	30	83	113
Bushbuck*	0	0	1	3	0	1	0	0	0	0	-
Buffalo	12	5	7	25	21	18	13	18	22	17	153
Total	45	-	46	47	43	50	63	36	136	179	645

*Omitted from both χ^2-analysis and canonical analysis; not entered into marginal or grand totals.

Source: Field and Laws (1970), Table 2 with modifications.

The analysis is performed largely on a matrix which we may call $\underset{\sim}{T}$ calculated from the matrix of observed frequencies $\underset{\sim}{A}$ in a manner shortly to be described. In order to relate Williams' method to the more familiar alternatives based on (17.2), however, we first observe that the quantity (17.2) can be re-expressed as

$$\chi_0^2 = n \left[\sum_i \sum_j \frac{a_{ij}^2}{a_{i.}a_{.j}} - 1 \right] . \qquad (17.3)$$

Williams (1952) has shown that the first term within the brackets of (17.3) is equal to the trace of the $r \times r$ symmetric matrix $\underset{\sim}{T}$, where

$$\underset{\sim}{T} = \underset{\sim}{G}\underset{\sim}{G}'$$

and $\underset{\sim}{G} = [g_{ij}]$ is an $r \times c$ matrix formed from the elements of $\underset{\sim}{A}$ by setting

$$g_{ij} = a_{ij}/(a_{i.}a_{.j})^{\frac{1}{2}} .$$

Equation (17.3) can therefore be expressed

$$\chi_o^2 = n[\text{tr}(\underset{\sim}{T}) - 1],\qquad\qquad(17.4)$$

where $\text{tr}(\underset{\sim}{T})$ stands for the trace of $\underset{\sim}{T}$.

In the absence of both association and sampling errors, χ_o^2 is zero. In such cases, it follows from (17.4) that

$$\text{tr}(\underset{\sim}{T}) = 1 \ . \qquad\qquad(17.5)$$

It can be shown (e.g. Maxwell, 1973, p. 156) that $\underset{\sim}{T}$ is a matrix of unit rank; consequently $\underset{\sim}{T}$ has only a single latent root (Timm, 1975, p. 80), the value of which must therefore be unity. It is now apparent that the term -1 in the overall criterion (17.4) refers to a latent root of $\underset{\sim}{T}$ which arises solely from the expected values of the observations in $\underset{\sim}{A}$ *and*, moreover, that a non-zero value of (17.4) must depend on the other latent roots of $\underset{\sim}{T}$.

Where, in contrast, the row and column attributes are asso- ciated, the rank of $\underset{\sim}{T}$ will exceed 1 and may at most reach r. Thus the number of latent roots will exceed one. In such cases the largest latent root of $\underset{\sim}{T}$ will be unity. Denoting the latent roots after the largest $(\lambda_1 = 1)$ by λ_i, $i = 2,\cdots,r$ and recalling that the sum of the latent roots of a matrix is equal to the trace of the matrix, we may therefore write

$$\sum_{i=2}^{r} \lambda_i = (\text{tr}(\underset{\sim}{T}) - 1).$$

Substituting this expression into (17.4) we arrive at

$$\chi_o^2 = n[\sum_{i=2}^{r} \lambda_i] \ . \qquad\qquad(17.6)$$

Expression (17.6) shows how the total chi-square, χ_o^2, given by (17.4) can be partitioned into additive components in terms of the latent roots of $\underset{\sim}{T}$. Observe carefully that the partition of χ_o^2 does not depend on the largest root, $\lambda_1 = 1$. The $(r-1)(c-1)$ degrees of freedom associated with the overall χ_o^2 for the table may similarly be partitioned (Maxwell, 1973, p. 156).

It is widely believed that each component $n\lambda_i (i > 1)$ of χ_o^2 given by (17.6) is a chi-squared variate based on a known number of degrees of freedom. Lancaster (1963) however has shown this supposition to be ill-founded. It would be well therefore to be wary of chi-squared tests based on such partitions of χ_o^2. In assessing the components reliance may accordingly have to be placed simply on their magnitude or on their percentage contribution to $[tr(\underset{\sim}{T}) - 1]$ or χ_o^2. It does however seem possible that the significance of individual components might perhaps be assessed by referring the roots, λ_i, directly to the gcr distribution, $\theta_\alpha(s,m,n)$.

TABLE 17.3: *Ruwenzori National Park, Uganda. Contingency table analysis of association between seven herbivore species and nine study areas.* $\underset{\sim}{T} = \underset{\sim}{G} \underset{\sim}{G}'$ *matrix, where the elements of* $\underset{\sim}{G}$ *are given by* $g_{ij} = a_{ij}/(a_{i.} a_{.j})^{\frac{1}{2}}$.

$$\underset{\sim}{T} = \underset{\sim}{G} \underset{\sim}{G}' = \begin{bmatrix} .0579 & .0551 & .1169 & .0686 & .0408 & .0000 & .1144 \\ .0551 & .0819 & .0887 & .0917 & .0624 & .0347 & .1335 \\ .1169 & .0887 & .3909 & .1287 & .0804 & .0318 & .2186 \\ .0686 & .0917 & .1287 & .1374 & .0426 & .0000 & .1602 \\ .0408 & .0624 & .0804 & .0426 & .4961 & .3179 & .2130 \\ .0000 & .0347 & .0318 & .0000 & .3179 & .3992 & .0968 \\ .1144 & .1335 & .2186 & .1602 & .2130 & .0968 & .3343 \end{bmatrix}$$

$$tr(\underset{\sim}{T}) = \Sigma t_{ii} = 1.8977.$$

The latent vectors $\underset{\sim}{x}_i$ associated with the roots $\lambda_i (i > 1)$ of $\underset{\sim}{T}$ are informative in interpreting the partitions of χ_o^2 given by (17.6). Moreover, the latent vectors $\underset{\sim}{y}_i$ of the related $c \times c$ matrix $\underset{\sim}{G}'\underset{\sim}{G}$ (which has the same r nonzero latent roots λ_i as $\underset{\sim}{T}$) in the singular value decomposition of $\underset{\sim}{G}$ are also useful for this purpose. In practice the $\underset{\sim}{y}_i$ are conveniently obtained from the latent vectors of $\underset{\sim}{T}$ by the relation $\underset{\sim}{y}_i = \underset{\sim}{G}'\underset{\sim}{x}_i$. The magnitude of the elements of $\underset{\sim}{x}_i$ and $\underset{\sim}{y}_i$ may be interpreted as weights which reflect the relative importance of the row and column attribute-states, respectively, in the *ith* component of the global association. More importantly, the latent vectors can be used to obtain 'scores' by means of which the attribute-states defining the contingency table can be scaled or metricized. The *sth* element, ξ_{si}, $(s=1,\cdots,r)$ of the $r \times 1$ vector of scores $\underset{\sim}{\xi}_i$ for the row attribute-states corresponding to λ_i is given by

$$\xi_{si} = x_{si}/m_i, \quad i = 2,\cdots,r \tag{17.7}$$

where x_{si} is the *sth* element of the *ith* latent vector of $\underset{\sim}{T}$ and $m_i = (a_{i.}/n)^{\frac{1}{2}}$. Similarly, the *hth* element, η_{hi}, $(h=1,\cdots,c)$ of the $c \times 1$ vector of scores η_i for the column attribute-states corresponding to λ_i is

$$\eta_{hi} = y_{hi}/n_i, \quad i = 2,\cdots,r \tag{17.8}$$

where y_{hi} is the *hth* element of the *ith* latent vector $y_i = \underset{\sim}{G}'\underset{\sim}{x}_i$ and $n_i = (a_{.j}/n)^{\frac{1}{2}}$. The resulting score vectors $\underset{\sim}{\xi}_i$ and $\underset{\sim}{\eta}_i$ are standardized to zero mean and unit variance. Other standardizations may however be preferred. For example, the latent vectors may be standardized so that the sum of squares of their elements are equal to λ_i before being used to calculate the scores. Alternatively, the score vectors $\underset{\sim}{\xi}_i$ and $\underset{\sim}{\eta}_i$ may themselves be standardized in a similar way so that:

$$\sum_{s=1}^{r} \xi_{si}^2 = \sum_{h=1}^{c} \eta_{hi}^2 = \lambda_i \ .$$

These procedures have the advantage that the standardized scores more clearly reflect the contribution of the latent roots to the global chi-squared.

Thus it turns out that the contingency table analysis enables metric information to be extracted from strictly nominally-scaled variables. The resulting metric scales often prove to be open to substantive interpretation.

TABLE 17.4: Ruwenzori National Park, Uganda. Contingency table analysis of association between seven herbivore species and nine study areas. Latent roots of $\underset{\sim}{T} = \underset{\sim}{G} \underset{\sim}{G}'$ and the partition of χ_0^2 into additive components by the relationship $\chi_0^2 = n\lambda_i$, $i = 2, \cdots, r$.

i	λ_i	df*	$n\lambda_i = \chi_i^2$	$\%\chi_0^2$
1	.999942	–	–	–
2	.563173	13	363.25	62.79
3	.169158	11	109.11	18.86
4	.109106	9	70.37	12.16
5	.037283	7	24.05	4.16
6	.013226	5	8.53	1.47
7	.005011	3	3.23	0.56
Total	1.896899	48	578.54	100.00

*Degrees of freedom: $r + c - (2i - 1)$

17.2.1 Results. The results of the contingency table analysis are summarized in Tables 17.3 – 17.6. The significance levels reported are purely nominal. They were obtained by treating the sample observations as though they were statistically independent. In the absence of valid tests, we shall be obliged to fall back on less formal procedures in assessing the results.

Table 17.3 reports the $\underset{\sim}{T}$ matrix. The sum of the diagonal elements, Σt_{ii}, yields the trace of the matrix; we see that $\text{tr}(\underset{\sim}{T}) = 1.8977$. In relation to unity, which is the value expected in the absence of association and errors of sampling, this represents a sizeable quantity in the present context. The magnitude of $\text{tr}(\underset{\sim}{T})$ therefore suggests that the herbivores and study areas are unlikely to be independent. A test of the overall hypothesis (17.1) of independence based on $\text{tr}(\underset{\sim}{T})$ may be made using (17.4) as follows:

$$\chi_o^2 = n[tr(\underset{\sim}{T}) - 1]$$

$$= 645(1.8977 - 1)$$

$$= 579.02, \text{ with } (6)(8) = 48 \text{ degrees of freedom.}$$

As $\chi_o^2(48) = 579.02 > \chi_\alpha^2(48) = 93.22$ for $\alpha = .0001$ the test would lead us to reject hypothesis (17.1). Clearly, the data do not support the assumption that the herbivores and study areas are independent; accordingly, there would be grounds for concluding that the entities are associated.

TABLE 17.5: Ruwenzori National Park, Uganda. Contingency table analysis of association between seven herbivore species and nine study areas. Normalized latent vectors $\underset{\sim}{x}_i$ and $\underset{\sim}{y}_i$ corresponding to the latent roots λ_2, λ_3 and λ_4.

$\underset{\sim}{x}_2$	$\underset{\sim}{x}_3$	$\underset{\sim}{x}_4$	$\underset{\sim}{y}_2$	$\underset{\sim}{y}_3$	$\underset{\sim}{y}_4$
-0.2144	0.0458	0.0075	-0.2675	-0.0560	0.0555
-0.1680	0.2258	0.3094	-0.4135	-0.5271	0.1464
-0.5319	-0.7020	-0.2360	-0.3224	0.5966	-0.5900
-0.3001	0.2738	0.3322	-0.2436	-0.0223	0.1151
0.4334	0.2120	-0.5977	-0.4047	-0.1901	-0.0378
0.5259	-0.4042	0.6973	0.1112	0.3560	0.5683
-0.2973	0.4132	0.1098	-0.1247	0.2312	0.1073
			0.3416	0.1556	0.2676
			0.5348	-0.3494	-0.4547

The latent roots λ_i of $\underset{\sim}{T}$ are reported in Table 17.4. The first root, λ_1, is, for all practical purposes unity, as expected. This root it will be recalled refers to the contribution of the expected frequencies to the trace of $\underset{\sim}{T}$. As λ_1 does not contribute to the association, it can henceforth be ignored. Of the remaining roots, λ_2 is clearly dominant; it absorbs 63 percent of the trace after the effect of λ_1 has been removed, while λ_3 and λ_4 account respectively for some 19 percent and 12 percent of the trace. Collectively, λ_2, λ_3 and λ_4 absorb 94 percent of the trace of $\underset{\sim}{T}$ after removing λ_1, or, equivalently, of the global chi-squared, χ_o^2. This figure provides a useful yardstick as to the dimensionality of the association, which for practical purposes can be taken to be three. Table 17.4 also shows the separate contributions of the partitions $n\lambda_i$ (i > 1) to the overall chi-squared, χ_o^2. It is easy to verify that, within rounding error, the latent roots do provide an additive decomposition of the total chi-squared for the table calculated above (χ_o^2 = 579.02). Moreover, the fundamental identity between the trace and the sum of the latent roots of a matrix which forms the algebraic basis of the method, is readily verified from the quantities given in Tables 17.3 and 17.4.

The normalized latent vectors $\underset{\sim}{x}_i$ and $\underset{\sim}{y}_i$ and the vectors of attribute-state scores $\underset{\sim}{\xi}_i$ and $\underset{\sim}{\eta}_i$ which correspond to the λ_i (i = 2,···,4) appear in Tables 17.5 and 17.6. Following Maxwell's (1973) recommendation, the score vectors have been standardized so as to reflect the magnitude of the latent root to which they correspond. Vectors of both kinds may be informative in interpreting the individual contributions to χ_o^2 of the partitions $n\lambda_i$ (i > 1). This point is true irrespective of any statistical significance which may or may not attach to the particular components $n\lambda_i$ themselves. Specifically, the sign and magnitude of the elements comprising the vectors enable inter-relationships between attribute-states to become apparent. In the absence of precise significance tests, in deciding on the number of ecologically informative components of χ_o^2, we have been guided principally by the substantive meaning, if any, which could be placed on these vectors, together with the percentage of $[tr(\underset{\sim}{T}) - 1]$ or χ_o^2 associated with the corresponding root, λ_i. Comparison of Tables 17.5 and 17.6 reveals an interesting feature concerning the latent vectors and the score vectors. This is that there is an appreciable degree of similarity between the numerical values of corresponding vectors. The similarity derives from the simple relationship between the elements of the two kinds of vector (see equations (17.7) and (17.8)). Where the similarity is appreciable, as here, obviously little would be gained from the interpretation of both sets of vectors.

TABLE 17.6: Ruwenzori National Park, Uganda. Contingency table analysis of association between seven herbivore species and nine study areas. Scores for row ξ_i and column η_i attribute-states corresponding to latent roots λ_i, $1 = 2, 3, 4$.

Row attribute-state	ξ_2	ξ_3	ξ_4	η_2	η_3	η_4	Column attribute-state
Elephant	-1.0216	.1194	.0159	-0.7599	-.0871	.0693	1
Warthog	-0.6537	.4816	.5298	-1.1617	-.8117	.1812	3
Hippo	-1.0628	-.7686	-.2077	-0.8964	1.0811	-.8586	4
Waterbuck	-1.0806	.5403	.5264	-0.7080	-.0356	.1472	5
Kob	0.5568	.1493	-.3380	-1.0909	-.2809	-.0449	6
Topi	0.9429	-.3970	.5502	0.2669	.4685	.6006	7
Buffalo	-0.4581	.3489	.0745	-0.3963	.4026	.1499	8
				0.5584	.1394	.1925	9
				0.7618	-.2728	-.2851	10

Further examination of the results and their ecological assessment is deferred until after consideration of the canonical analysis, to which we now turn.

17.3 Canonical Analysis. Canonical analysis provides an alternative means of investigating independence in 2-way tables. Recall that the formal objective of canonical analysis is to obtain linear transformations of two sets of variables such that the product-moment correlation coefficient between the transformed variables is maximized. In the present context, the null hypothesis of independence becomes the nullity of the population intercorrelation matrix $\underset{\sim}{\Sigma}_{12}$ between variables representing the two classifications of the frequency table. An equivalent statement of the null hypothesis in terms of population correlation coefficients ρ_k is also possible. We may therefore write

$$H_o: \underset{\sim}{\Sigma}_{12} = \underset{\sim}{0} \quad \text{or} \quad \rho_k = 0 \quad \text{for all } k, \quad k = 1, \cdots, s. \quad (17.9)$$

The largest sample squared canonical correlation coefficient, r_1^2, provides a suitable test criterion for hypothesis (17.9). The statistic is conveniently referred to gcr distribution, $\theta_\alpha(s,m,n)$.

Wilk's lambda $\Lambda = \Pi(1 - r_k^2)$ provides an alternative criterion, which, following Bartlett's transformation, may be referred to the χ^2-distribution. Rejection of H_o leads to acceptance of the alternative hypothesis - namely, that the attributes defining the frequency table are correlated. Where H_o can be rejected, examination of the remaining squared correlations and the latent vectors will generally provide insight as to the nature of the interrelationships present.

Before proceeding to the canonical analysis it is first necessary to express the data of the contingency table (Table 17.2) in a form suitable for canonical analysis. This calls for writing the r × c matrix $\underset{\sim}{A}$ of the 2-way table as an *incidence matrix*, $T(A)$, of order (r + c) × n. An incidence matrix consists of binary variables and has the special property that its row sums are equal to the row and column totals of A. A matrix of this kind is readily constructed from A by creating two sets of binary-valued dummy (or indicator) variables to represent the classifications of the 2-way table. Thus r dummy variables are initially required for the row categories (herbivores) of A and c dummy variables for the column (study area) categories. Each doubly-classified observation within the body of the frequency table is then coded appropriately with respect to both sets of dummy variables. For example, an observation of a particular herbivore in a particular study area would be given a score of one on the dummy variable of each set corresponding to the category in question and zero on all others. A simple, worked example using hypothetical data may help to clarify the procedure (Table 17.7). A 2-way frequency table is represented by the matrix $\underset{\sim}{A}$ of order 3 × 4, beneath which is the 7 × 10 incidence matrix $T(A)$ corresponding to it. The first three rows of $T(A)$ consist of dummy variables representing the row attribute-states or categories of $\underset{\sim}{A}$, while the four remaining rows represent the column categories of $\underset{\sim}{A}$. It is readily verified that the row sums of $T(A)$ are identically the row and column totals of $\underset{\sim}{A}$. Each column of $T(A)$ corresponds to an individual observation or frequency record in $\underset{\sim}{A}$. Thus in all $T(A)$ contains ten columns, ten being the total number of observations in $\underset{\sim}{A}$.

One further step is required before the canonical analysis can be undertaken. We have seen (Section 6.1) that k - 1 dummy variables are sufficient to represent any classificaiton into k mutually exclusive and exhaustive categories. Consequently, two rows of $T(A)$ are redundant, one belonging to the dummy variables corresponding to the row categories of $\underset{\sim}{A}$ and one belonging to the dummy variables for the column categories of $\underset{\sim}{A}$. The redundancy may be removed by deleting one row from each subset of variables in $T(\underset{\sim}{A})$. The choice as to the rows for deletion is unrestricted. In the example the dummy variables corresponding to the last row category and the last column category of $\underset{\sim}{A}$ have been deleted. In this way we arrive at the 'data' matrix $\underset{\sim}{X}$ of order 5×10 for canonical analysis. Each observation of the original 2-way table is represented in $\underset{\sim}{X}$ by a *partitioned* 5×1 column vector $\underset{\sim}{x}_j$, $j = 1, \cdots, 10$, which completely specifies the row and column attribute-states which correspond to a particular observation. Our interest focuses on the correlations within and between the two sets of variables which collectivley comprise $\underset{\sim}{X}$. The internal structure of the 5×5 correlation matrix of these variables is readily investigated by canonical analysis.

A 'data' matrix of precisely the kind described was constructed from the 2-way table of frequencies (Table 17.2) for the rangeland data. Buffalo and site 1 (short grass/thicket) were the herbivore and study area attribute-states deleted in arriving at the $\underset{\sim}{X}$ matrix. Consequently, observations falling in either or both categories came to be specified by a score of zero on *all* members of the corresponding subset of subsets of dummy variables.

For the canonical analysis we therefore had a biased sample of $N = 645$ observations, $p = 8$ indicator variables designating nine study areas, and $q = 6$ indicator variables designating seven herbivore species. The ecological objective of the analysis was to assess the evidence for association between the herbivores and study areas and to characterize the association, if present. There were two further, methodological objectives. First to show how canonical analysis can be used to analyse the familiar $r \times c$ contingency table; and secondly, to draw attention to relationships which exist between canonical analysis and contingency table analysis.

The results of the canonical analysis appear in Tables 17.8 – 17.11.

TABLE 17.7: Construction of an incidence matrix, $T(A)$, and 'data' matrix, X, for canonical analysis from a 2-way table of frequencies represented by A. Dotted line separates indicator variables corresponding to the rows (above) and columns (below) of A, respectively.

$$A = \begin{bmatrix} 1 & 2 & 1 & 0 \\ 2 & 1 & 0 & 1 \\ 1 & 0 & 0 & 1 \end{bmatrix}$$

$$T(A) = \left[\begin{array}{cccccccccc} 1 & 1 & 1 & 1 & 0 & 0 & 0 & 0 & 0 & 0 \\ 0 & 0 & 0 & 0 & 1 & 1 & 1 & 1 & 0 & 0 \\ 0 & 0 & 0 & 0 & 0 & 0 & 0 & 0 & 1 & 1 \\ - & - & - & - & - & - & - & - & - & - \\ 1 & 0 & 0 & 0 & 1 & 1 & 0 & 0 & 1 & 0 \\ 0 & 1 & 1 & 0 & 0 & 0 & 1 & 0 & 0 & 0 \\ 0 & 0 & 0 & 1 & 0 & 0 & 0 & 0 & 0 & 0 \\ 0 & 0 & 0 & 0 & 0 & 0 & 0 & 1 & 0 & 1 \end{array}\right]$$

$$X = \left[\begin{array}{cccccccccc} 1 & 1 & 1 & 1 & 0 & 0 & 0 & 0 & 0 & 0 \\ 0 & 0 & 0 & 0 & 1 & 1 & 1 & 1 & 0 & 0 \\ - & - & - & - & - & - & - & - & - & - \\ 1 & 0 & 0 & 0 & 1 & 1 & 0 & 0 & 1 & 0 \\ 0 & 1 & 1 & 0 & 0 & 0 & 1 & 0 & 0 & 0 \\ 0 & 0 & 0 & 1 & 0 & 0 & 0 & 0 & 0 & 0 \end{array}\right]$$

17.3.1 Results.

Dimensionality. The canonical correlation coefficients (Table 17.8 (a)) range in size from r_1 = .75 to r_6 = .07. While, with the exception of r_1, the correlations are not large the effect of the relatively large sample (N = 645; v/s = .022) on their size must be taken into account in their assessment. Bearing this point in mind, it appears that the first three or even four canonical correlations are not negligible. From the magnitude of r_1, in particular, there can be little doubt that the herbivores and study areas are indeed associated.

Table 17.8 (a) and (b) also contains the results of the nominal tests of the overall null hypothesis (17.9). Roy's criterion yields $\theta_s(6, \frac{1}{2}, 314\frac{1}{2})$ = .563 > $\theta_\alpha(6, \frac{1}{2}, 314\frac{1}{2})$ = .099 for α = .01, while Bartlett's chi-squared approximation of Λ provides $\chi_s^2(48)$ = 753.72 > $\chi_\alpha^2(48)$ = 93.22 for α = .0001. Both tests lead unequivocally to be rejection of H_o; the tests therefore indicate that the herbivores and study areas are likely to be associated. Accordingly, interest now turns to the *number* of identifiable 'components' which together make up the overall association and to the nature of these components. The results of the nominal tests of the smaller roots, r_k^2 (k > 1) using Roy's criterion and of Bartlett's test of dimensionality are shown in Tables 17.8 (a) and 17.9. The tests broadly agree in suggesting that three or four components should be sufficient to fully resolve the interrelationships among the herbivores and study areas. Notice, however, that a discrepancy does exist between the two procedures concerning the apparent significance of r_4^2. Bartlett's test is strictly a test of the residuals after the roots presumed to be significant have been removed, not a test of the roots themselves (Section 4.2). From Table 17.9 it appears that r_5^2 and r_6^2 do not differ significantly from zero; this in turn suggests that the dimensionality of the association is four. Roy's criterion, on the other hand (Table 17.8 (a)), points to the significance of only the first three canonical roots, and hence a dimensionality of three. In order to obtain insight into the substantive nature of some or all of these separate components of the total association, we need to examine the canonical variates associated with the r_k^2. Table 17.10 reports the scores of samples on the canonical variates U_k and V_k (k = 1,$\cdots$,3), while Figure 17.1 displays these scores graphically.

TABLE 17.8: Ruwenzori National Park, Uganda. Canonical analysis of relationships among seven herbivore species and nine study areas. Canonical correlation coefficients and related significance tests.

(a) The canonical correlation coefficients r_k and their nominal significance. Approximate critical values of Roy's largest-root criterion, $\theta_\alpha(s,m,n)$, are shown for $\alpha = .05$ $\alpha = .01$. ($m = \frac{1}{2}$, $n = 314\frac{1}{2}$).

k	s	r_k	r_k^2	$\theta_{.05}(s,m,n)$	$\theta_{.01}(s,m,n)$	p
1	6	.750	.563	.083	.099	<.01
2	5	.411	.169	.072	.085	<.01
3	4	.330	.109	.061	.074	<.01
4	3	.193	.037	.049	.061	>.05
5	2	.115	.013	.038	.048	>.05
6	1	.071	.005	.024*	.031*	>.05

*From the U-distribution: $\theta_\alpha(1,m,n) = 1 - U(p,1,n)$.

(b) Bartlett's approximate test of the overall hypothesis $H_o: \rho_k = 0$ for all k.

$$\Lambda = \prod_{k=1}^{s} (1 - r_k^2)$$

$$= (1 - .563)(1 - .169) \cdots (1 - .005) = 0.306 \ .$$

$$\chi^2 = -\{(N - 1) - \tfrac{1}{2}(p + q + 1)\} \log_e \Lambda$$

$$= -\{644 - \tfrac{1}{2}(8 + 6 + 1)\}(-1.184)$$

$$= 753.717, \text{ with } p.q = 48 \quad df; \quad p < .0001 \ .$$

The canonical variates. The sample scores reported in Table 17.10 and graphed in Figure 17.1 have been standardized to reflect the magnitude of the *kth* squared canonical correlation coefficient, r_k^2. This is the same convention as that adopted in connection with

the scores of the contingency table analysis and will facilitate
the comparison of the two sets of results in Section 17.4. The
scores for items omitted from the analysis (bushbuck, reedbuck
and site 2) for the reasons stated above, were obtained by applying
the vectors of weights (not reported here) from the canonical analysis
to the standardized observed sample-vectors of the entities in
question and post-normalizing. Figure 17.1 displays the sample
(N = 645) mapped into a number of distinct subspaces defined by
the canonical variates. Figure 17.1 (a) and (b) shows the sample
projected into subspaces defined by the first two canonical
variates of study area space (V_1, V_2) and herbivore space (U_1, U_2),
respectively, while Figure 17.1 (c) displays the sample simultaneously mapped into the subspace of the first three canonical
variates, V_k and U_k (k = 1,$\cdots$,3), of each domain. The most
notable feature of Figure 17.1 is the *scaling* of the attribute-
states which the canonical variates achieve. This property enables
the structural relationships within and between the various
attribute-states to be communicated with considerable efficiency.
We postpone the ecological interpretation of these relationships
until the connections between the contingency table analysis and
canonical analysis have been examined (Section 17.4).

Some further results appear in Table 17.11, notably the
structure correlations and the redundancies. We shall refer to
these results again in Section 17.5.

Before embarking on the comparison of methods some further
explanation of the sample scores reported in Table 17.10 is
called for. Notice that although N, the total number of samples,
is 645 the dimensions of the vectors U_k and V_k of what in
Table 17.10 purport to be sample scores are 7 × 1 and 9 × 1,
respectively - rather than 645 × 1 as might be expected. (We
choose to ignore for the present the two herbivores and the study
area omitted from the canonical analysis.). Moreover, the elements
of the vectors are identified in terms of *attribute-states* rather
than samples. Table 17.10 in fact consists of a *summary* of the
sample scores on the canonical variates. The reason for this is
as follows. The scores for the total sample (N = 645) when mapped
into U-space (herbivores) fall into seven discrete groups; similarly,
the scores mapped into V-space (study areas) fall into nine discrete
groups. What Table 17.10 does is to report the coordinates or
scores of these *sample-groups* on the canonical variates. The
number of samples belonging to each group is not shown in the
Table, but if required are simply the row and column marginal
totals, respectively, of the original 2-way table (Table 17.2).
Little would be gained by reporting the scores for all 645 samples
in extenso here. It is nevertheless worth bearing in mind that
Table 17.12 does in summary form refer to all N = 645 samples.

The feature described stems from the fact that the two attributes underlying the frequency table were themselves discrete variables with seven and nine states, respectively, rather than continuous variables *and* that each sample is distinguishable only up to two attribute-states. As Figure 17.1 is a graph of the numerical entries of Table 17.10, the above remarks apply equally to the Figure.

TABLE 17.9: Ruwenzori National Park, Uganda. Canonical analysis of relationships among seven herbivore species and nine study areas. Bartlett's approximate χ^2 test of dimensionality; $m = \{(N - 1) - \frac{1}{2}(p + q + 1)\} = 636.5$.

Source	df*	$-m \ln(1 - r_k^2)$	χ^2	p
r_1^2	13	$-m \ln(1 - r_1^2)$	526.91	–
r_2^2	11	$-m \ln(1 - r_2^2)$	117.83	–
r_3^2	9	$-m \ln(1 - r_3^2)$	73.46	–
r_4^2	7	$-m \ln(1 - r_4^2)$	24.00	–
Remaining roots	8	$-m \ln\{\Lambda/(1-r_1^2)\cdots(1-r_4^2)\}$	11.52	.174
Total	p.q = 48	$-m \ln\Lambda$	753.72	<.0001

*Degrees of freedom: $p + q - (2k - 1)$.

17.4 Relationships Between Contingency Table Analysis and Canonical Analysis. The relationship between the methods is perhaps most easily demonstrated by comparing the statistics on which the tests of hypothesis (17.1) or (17.9), as well as the subsidiary hypotheses which arise when (17.1) and (17.9) can be rejected, are based. We will therefore first compare the test criteria of the two methods at and beyond the acceptance decision (Section 17.4.1) before going on to sketch some of their wider relationships (Section 17.4.2).

17.4.1 Relationships between criteria for testing H_o. In Section 17.2 we saw that the classical chi-squared statistic (17.2) for testing independence in $r \times c$ tables can be calculated from the trace of a matrix $\underset{\sim}{T}$ derived from the contingency table:

$$\chi_o^2 = n[tr(\underset{\sim}{T}) - 1] \quad .$$

The quantity in brackets on the right-hand side of this expression is closely related to the trace of the quadruple matrix product $\underset{\sim}{R}_{22}^{-1}\underset{\sim}{R}_{21}\underset{\sim}{R}_{11}^{-1}\underset{\sim}{R}_{12}$ which arises in canonical analysis (McKeon, 1965, p. 32). Specifically

$$tr(\underset{\sim}{T}) - 1 = tr(\underset{\sim}{R}_{22}^{-1}\underset{\sim}{R}_{21}\underset{\sim}{R}_{11}^{-1}\underset{\sim}{R}_{12}) \quad . \tag{17.10}$$

Now, as the trace of a matrix is identically the sum of its latent roots, and recalling that the largest root λ_1 of $\underset{\sim}{T}$ is unity, we may re-write (17.10) as

$$\sum_{i=2}^{r} \lambda_i = \sum_{k=1}^{\Sigma} r_k^2 \quad . \tag{17.11}$$

This result shows that if we ignore the largest root $\lambda_1 = 1$ of $\underset{\sim}{T}$, the sums of the latent roots of the matrices in question are identical (cf. Kshirsagar, 1972, p. 383). Equation (17.11) specifies the fundamental connection between the classical χ^2-test of the independence and canonical analysis in relation to a test of the overall hypothesis (17.1) or (17.9).

TABLE 17.10: Ruwenzori National Park, Uganda. Canonical analysis of relationships among seven herbivore species and nine study areas. Sample scores on the canonical variates $U_k = \underset{\sim}{a}_k'\underset{\sim}{X}$ and $V_k = \underset{\sim}{b}_k'\underset{\sim}{Y}$, $k = 1, \cdots, 3$. Attribute-states corresponding to sample-groups are indicated.

Herbivore	U_1	U_2	U_3	V_1	V_2	V_3	Site
Elephant	−1.021	.119	.016	−0.758	−.086	−.069	1
Warthog	−0.653	.481	.530	−0.629	.648	−.025	2*
Hippo	−1.066	−.769	−.208	−1.163	−.810	−.181	3
Reedbuck*	0.744	.337	−.571	−0.893	.909	.722	4
Waterbuck	−1.080	.539	.526	−0.705	−.037	−.147	5
Kob	0.555	.148	−.338	−1.088	−.280	.045	6
Topi	0.946	−.399	.479	0.270	.469	−.601	7
Bushbuck*	−1.317	.795	1.229	−0.398	.403	−.150	8
Buffalo	−0.458	.350	.075	0.555	.140	−.193	9
				0.758	−.271	.285	10

*Sample scores calculated subsequent to main analysis.

Multiplication of the left-hand side of (17.11) by n yields the chi-squared statistic of the classical test, the significance of which is normally assessed by reference to the χ^2-distribution with (r-1)(c-1) degrees of freedom. The significance of the corresponding quantity in canonical analysis, i.e. the right-hand side of (17.11), could be assessed in the same way. In canonical analysis, however, it is customary to employ functions of the roots r_k^2 other than $n\Sigma r_k^2$ as criteria for testing (17.9). Most often $\Lambda = \Pi(1 - r_k^2)$ is used for this purpose; alternatively, the quantity Σr_k^2 may itself be referred to a table of percentage points of Pillai's trace criterion (Pillai, 1960), or r_1^2 may be referred to Roy's largest-root distribution. Thus canonical analysis may lead to distributions other than the χ^2-distribution in conducting the test of (17.9). These variations on the conduct of the test, however, are comparatively unimportant in relation to the fundamental identity (17.10). Consequently, correspondence (17.11) effectively establishes the *formal equivalence* of both the classical chi-squared test of independence and of Williams' (1952) analysis to a canonical analysis in which both sets of variables are binary-valued dummy variables with respect to hypothesis (17.1) and hypothesis (17.9).

We turn now to consider affinities between contingency table analysis and canonical analysis beyond the acceptance decision stemming from (17.1) or (17.9). As in multivariate analysis of variance, a formidable range of options exist for pursuing the analysis beyond H_o (e.g. see Bock, 1975, Ch. 8; Bishop, Fienberg and Holland, 1975; Nelder, 1975; Gokhale and Kullback, 1978). For our purpose is sufficient to note that of the various alternatives, Williams' (1952) method has the closest affinity with canonical analysis. For this reason we shall confine our attention to this particular form of contingency table analysis. The affinity between the methods arises in large part from the eigenstructure basis of both methods. Inspection of the latent roots λ_i, i = 2,$\cdots$,r of the matrix $\underset{\sim}{T}$ of the Williams analysis (Table 17.4) and of the squared canonical correlation coefficients, r_k^2, k = 1,$\cdots$,s of the canonical analysis (Table 17.8) shows corresponding roots to be identical. Thus, not only are the traces of the matrices of interest identical, but the identity extends to the individual roots λ_i (i > 1) and r_k^2 which combine to form tr($\underset{\sim}{T}$) and tr($\underset{\sim}{R}_{22}^{-1}\underset{\sim}{R}_{21}\underset{\sim}{R}_{11}^{-1}\underset{\sim}{R}_{12}$), respectively.

The significance of the individual roots λ_i in contingency table analysis is customarily tested by first applying the transformation $n\lambda_i$ and then referring the resulting variate to a chi-squared distribution. The proscription mentioned in connection with partitioned Λ-tests (Section 4.2), however, is relevant here. In practice, difficulties are encountered in attempting to associate with each term $n\lambda_i$ in the partition (17.6) of χ_o^2 its corresponding term in the parallel decomposition of the total $(r-1)(c-1)$ degrees of freedom. Thus the appropriate chi-squared reference distribution cannot be identified with certainty. Consequently, the tests of the $n\lambda_i$ may be invalid.

In canonical analysis, on the other hand, we have seen that the individual roots r_k^2 can be referred directly to the gcr distribution. This procedure effectively circumvents the inherent difficulty of assessing the partitions $n\lambda_i$ in contingency table analysis and might well prove worthwhile in the latter. Nevertheless, notwithstanding these differences in the assessment of particular components of the global association, the identity of the individual roots largely establishes the equivalence of the methods beyond the initial acceptance decision.

The identities to which attention has been drawn are for the most part readily verified from the numerical entries of Tables 17.4 and 17.8 (a) and (b). A discrepancy is however apparent concerning the respective criteria for testing the overall hypothesis (17.1) or (17.9). In the Williams' (1952) analysis the test criterion yielded a chi-squared value of $\chi_o^2 = 579.02$ based on 48 degrees of freedom (p. 000). The corresponding criterion in the canonical analysis was Wilks' Λ; following Bartlett's transformation a chi-squared value of $\chi_o^2 = 753.72$ based on 48 degrees of freedom was obtained (Table 17.8 (b)). Thus, while the test criteria are alike in leading to the decisive rejection of H_o, they differ numerically. The discrepancy arises because in finite samples each criterion is only an *approximation* of a single, underlying chi-squared variate; the two statistics are however asymptotically equivalent (Bartlett, 1947; Rao, 1952, p. 372).

In addition to the identities considered above, the two methods share other features in common. Reference to Tables 17.6 and 17.10 shows the attribute-state scores of the contingency table analysis and the sample scores of the canonical analysis also to be substantial equal, at least within rounding error and a possible reflection in sign. Given the identity of the roots λ_i (i > 1) and r_k^2, we may anticipate that the corresponding latent vectors will be identica up to a scale factor, at least. Consequently, the identity of the

two sets of scores, the evaluation of which depends on the latent
vectors, comes as no surprise. Thus we see that the formal equivalence of the methods is virtually complete.

The relationships to which we have drawn attention explain why
the contingency table analysis of Williams' (1952) is often referred
to as canonical analysis (McKeon, 1965, p. 11; Hatheway, 1971;
Kendall and Stuart, 1967, p. 568). More specifically, we can see
that Williams' (1952) method for the analysis of independence in
r × c tables may be regarded as equivalent to the particular case
of canonical analysis which arises under specialization of both
sets of variables to binary-valued dummy variables.

17.4.2 Further relationships: correspondence analysis. Hill (1974)
has proposed subsuming the two procedures applied in Sections 17.2
and 17.3 under the more general term *correspondence analysis;* see
also David, Campiglio and Darling (1974) and David, Dagbert and
Beauchemin (1977). Correspondence analysis is a method for
analysing both frequency and incidence data which stresses the
scaling properties of the solution. The technique is closely related
to those classical methods for the analysis of 2-way tables which
are motivated by a search for scores with certain optimal properties
(Hirschfeld, 1935; Fisher, 1938; Williams, 1952) as opposed to
methods which seek a meaningful partition of some overall measure
of association (e.g. Lancaster, 1949; Irwin, 1949). Two forms of
correspondence analysis were distinguished by Hill (1974) - zero-
order and first-order correspondence analysis, denoted C_0 and
C_1 respectively.

The analysis C_0 corresponds closely to the scoring method of
analysis for 2-way frequency tables associated particularly with
Fisher and Williams. Hill (1974) has given an algebraic proof which
shows that contingency table analysis of this kind and canonical
analysis of an incidence matrix $T(A)$ derived from A are in fact
equivalent forms of C_0. Hill further pointed out that an incidence
matrix can itself be treated *as if* it were a 2-way table of frequencies
and analysed accordingly by C_0. It is this procedure which constitutes first-order correspondence analysis. This aspect of the
relationship between C_0 and C_1 may therefore be specified as
follows: $C_1(A) = C_0(T(A))$. The analysis $C_1(A)$ simultaneously
scales the attribute-states (rows of $T(A)$) and the samples themselves
(columns of $T(A)$) and is regarded by Hill as essentially a form of
multidimensional scaling. The practical significance of first-
order correspondence analysis is that an incidence matrix can be
defined when A is an m-way table of frequencies. Thus, C_1
is defined for *all* contingency tables, not just 2-way tables.

TABLE 17.11: *Ruwenzori National Park, Uganda. Canonical analysis of relationships among seven herbivore species and nine study areas. Correlations between the original variables and canonical variates.*

Canonical variate	U_1	U_2	U_3	h^2_w	V_1	V_2	V_3	h^2_b
Herbivore								
Elephant	-.217	.076	.008	.049	-.163	.019	.003	.027
Warthog	-.171	.230	.315	.181	-.128	.095	.104	.036
Hippopotamus	-.574	-.757	-.255	.968	-.431	-.311	-.084	.290
Reedbuck*	.052	.078	-.206	.051	.039	.032	-.068	.007
Waterbuck	-.307	.280	.340	.288	-.230	.115	.112	.079
Kob	.534	.261	-.736	.895	.401	.107	-.243	.231
Topi	.579	-.445	.669	.981	.435	-.138	.221	.272
Bushbuck*	-.206	.416	.883	.995	-.154	.171	.293	.139
Buffalo	-.340	.474	.126	.356	-.255	.195	.042	.105
Variance Extracted	.142	.153	.234	.529	.080	.026	.026	.132
Redundancy	.080	.026	.132	.132	.080	.026	.026	.132

TABLE 17.11, Continued.

Canonical variate	V_1	V_2	V_3	h_w^2	U_1	U_2	U_3	h_b^2
Study area								
1 Grass/thicket	-.277	-.057	-.057	.083	-.207	-.024	-.019	.044
2* Tall grass/thicket	-.125	.429	-.026	.200	-.094	.177	-.008	.040
3 Grass/thicket	-.429	-.547	-.152	.506	-.322	-.225	-.050	.157
4 Grass/thicket	-.335	.619	.613	.871	-.251	.255	.202	.169
5 Grass/thicket	-.252	-.023	-.119	.078	-.189	-.009	-.039	.037
6 Grass/thicket	-.421	-.198	.039	.218	-.316	-.081	.013	.107
7 Short grassland	.117	.375	-.598	.512	.088	.154	-.198	.071
8 Grass/thicket	-.128	.238	-.110	.085	-.096	.098	-.036	.020
9 Short grassland	.384	.175	-.301	.269	.288	.072	-.099	.098
10 Short grassland	.629	-.411	.535	.851	.472	-.169	.177	.283
Variance Extracted	.120	.131	.117	.367	.068	.022	.013	.103
Redundancy	.068	.022	.013	.103	.068	.022	.013	.103

*Structure correlations calculated subsequent to main analysis.

The incidence matrix of an m-way table is distinguished by
m rather than just two 1's in each column. The analysis
of an incidence matrix of this kind could be undertaken equally
or by either (a) zero-order correspondence analysis, C_0; or
(b) generalized canonical analysis using m sets of dummy variables.
While alternative methods for analyzing multi-dimensional contingency
tables are available (Fienberg, 1970; Bishop, Fienberg and Holland,
1975; Nelder, 1975; Gokhale and Kullback, 1978), these alternatives
in general lack the scaling properties which are an attractive feature
correspondence analysis (cf. however, Healy and Goldstein, 1976).
Thus the extention of the procedures for analyzing r × c tables
represented by correspondence analysis greatly enhances the opportunities for the analysis of frequency data in ecology, where
interest is rarely confined to just two variables, whether discrete
or continuous.

17.5 Ecological Interpretation of the Results. Having shown
the results of the contingency table analysis and the canonical
analysis to be substantially the same we have in effect only one
set of results to deal with. We choose to confine our attention
here to the results of the canonical analysis. Before proceeding,
however, it will be useful to recall the salient points to have
emerged so far. It has been shown that there are grounds for
believing:

(a)   that the herbivores and study areas are associated;

(b)   that the association may be resolved into three elements
      or components; and that

(c)   of the three components, one is appreciably stronger than
      the remaining two, which themselves are of roughly equal
      importance.

Our concern now is to enquire into the *nature* of the components.
In so doing we shall focus our attention on the scores of samples
on the canonical variates (Table 17.10). The canonical weights
have little intrinsic interest in the present analysis because
the assignment of dummy variables among the attribute-states is
to some extent arbitrary, while the structure correlations (Table
17.11) are highly correlated with the sample scores and therefore
add little to the information contributed by the scores themselves.
Table 17.11 is presented here mainly for completeness and to allow
comparison with the canonical analyses of previous Sections. Two
general points are however worth making. First, it is apparent
that the numerical entries of Table 17.11 are in general smaller
than their counterparts in previous analyses. Such differences
are attributable at least in part to differences in sample size,

the present analysis being based on an appreciably larger sample
than any of the earlier analyses. Secondly, the two sets of interset structure correlations merit equal attention. This follows
because ecological interest could well embrace both possible effects
of site-properties on the occurrence of the herbivores as well as
possible effects of the herbivores on the vegetation and other
physical site characteristics. It is convenient to draw attention
at this point also to the redundancy associated with the canonical
variates corresponding to the fourth canonical root r_4^2, although
the quantities in question are not reported in Table 17.11. The
redundancies of V_4 and U_4 are $v_{x|v_4}^2 < .004$ and $u_{y|u_4}^2 < .004$,
respectively. The explanatory power of these canonical variates
across domains is evidently small – a finding which contributed
to the decision to adopt the rank 3 model for the analysis which is
reported in Tables 17.10 and 17.11.

The sample scores are graphed in Figure 17.1 (a), (b) and (c).
As previously noted (Section 17.3.1), the scores fall into discrete
groups which are characterized by attribute-states, either study
areas or species, which distinguish particular subsets of samples.
For this reason, although the various mappings of Figure 17.1 are
mappings of sample scores they appear to emphasize attribute-states;
it happens to be more convenient to refer to a group of samples by
the single attribute-state which characterizes it in U_k - or
V_k-space than by enumerating the samples which together make up the
group. In interpreting the sample scores, below, Field and Laws'
(1970) descriptive account has been drawn on extensively. It
should be clearly understood, however, that the views expressed
here are not necessarily those of Field and Laws.

Inspection of Figure 17.1 (a) shows the first canonical variate
of site-space, V_1, to contrast sites 7, 9 and 10, which have
positive scores, against sites 1, 2, 3, 4, 5, 6 and 8 with negative
scores. Reference to the summary of site characteristics (Table 17.1)
shows the sites with positive scores to support open grassland vegetation at some distance from standing water, while sites with negative scores for the most part support a mosaic of grassland and
woody, thicket vegetation adjacent to water. Furthermore, the grass/
thicket sites with the smallest negative scores (sites 3, 4 and 6)
are heavily overgrazed while in contrast sites 7 and 9 towards
the positive pole bear fire-climax grassland. These site relationships can be comprehended in terms of the effects of grazing and
fire on the vegetation together with the partial control of both
factors by standing water acting through the herbivores themselves.
We shall defer further consideration of the interrelations
involved until something has been said of the structure of the
herbivore community (p. 000). For the present it is sufficient

to note that V_1 appears to express the direct or indirect effects of an interplay between grazing, fire and standing water on the vegetation. Sites with negative scores on V_1 tend in general

to be (a) close to water, (b) overgrazed, (c) relatively fire-free, and (d) to support a mosaic of herbaceous and woody vegetation. The converse is true of sites with positive scores. There is also a large geographical component to this canonical variate, as can be seen by comparing the site scores with the spatial position of the sites as shown on Field and Laws' map of the Park (Field and Laws, 1970, Fig. 1). Sites centered about the Mweya Peninsular are on the whole rather sharply differentiated by their negative scores on V_1 from sites situated south of the Maramagambo Forest or

otherwise at some distance from standing water (sites 7, 9 and 10).

The second canonical variate, V_2, operates chiefly within the grass/thicket sites, among which site 3 (-.81) is contrasted against site 4 (.91). Reference to Table 17.1 and to Field and Laws' account shows site 4 to have been subjected to continuous hippopotamus cropping over a lengthy period, while site 3 has not; on the contrary, there has been an influx of hippopotamus into site 3 from the surrounding area during the period of the survey. Moreover, the virtual disappearance of hippopotamus from site 4 has been accompanied by an increase in the density of other species in this site, especially buffalo and waterbuck. Thus V_2 may well correspond to the effects of cropping or management on the herbivore complement of the sites. The scores of other sites on V_2 are broadly consistent with such an interpretation. In addition there is evidence to show that in general such dynamic changes in herbivore density have repercussions on the vegetation and other physical characteristics of the sites (Field, 1968, 1972; Thornton, 1971; Lock, 1972). Thus there may well be a corresponding physical differentiation of the sites in response to the cropping of hippopotamus also.

The third canonical variate, V_3, (Figure 17.1 (c)), chiefly affects sites 4 (.72) and 7 (-.60); the remaining sites with the exception of site 10 (.28) have scores < $|.20|$. While the ecological implications of this canonical variate, if any, are not clear it seems possible that it may represent a distinction between fire-climax grassland (sites 5, 7 and 9), on the one hand, and the overgrazed, comparatively fire-free woody, thicket vegetation of sites 4 and 6, on the other. Thus V_3 may correspond to some aspect of the influence of fire on the vegetation.

In summary, it seems that of the site characteristics examined, vegetation contributes most directly to the physical differentiation

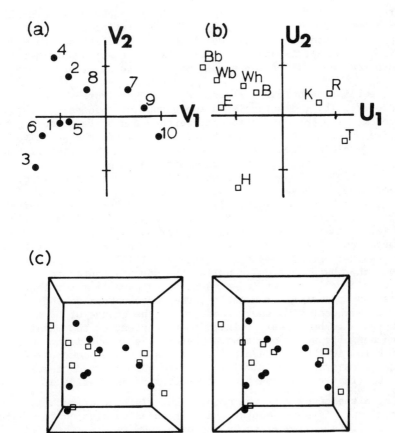

FIG. 17.1: *Ruwenzori National Park, Uganda. Canonical analysis of association between nine study areas and seven species of herbivore.* N = 645 *observations mapped into subspaces defined by canonical variates (a)* V_1 *and* V_2 *of the site domain; (b)* U_1 *and* U_2 *of the herbivore domain; and (c) simultaneously mapped into* V_k *and* U_k, k = 1, $\cdots$, 3. *In (c), identification of points as in (a) and (b); orientation of axes as in FIG. 13.1. The canonical variates normalized to* r_k^2 *in each diagram.* B: Buffalo; Bb: Bushbuck; E: Elephant; H: Hippopotamus; K: Kob; R: Reedbuck; T: Topi; Wb: Waterbuck; Wh: Warthog.

of the sites, while grazing, fire, proximity to water and management
have more remote though nevertheless appreciable effects. We note
also that the site-configuration in the (V_1, V_2)-plane largely recovers the habitat categories into which Field and Laws initially
grouped the sites (p. 000 above); only the distinctiveness of
Group 1 (composed of site 2), which here is affiliated with Group
4 (sites 4 and 8), is unsubstantiated by the results obtained.
We turn now to consider the sample mapped into herbivore-space.

The first canonical variate, U_1, of herbivore-space
(Figure 17.1 (b)), distinguishes two groups of species - bushbuck,
waterbuck, hippopotamus, elephant, warthog and buffalo towards the
negative pole, and kob, reedbuck and topi towards the positive
pole. A striking ecological difference between these two groups
of species concerns the extent of their dependence on free standing
water. Buffalo, elephant, hippopotamus and warthog are all dark-
skinned species whose reduced hair renders them particularly
sensitive to heat. Consequently, they depend for their survival
on water-containing wallows (Field and Laws, 1970); in a similar
way, waterbuck is known to require water for drinking while
bushbuck has an affinity for swamps and is seldom found far from
water (Walker, *et al.*, 1968, pp. 1437 and 1414). Kob, reedbuck and topi
on the other hand, are quite unlike the species mentioned in being
to a considerable extent independent of both wallows for thermo-
regulation and water for drinking; they obtain much of their
water requirement with their food. Thus the change of sign along
U_1 appears to be closely related to the extent of the physiological
dependence of species on standing water. For convenience we shall
refer below to the two species' groups as *water-dependent* and
water-independent respectively. We also observe in connection
with U_1 that the disposition of species is suggestive of a

partial discontinuity in the data. The existence of such a
discontinuity is borne out by the corresponding site configuration
(Figure 17.1(a)) as well as by the magnitude of r_1^2 (cf. Teil,
1975).

U_2 is characterized principally by a contrast within the water-
dependent species. This distinguishes hippopotamus (-.77) from a
majority of the remaining species, particularly bushbuck (.79),
waterbuck (.54) and buffalo (.35). A singular feature of
hippopotamus is that this was the only species subjected to cropping.
Hippopotamus was substantially eliminated from one study area (site
4) and its numbers drastically reduced in two others (sites 1 and
8). Moreover, as noted above, the cropping program had an effect
also on the population densities of several other species, notably
waterbuck and buffalo, not themselves directly involved. Thus U_2
appears to correspond to changes in the dynamic balance between the
water-dependent species as a result of the cropping program directed

against hippopotamus. Notice, however, that U_2 also distinguishes topi (-.40) from reedbuck (.34) and kob (.15) among the water-independent species. It seems unlikely that this effect is attributable to the hippopotamus cropping program, however, and to this extent the ecological significance of U_2 cannot be considered to have been completely resolved. The inability to fully account for U_2 may well be a consequence of the partial discontinuity in the data referred to above.

The third canonical variate, U_3, (Figure 17.1 (c)) is characterized principally by bushbuck (1.23), towards the positive pole, and hippopotamus (-.21), kob (-.34) and reedbuck (-.57) towards the negative pole. In view of the rarity of reedbuck (see Table 17.1), we may for convenience regard V_3 as essentially a contrast between bushbuck on the one hand and hippopotamus and kob on the other. One marked ecological difference between these species concerns their feeding habit. Bushbuck is basically a *browser*, having a decided preference for the woody, dicotyledenous shrub *Capparis tormentosa* Lam. (Field and Laws, 1970), while hippopotamus and kob are both exclusive *grazers* (Field, 1969). Moreover, waterbuck (.53) and Buffalo (.10), which are distinguished by positive scores on V_3, have also been reported as extensive browsers of *Capparis* thicket (Lock, 1977, p. 399). Thus U_3 appears to correspond to a difference in the feeding habit of certain species. Like U_2, this canonical variate also draws attention to a distinction between kob (-.34) and topi (.48). Whether the distinction in this case reflects a difference in feeding habit, however, is far from clear though it seems rather unlikely.

We may summarize our ecological assessment of the canonical variates of herbivore-space as follows. The variates examined appear to reflect: (a) a major distinction among species related primarily to the extent of their dependence on standing water; (b) dynamic changes among species in response to a management program directed at hippopotamus; and (c) the contrasting food habits of grazers and browsers.

As a final step, it is necessary to integrate the interpretations placed on each set of canonical variates. The stereogram of Figure 17.1 (c), which is a representation of the covariance structure of the data, is particularly helpful for this purpose. A striking feature of the Figure is the spatial proximity of grass/thicket sites adjacent to water and water-dependent species, on the one hand, and of grassland sites at some distance from water and water-independent species on the other. Observe, for example, the proximity of kob, reedbuck and topi to sites 7, 9 and 10 in the stereogram (cf. also Table 17.1). The interrelationships between sites and herbivores specified jointly by U_1 and V_1, in particular, can be accounted for as follows. The present aspect of the vegetation of the Park, which varies from open grassland to thicket, woodland and forest, is believed to be due in large measure to the effects of grazing and fire (Osmaston, 1965; Lind and Morrison, 1974, p. 68). Together, these factors largely determine the balance between the woody and non-woody components of the vegetation. Over-grazing tends to shift the balance between the two towards woody vegetation; woody vegetation itself is resistent to fire, which in turn tends to maintain or shift the balance further towards the woody component. A reduction in grazing intensity, on the other hand, enables grassland to expand at the expense of woody vegetation, especially in the presence of fire. Fire itself has the effect of stimulating the development of grassland and so tends to stabilize the vegetation in the grassland state. In addition, standing water appears to play an indirect though decisive part in determining the balance between woody and non-woody vegetation. This arises from the control exerted by open water on the distribution of the water-dependent herbivores. Such species are to a greater or lesser degree confined to the vicinity of free standing water; this in turn leads to the association between over-grazing, thicket development, reduced incidence of fire and water-dependent herbi-vores of sites adjacent to water which is evident in Figure 17.1 (c). In sites at a greater distance from water, the situation differs. The mobility of the herbivores of such sites is not restricted by a need for standing water. Consequently, the intensity of grazing is less, grassland thrives at the expense of woody vegetation and is maintained as such by the frequency of fire. In this way the association between the fire-climax grassland of sites 7, 9 and to a lesser extent 10, all of which are located at some distance from water, and the water-independent species kob, topi and reedbuck appears to have arisen. Thus it seems that the effects of grazing, fire, standing water and the physiological requirements of the herbivores together underlie the relationship between sites and species jointly specified by U_1 and V_1. Of the factors involved, standing water appears to exert a direct influence on the distri-bution of the herivores, while grazing and fire seem to be among the direct controls of site vegetation. In addition, standing water seems likely to exert an indirect effect on the vegetation through its influence on the herbivores.

The stereogram also draws attention to the contrasting manage-ment regimes of sites 3 and 4, particularly in the opposition of these sites in relation to V_2. The association between hippopotamus and each of these sites (positive and negative respectively) is clearly implied by the spatial relationships involved (cf. also

Table 17.1). In a similar way, the proximity of waterbuck and buffalo to site 4 reflects the positive association between these entities, which is, presumably, to some extent a consequence of the elimination of hippopotamus from site 4.

The distinction between the comparatively fire-free, grassland/ thicket vegetation of site 4 and the fire-climax grassland of sites 5, 7 and 9 is a further feature of interest in Figure 17.1 (c). The positive association of the browser, bushbuck, with site 4 and between the grazers kob and reedbuck and sites 7 and 9 particularly, is also plain. Similarly, the distinction (i.e. negative association) between the vegetation of site 4 on the one hand, and sites 7 and 9 on the other, and also between their respective herbivores, is also clearly implied by the spatial opposition of the symbols representing these entities.

There are several further features of interest in Figure 17.1 (c). Notice, for example, the strong positive association between hippopotamus and site 3, bushbuck and site 4, kob and site 9 and between topi and site 10, as well as the negative association between elephant and topi (cf. also Table 17.1). The specialized require-ments and narrow ecological range of a majority of these species is clearly implied by their peripheral position with respect to the configuration as a whole. Buffalo on the other hand, which is a broadly and rather evenly distributed species, is located towards the center of the configuration. Its central position reflects the wide ecological amplitude of this species. The stereogram also shows kob and topi to be alike at least to the extent of being essentially water-independent species of open grassland. Yet the considerable spatial separation of the species which is also apparent is indicative of the appreciable ecological 'dis-tance' between them. The nature of the separation in ecological terms, however, remains unclear; in view of the dispositon of the species, it is tempting to speculate that it may perhaps in some way correspond to the distinction between browsers and grazers attributed to U_3. Furthermore, the peripheral and rather isolated position of topi is suggestive of the control of this species by some overriding ecological requirement or characteristic (cf. for example hippopotamus). The identity of this hypothesized control, however, is not resolved. Another point of interest concerns the extent of the spatial separation of the species as a whole, which in general tend to be comparatively distant from one another. This is an expression of the *ecological separation* of the species in the field, to which Field (1968) and Field and Laws (1970) have drawn attention, the effect of which is to minimize competition between them.

The salient points of ecological interest may now be drawn together. It appears that the degree of physiological dependence of the herbivores on standing water may be the primary mechanism of ecological separation among them, while the cropping of hippopotamus and feeding habit have subsidiary though still appreciable effects. The influence of these factors seems to be as follows.

(a) *Standing water* generates or contributes in sites close to water to association between water-dependent species, over-grazing and grass/thicket vegetation; and in sites relatively distant from water to association between water-independent species, fire and open grassland.

(b) *Management cropping of hippopotamus* leads to negative association between hippopotamus and sites 4 and 8 on the one hand, and to positive association between these sites and buffalo and waterbuck, among other species, on the other.

(c) Differences in *feeding habit* apparently give rise to positive association between sites bearing woody, thicket vegetation and browsers such as bushbuck on the one hand, and between sites of fire-climax grassland and grazing species exemplified by kob on the other.

Another point to emerge is the contrast between specialized species such as bushbuck and between species of wide ecological amplitude such as buffalo. A distinction between kob and topi was also noted, though it was not accounted for. Although other relationships of interest are suggested by Figure 17.1 (c), sufficient has been said to identify at least in a provisional way some of the ecological processes which appear to govern the distribution of the herbivore species examined.

17.6 Conclusions. The analysis had two principal objectives, one ecological the other theoretical. Ecologically, the aim of analysis was to examine data on the distribution of nine herbivore species in ten study areas for evidence of association. The second objective was to explore connections between canonical analysis and contingency table analysis, the underlying motivation being to throw light on the extent of the applicability of canonical analysis in ecology.

The results of the canonical analysis point to the herbivores and study areas being associated. In so doing, analysis enabled a number of the operative environmental or other controls of the herbivores to be identified, at least in a provisional way, as well as allowing the apparent effects of these controls on particular species to be recognized. Broadly speaking, the results obtained are consistent with those of Field and Laws (1970), who used a quite different approach. This independent substantiation of the results of the canonical analysis lends weight to the view that canonical analysis does provide a useful means of arriving at sound ecological insight. Furthermore, it could be argued that the canonical analysis provides a more convenient summarization of the inherent structure of the field data which is at once more penetrating the better integrated than that of earlier studies. The analysis unquestionably generates considerable insight into relationships within and between the sites and herbivores of the sample data examined. Just how rewarding such insight proves to be in ecological terms, however, remains to be critically assessed. On the basis of the evidence presently available, though, the canonical analysis does appear to have been successful in contributing towards a better understanding of the distribution of the larger herbivores of the Ruwenzori National Park.

In connection with the second objective, the results obtained show the analysis of r × c contingency tables by a latent root and vector method to be formally equivalent to a canonical analysis based on two sets of dummy variables derived from a 2-way table. Attention was also drawn to the generalization of this result to the case of m-way contingency tables (m > 2). The significance of these relationships to us is that they throw light on the extent of the opportunities offered by canonical analysis in the analysis of ecological data. In view of the connections established, it appears that canonical analysis could have wide application in relation to the analysis of frequency data in ecology.

18. REAL APPLICATIONS: APPRAISAL AND CONCLUSIONS

The analyses of Sections 13 to 17 were motivated primarily by a desire to assess the performance of canonical analysis in a variety of ecological contexts. The applications dealt with (a) relationships between soils and vegetation and (b) between trees and seedlings in a lowland tropical rain forest, (c) the structure of a limestone grassland community in Anglesey, North Wales, (d) the comparative responses of eight grass species in a mineral nutrition experiment and (e) factors affecting the distribution of large herbivores in the Ruwenzori National Park, Uganda. A minimum prerequisite for analysis in each case was that the data structure determined by the observations or measurements was essentially linear and continuous. The chief justification for supposing this requirement to be not unrealistic was the small scale of the investigations in terms of either geographical extent or of the number of species or experimental treatments involved. It is in applications of precisely this kind that grounds for expecting the precondition to be at least reasonably closely satisfied are strongest and the application of linear models generally to be most productive. The analyses were also intended to serve a rather different purpose, namely to draw attention to the remarkable flexibility of canonical analysis. A related objective specifically of the analyses of Sections 16 and 17 was to shed light on the extent of the applicability of canonical analysis in ecology. This objective was pursued by examining connections between canonical analysis and several other analytical methods familiar to ecologists.

In this Section we examine the results of the five analyses with the declared objectives in mind. The conclusions reached on the basis of this examination are then stated.

18.1 Assessment.   Table 18.1 summarizes some characteristics in terms of which the analyses can be compared. Such a comparison is instructive because it throws light on the process of assessment. One of the most notable features of the Table is that, with the exception of the rangeland analysis, the analyses are all based on small samples ($N \leqslant 45$). Given the scale of the investigations this in itself need not necessarily give rise to undue concern. In relation to the number of variables involved (i.e. to the *subject* domain), however, the adequacy of the samples may seriously be questioned. From the Table we see that the total number of variables used ranged from 12 to 33, and that, with the exception once again of the rangeland analysis, the variable to sample ratios (v/s) are disturbingly high ($\geqslant .30$). Barcikowski and Stevens (1975) and Thorndike (1978, p. 184) have provided guidelines which suggest that for a total of, say, n = 15 variables, a sample of the order of 200 to 275 should prove sufficient for predictive purposes (v/s = .07 to v/s = .05). These estimates provide a rough yardstick as to just how inadequate the sample sizes might be. Two consequences follow immediately: (a) the results are best regarded as intrinsic features of the samples actually examined; and (b) indices of strength of relationship are likely to be exaggerated to an unknown extent. These points will need to be heeded in assessing the analyses. The first point, in particular, has an obvious bearing on the possible wider validity of the results and hence on the uses to which they might be put.

In assessing the analyses criteria intrinsic and external to canonical analysis will be used. Explained or predictable variance is used as an intrinsic measure of success while the extent of ecological insight afforded provides an external yardstick. Squared canonical correlation coefficients and redundancy are both expressions of explained variance. While there are difficulties in using these quantities as measures of success,

TABLE 18.1: Some sample characteristics of the canonical analyses of Sections 13-17.

	13. TRF (a) Guyana	14. TRF (b) Guyana	15. Grassland structure	16. Nitrate nutrition	17. Rangeland Uganda
N	25	25	45	40	645
p	6	6	31	7*	8*
q	9	6	2*	5	6*
v/s	.60	.48	.73	.30	.02

*Binary-valued dummy variables.

used circumspectly they may be helpful in this way. Ultimately, of course, the worth of an analysis is entirely a matter for ecological judgement.

The squared canonical correlations (Table 18.2) express the proportion of the variance of one member of a pair of canonical variates which can be accounted for or predicted by the other. As the significance tests of Sections 13-17 for the most part have only nominal value, we are forced to rely on the magnitude of the squared correlations in assessing this aspect of the analyses. In four of the analyses it can be seen that the magnitude of the r_k^2 (k = 1,2) is appreciable ($r_1^2 \geqslant .86$; $r_2^2 \geqslant .72$). Thus, even allowing for some inflationary effect stemming from the large v/s ratios, i.e. from *overfitting*, these analyses appear to have been successful in terms of the explained variance of the canonical variates U_k and V_k (k = 1, 2). In the rangeland analysis, the r_k^2 are smaller ($r_1^2 = .56$; $r_2^2 = .17$) and hence relationships between the canonical variates weaker. The difference to some extent, however, can be attributed to the smaller v/s ratio (.02) of this analysis. This in turn places our estimate of the relative value of the rangeland analysis on a somewhat different footing; it seems that in this case also useful portions of the variance of the U_k and V_k (k = 1, 2) are accounted for.

The percentages of the trace of the matrix product $R_{22}^{-1}R_{21}R_{11}^{-1}R_{12}$ absorbed by the canonical roots r_k^2 (k = 1,···,s) individually and cummulatively provide useful criteria of goodness

of fit in canonical analysis. The relevant quantities are shown
in Table 18.2 (a) for the five analyses, where the cumulative
percentage trace or canonical variance associated with the reduced-
rank solution adopted in each case is underlined. It can be seen
that the explained variance of the reduced-rank models varies from
80.18 percent for the nitrogen experiment, through 93.86 percent
for the rangeland analysis to 100 percent in the limestone grassland analysis. Thus the fit of the models in this respect is
good. It is helpful to bear in mind, however, that the adequacy
of the fitted model expressed in this way is to some extent a
function of $R(\underset{\sim}{R}_{22}^{-1}\underset{\sim}{R}_{21}\underset{\sim}{R}_{11}^{-1}\underset{\sim}{R}_{12})$, where R denotes the rank of the
indicated matrix, and is at most $s = \min(p,q)$, the number of
variables of the smaller set. In each analysis, a useful low-
dimensional solution was achieved, the solutions all being of
rank $\leqslant 4$. Moreover, largely because the canonical roots can be
ordered according to their contribution to overall explanatory
power, even in those analyses where the reduction in dimensionality
was least, the fitted models contributed enormously to the comprehension of salient relationships. Thus, by efficiently
summarizing and giving concreteness to the covariance structures,
there can be little doubt as to effectiveness of the analyses
in describing the relationships of interest.

Canonical roots are not necessarily reliable indices of the
strength of the relationship between two measurement domains,
partly because the canonical variates themselves are not observed
phenomena operationally defined through their measurement. For
this reason redundancy provides a useful supplementary index of
relationship. The redundancies $V_{x|v_1 \cdots v_k}^2$ and $U_{y|u_1 \cdots u_k}^2$
reported in Table 18.2 (b) vary between .103 and .947. It is
clear that in general the explanatory power of the analyses as
expressed by the redundancies is modest, especially if an attempt
is made to allow for a probable upward bias stemming from the
high v/s ratios in four analyses. It is well to recall, however,
that quite small redundancy values can nevertheless reflect
relationships of considerable ecological significance (cf.
Sections 11.2.5 and 12). Thus it is difficult to arrive at a
clear perception of the success of the analyses on the basis of
the unsupported redundancy indices alone.

Ecologically, the results of all five analyses can be
characterized in general terms as follows. The results are:

(a) substantiated in their broad features by the results
    of independent studies; and

(b) appear to provide a more informative and more convenient
    statement of the interrelationships in question than
    existed hitherto.

TABLE 18.2: *Goodness of fit of canonical analyses of Sections 13-17.*

(a) Percentage $tr(R_{22}^{-1}R_{21}\sim R_{11}^{-1}R_{12})$ absorbed by the canonical roots, r_k^2 (k = 1,···,s), individually (% Tr) and cummulatively (C %).

k	13. TRF (a) r_k^2	%Tr	C%	14. TRF (b) r_k^2	%Tr	C%	15. Grassland r_k^2	%Tr	C%	16. N nutrition r_k^2	%Tr	C%	17. Rangeland r_k^2	%Tr	C%
1	.857	29.08	29.08	.977	24.60	24.60	.963	51.01	51.01	.910	44.65	44.65	.563	62.83	62.83
2	.817	27.72	56.80	.914	23.01	47.61	.925	48.99	100.00	.724	35.53	80.18	.169	18.86	81.69
3	.556	18.87	75.67	.894	22.51	70.12				.298	14.62	94.80	.109	12.17	93.86
4	.427	14.49	90.16*	.760	19.13	89.25				.093	4.56	99.36	.037	4.13	97.99
5	.220	7.47	97.63	.424	10.67	99.92				.013	0.64	100.00	.013	1.45	99.44
6	.070	2.38	100.01	.003	0.08	100.00							.005	0.56	100.00
Trace	2.947	-	-	3.972	-	-	1.888	-	-	2.038	-	-	0.896	-	-

*For the roots r_k^2 (k = 1,2,3,5), C% = 83.14.

(b) Total redundancies of the fitted models.

	13. TRF (a)	14. TRF (b)	15. Grassland	16. N nutrition	17. Rangeland
$V_{x|v_1\cdots v_k}^2$	.501	.757	.392	.233	.132
$U_{y|u_1\cdots u_k}^2$	.391	.767	.947	.732	.103

A fundamental difficulty which arises in attempting to assess the analyses in purely ecological terms is that the *true* relationships in question are not known. In assessment, we are therefore forced to fall back on preconceived notions as to what these relationships might be. Notions of this kind are acquired by substantive experience or on the basis of what some other analytic procedure has led one to expect. Neither standard is completely trustworthy. However, provided one is alert to the danger of bias, this form of assessment is reasonably satisfactory. Notwithstanding difficulties of this kind, external confirmation of the broad features of the results does lend weight to the view expressed in Section 12 that canonical analysis can yield results that are ecologically sound (cf. Tukey, 1969). Since the major features of each analysis, however, could also almost certainly be recovered by other, possibly less elaborate means, any ecological worth of canonical analysis must depend ultimately on the insight it yields over and above that relating to the most obvious features of the data. Just how useful the present analyses will eventually prove to be in this way remains to be seen. On the strength of the available evidence, the descriptions of the covariance structures of interest were consistently found to be informative. Numerous previously unsuspected or dimly perceived relationships emerged or were specified with enhanced precision. These relationships could well provide starting points for future work.

We have seen that the fit achieved in all five analyses, as expressed by the cummulative percentage of the canonical variance associated with the first few canonical roots of each analysis, is good (C% ≥ 80.18). This remark is no less true of the analyses based on small samples (analyses 13-16) as of the rangeland analysis, in which N = 645. However, the small-sample analyses are precisely those most vulnerable to the consequences of overfitting (p. 000) and to sample-specific variation and covariation. Thus, while the results of these analyses may be optimal for the particular data sets examined, they are unlikely to have good *general* validity. Accordingly, in the absence of cross-validation of some kind, the value of their results for *predictive* purposes would be practically nil. By the same token it also follows that the results of the rangeland analysis would have considerably greater predictive value. For the rather different goal of *describing* ecological relationships provisionally regarded as properties of the samples actually examined, on the other hand, the results of all five analyses would be more uniformly useful. Descriptively, the results could be useful in themselves, as for example in vegetation survey, or in guiding the direction of further work. Thus, in coming to an appraisal of the worth of the analyses in ecological terms we should recognize that, while the value of the results for descriptive purposes can with some justification be said to have been established, their predictive value would, in general, be very small.

In summarizing, we note that in terms of the various assessment criteria employed, the overall success of the analyses can scarcely be doubted. Differences between analyses in their value for different purposes were, however, pointed out and related to differences in the design of the investigations. Furthermore, it is not difficult to see how cross-validation, by sharpening interpretation and clarifying the extent of the applicability of the results, might have added considerably to their ecological value.

18.2 Flexibility of canonical analysis. In attempting to clarify what canonical analysis might contribute to ecological data analysis, the flexibility of the method derserves to be recognized. One objective of the analyses reported in Sections 13-17 was to draw attention to this property of canonical analysis. The varied ecological content of the applications, which embraced both exploratory field survey and experimental situations, in itself gives some indication of the flexibility of the method. The analyses also show how by the use of variables of different kinds the applicability of canonical analysis can be widened and exemplify some of the specific purposes for which canonical analysis may be employed in data analysis.

Use was made of sets of continuous, nominal-scaled and binary variables in various combinations. Binary variables were employed as a means of handling nominal-scaled variables. This device was shown to increase the utility of canonical analysis very considerably. One-way multivariate analysis of variance, discriminant analysis and a latent root and vector method of contingency table analysis were all shown to be special cases of canonical analysis which arise when either one or both sets of variables in the latter are binary. It was also demonstrated that a particular class of univariate analysis of variance (split-plot designs) can be undertaken by canonical analysis. The bearing of these relationships on our objective of clarifying the domain of canonical analysis in ecology will largely be self-evident. We are now able to appreciate just how extensive the potential role of the method is. The analyses also illustrate the use of variables of other kinds. In the mineral nutrition experiment *all* q response variables were identical and were measured in the same metric and had the same origin and unit; in the more usual case, the variables used in canonical analysis are *qualitatively* distinct, at least. The nutrition analysis is an example of a *repeated-measures* design. In the rain forest analyses principal components were used as the input variables with the specific aim of circumventing difficulties which arise when the number of observed variables exceed sample size.

Sections 13-17 also illustrate different purposes for which
canonical analysis may be used in data analysis. In the applications considered, stress was placed in general on canonical analysis
as a means of *exploratory data analysis*. In such cases, the
general objective was to summarize and expose anticipated and
unanticipated features of the covariance structure of interest.
The use of canonical analysis in this way minimized the requirements
placed on the data for analysis. Where stronger assumptions could
reasonably be made, we saw how canonical analysis can be used for
inferential purposes. In this connection the method was used for
parameter estimation and for testing hypotheses of (a) independence
and (b) rank. Tests of the independence of two sets of
continuous variables, of the equality of k vector-means and
of independence in 2-way contingency tables are all fundamentally
tests of independence. Tests of rank evaluate the hypothesis
that the first s canonical variates are sufficient to account
for the association or discrimination between the respective sets
of variables.

18.3 Choice of method. We consider here some subsidiary points
which arise specifically from the analyses of Sections 16 and 17.
From the formal equivalence of certain special cases of canonical
analysis to multivariate analysis of variance and contingency
table analysis, which was demonstrated in these Sections, we can
envisage something of the extent of the potential domain of
canonical analysis in ecology. Formal equivalence in itself,
however, does not necessarily mean that two methods would be
equally rewarding in practice. We shall see, for example, that
in relation to one important class of applications, canonical
variates analysis has certain advantages over multivariate
analysis of variance.

The algebraic equivalence of multivariate analysis of
variance and canonical variates analysis in relation to a test
of the hypothesis the equality of k p-variate means (Section
16.4.1) does not imply that the two methods would be of equal
practical value in all situations. The distinction between the
methods is sharpest when, in the multivariate analysis of variance,
the test of hypothesis (16.1) is conducted using determinants, as
is customary, rather than eigenvalues (Section 16.2). For
simplicity, we shall therefore confine our attention to this
case. Canonical variates analysis it will be recalled, is almost
invariably treated as an eigenproblem (Section 16.3). Thus, it
turns out that a choice between multivariate analysis of
variance and canonical analysis, may, and in practice commonly
does, amount to a choice between the adoption of a likelihood-ratio
or a union-intersection test procedure. There is unfortunately
no general answer to the question as to which procedure leads to

the more powerful test (Kshirsagar, 1972, p. 331-334). It has
been argued that where the deviation from the null hypothesis is
unidimensional or nearly so, as is not uncommon in ecological
applications, the largest-root criterion may be expected to have
somewhat greater power and is therefore to be preferred (Bock,
1975, p. 154; cf. Kshirsagar, 1972, p. 333); but see Olson
(1976, 1979) and Stevens (1979) for different viewpoints. In
other respects, however, differences between the likelihood-ratio
and union-intersection procedures are clear-cut. In particular,
we note that the union-intersection test has the advantage that,
when it is possible to reject the null hypothesis, it is automatically known which combination of the response variables provides
the strongest evidence against H_o (Harris, 1975, p. 6). This
information is specified by the eigenvector associated with the
root r_1^2 on which the union-intersection test depends. Rejection
of H_o in the multivariate analysis of variance, on the other hand,
is generally followed by univariate tests of the effect of the
response variables individually. In such an analysis there is
therefore no hope of detecting the effect of critical *combinations*
of variables. For this reason there are strong grounds for
preferring the union-intersection approach over the likelihood-
ratio alternative, particularly in exploratory work. Moreover,
the axes of the canonical representation derived from the eigenvectors corresponding to the r_k^2, which function as synthetic
contrasts with which to search for clustering or structure among
the sample centroids, have no counterpart in multivariate analysis
of variance. Yet these contrasts are precisely those which in
exploratory work are most likely to prove rewarding. In exploratory
studies at least, canonical analysis would therefore appear to
offer a more unified and penetrating approach than the alternative
represented by the multivariate analysis of variance.

Turning to the contingency table analyses, the equivalence
shown between canonical anlaysis and Williams' (1952) method in
relation to a test of the independence of two sets of categorized
variables (Section 17.4.1), is not accompanied by any distinction
in practical usage comparable to that described above. Both procedures are fundamentally eigenanalyses, and, as such, the solutions
they provide are characterized by the scaling properties which we
have seen are associated with canonical analysis generally. The
two methods in fact are completely interchangeable. In contingency
table analysis, a distinction comparable to that considered above
can be drawn between scaling methods and those alternatives which
set out to partition the overall test criterion χ_o^2, in some
way other than by eigenanalysis.

496   R. GITTINS

The above remarks have focussed on those aspects of the
equivalence between methods which have relevance to practical
applications. But the relationships could be made use of in
other ways, notably perhaps in the teaching of ecology. The
relationships described in Sections 16 and 17, as well as those
alluded to previously (Section 6.2), show canonical analysis to
provide a means of (a) clarifying relationships between many
familiar methods of data analysis, both univariate and multivariate;
and of (b) unifying the presentation of a substantial part of
parametric data analysis. As a consequence, it is largely self-
evident that canonical analysis could be employed to good effect
in the teaching of ecology.

Based on the applications of Sections 13-17 the following
conclusions are drawn. First that canonical analysis is able
to produce meaningful ecological results, at least under certain
scale conditions. Secondly, that canonical analysis is a remarkably
versatile procedure of wide applicability. Finally, that the role
of canonical analysis in ecology is potentially an extensive one.

CHAPTER IV

DISCUSSION

19. INTRODUCTION

Methods for studying interrelationships among variables in
all but the simplest cases are surprisingly few. Of the methods
which are available, canonical analysis appears on theoretical
gounds to be well-suited to the analysis of one class of problem
encountered in ecology, namely that in which the joint distribution
of the variables is essentially linear and continuous. Assessments
of canonical analysis in practice however have been conflicting.
Several workers have concluded that the method has comparatively
little to contribute to data analysis in ecology (Austin, 1968;
Cassie and Michael, 1968; Cassie, 1969; Barkham and Norris, 1970;
Goodall, 1970; Gauch and Wentworth, 1976; Kessel and Whittaker,
1976; W. T. Williams, 1976, pp. 66-67; Gauch and Stone, 1979).
Furthermore, the practical value of canonical analysis has
frequently been questioned by statisticians (E. J. Williams, 1967;
Kendall and Stuart, 1967, p. 305; Dempster, 1969, p. 179;
Marriott, 1974, p. 31; Kendall, 1975, pp. 64 and 69; see also
Kempthorne, 1966). On the other hand numerous apparently
successful applications of the technique in ecology or in related
areas have been reported (Rao, 1952, pp. 364-378; Ashton, Healy
and Lipton, 1957; Delany and Healy, 1964, 1966; Glahn, 1968;
Delany and Whittaker, 1969; Lee, 1969; Pearce, 1969; Calhoon and
Jameson, 1970; Vogt and Jameson, 1970; Fritts, Blasing, Hayden
and Kutzbach, 1971; Grigal and Goldstein, 1971; Hatheway, 1971;
Webb, Tracey, Williams and Lance, 1971; Goldstein and Grigal, 1972;
Gould and Johnson, 1972; Smouse, 1972; Webb and Bryson, 1972; Hill,
1973; Ray and Lohnes, 1973; Thaler and Plowright, 1973; Zuckerman,
Ashton, Flinn, Oxnard and Spence, 1973; Bryson and Kutzbach, 1974;
Gipson, Sealander and Dunn, 1974;Hills, Klovan and Sweet, 1974; Spain,
1974a,b; Ashton, Flinn and Oxnard, 1975; Birks, Webb and Berti, 1975;
Dancik and Barnes, 1975a,b; Grigal and Ohmann, 1975; Karr and James,
1975; Birks, 1976; Kowal, Lechowicz and Adams, 1976; Thorpe, 1976;
Kercher and Goldstein, 1977; Webster, 1977a,b, pp. 196-200; Albrecht,
1978; Blasing, 1978; Chacon, Stobbs and Dale, 1978; Cody and Mooney,

1978; Jeffers, 1978, pp. 128–138; Bissett and Parkinson, 1979;
Fritts, Lofgren and Gordon, 1979; Hull, 1979). It is of some
interest to observe that the great majority of these applications
involve a particular case of canonical anlaysis, namely canonical
variates. In view of conflicting evaluations of the method, the
present paper set out to clarify what can in fact be accomplished
by means of canonical analysis in ecology. In pursuit of this aim,
seven applications of canonical analysis were made. It was concluded
from the results obtained that the role of the method in ecology
is potentially an extensive one. It has been suggested above
that a comparatively narrow range of variation and covariation
in the material examined may partly account for the apparent
success of the analyses described; what for convenience we may
characterize as their small scale may have ensured that the
fundamental requirements of linearity and continuity were not
grossly violated. In addition, the scant attention paid to the
canonical weights together with reliance on several comparatively
novel interpretive indices may have made a more incisive interpretation of the results possible than has often been the case.

The linear, orthogonal character of the algebraic model
underlying canonical analysis as well as the high error variance
of ecological data have both been advanced to account for what
has been seen as either the inappropriateness or the ineffectiveness
of canonical analysis in ecological studies. There is undoubtedly
substance to these remarks in relation to certain applications, at
least. What has not always been sufficiently recognized, however,
is that deficiencies in design, execution and interpretation have
also been at least partly responsible for the lack of success
sometimes encountered. It is not difficult in retrospect, for
example, to identify deficiencies in technique in connection with
the seven applications of canonical analysis reported above.
Shortcomings in technique are also by no means uncommon in
applications reported in the literature generally. In order to
arrive at a balanced appraisal of canonical analysis it is
obviously necessary to identify and as far as possible distinguish
deficiencies of this kind from possible limitations of the
method itself. In Section 20, therefore, we draw attention to
several factors bearing on the conduct of an analysis which can
profoundly influence the quality of the results obtained. In
this way we may perhaps contribute towards both a better
appreciation of the potentialities of canonical analysis and
towards more responsible use of the method. In addition to
questions of technique, there are also a number of outstanding
problems associated with canonical analysis itself, the solution
or clarification of which could add appreciably to the value of
the method. In Section 21 four such problems are identified and
their implications for the prospect of canonical analysis in ecology
discussed.

20   Conduct of canonical analysis. Having identified canonical analysis as an appropriate means of attaining a specific ecological objective, the question of the design of the investigation immediately arises. Orlóci (1978, pp. 1-41) has provided a useful general review of issues which require consideration at this stage in the context of vegetation survey. In Section 20.1 we address specific points which arise in the initial stages of applications of canonical analysis. Secondly, having acquired data, it is good practice to validate these against the specification of the model. It may, as a consequence, in some cases be found necessary to adjust the data to minimize any discrepancy before proceeding to canonical analysis. These aspects of an investigation are taken up in Section 20.2. Other features of the data unrelated to the specification of the model also deserve attention at this stage. The most notable of these are measurement error and multicollinearity. These features and means of combating their effects are discussed in Section 20.3. Finally, there are a number of steps which can be taken following an analysis to sharpen interpretation; some of these are reviewed in Section 20.4.

20.1 Design. Two decisions with far reaching effects on the outcome of an analysis concern the selection of variables and the number of samples to employ. These are matters which therefore deserve careful attention.

20.1.1 The selection of variables. The need for substantive judgement in the selection of variables is self-evident. It is plain that the omission of variables which play a significant role in the system investigated will have serious and irredeemable consequences. On the other hand, the overinclusion of variables also has harmful effects. In such cases there will be loss of precision in estimation of the canonical weights; consequently, the overfitted model will be sensitive to sampling error. Further, the power of any statistical tests which might be applied is strictly a decreasing function of the number of variables. Thus, as more variables are acquired, the greater the danger that interesting relationships or differences will go undetected (Dempster, 1971). Quite different considerations which also bear on the selection of variables concern the relative number of predictor and criterion variables and the extent to which the variables of each set are correlated among themselves. The former is important in predictive or directed analyses because of its effect on the size of the explained variance of a measurement domain which can be achieved. Consider, for example, the effect of increasing the number of criterion variables relative to a fixed set of predictors on the explained criterion variance. Highly correlated

variables, on the other hand, add little to explanatory power
and frequently create difficulties in the numerical estimation
of the canonical weights, rendering them unstable and sensitive
to sampling error.

Once the essential variables have been identified, the
question of the level or scale of their measurement arises. While
interval- or ratio-scaled measurements are frequently employed in
ecology, the use of nominal- or ordinal-scaled measurements
might often be more appropriate. Goodall (1970) has drawn
attention to the enormous sampling variance characteristic of
many kinds of ecological data and pointed out further that
variance is more effectively reduced by increased intensity of
sampling than by improved precision. Thus nominal- or ordinal-
scaled measurement may well be more rewarding than higher-scale
measurement in many studies. The advantages of the larger
samples which the use of such measurement would permit will become
apparent in the following subsection. Maxwell (1961), Burnaby
(1971), Buzas (1972), Harris (1975 , pp. 137-140) and Green (1978,
pp. 266-279) all describe canonical analyses based on nominal- or
ordinal-scaled response variables.

20.1.2 Sample size. Because of the profound consequences of
sample size on the use to which the results of an analysis may
be put, in the design stage it is of the utmost importance that
the primary purpose of analysis - whether *descriptive* or *predictive*
- be explicitly recognized and stated. The choice has far reaching
ramifications on the design of a study, which influence not only
the size of the sample but also the sampling method (see Orlóci,
1978, pp. 21-25) and the measurement scale or scales adopted
for the variables.

A prime requisite of a sample is that it be unbiased. Moreover,
as Guttman (1968) has observed: "It is the task of experimental
design - before the data are gathered - to ensure that *enough* points
of [sample] space are observed, with sufficient *spread*, to reveal
the basic lawfulness sought; a computing algorithm by itself need
not always be able to compensate for deficiencies in the design of
observations". In this connection it is well to be aware that
a multidimensional space needs surprisingly many samples to avoid
quite large holes (Gower, 1973). Guidelines as to what might
constitute a sample of adequate size in relation to the number of
variables in predictive applications of canonical analysis are
given by Barcikowski and Stevens (1975) and by Thorndike (1978,
p. 184). Samples fulfilling the guidelines proposed, however, would
often be wholly beyond realization in practice in ecological applications particularly in exploratory surveys to remote areas. In such
cases, there would be no alternative but to work with the data base tha

could be realized. Nevertheless, a serious effort to obtain sufficient samples to permit cross-validation or double cross-validation (Mosier, 1951) would be a worthwhile goal in many studies, the latter providing the most generally suitable means of gauging the extent of the sample-specific dependence of the results. The scale of measurement adopted for the variables might well be considered with the benefits stemming from cross-validation of some kind in mind, as indicated earlier.

Having acquired data for analysis, it is desirable to check that the minimum requirements of linearity and continuity are broadly satisfied. If the requirements are not satisfied, then steps to appropriately transform the data so that the discrepancy is minimized would generally be worthwhile. It is to these matters that we now turn.

20.2 Data validation. A minimal condition for the proper use of canonical analysis is that the structure of the data to be analysed is substantially linear and continuous. More specifically, the requirement is that the conditional mean of $\underset{\sim}{Y}$, given $\underset{\sim}{X}$, is linear in $\underset{\sim}{X}$ (cf. Timm, 1975, eq. (2.55)). The purpose of validation is to as far as possible assess the extent to which this condition, at least, is satisfied. The need for validation increases with the scale of the investigation. As scale increases we can anticipate that in general the distribution of individual variables will depart increasingly from unimodality, perhaps becoming truncated, while bivariate distributions will become increasingly nonlinear. Similarly, the p- and q-variate joint distributions of the variables of each set will also depart increasingly from linearity and continuity, features which will be reflected in the corresponding data structures. Hence, in validation interest centers on the distribution of the variables for the information conveyed about the data structure.

20.2.1 Distributional properties. While in practice it is impossible to exactly determine and specify the joint distribution of a set of multivariate observations, partial insight can be obtained by examining the distribution of the variables individually, pairwise, and, after mapping the sample into a linear subspace, simultaneously. Graphical analysis is particularly helpful for this purpose provided the data are moderately extensive (Gnanadesikan and Wilk, 1969; Gnanadesikan, 1977, pp. 150-195; Albrecht, 1978, pp. 40-50; Everitt, 1978, pp. 65-81). Moment statistics may also be informative. To examine *univariate* distributions, histograms, rankit plots (Bliss, 1967, p. 108), normal probability plots (Bock, 1975, pp. 161-168; Gnanadesikan, 1977, p. 167) and coefficients of

skewness and kurtosis are useful, while for *bivariate* distributions
scatter plots (Gnanadesikan, 1977, pp. 177-182) and simple
regression are available. Procedures which directly address the
simultaneous distribution of the variables include mapping and
display of the sample in the space of the first two or three
principal components (Albrecht, 1978, pp. 42-45), chi-squared
probability plots (Gnanadesikan, 1977, pp. 168-195; Albrecht, 1978,
pp. 42-45; Everitt, 1978, pp. 67-81), and evaluation of multivariate skewness and kurtosis (Mardia, 1970). It is worth
noticing that several of the procedures mentioned here are
strictly speaking concerned with assessing normality or joint-
normality of distribution rather than linearity or continuity
as such. Thus they may be sensitive to features of the data
which might perhaps fall outside our immediate concern. In
employing these procedures we are therefore exploiting the
serendipitous value of the methods for summarizing the structure
of data, to which Gnanadesikan (1977, p. 137) has referred.

Nonlinearity, heterogeneity and the presence of deviant
observations (outliers) can largely nullify a canonical analysis.
Thus validation is directed largely to the detection of these
features. If, after examining distributional properties of the
data, it is considered that the assumptions of the model are not
sufficiently met, then efforts directed towards improving
conformity are indicated. In general, the conditions of non-
linearity, heterogeneity and the presence of outliers can be
expected to increase as the scale of the investigation increases,
or, roughly speaking, with the size of the data set (Nelder,
1977). Procedures for ameliorating the effects of such features
on the course of analysis are available, some of which we now
sketch.

20.2.2 Outliers. Outliers may be revealed by the screening
procedures described above. In addition, techniques have also
been developed specifically to aid the detection of multivariate
outliers (Gnanadesikan and Kettenring, 1972; Gnanadesikan, 1977,
pp. 271-284; Barnett and Lewis, 1978; Orlóci, 1978, pp. 38-41).
While outliers can contribute useful information about the
processes which generated the data, in view of their drastic
effects on the outcome of analysis they are best discarded before
canonical analysis is undertaken (cf. Gnanadesikan, 1977, p. 271).

20.2.3 Nonlinearity. The golden rule when confronted by non-
linearity is to linearize (Kendall, 1975, p. 28), although other
strategies exist. There are several ways in which linearization
can be effected.

Where the number of variables is relatively small it may be possible to transform the affected variables individually to achieve linearity (Kendall, 1975, p. 28; Gnanadesikan, 1977, pp. 137-150; Howarth and Earle, 1979). Known functional relationships are obviously worth exploiting whenever possible (Tukey and Wilk, 1966; Austin, 1971, 1972; Nelder, 1977). In other cases polynomial functions of the variables of one or other set may be helpful (cf. Section 10 above). Frequently, however, the number of variables will be too large to handle in these ways. In such cases linearization of the *data structure* itself may provide an effective means of dealing with the problem. Noy-Meir (1974a,b) has shown how Shepard and Carroll's (1966) method for the parametric representation of nonlinear data structures can be used for precisely this purpose in ecology. A simpler and quite different approach to nonlinearity is possible where the sample can be clustered meaningfully on the basis of the variables of one or other domain. After clustering, canonical variates analysis can be used to obtain a reduced-dimensional representation of the group-centroids in the space of the variables of the 'second' domain. This procedure can readily be extended by replacing the binary-valued dummy variables with orthogonal polynomials as contrasts, thus enhancing its ability to deal with more extreme forms of nonlinearity. Yet another alternative would be to employ a resemblance function with linearizing properties in the canonical analysis itself, in place of the more usual covariance or correlation coefficient (cf. Orlóci, 1978, pp. 93-95).

20.2.4 Heterogeneity. Where the sample is or appears to be heterogeneous, several steps can be taken to render the data more suitable for canonical analysis. Care however must be exercised against the spurious appearance of heterogeneity in small samples (Day, 1969). In small samples apparent heterogeneity may indicate simply the need for a larger sample; in such cases steps to meet this need might then be taken. On the other hand, where heterogeneity is pronounced and the data sufficiently extensive, the sample could be clustered, leading to a number of smaller, more homogeneous sample-groups or clusters. Separate canonical analyses might then be performed on at least the larger clusters. With less extensive data, an initial cluster analysis could be followed by a canonical variates analysis performed over *all* groups as described in Section 20.2.3 above.

Methods of assessing and testing *multivariate normality* as such have been discussed by Andrews, Gnanadesikan and Warner (1973), Gnanadesikan (1977, pp. 161-195) and Cox and Small (1978), while transformations designed specifically to improve joint normality have been developed or used by Andrews, Gnanadesikan and Warner (1971), Gnanadesikan (1977, pp. 137-150) and Malmgren (1979).

Conditions other than nonlinearity, heterogeneity and, perhaps, multivariate non-normality can also jeopardize an analysis. Measurement error and multicollinearity, in particular, deserve attention during data screening. In the following Section we mention steps which can be taken to combat these conditions.

20.3 Measurement error; multicollinearity. Most ecological data are characterized by large error variance (Goodall, 1970; W. T. Williams, 1976, p. 67). In view of the known effects of error variance on canonical analysis (Meredith, 1964), it is likely to be worthwhile to correct for its presence whenever possible. Correction may be accomplished in either of two ways. First, where the data are sufficiently extensive, error variance might be estimated from the data (e.g. see Maxwell, 1977a, pp. 15 and 48). The estimates could then be subtracted from the diagonal elements of the correlation or variance-covariance matrix, R or V, before proceeding to canonical analysis, as described by Meredith (1964). Alternatively, factor analysis might be employed to determine orthogonal factor matrices and communalities for each set of variables and then canonical analysis performed across the common factor spaces of each domain.

Strong linear relationships among the variables of one or both sets lead to correlation matrices, R_{11} and R_{22}, that are singular or almost so (e.g. see Webster, 1977a). The condition of multicollinearity results in unstable, meaningless canonical weights. Multicollinearity or near-multicollinearity can be extremely difficult to detect; the problem of detection has been considered by Willan and Watts (1978), who provide valuable guidelines for this purpose. One means of nullifying multicollinearity is to perform a prior factor analysis or similar orthogonal transformation on the original variables and then to base the canonical analysis on the transformed variables. A related problem which also results in singular correlation matrices, arises when the number of variables of either or both sets exceeds sample size (Barkham and Norris, 1970; Gauch and Wentworth, 1976; Gauch and Stone, 1979). Here again the problem may be combated by a prior orthogonal transformation with the objective in this case of condensing the variables. Canonical analysis may then be performed on some subset or subsets of the transformed variables. Alternatively, a number of the original variables might themselves be discarded; McHenry (1978) and Orlóci (1978, pp. 25-38) have each described algorithms for obtaining optimal subsets of variables in multivariate analysis which would be useful for this purpose. An effective means of dealing with the same problem specifically in canonical variates analysis is provided by the Q-technique for calculating canonical variates described by Gower (1966).

Yet another strategy is available. This consists quite simply
of using the generalized inverses, $R_{\sim 11}^{-}$ and $R_{\sim 22}^{-}$ in place of
$R_{\sim 11}^{-1}$ and $R_{\sim 22}^{-1}$ (Marriott, 1974, pp. 93–94; Khatri, 1976).

We note in passing that a number of the procedures for dealing
with an excess of variables might also prove useful in connection
with the related problem of overfitting.

20.4 The interpretation of results. We have seen that a variety
of interpretive indices exist in canonical analysis (cf. Finn,
1974, p. 193). Among these we mention canonical correlation
coefficients, structure correlations, within-set variance,
redundancy and variable communalities. The canonical weights,
on the other hand, on which previously reliance has often been
placed, are considered here in general to be of little interpretive
value. This is because of their instability in response particularly
to multicollinearity, overfitting and sample-specific error. We
have also seen that graphic presentation can contribute significantly
to the description and interpretation of results. In applications
where a model of rank two or three can be fitted, scatter plots
or stereograms can have enormous interpretive value. In other
cases Andrews' (1972) technique offers a more flexible approach
unrestricted by the dimensionality of the solution as Oxnard (1973),
Gnanadesikan (1977, pp. 213–225) and Everitt (1978, pp. 81–87) have
shown. The minimum spanning tree of Gower and Ross (1969) is
another useful graphic device. Steps to further enhance substantive
interpretation can be taken. Among these we mention rotation of the
canonical variates (Cliff and Krus, 1976), the discarding of noninformative variables (McHenry, 1978) and the canonical prediction
strategy of Thorndike (1977). Rotation in canonical analysis has
generally involved identical orthogonal transformations of both
sets of canonical variates (Hall, 1977; Perreault and
Spiro 1978; Fornell, 1979; Wingard, Huba and Bentler, 1979); Scott
and Koopman (1977), however, have shown that nonidentical transformations of the variates are also sometimes useful. Double
cross-validation is almost always worthwhile in canonical analysis,
sharpening interpretation appreciably, and is indeed obligatory
in predictive studies in order that the extent of the sample-
dependency of the results can be gauged. Finally, radomization
or more formal statistical tests may each strengthen interpretation
in particular circumstances.

Few ecological applications of canonical analysis appear to
have been designed from the outset with canonical analysis
specifically in mind. Similarly, very few if any reported
applications have fully exploited the range of interpretive
devices available. For these reasons, opportunities for incisive

analysis frequently appear to have been lost. The uncritical
selection and use of computer programs seems only to have exacer-
bated this situation. Altogether, it is hardly to be wondered at
that reactions to canonical analysis have often been unfavorable.
Fortunately, there are grounds for expecting that increased
attention to the conduct of analysis would be repaid by more
uniformly rewarding results.

21.   UNRESOLVED PROBLEMS

Among the unsolved problems of canonical analysis we draw
attention to those associated with (a) overfitting, (b) overly
restrictive assumptions, (c) the extraction of information from
residuals and (d) the properties of solutions associated with
different resemblance structures.

21.1 Overfitting. Instances in which adequate samples are quite
beyond realization are always likely to be with us in ecology.
Thus, almost inevitably, the problem of overfitting has to be
faced. It is in situations of precisely this kind that cross-
validation, which might otherwise be used to shapren interpretation, is unable to offer a solution. Where statistical tests for
one reason or another are also unavailable, interpretation is bound
to be to some extent equivocal. This indeed proved to be the
case in a majority of the applications considered above. Overfitting results in the explanatory power of a fitted model being
exaggerated and in the sensitivity of the canonical weights and of
other interpretive indices to sampling error. Consequently, the
model is likely to poorly approximate the functional dependencies
among the variables. Examples of overfitting in ecological uses
of canonical analysis are frequent (e.g. Hedges, Wheeler and
Williams, 1973; Trenbath and Harper, 1973; van der Aart and
Smeenk-Enserink, 1975; Auclair, 1975; Gauch and Wentworth, 1976;
Herrera, 1978). Indeed, instances in which one or more canonical
correlation coefficients of unity are reported are by no means
uncommon. The question therefore arises as to whether the
"fitted parameters are meaningful or, conversely, whether the
numerical processes which produce them are not empty exericses"
(Dempster, 1971, p. 336). Sometimes evidence of the reasonableness
of a result can be put forward following analysis by recognizing
substantive meaning in the canonical variates. Suggestions of
a preliminary kind as to how the problem of overfitting in
canonical analysis might be tackled have been made by Weinberg
and Darlington (1976). However, as Dempster (1971) has pointed
out in connection with overfitting generally, "a crucial problem
for theoretical statisticians is to assess the evidence [for substantive meaning] internally during the course of analysis, and

to alter the course where necessary so that the outputs of analysis have high signal-to-noise ratio". Progress in this direction would clearly be of immense benefit to ecologists. In the meantime it would be advisable either (a) to work whenever possible with measurements of a lower scale than might otherwise have been chosen, with the aim of securing a larger sample; or (b) to anticipate the consequences of overfitting and take account of them in interpretation. We note also that Dempster (1966) has addressed the problem of estimation in canonical analysis and has provided expressions for correcting bias in sample estimates of canonical correlation coefficients and canonical variates of use with moderately large samples.

21.2 Overly restrictive assumptions. Even as a method of exploratory data analysis, canonical analysis places constraints on field data which are unlikely to be met in a wide class of ecological endeavors. The question naturally arises as to whether the constraints specifically of linearity and continuity might somehow be relaxed. This is the converse of the question tackled above when techniques to enhance conformity between the data and the model requirements by adjustments to the data were outlined (Sections 20.2.3 and 20.2.4). We refer here to lines along which current work is proceeding in order to remove the constraints of linearity and continuity and mention further possible developments.

21.2.1 Linearity. A technique which would unquestionably be of value in this connection is a nonmetric analog of canonical analysis comparable to nonmetric analogs of, say, principal components analysis. Nonmetric scaling methods which preserve either (a) a *monotonic* or (b) a yet more general *continuity* relationship with observed distances between objects have been available for some years (e.g. Shepard, 1962; Kruskal, 1964; Shepard and Carroll, 1966). Similarly, nonlinear, monotonic analogs of multiple regression analysis and the analysis of variance have also been developed (Kruskal, 1965; Carroll, 1972), while Carroll (1972) has described a nonlinear counterpart of multiple regression based on the continuity approach. The extention of these concepts to canonical analysis could, in principle, provide a nonmetric analog of the method. The first significant steps towards realization of this goal appear to have been taken. Recently, Young, de Leeuw and Takane (1976, 1978) have described a nonmetric version of canonical correlation analysis which places virtually no constraints on the data, and, in additon, also provide a computer program (CORALS: Canonical Optimal Regression by Alternating Least Squares) to perform the analysis. Reports of the performance of the method in practice would obviously be of more than ordinary interest and significance to ecologists.

Thus it seems that it may already be possible to completely bypass the problem of nonlinearity in canonical analysis. Apart from removing an irksome difficulty, we may anticipate that such developments will provide us with lower dimensional solutions than the familiar metric analysis, yet with no loss in interpretability. It would, however, would be a mistake to suppose that advantages of this kind could be gained without cost. It is, for example, well-known that as prior assumptions of the data are weakened, in general the power of analysis diminishes also. In the present context, inferential aspects of canonical analysis for instance will almost surely have to be forfeited. The greatest obstacle to the rapid implementation of nonmetric forms of canonical analysis, however, given the likely complexity of the algorithms, may well be the programming effort required to get computer programs operational at local installations. Moreover, the programs in this field tend to become obsolete very rapidly and the algorithms underlying them to be prone to problems of local minima and solution degeneracy. Nevertheless, ecologists will take a keen interest in continuing efforts to perfect nonmetric counterparts of canonical analysis.

21.2.2 Continuity. As with nonlinearity, recent developments seem likely to make the analysis of heterogeneous data structures in canonical analysis much more tractable than has hitherto been the case. Kercher and Goldstein (1977), for example, have developed a technique (canonical group analysis) which, in cases where the data structure corresponding to one set of variables consists of well-defined, discrete groups, enables the requirement of continuity to be relaxed. From the results obtained, it appears that the method offers an effective means of tackling the problem in cases of this kind. In practice, the data are usually clustered prior to the main analysis in order to yield the strictly discontinuous data structure assumed.

The need for a completely discontinuous data structure, however, would in itself be unrealistic in many ecological contexts. Often, ecological processes are known to generate data structures which hover tantalizingly between the continuous and the discontinuous (cf. Webb, 1954). Prior classification in such cases would seriously distort the intrinsic structure of the data. There exists a need, therefore, for a procedure, which, while preserving intrinsic structure, would be entirely free from considerations as to the extent of the continuity or otherwise of the data. An approach of this kind, which has not so far been tried, would involve only a slight modification of the usual procedure. Indeed, it seems that analysis need differ from the standard procedure of canonical analysis only in being based on the raw sums of squares and crossproducts matrix, rather than the correlation of variance-covariance matrix, as is customary. In this way *noncentered*

canonical variates would result. These linear composites could
then be rotated, for example by the varimax criterion, towards
phases of higher density in each sample space. Identical
transformations might be applied to the two sets of canonical
variates for this purpose, but there would be no necessity for
this to be the case (cf. Scott and Koopman, 1977). Post-
normalization of the canonical variates following rotation might
also prove worthwhile. The effects of modifications of the kind
described on principal components analysis have been examined
in some detail by Noy-Meir (1970, 1973b). Noncentered components
analysis has in fact been applied in connection with large-scale
descriptive surveys of heterogeneous vegetation with excellent
results (e.g. Noy-Meir, 1971). It seems possible that
canonical analysis of noncentered data followed by varimax
rotation and post-normalization of the canonical variates might
similarly extend the applicability of canonical analysis to
heterogeneous data.

Thus recent developments hold out the promise that non-
linearity and heterogeneity will not always be the stumbling blocks
to the applicability of canonical analysis that they have sometimes
been in the past. Moreover, there is hope that in the not too
distant future intelligent programs, which will use the data to
check if the assumptions underlying an analysis are met, and
then perhaps adjust course accordingly, may be developed (Nelder,
1977). A related task which we raise but do not pursue here,
centers on the *robustness* of canonical analysis itself. Very
little is known of the robustness of parametric forms of multivariate analysis generally. Harris (1975, pp. 231-233), however,
has expressed the view that there are grounds for believing
that on investigation these methods will in fact be shown to be
reasonably robust. Obviously, there is scope for further work
by theoretical statisticians here, the results of which could
prove to have considerable practical importance for ecologists.

21.3 Residuals. The analysis of residuals in canonical analysis
has scarcely begun, although Brillinger (1967) drew attention
to some of the opportunities several years ago.
Yet there is no reason to suppose that canonical residuals will
prove any less informative in exposing inadequacies in a fitted
canonical model than has been the case with residuals in statistical
data analysis generally (cf. Tukey and Wilk, 1966). Brillinger (1967)
Glahn (1968), and Skinner (1977, eq. (23)) have shown how
residuals from solutions of less than full rank can be calculated
in canonical analysis (see also Gleason, 1976; Webster, 1977a).
Moreover, Gnanadesikan and his colleagues have shown how in
related forms of analysis, multivariate residuals can profitably
be examined. Specifically, they advocate (a) normal probability
plots of the residuals and gamma probability plots of quadratic

forms of the residuals; and (b) principal components analysis of
the residuals. (See Gnanadesikan and Kettenring, 1972; Devlin,
Gnanadesikan and Kettenring, 1975; Gnanadesikan, 1977, pp. 259-
270). Such procedures applied to residuals in canonical analysis
could well prove informative. It is also clear that additional
work directed towards detecting structure specifically in canonical
residuals might well prove rewarding.

21.4 Resemblance structures. Little attention has been given to
the effect of the resemblance structure on the properties of the
resulting solution in canonical analysis. The resemblance
structure is itself partly a function of the selected resemblance
function (cf. Orlóci, 1978, pp. 42-43). In canonical analysis
the correlation coefficient or the covariance (calculated between
variables) have traditionally been the preferred resemblance
functions (cf. however Tatsuoka, 1971, p. 183). Yet we have
seen that solutions resulting from the choice of other scalar-
products may well have interesting and useful properties (p. 000);
see also Green (1978, p. 280). Moreover, Gower (1966) has shown
how a matrix of Mahalanobis distances, D^2, calculated between
samples can provide a starting point for canonical variates
analysis and that such an analysis has practical and statistical
properties of considerable value. A wide range of resemblance
functions exist, both metric and nonmetric, the properties of
many of which have been described by Orlóci (1978, pp. 42-101).
Some of these functions for example are known to have a linearizing
effect on the data structure and therefore could conceivably prove
helpful in combating nonlinearity. Orlóci (1978, pp. 141-142) has
also shown how, given certain assumptions concerning the form of
the functional relationship between variables of two kinds, a
resemblance function which will optimize a particular reduced-
dimensional representation of the data can be derived. From
these remarks, it will be clear that much remains to be learnt
concerning the role of the resemblance structure in canonical
analysis. Further work in this area could have far-reaching
effects for ecologists.

The foregoing discussion has centered on considerations which
arise in the planning and execution of canonical analysis and on
some outstanding issues concerning the method itself. As the
standard of application rises and solutions to unresolved problems
are found, we may perhaps anticipate that the use of canonical
analysis generally will become more rewarding. Thus, there is
every likelihood of canonical analysis coming to fulfill the
role in ecological data analysis which it appears well-fitted
to perform but which has yet to be fully realized. Should this
hope prove to be ill-founded, however, then an urgent need for
methods which will render relationships between sets of variables
analytic will be abundantly clear.

CHAPTER V

CONCLUSIONS

Relationships among sets of variables have always occupied a central place among the interests of ecologists. Canonical analysis appears to be one of the most promising procedures for the analysis of such relationships. Yet the method is largely unknown to ecologists. Our aim has been to draw canonical analysis and some of the opportunities afforded by the method to the attention of ecologists.

Applications of the method were confined to data structures thought to be broadly linear and continuous. Within this class, canonical analysis was consistently found to yield useful results. The remarkable flexibility of the method in connection with applications of this kind was also demonstrated. This aspect of canonical analysis however was not examined exhaustively. For completeness, we note here three generalizations of the procedure about which little or nothing was said which significantly extend the utility of the method. First, by the use of suitably coded dummy variables, the equivalence shown to exist between canonical analysis and one-way multivariate analysis of variance can be extended to other more complex experimental designs (E. J. Williams, 1967; Hope, 1968, pp. 129-137; Dempster, 1969, pp. 222-241; Kshirsagar, 1972, pp. 372-374; cf. also Timm, 1975, pp. 381, 392 and 415). The chief consequence of this wider equivalence is that, while canonical analysis may have much to contribute in connection with exploratory studies, its usefulness is by no means confined to investigations of this sort. In particular, it will now be apparent that the analysis of designed experiments generally can also be undertaken by canonical analysis. Secondly, the extension of canonical analysis to m sets of variables (m > 2) has for some time been a practical proposition, due largely to the computer programs CANCOR and CANDECOMP developed by Carroll and Chang (Carroll, 1968; Carroll and Chang, 1970; Green, 1978, pp. 283-285). Finally, we note that canonical variates find extensive application in the optimal allocation of unidentified samples among k pre-existing categories (cf. Section 15.1;

Bock, 1975, pp. 395-415). In this context, the canonical variates
are generally referred to as discriminant functions. Thus, the
domain of canonical analysis in ecology is even wider than the
applications discussed might have led us to suppose.

The constraints of linearity and continuity are undoubtedly
troublesome in many ecological situations. We have however seen
how by (a) the use of transformations, (b) adjustments to the
specification of the model, or by (c) the use of nonmetric counterparts, steps can be taken to relax the constraints. Such freedom
is bought at a price - the need for added vigilance in data
validation and preprocessing as well as a need to rely on
complicated algorithms in computation which may embody unfamiliar
procedures of numerical analysis. Nevertheless, it remains true
that much can be done to widen the class of problem whose
analysis may profitably be undertaken by canonical analysis.

There are, on the other hand, unresolved issues which curtail
the applicability and effectiveness of canonical analysis. As
these problems become the focus of increased attention by theoretical
statisticians we may anticipate that their eventual solution in
some cases at least will add to the range and penetration of the
method.

The principal conclusions drawn from this study are:

(a)   that canonical analysis can contribute usefully to
      data analysis in ecology;

(b)   that the role of canonical analysis within the class
      of applications characterized by broadly linear and
      continuous data structures is potentially an extensive
      one; and that

(c)   provided adequate steps are taken to ensure conformity
      between the specification of the model and the structure
      of the data to be analyzed, the utility of canonical
      analysis is not restricted to data structures which are
      substantially linear and continuous.

Three corollaries are worth stating. First, that opportunities for
ill-considered application or misuse of the method are likely to
increase as the scale of investigation increases. Secondly, that
whenever a choice can be made, there are likely to be advantages in
minimizing the scale of application, as far as is consistent
with realization of the overall objective. Finally, that in large
scale studies it would be well to recognize a need to allocate
a proportionately greater expenditure of time and effort to data
validation and preprocessing than exists in smaller scale studies.

Obstacles to the wider and more effective use of canonical analysis in ecology are attributable in part not so much to the method itself as to such external matters as the slow diffusion of computer software and the need for a more determined effort on the part of ecologists in the field. Much of the software necessary to implement canonical analysis, as well as for data validation and preprocessing, and for the effective display of the results including residuals, already exists. Programs however not only vary greatly in quality but are also scattered and access generally is far from easy, particularly in the case of the better versions. We may look forward in the not too distant future to remote access, interactive systems of programs for canonical analysis which support facilities for multi-stage analysis. These would include opportunities for probing and exploring data structures, metric and nonmetric algorithms for canonical analysis, procedures for the analysis of residuals and flexible means of graphical display. The development of systems of this kind is, however, still some way off. With respect to field work, it is something of a truism that the advancement of ecological understanding is founded on a thorough working knowledge of ecosystems in the field. What is not sufficiently recognized is that this remark is particularly applicable to studies which are analytic rather than verbal in nature. If the wealth of ecological insight that canonical analysis is potentially able to yield is to be recognized for what it is and be used to best advantage, then it occurs to me that ecologists may need to consider whether they ought to devote a greater proportion of their time and overall effort to field work than is customary at present.

In short, it will have emerged that the theory of canonical correlation analysis which was developed originally to identify the most predictable criterion has since proved to have other applications. I incline towards the view expressed by Kshirsagar 1972, p. 284) that canonical analysis is both more widely applicable and richer than comparable procedures such as principal components analysis, which are fairly well-known to ecologists. As multi-stage analysis becomes computationally more convenient and the standard of application arises, I believe there are grounds for optimism that the promise which canonical analysis appears to hold for ecological data analysis will come to be fully realized.

ADDENDUM

THE SAMPLING DISTRIBUTION AND A TEST OF SIGNIFICANCE FOR REDUNDANCY

Since preparing the above review, Miller's (1975) derivation of the sampling distribution of redundancy has come to my notice. Arguing partly by analogy from the distribution of the coefficient of multiple determination R^2 in multiple regression analysis and on the basis of Monte Carlo studies, Miller (1975) has shown a simple multiple of the ratio $[U^2_{y|u_1 \cdots u_s}] / [1 - U^2_{y|u_1 \cdots u_s}]$ to be a ratio of two χ^2 variates and hence to have an F distribution. Specifically,

$$\{ [U^2_{y|u_1 \cdots u_s}] / [1 - U^2_{y|u_1 \cdots u_s}] \} \cdot \{ [N - p - 1]q / pq \} \sim F_{\nu_1, \nu_2}.$$

For the model of full rank $(r = s,$ where $s = \min(p,q))$

$$\nu_1 = pq$$

$$\nu_2 = (N - p - 1)q .$$

For the reduced rank model $(r < s)$, the total redundancies of the leading term of the left-hand side are replaced by the redundancies of the reduced rank model, $U^2_{y|u_1 \cdots u_r}$ and ν_1 and ν_2 by $\nu_1 = pq'$ and $\nu_2 = (N - p - 1)q'$, respectively, where $q' = r$, the number of retained canonical roots.

The practical importance of this result lies in making a test of the null hypothesis

$$H_o : P^2_{y|u_1 \cdots u_s} = 0$$

against the alternative

$$H_1 : P^2_{y|u_1 \cdots u_s} \neq 0$$

possible, where we have written $P^2_{y|u_1 \cdots u_s}$ for the redundancy in the population. Miller's result represents a very considerable step forward in strengthening the interpretation of statistical applications of canonical analysis.

ACKNOWLEDGEMENTS

The completion of this work was made possible by a
Natural Sciences and Engineering Research Council of Canada
award to L. Orlóci. It is a pleasure to acknowledge my
indebtedness both to Dr. Orlóci for his support and encouragement
and to the Department of Plant Sciences at the University of
Western Ontario for hospitality and help.

To Dr. J. Ogden of the Australian National University,
Canberra appreciation is expressed for the rain forest data on
which the analyses of Sections 13 and 14 were based and for his
invaluable cooperation in the interpretation of the results of
these analyses. Dr. A. D. Bradshaw, formerly of the University
College of North Wales, Bangor provided valued guidance and
support in connection with the nitrate nutrition experiment
of Section 16, while Miss G. K. Crossley shared with me in the
running of the experiment. Thanks are expressed also to my
friend Robert Morris who shared in much of the early computing
work. The stereograms of Figures 13.1 and 17.1 were prepared
with the aid of a computer program kindly placed at my disposal by
my former colleague, Dr. P. A. Burrough.

REFERENCES

Aart, P. J. M. van der and Smeenk-Enserink, N. (1975).   Correla-
     tions between distributions of hunting spiders (Lycosidae,
     Ctenidae) and environmental characteristics in a dune area.
     Netherlands Journal of Zoology, 25, 1-45.

Albrecht, G. H. (1978).   The craniofacial morphology of the
     Sulawesi macaques: multivariate approaches to biological
     problems.   *Contributions to Primatology*, 13. Karger, Basel.

Anderberg, M. R. (1973).   *Cluster analysis for applications*.
     Academic Press, New York.

Andrews, D. F. (1972).   Plots of high-dimensional data.
     Biometrics, 28, 125-136.

Andrews, D. F., Gnanadesikan, R., and Warner, J. L. (1971).
     Transformations of multivariate data.   *Biometrics*, 27,
     825-840.

Andrews, D. F., Gnanadesikan, R., and Warner, J. L. (1973).
     Methods for assessing multivariate normality.   In *Multi-
     variate Analysis III*, P. R. Krishnaiah, ed.   Academic
     Press, New York.   95-116.

Ashton, E. H., Healy, M. J. R., and Lipton, S. (1957). The descriptive use of discriminant functions in physical anthropology. *Proceedings of the Royal Society, London B*, 146, 552–572.

Ashton, E. H., Flinn, R. M., and Oxnard, C. E. (1975). The taxonomic and functional significance of overall body proportions in Primates. *Journal of Zoology, London*, 175, 73–105.

Auclair, A. N. (1975). Sprouting response in *Prunus serotina* Erhr.: multivariate analysis of site, forest structure and growth rate relationships. *American Midland Naturalist*, 94, 72–87.

Austin, M. P. (1968). An ordination study of a chalk-grassland community. *Journal of Ecology*, 89, 408–425.

Austin, M. P. (1971). Role of regression analysis in plant ecology. *Proceedings of the Ecological Society of Australia*, 6, 63–75.

Austin, M. P. (1972). Models and analysis of descriptive vegetation data. In *Mathematical Models in Ecology*, J. N. R. Jeffers, ed. Blackwell, Oxford, 61–86.

Austin, M. P. (1977). Use of ordination and other multivariate descriptive methods to study succession. *Vegetatio*, 35, 165–175.

Austin, M. P. and Noy-Meir, I. (1971). The problem of non-linearity in ordination: experiments with two-gradient models. *Journal of Ecology*, 59, 763–773.

Barcikowski, R. S. and Stevens, J. P. (1975). A Monte Carlo study of the stability of canonical correlations, canonical weights and canonical variate-variable correlations. *Multivariate Behavioral Research*, 10, 353–364.

Bargmann, R. E. (1962). *Representative ordering and selection of variables*. Part A. Virginia Polytechnic Institute. Cooperative Research Project No. 1132. U.S. Office of Education.

Barkham, J. P. and Norris, J. M. (1970). Multivariate procedures in an investigation of vegetation and soil relations of two beech woodlands, Cotswold Hills, England. *Ecology*, 51, 630–639.

Barnett, V. D. and Lewis, T. (1978). *Outliers in Statistical Data*. Wiley, Chichester.

Barrett, J. C. and Healy, M. J. R. (1978). A remark on algorithm AS 6: triangular decomposition of a symmetric matrix. *Applied Statistics*, 27, 379-380.

Bartlett, M. S. (1947). Multivariate analysis. *Journal of the Royal Statistical Society, Series B*, 9, 176-197.

Bartlett, M. S. (1951). The goodness of fit of a single hypothetical discriminant function in the case of several groups. *Annals of Eugenics, London*, 16, 199-214.

Bartlett, M. S. (1965). Multivariate statistics. In *Theoretical and Mathematical Biology*, T. H. Waterman and H. J. Morowitz, eds. Blaisdell, New York, 201-224.

Birks, H. J. B. (1976). Late-Wisconsinan vegetational history at Wolf Creek, Minnesota. *Ecological Monographs*, 46, 395-429.

Birks, H. J. B., Webb, T., and Berti, A. A. (1975). Numerical analysis of surface samples from central Canada: a comparison of methods. *Review of Palaeobotany and Palynology*, 20, 133-169.

Bishop, Y., Fienberg, S. E., and Holland, P. (1975). *Discrete Multivariate Analysis: Theory and Practice*. Massachusetts Institute of Technology Press, Cambridge, Massachusetts.

Bissett, J. and Parkinson, D. (1979). Functional relationships between soil fungi and environment in alpine tundra. *Canadian Journal of Botany*, 51, 1642-1659.

Blackith, R. E. and Reyment, R. A. (1971). *Multivariate Morphometrics*. Academic Press, New York.

Blasing, T. J. (1978). Time series and multivariate analysis in palaeoclimatology. In *Time Series and Ecological Processes*, H. H. Shugart, Jr., ed. Proceedings of the Society for Industrial and Applied Mathematics Conference, Utah. SIAM, Philadelphia, 211-226.

Bliss, C. I. (1967). *Statistics in Biology: Statistical Methods for Research in the Natural Sciences*. Volume 1. McGraw-Hill, New York.

Bock, R. D. (1975). *Multivariate Statistical Methods in Behavioral Research*. McGraw-Hill, New York.

Bourlière, F. (1965). Densities and biomass of some ungulate populations in Eastern Congo and Rwanda, with notes on population structure and lion/ungulate ratios. *Zoologica Africana*, 1, 199-207.

Box, G. E. P. and Jenkins, G. M. (1976). *Time Series Analysis: Forecasting and Control*. Holden-Day, San Francisco.

Brent, R. P. (1970). Error analysis of algorithms for matrix multiplication and triangular decomposition using Winograd's identity. *Numerische Mathematik*, 16, 145-156.

Brillinger, D. R. (1967). Contribution to discussion of E. J. Williams' (1967) paper 'Analysis of association among many variates'. *Journal of the Royal Statistical Society, Series B*, 29, 199-242.

Brillinger, D. R. (1975). *Time Series: Data Analysis and Theory*. Holt, Rinehart and Winston, New York.

Bryson, R. A. and Kutzbach, J. E. (1974). On the analysis of pollen-climate canonical transfer functions. *Quaternary Research*, 4, 162-174.

Burnaby, T. P. (1971). The skeletal form of some African shrews. Quoted from R. E. Blackith and R. A. Reyment (1971), *Multivariate Morphometrics*. Academic Press, London, 105-107.

Buzas, M. A. (1972). Biofacies analysis of presence or absence data through canonical variate analysis. *Journal of Paleontology*, 46, 55-57.

Calhoon, R. E. and Jameson, D. L. (1970). Canonical correlation between variation in weather and variation in size in the Pacific tree frog, *Hyla regilla*, in Southern California. *Copeia*, 1970 (1), 124-134.

Carney, E. J. (1975). Ridge estimates for canonical analysis. In *Computer Science and Statistics: Eigth Annual Symposium on the Interface*, J. W. Frane, ed. University of California, Los Angeles, 252-256.

Carroll, J. D. (1968). A generalization of canonical correlation analysis to three or more sets of variables. *Proceedings of 76th Annual Convention of the American Psychological Association*, 227-228.

Carroll, J. D. (1972). Individual differences and multi-dimensional scaling. In *Multidimensional Scaling: Theory and Application in the Behavioral Sciences, Vol. 1,* R. N. Shepard, A. K. Romney , and S. Nerlove, eds. Seminar Press, New York.

Carroll, J. D. and Chang, J. J. (1970). Analysis of individual differences in multidimensional scaling via an N-way generalization of 'Eckart-Young' decomposition. *Psychometrika,* 35, 283-319.

Cassie, R. M. (1969). Multivariate analysis in ecology. *Proceedings of the New Zealand Ecological Society,* 16, 53-57.

Cassie, R. M. and Michael, A. D. (1968). Fauna and sediments of an intertidal mudflat: a multivariate analysis. *Journal of Experimental Marine Biology and Ecology,* 2, 1-23.

Chacon, E. A., Stobbs, T. H. and Dale, M. B. (1978). Influence of sward characteristics on grazing behaviour and growth of Hereford steers grazing tropical grass pastures. *Australian Journal of Agricultural Research,* 29, 89-102.

Cliff, N. and Krus, D. J. (1976). Interpretation of canonical analysis: rotated *vs.* unrotated solutions. *Psychometrika,* 41, 35-42.

Cody, M. L. and Mooney, H. A. (1978). Convergence *versus* non-convergence in Mediterranean-climate ecosystems. *Annual Review of Ecology and Systematics,* 9, 265-321.

Coe, M. J., Cumming, D. H., and Phillipson, J. (1976). Biomass and production of large African herbivores in relation to rainfall and primary production. *Oecologia,* 22, 341-354.

Cooley, W. W. and Lohnes, P. R. (1971). *Multivariate Data Analysis.* Wiley, New York.

Cox, D. R. and Small, N. J. H. (1978). Testing multivariate normality. *Biometrika,* 65, 263-272.

Coxhead, P. (1974). Measuring the relationship between two sets of variables. *British Journal of Statistical and Mathematical Psychology,* 27, 205-212.

Cramer, E. M. and Nicewander, W. A. (1979). Some symmetric, invariant measures of multivariate association. *Psychometrika,* 44, 43-54.

Cronbach, L. J. (1971). Validity. In *Educational Measurement*, R. L. Thorndike, ed. American Council on Education, Washington.

Dale, M. (1975). On objectives of methods of ordination. *Vegetatio*, 30, 15-32.

Dancik, B. P. and Barnes, B. V. (1975a). Leaf variability in Yellow Birch (*Betula alleghaniensis*) in relation to environment. *Canadian Journal of Forest Research*, 5, 149-159.

Dancik, B. P. and Barnes, B. V. (1975b). Multivariate analysis of hybrid populations. *La Naturaliste Canadien*, 102, 835-843.

David, M., Campiglio, C., and Darling, R. (1974). Progresses in R- and Q-mode analysis: correspondence analysis and its application to the study of geological processes. *Canadian Journal of Earth Sciences*, 11, 131-146.

David, M., Dagbert, M., and Beauchemin, Y. (1977). Statistical analysis in geology: correspondence analysis method. *Quarterly of the Colorado School of Mines*, 72, 1-60.

Davis, T. A. W. and Richards, P. W. (1933). The vegetation of Moraballi Creek, British Guiana: an ecological study of a limited area of tropical rain forest. Part I. *Journal of Ecology*, 21, 350-384.

Davis, T. A. W. and Richards, P. W. (1934). The vegetation of Moraballi Creek, British Guiana: an ecological study of a small area of tropical rain forest. Part II. *Journal of Ecology*, 22, 106-155.

Day, N. R. (1969). Divisive cluster analysis and a test for multivariate normality. *Bulletin of the International Statistical Institute*, 43, 110-112.

Delany, M. J. and Healy, M. J. R. (1964). Variation in the long-tailed field mouse (*Apodemus sylvaticus L.*) in north-west Scotland. II. Simultaneous examination of all characters. *Proceedings of the Royal Society, London B*, 161, 200-207.

Delany, M. J. and Healy, M. J. R. (1966). Variation in the white-toothed shrews (*Crocidura* spp.). *Proceedings of the Royal Society, London B*, 164, 63-74.

Delany, M. J. and Whittaker, H. M. (1969). Variation in the skull of the long-tailed field mouse, *Apodemus sylvaticus* in mainland Britain. *Journal of Zoology, London*, 157, 147–157.

Dempster, A. P. (1966). Estimation in multivariate analysis. In *Multivariate Analysis*, P. R. Krishnaiah, ed. Academic Press, New York.

Dempster, A. P. (1969). *Elements of Continuous Multivariate Analysis*. Addison-Wesley, Reading, Massachusetts.

Dempster, A. P. (1971). An overview of multivariate data analysis. *Journal of Multivariate Analysis*, 1, 316–346.

Devlin, S. J., Gnanadesikan, R., and Kettenring, J. R.   (1975). Robust estimation and outlier detection with correlation coefficients. *Biometrika*, 62, 531–545.

Everitt, B. S. (1978). *Graphical Techniques for Multivariate Data*. Heinemann, London.

Fanshawe, D. B. (1952). *The vegetation of British Guiana: a preliminary review*. Imperial Forestry Institute Paper 29. Oxford.

Field, C. R. (1968). A comparative study of the food habits of some wild ungulates in the Queen Elizabeth National Park, Uganda. *Symposium of the Zoological Society of London*, 21, 135–151.

Field, C. R. (1972). The food habits of wild ungulates in Uganda by analyses of stomach contents. *East African Wildlife Journal*, 10, 17–42.

Field, C. R. and Laws, R. M. (1970). The distribution of the larger herbivores in the Queen Elizabeth National Park, Uganda. *Journal of Applied Ecology*, 7, 273–294.

Fienberg, S. E. (1970). The analysis of multidimensional contingency tables. *Ecology*, 51, 419–433.

Finn, J. D. (1974). *A General Model for Multivariate Analysis*. Holt, Rinehart and Winston, New York.

Fisher, R. A. (1938). *Statistical Methods for Research Workers*. Seventh Edition. Oliver and Boyd, Edinburgh.

Fornell, C. (1979). External single-set components analysis of multiple criterion/multiple predictor variables. *Multivariate Behavioral Research*, 14, 323-338.

Fritts, H. C., Blasing, T. J., Hayden, B. P., and Kutzbach, J. E. (1971). Multivariate techniques for specifying tree-growth and climate relationships and for reconstructing anomalies in paleoclimate. *Journal of Applied Meteorology*, 10, 845-864.

Fritts, H. C., Lofgren, G. R., and Gordon, G. A. (1979). Variations in climate as reconstructed from tree rings. *Quaternary Research*, 12, 18-46.

Gauch, H. G. and Wentworth, T. R. (1976). Canonical analysis as an ordination technique. *Vegetatio*, 33, 17-22.

Gauch, H. G. and Stone, E. L. (1979). Vegetation and soil pattern in a mesophytic forest at Ithaca, New York. *American Midland Naturalist*, 102, 332-345.

Gipson, P. S., Sealander, J. A., and Dunn, J. E. (1974). The taxonomic status of wild *Canis* in Arkansas. *Systematic Zoology*, 23, 1-11.

Gittins, R. (1965). Multivariate approaches to a limestone grassland community. I. A stand ordination. *Journal of Ecology*, 53, 385-401.

Gittins, R. (1969). The application of ordination techniques. In *Ecological Aspects of the Mineral Nutrition of Plants*, I. H. Rorison, ed. Blackwell, Oxford, 37-66.

Gittins, R. (1975). *A comparative study of the phosphate nutrition of nine grass species*. Unpublished manuscript.

Gittins, R. and Ogden, J. (1977). *A reconnaissance survey of lowland tropical rain forest in Guyana*. Unpublished manuscript.

Glahn, H. (1968). Canonical correlation and its relation to discriminant analysis and multiple regression. *Journal of the Atmospheric Sciences*, 25, 23-31.

Gleason, T. C. (1976). On redundancy in canonical analysis. *Psychological Bulletin*, 83, 1004-1006.

Gnanadesikan, R. (1977). *Methods for Statistical Data Analysis of Multivariate Observations*. Wiley, New York.

Gnanadesikan, R. and Wilk, M. B. (1969). Data analytic methods in multivariate statistical analysis. In *Multivariate Analysis II*, P. R. Krishnaiah, ed. Academic Press, New York, 593-638.

Gnanadesikan, R. and Kettenring, J. R. (1972). Robust estimates, residuals and outlier detection with multiresponse data. *Biometrics*, 28, 81-124.

Gokhale, D. V. and Kullback, S. (1978). *The Information in Contingency Tables*. Marcel Dekker, New York.

Goldsmith, B. (1973). The vegetation of exposed sea cliffs at South Stack, Anglesey. I. The multivariate approach. *Journal of Ecology*, 61, 787-818.

Goldstein, R. A. and Grigal, D. F. (1972). Definition of vegetation structure by canonical analysis. *Journal of Ecology*, 60, 277-284.

Goodall, D. W. (1970). Statistical plant ecology. *Annual Review of Ecology and Systematics*, 1, 99-124.

Gould, S. J. and Johnston, R. F. (1972). Geographic variation. *Annual Review of Ecology and Systematics*, 3, 457-498.

Gower, J. C. (1966). A Q-technique for the calculation of canonical variates. *Biometrika*, 53, 588-590.

Gower, J. C. (1973). Contribution to discussion of R. Gnanadesikan's (1973) paper 'Graphical methods for informal inference in multivariate data analysis'. *Bulletin of the International Statistical Institute, Proceedings*, 39, 195-205 and 233-235.

Gower, J. C. and Ross, G. J. S. (1969). Minimum spanning trees and single linkage cluster analysis. *Applied Statistics* , 18, 54-64.

Green, P. E. (1976). *Mathematical Tools for Applied Multivariate Analysis*. Academic Press, New York.

Green, P. E. (1978). *Analyzing Multivariate Data*. The Dryden Press, Illinois.

Grigal, D. F. and Goldstein, R. A. (1971). An integrated ordination classification analysis of an intensively sampled oak-hickory forest. *Journal of Ecology*, 59, 481-492.

Grigal, D. F. and Ohmann, L. F. (1975). Classification, description and dynamics of upland plant communities within a Minnesota wilderness area. *Ecological Monographs*, 45, 389-407.

Guttman, L. (1968). A general non-metric technique for finding the smallest coordinate space for a configuation of points. *Psychometrika*, 33, 469-506.

Hall, C. E. (1977). Some elements of the design of multivariate experiments. *Journal of Experimental Education*, 45, 26-37.

Harris, R. J. (1975). *A Primer of Multivariate Statistics*. Academic Press, New York.

Harris, R. J. (1976). The invalidity of partitioned-U tests in canonical correlation and multivariate analysis of variance. *Multivariate Behavioral Research*, 11, 353-365.

Hatheway, W. H. (1971). Contingency-table analysis of rain forest vegetation. In *Statistical Ecology, Vol. 3*, G. P. Patil, E. C. Pielou, and W. E. Waters, eds. Pennsylvania State University Press, University Park, Pennsylvania, 217-313.

Healy, M. J. R. (1968). Triangular decomposition of a symmetric matrix. *Applied Statistics*, Algorithm AS 6, 17, 195-197.

Healy, M. J. R. and Goldstein, H. (1976). An approach to the scaling of categorized variables. *Biometrika*, 63, 219-229.

Hedges, D. A., Wheeler, J. L.   and Williams, W. T. (1973). The efficiency of utilization of forage oats in relation to the quantity initially available. *Australian Journal of Agricultural Research*, 24, 257-270.

Herrera, C. M. (1978). Ecological correlates of residence and non-residence in a mediterranean passerine bird community. *Journal of Animal Ecology*, 47, 871-890.

Hill, M. O. (1973). Reciprocal averaging: an eigenvector method of ordination. *Journal of Ecology*, 61, 237-249.

Hill, M. O. (1974). Correspondence analysis: a neglected multivariate method. *Applied Statistics*, 23, 340-354.

Hills, L. V., Klovan, J. E., and Sweet, A. R. (1974). *Juglans eocinerea* n. sp., Beaufort Formation (Tertiary), southwestern Banks Island, Arctic Canada. *Canadian Journal of Botany*, 52, 65-90.

Hirschfeld, H. O. (1935). A connection between correlation and contingency. *Proceedings of the Cambridge Philosophical Society*, 31, 520-524.

Hoerl, A. E. and Kennard, R. W. (1970). Ridge regression: biased estimation for nonorthogonal problems. *Technometrics*, 12, 55-67.

Hope, K. (1968). *Methods of Multivariate Analysis*. University of London Press, London.

Horst, P. (1961). Relations among m sets of measures. *Psychometrika*, 26, 129-150.

Hotelling, H. (1935). The most predictable criterion. *Journal of Educational Psychology*, 26, 139-142.

Hotelling, H. (1936). Relations between two sets of variates. *Biometrika*, 28, 321-377.

Howarth, R. J. and Earle, S. A. M. (1979). Application of a generalized power transformation to geochemical data. *Mathematical Geology*, 11, 45-62.

Hubbard, C. E. (1954). *Grasses: a guide to their structure, identification, uses and distribution in the British Isles*. Penguin Books, Harmondsworth.

Hull, D. B. (1979). A craniometric study of the black and white *Colobus* IIIiger 1811 (Primates: Cercopithecoidea). *American Journal of Physical Anthropology*, 51, 163-182.

Irwin, J. O. (1949). A note on the subdivision of χ^2 into components. *Biometrika*, 36, 130-134.

Jeffers, J. N. R. (1978). *An Introduction to Systems Analysis: with Ecological Applications*. University Park Press, Baltimore, Maryland.

Karr, J. R. and James, F. C. (1975). Ecomorphological configurations and convergent evolution in species and communities. In *Ecology and Evolution of Communities*, M. L. Cody and J. M. Diamond, eds. Belknap Press, Cambridge, Massachusetts, 258-291.

Kempthorne, O. (1966). Multivariate responses in comparative experiments. In *Multivariate Analysis*, P. R. Krishnaiah, ed. Academic Press, New York, 521-540.

Kendall, M. G. (1975). *Multivariate Analysis*. Griffin, London.

Kendall, M. G. and Stuart, A. (1967). *The Advanced Theory of Statistics, Vol. 2: Inference and Relationship*. Second Edition. Griffin, London.

Kendall, M. G. and Stuart, A. (1968). *The Advanced Theory of Statistics, Vol. 3: Design and Analysis, and Time-Series*. Second Edition. Griffin, London.

Kercher, J. R. and Goldstein, R. A. (1977). Analysis of an east Tennessee oak hickory forest by canonical correlation of species and environmental parameters. *Vegetatio*, 35, 153-163.

Kerlinger, F. N. and Pedhazur, E. J. (1973). *Multiple Regression in Behavioral Research*. Holt, Rinehart and Winston, New York.

Kessel, S. R. and Whittaker, R. H. (1976). Comparisons of three ordination techniques. *Vegetatio*, 32, 21-29.

Kettenring, J. R. (1971). Canonical analysis of several sets of variables. *Biometrika*, 58, 433-451.

Khatri, C. G. (1976). A note on multiple and canonical correlation for a singular covariance matrix. *Psychometrika*, 41, 465-470.

Knapp, T. R. (1978). Canonical correlation analysis: a general parametric significance-testing system. *Psychological Bulletin*, 85, 410-416.

Kowal, R. R., Lechowicz, M. J., and Adams, M. S. (1976). The use of canonical analysis to compare response curves in physiological ecology. *Flora*, 165, 29-46.

Kruskal, J. B. (1964). Multidimensional scaling by optimizing goodness of fit to a nonmetric hypothesis. *Psychometrika*, 29, 1-27.

Kruskal, J. B. (1965). Analysis of factorial experiments by estimating monotone transformations of the data. *Journal of the Royal Statistical Society, Series B*, 27, 251-263.

Krzyśko, M. (1979). Discriminant variables. *Biometrical Journal*, 21, 227-241.

Kshirsagar, A. M. (1972). *Multivariate Analysis*. Marcel Dekker, New York.

Lancaster, H. O. (1949). The derivation and partition of χ^2 in certain discrete distributions. *Biometrika*, 36, 117–129.

Lancaster, H. O. (1963). Canonical correlations and partitions of χ^2 . *Quarterly Journal of Mathematics, Oxford*, 14, 220–224.

Lawley, D. N. and Maxwell, A. E. (1973). Regression and factor analysis. *Biometrika*, 60, 331–338.

Lee, P. J. (1969). The theory and application of canonical trend surfaces. *Journal of Geology*, 77, 303–318.

Lee, Sik-Yum (1978). Generalizations of the partial, part and bipart canonical correlation analysis. *Psychometrika*, 43, 427–431 and, 44, 131.

Lind, E. M. and Morrison, M. E. S. (1974). *East African Vegetation*. Longman, London.

Lock, J. M. (1972). The effects of hippopotamus grazing on grasslands. *Journal of Ecology*, 60, 445–467.

Lock, J. M. (1977). The vegetation of the Ruwenzori National Park, Uganda. *Botanische Jahrbücher*, 98, 372–448.

Macnaughton-Smith, P. (1965). *Some statistical and other numerical techniques for classifying individuals*. HMSO, London.

McDonald, R. P. (1968). A unified treatment of the weighting problem. *Psychometrika*, 33, 351–381.

McHenry, C. E. (1978). Computation of a best subset in multivariate analysis. *Applied Statistics*, 27, 291–296.

McKeon, J. J. (1965). *Canonical Analysis: Some Relations Between Canonical Correlation, Factor Analysis, Discriminant Function Analysis and Scaling Theory*. Psychometric Monographs No. 13. The Psychometric Society, University of Chicago Press.

Maindonald, J. H. (1977). Least squares computations based on the Cholesky decomposition of the correlation matrix. *Journal of Statistical Computation and Simulation*, 5, 247–258.

Malmgren, B. A. (1979). Multivariate normality tests of plank-
tonic foraminiferal data. *Mathematical Geology*, 11, 285-
297.

Mardia, K. V. (1970). Measures of multivariate skewness and
kurtosis with applications. *Biometrika*, 57, 519-530.

Marriott, F. H. C. (1974). *The Interpretation of Multiple
Observations*. Academic Press, London.

Maxwell, A. E. (1961). Canonical variate analysis when the
variables are dichotomous. *Educational and Psychological
Measurement*, 21, 259-271.

Maxwell, A. E. (1973). Tests of association in terms of matrix
algebra. *British Journal of Statistical and Mathematical
Psychology*, 26, 155-166.

Maxwell, A. E. (1977a). *Multivariate Analysis in Behavioural
Research*. Halsted, New York.

Maxwell, A. E. (1977b). Multiple regression and poorly
conditioned matrices. *British Journal of Mathematical
and Statistical Psychology*, 30, 210-212.

Meredith, W. (1964). Canonical correlation with fallible data.
Psychometrika, 29, 55-65.

Miller, J. K. (1975). The sampling distribution and a test for
the significance of the bimultivariate redundancy statistic:
a Monte Carlo study. *Multivariate Behavioral Research*, 10,
233-244.

Miyata, M. (1970). Complex generalization of canonical correlation and its application to a sea-level study. *Journal
of Marine Research*, 28, 202-214.

Morrison, D. F. (1976). *Multivariate Statistical Methods*.
Second Edition. McGraw-Hill, New York.

Mosier, C. I. (1951). Problems and designs of cross-validation.
Educational and Psychological Measurement, 11, 5-11.

Mulaik, S. A. (1972). *The Foundations of Factor Analysis*.
McGraw-Hill, New York.

Nelder, J. A. (1975). *General Linear Interactive Modelling*.
Numerical Algorithms Group, Oxford.

Nelder, J. A. (1977). Intelligent programs, the next stage in statistical computing. In *Recent Developments in Statistics*, J. R. Barra, et al., eds. North Holland, Amsterdam, 79-86.

Noy-Meir, I. (1970). *Component analysis of semi-arid vegetation in southeastern Australia.* Ph.D. thesis, Australian National University, Canberra.

Noy-Meir, I. (1971). Multivariate analysis of the semi-arid vegetation in southeastern Australia: nodal ordination by components analysis. *Proceedings of the Ecological Society of Australia*, 6, 159-193.

Noy-Meir, I. (1973b). Data transformations in ecological ordination. I. Some advantages of non-centering. *Journal of Ecology*, 61, 329-341.

Noy-Meir, I. (1974a). Multivariate analysis of the semi-arid vegetation in southeastern Australia. II. Vegetation catenae and environmental gradients. *Australian Journal of Botany*, 22, 115-140.

Noy-Meir, I. (1974b). Catenation: quantitative methods for the definition of coenoclines. *Vegetatio*, 29, 89-99.

Noy-Meir, I. and Anderson, D. J. (1971). Multiple pattern analysis, or multiscale ordination: towards a vegetation hologram? In *Statistical Ecology, Vol. 3*, G. P. Patil, E. C. Pielou, and W. E. Waters, eds. The Pennsylvania State University Press, University Park, Pennsylvania, 207-225.

Noy-Meir, I., Orshan, G., and Tadmor, N. H. (1973). Multivariate analysis of desert vegetation. III. The relation of vegetation units to habitat classes. *Israel Journal of Botany*, 22, 239-257.

Noy-Meir, I., Walker, D., and Williams, W. T. (1975). Data transformations in ecological ordination. II. On the meaning of data standardization. *Journal of Ecology*, 63, 779-800.

Ogden, J. (1966). *Ordination studies on a small area of tropical rain forest.* M.Sc. thesis, University of Wales.

Ogden, J. (1977). Personal communication.

Olson, C. L. (1976). On choosing a test statistic in multivariate analysis of variance. *Psychological Bulletin*, 83, 579-586.

530   R. GITTINS

Olson, C. L. (1979). Practical considerations in choosing a
    MANOVA test statistic: a rejoinder to Stevens. *Psycho-
    logical Bulletin*, 86, 1350-1352.

Orlóci, L. (1975). *Multivariate Analysis in Vegetation Research*.
    Junk, The Hague.

Orlóci, L. (1978). *Multivariate Analysis in Vegetation Research*.
    Second Edition. Junk, The Hague.

Ortega, J. (1967). The Givens-Householder method for symmetric
    matrices. In *Mathematical Methods for Digital Computers*,
    Vol. II, A. Ralston and H. S. Wilf, eds. Wiley, New York.
    94-115.

Osmaston, H. A. (1965). The vegetation. In *Uganda National
    Parks Handbook*. Trustees of the Uganda National Parks,
    Kampala.

Oxnard, C. E. (1973). Some locomotor adaptations among lower
    primates: implications for primate evolution. *Symposium of
    the Zoological Society of London*, 33, 255-299.

Pearce, S. C. (1969). Multivariate techniques of use in
    biological research. *Experimental Agriculture*, 5, 67-77.

Perreault, W. D. and Spiro, R. L. (1978). An approach for
    improved interpretation of multivariate analysis. *Decision
    Sciences*, 9, 402-413.

Pielou, E. C. (1977). *Mathematical Ecology*. Wiley-Interscience,
    New York.

Pillai, K. C. S. (1960). *Statistical Tables for Tests for
    Multivariate Hypotheses*. Statistical Center, University
    of the Philippines, Manila.

Rao, C. R. (1952). *Advanced Statistical Methods in Biometric
    Research*. Wiley, New York.

Ray, D. M. and Lohnes, P. R. (1973). Canonical correlation in
    geographical analysis. *Geographia Polonica*, 25, 49-65.

Rohlf, F. J. (1968). Stereograms in numerical taxonomy. *System-
    atic Zoology*, 17, 246-255.

Roy, J. (1958). Step-down procedures in multivariate analysis.
    Annals of Mathematical Statistics, 29, 1177-1187.

Roy, S. N. (1957). *Some Aspects of Multivariate Analysis.* Wiley, New York.

Schall, J. J. and Pianka, E. R. (1978). Geographic trends in numbers of species. *Science*, 201, 679-686.

Schultz, J. P. (1960). *Ecological studies on rain forest in northern Surinam.* Amsterdam.

Scott, D. W. and Koopman, R. F. (1977). *Rotation of canonical variates using nonidentical orthogonal transformations.* Psychology Department, Simon Fraser University. (Mimeograph). Unpublished manuscript.

Seal, H. (1964). *Multivariate Statistical Analysis for Biologists.* Methuen, London.

Shepard, R. N. (1962). The analysis of proximities: multidimensional scaling with an unknown distance function. (I and II). *Psychometrika*, 27, 125-139 and 219-246.

Shepard, R. N. and Carroll, J. D. (1966). Parametric representation of nonlinear data structures. In *Multivariate Analysis*, P. R. Krishnaiah, ed. Academic Press, New York, 561-592.

Skinner, H. A. (1977). Exploring relationships among multiple data sets. *Multivariate Behavioral Research*, 12, 199-220.

Smouse, P. E. (1972). The canonical analysis of multiple species hybridization. *Biometrics*, 28, 361-371.

Spain, A. V. (1974a). Litter fall in a New South Wales conifer forest: a multivariate comparison of plant nutrient element status and return in four species. *Journal of Applied Ecology*, 10, 527-556.

Spain, A. V. (1974b). A preliminary study of spatial patterns in the accession of conifer litter. *Journal of Applied Ecology*, 10, 557-567.

Sparks, D. N. and Todd, A. D. (1973). Latent roots and vectors of a symmetric matrix. *Applied Statistics*, Algorithm AS 60, 22, 260-265.

Stevens, J. P. (1979). Comment on Olson: choosing a test statistic in multivariate analysis of variance. *Psychological Bulletin*, 86, 355-360.

Stewart, D. K. and Love, W. A. (1968). A general canonical correlation index. *Psychological Bulletin*, 70, 160–163.

Stewart, G. W. (1970). Algorithm 348: eigenvalues and eigenvectors of a real symmetric matrix. *Communications of the Association for Computing Machinery*, 13, 369–371; Errata 13, 750.

Stewart, G. W. (1973). *Introduction to Matrix Computations*. Academic Press, New York.

Swan, J. M. A. (1970). An examination of some ordination problems by use of simulated vegetation data. *Ecology*, 51, 89–102.

Tansley, A. G. (1939). *The British Islands and Their Vegetation*. Cambridge University Press.

Tatsuoka, M. M. (1971). *Multivariate Analysis: Techniques for Educational and Psychological Research*. Wiley, New York.

Teeri, J. A., Stowe, L. G., and Murawski, D. A. (1978). The climatology of two succulent plant families: Cactaceae and Crassulaceae. *Canadian Journal of Botany*, 56, 1750–1758.

Teil, H. (1975). Correspondence factor analysis: An outline of its method. *Mathematical Geology*, 7, 3–12.

Thaler, G. R. and Plowright, R. C. (1973). An examination of the floristic zone concept with special reference to the northern limit of the Carolinian zone in southern Ontario. *Canadian Journal of Botany*, 51, 1765–1789.

Thorndike, R. M. (1976). Studying canonical analysis: Comments on Barcikowski and Stevens. *Multivariate Behavioral Research*, 11, 249–253.

Thorndike, R. M. (1977). Canonical analysis and predictor selection. *Multivariate Behavioral Research*, 12, 75–87.

Thorndike, R. M. (1978). *Correlational Procedures for Research*. Gardner Press, New York.

Thorndike, R. M. and Weiss, D. J. (1973). A study of the stability of canonical correlations and canonical components. *Educational and Psychological Measurement*, 33, 123–134.

Thornton, D. D. (1971). The effect of complete removal of hippopotamus on grassland in the Queen Elizabeth National Park, Uganda. *East African Wildlife Journal*, 9, 47–55.

Thorpe, R. S. (1976). Biometric analysis of geographical variation and racial affinities. *Biological Reviews*, 51, 407–452.

Timm, N. H. (1975). *Multivariate Analysis with Applications in Education and Psychology*. Wadsworth, Belmont, California.

Timm, N. H. and Carlson, J. E. (1976). Part and bipartial canonical correlation analysis. *Psychometrika*, 41, 159–176.

Trenbath, B. R. and Harper, J. L. (1973). Neighbour effects in the genus *Avena*. I. Comparison of crop species. *Journal of Applied Ecology*, 10, 379–400.

Tukey, J. W. (1969). Analyzing data: Sanctification or detective work? *American Psychologist*, 24, 83–91.

Tukey, J. W. and Wilk, M. B. (1966). Data analysis and statistics: An expository overview. *AFIPS Conference Proceedings, Fall Joint Computer Conference*, 29, 695–709.

Van de Geer, J. P. (1971). *Introduction to Multivariate Analysis for Social Sciences*. Freeman, San Francisco.

Vogt, T. and Jameson, D. L. (1970). Chronological correlation between change in weather and change in morphology of the Pacific tree frog in Southern California. *Copeia*, 1970(1), 135–144.

Vuilleumier, F. (1970). Insular biogeography in continental regions. I. The northern Andes of South America. *The American Naturalist*, 104, 373–388.

Walker, E. P., Warnick, F., Lange, K. I., Uible, H. E., Hamlet, S. E., Davis, M. A., and Wright, P. F. (Second edition revised by J. L. Paradiso) (1968). *Mammals of the World*. Second Edition. The Johns Hopkins Press, Baltimore.

Wall, F. J. (1968). *The Generalized Variance Ratio of the U-Statistic*. The Dikewood Corporation, Albuquerque, New Mexico.

Webb, D. A. (1954). Is the classification of plant communities either possible or desirable? *Botanisk Tidsskrift*, 51, 362–370.

Webb, L. J., Tracey, J. G., Williams, W. T., and Lance, G. N. (1971). Prediction of agricultural potential from intact forest vegetation. *Journal of Applied Ecology*, 8, 99-121.

Webb, T. III and Bryson, R. A. (1972). The late- and post-glacial sequence of climatic events in Wisconsin and east-central Minnesota: Quantitative estimates derived from fossil pollen spectra by multivariate statistical analysis. *Quaternary Research*, 2, 70-115.

Webster, R. (1977a). Canonical correlation in pedology: How useful? *Journal of Soil Science*, 28, 196-221.

Webster, R. (1977b). *Quantitative and Numerical Methods in Soil Classification and Survey*. Clarendon Press, Oxford.

Weinberg, S. L. and Darlington, R. B. (1976). Canonical analysis when the number of variables is large relative to sample size. *Journal of Educational Statistics*, 1(4), 313-332.

Willan, A. R. and Watts, D. G. (1978). Meaningful multicollinearity measures. *Technometrics*, 20, 407-412.

Williams, E. J. (1952). Use of scores for the analysis of contingency tables. *Biometrika*, 13, 274-289.

Williams, E. J. (1967). The anlaysis of association among many variates (with discussion). *Journal of the Royal Statistical Society, Series B*, 29, 199-242.

Williams, W. T. (1971). Strategy and tactics in the acquisition of ecological data. *Proceedings of the Ecological Society of Australia*, 6, 57-62.

Williams, W. T. (1976). *Pattern Analysis in Agricultural Research*. CSIRO, Melbourne and Elsevier, Amsterdam.

Williams, W. T. and Lance, G. N. (1968). Choice of strategy in the analysis of complex data. *The Statistician*, 18, 31-43.

Wingard, J. A., Huba, G. J., and Bentler, P. M. (1979). The relationship of personality structure to patterns of adolescent substance use. *Multivariate Behavioral Research*, 14, 131-143.

Wollenberg, A. L. van den. (1977). Redundancy analysis an alternative for canonical correlation analysis. *Psychometrika*, 42, 207-219.

Young, F. W., de Leeuw, J., and Takane, Y. (1976). Regression with qualitative and quantitative variables: An alternating least squares method with optimal scaling features. *Psychometrika*, 41, 505–529.

Young, F. W., de Leeuw, J., and Takane, Y. (1978). *Quantifying qualitative data*. Report 149, L. L. Thurstone Psychometric Laboratory, University of North Carolina, Chapel Hill.

Zuckerman, S., Ashton, E. H., Flinn, R. M., Oxnard, C. E., and Spence, T. F. (1973). Some locomotor features of the pelvic girdle in Primates. *Symposium of the Zoological Society of London*, 33, 71–168.

[*Received October* 1979]

AUTHOR INDEX

Aart, P. J. M. van der, 506, 515
Abderson, T. W., 277
Adams, M. S., 322, 350, 362, 497, 526
Agrell, I., 122
Albrecht, G. H., 497, 501, 502, 515
Anderberg, M. R., 454, 515
Anderson, D. J., 314, 529
Anderson, T. W., 299
Andrewartha, H. G., 52
Andrews, D. F., 336, 503, 505, 515
Arnold, G. W., 268
Ashton, E. H., 360, 497, 516, 535
Auclair, A. N., 506, 516
Austin, M. P., 102, 123, 314, 315, 390, 403, 413, 497, 503, 516

Baker, F., 5, 8
Bakker, D., 82, 85
Banfield, C., 229, 232, 233, 235
Barcikowski, R. S., 488, 500, 516
Bargmann, R. E., 1, 2, 5, 8, 9, 354, 356, 516
Barkham, J. P., 314, 315, 497, 504, 516
Barnes, B. V., 299, 497, 520
Barnett, V. D., 502, 517
Barra, J. R., 529
Barrett, J. C., 361
Bartlett, M. S., 322, 329, 337, 338, 351, 436, 474, 517
Baskett, T. S., 53
Beals, E. W., 198, 201
Beauchemin, Y., 475, 520
Bedford, B. D., 52

Bednarek, A. R., 17, 19
Bentler, P. M., 505, 534
Benzecri, J. P., 129, 138, 172
Beron, M. L., 234
Berti, A. A., 297, 517
Beverton, R. J. H., 52
Billings, W. D., 123
Birks, H. J. B., 214, 233, 497, 517
Bishop, Y. M. M., 52, 62, 473, 478, 517
Bissett, J., 498, 517
Blackith, R. E., 180, 188, 227, 233, 517, 518
Blanc, F., 128, 148, 172
Blasing, T. J., 497, 517, 522
Bliss, C. I., 501, 517
Blum, H., 176, 188
Blumer, M., 260, 262
Bock, R. D., 314, 332, 352, 429, 473, 495, 501, 512, 517
Bookstein, F. L., 176, 188
Booth, T. L., 16, 19
Bourliere, F., 452, 518
Bouxin, G., 123
Box, G. E. P., 315, 518
Bradbury, R. M., 15, 19
Brandsma, M., 82, 86
Bray, J. R., 123, 266, 277
Brayer, J. M., 17, 19
Brent, R. P., 361, 518
Brillinger, D. R., 139, 151, 172, 360, 509, 518
Brown-Leger, L. S., 260, 262
Bryson, R. A., 497, 518, 534
Bunce, R. G. H., 90, 99
Burnaby, T. P., 229, 233, 500, 518
Burton Barnes, V., 289
Buzas, M. A., 500, 518

Caillez, F., 134, 172

SUBJECT INDEX

Ridge regression, 354, 356
Roy's largest-root distribu-
    tion, 442, 473

Salinity, 229
Sample-specific covariation,
    357
Sample-specific validation,
    492
Sample-specific variation,
    332
Sampling design, 403
Scaling, 351
Sedimentary components, 220
Seedling domain, 411
Shannon-Weaver index, 214
Shape vector, 175
Shrunken estimators, 224
Significance tests, 336
Similarity measures, 253
Simulation, 301
Singular value decomposition,
    360, 459
Size and shape variables, 180
Size component, 227
Size variables, 175
Skewness, 502
Soil factor-complex, 396,
    397, 399, 401, 421
Soil-species relationships,
    379
Soil-vegetation relationships,
    309, 391, 392, 401-403
Spatial distribution, 378, 389
Spatial domain, 373, 374
Spatial measures, 371, 375
Spatial paleoecology, 214
Spatial point distribution,
    303
Spatial relationships, 421
Spatial variables, 368
Spatial variation, 309, 366,
    367
Species clumping, 102
Species domain, 373, 374,
    377, 378, 388
Species weighting techniques,
    97
Species weights, 87

Specific factor, 297
Specificity, 298
Split-plot analysis, 452
Split-plot designs, 493
Spurious correlation, 237
Stability of vegetation, 403
Stabilized canonical variates,
    223
Step-down analysis, 358
Stereogram, 336, 399, 483
Stratification, 62
Structural simultaneity, 302
Structure coefficients, 219
Structure correlations, 342,
    356, 357, 394-397, 410,
    420, 423, 438, 443, 449,
    470, 479, 505
Synthetic contrast, 495
Systematic sample, 405

Thanatocoenetic, 218
Trampling effect, 306
Tree domain, 411
Tree-seedling relationship,
    406, 408, 409, 411, 413
Turtles, 58, 59

Uganda kob, 453
Union-interesection criterion,
    1, 3, 339, 432
Union-intersection tests, 426,
    428, 431, 432, 434, 436,
    438, 441, 494, 495

Vagility, 66
Variance extracted, 370, 398
Variance-covariance matrix,
    325, 330, 334, 355
Varimax criterion, 357
Vegetation, 191, 301
    grassland, 416, 419
Vegetation classification, 415
Vegetation domain, 400
Vegetation structure, 302,
    421, 423, 424
    grassland, 415-421
Vegetation survey, 402, 414,
    424, 492, 499
Vegetation-herbivore relation-
    ships, 309

INTERNATIONAL STATISTICAL ECOLOGY PROGRAM

The International Statistical Ecology Program (ISEP) consists of the activities of the Statistical Ecology Section of the International Association for Ecology and of the Liaison Committee on Statistical Ecology of the International Statistical Institute, the Biometric Society, and the International Association for Ecology. The ISEP is a non-profit program formulated to serve the needs of interdisciplinary research and training in the newly emerging fields of Statistical Ecology and Ecological Statistics.

SATELLITE PROGRAM IN STATISTICAL ECOLOGY

The Second International Congress of Ecology was held in Jerusalem during September 1978. In this connection, ISEP organized a Satellite Program in Statistical Ecology during 1977 and 1978. The emphasis was on research, review, and exposition concerned with the interface between quantitative ecology and relevant quantitative methods. Both theory and application of ecology and ecometrics received attention. The Satellite Program consisted of instructional coursework, seminar series, thematic research conferences, and collaborative research workshops.

Research papers and research-review-expositions were specially prepared for the program by concerned experts and expositors. These materials have been refereed and revised, and are now available in a series of ten edited volumes listed on page ii of this volume.

The Satellite Program takes as its theme the better melding of fundamental ecological concepts with rigorous empirical quantification. The overall result should be progress toward a stronger body of general ecologic and ecometric theory and practice.

FUTURE DIRECTIONS

The satellite-like-programs help create and sustain enthusiasm, inward strength, and working efficiency of those who desire to meet a contemporary social need in the form of some interdisciplinary work. It should be only proper and rewarding for everyone involved that such programs are planned from time to time.

Plans are being made for a satellite program in conjunction with the next Biennial Conference of the International Statistical Institute and the next International Congress of Ecology. Care should be exercised that the next program not become a mere replica of the present one, however successful it has been. Instead, the next program should be organized so that it helps further the evolution of statistical ecology as a productive field.

The next program is being discussed in terms of subject area groups. Each subject group is to have a coordinator assisted by small committees, such as a program committee, a research committee, an annual review committee, a journal committee, and an education committee. This approach is expected to respond to the need for a journal on statistical ecology, and also to the need of bringing out well planned annual review volumes. The education committee would formulate plans for timely modules and monographs. Interested readers may feel free to communicate their ideas and interests to those involved in planning the next program. The mailing address is: International Statistical Ecology Program, P. O. Box 218, State College, PA 16801, USA.